CONTEMPORARY MATHEMATICS

734

Differential Equations, Mathematical Physics, and Applications: Selim Grigorievich Krein Centennial

Peter Kuchment
Evgeny Semenov
Editors

AMS AMERICAN MATHEMATICAL SOCIETY

2010 *Mathematics Subject Classification.* Primary 01Axx, 14E20, 30C65, 34C23, 35L05, 37E35, 37G99,47B44,83C57, 92C37.

Library of Congress Cataloging-in-Publication Data

Names: Kuchment, Peter, 1949– editor. | Semenov, E. M. (Evgeniĭ Mikhaĭlovich), editor.
Title: Differential equations, mathematical physics, and applications : Selim Grigorievich Krein centennial / Peter Kuchment, Evgeny Semenov, editors. Other titles: Selim Grigorievich Krein centennial
Description: Providence, Rhode Island : American Mathematical Society, [2019] | Series: Contemporary mathematics ; volume 734 | Includes bibliographical references.
Identifiers: LCCN 2019008849 | ISBN 9781470437831 (alk. paper)
Subjects: LCSH: Differential equations. | Mathematical physics. | Kreĭn, S. G. (Selim Grigor′evich), 1917– | Festschriften. | AMS: History and biography – History of mathematics and mathematicians – History of mathematics and mathematicians. msc | Algebraic geometry – Birational geometry – Coverings. msc | Functions of a complex variable – Geometric function theory – Quasiconformal mappings in ... msc | Ordinary differential equations – Qualitative theory – Bifurcation. msc | Partial differential equations – Hyperbolic equations and systems – Wave equation. msc | Dynamical systems and ergodic theory – Low-dimensional dynamical systems – Flows on surfaces. msc | Dynamical systems and ergodic theory – Local and nonlocal bifurcation theory – None of the above, but in this section. msc | Operator theory – Special classes of linear operators – Accretive operators, dissipative operators, etc. msc | Relativity and gravitational theory – General relativity – Black holes. msc | Biology and other natural sciences – Physiological, cellular and medical topics – Cell biology. msc.
Classification: LCC QA371 .D44745 2019 | DDC 515–dc23
LC record available at https://lccn.loc.gov/2019008849

Contemporary Mathematics ISSN: 0271-4132 (print); ISSN: 1098-3627 (online)

DOI: https://doi.org/10.1090/conm/734

∞ The paper used in this book is acid-free and falls within the guidelines
established to ensure permanence and durability.
Visit the AMS home page at `https://www.ams.org/`

10 9 8 7 6 5 4 3 2 1 24 23 22 21 20 19

Dedicated to the centennial of Professor Selim Grigorievich Krein

Photograph provided by Dr. T. Voronina (nee Krein).

Selim Grigorievich Krein

Contents

Preface

This is the second of the two volumes dedicated to the centennial of distinguished mathematician Professor Selim Grigorievich Krein. Krein has made numerous major contributions to functional analysis, operator theory, partial differential equations, fluid dynamics, and other areas. He had written several influential monographs in these areas. He had also been a prolific teacher, graduating 83 PhD students, including quite a few well known mathematicians currently scattered around the globe. Another his major activity was creation and running the annual Voronezh Winter Mathematical Schools, which has influenced significantly the mathematical life in the former Soviet Union since 1967.

The articles in this volume are written by former students and colleagues of Selim Krein, as well as lecturers and participants of Voronezh Winter Schools. They are devoted to a variety of contemporary problems of differential equations, fluid dynamics, and applications. The papers present new or survey recent results in these areas.

The editors express their sincere gratitude to all the contributors to the volume and to the AMS publishing staff, in particular to Ms. Christine Thivierge and production editor Mike Saitas, for great help provided.

Peter Kuchment, Mathematics Department, Texas A&M University, College Station, TX 77843-3368, USA

Evgeny Semenov, Mathematics Department, Voronezh State University, Voronezh, Russia

November 24th, 2018

Contemporary Mathematics
Volume **734**, 2019
https://doi.org/10.1090/conm/734/14757

Introduction

Peter Kuchment and Evgeny Semenov

Dedicated to the centennial of Professor Selim G. Krein

1. Selim Grigorievich Krein and his mathematics

We consider it a privilege and an honor to edit the two volumes dedicated to the centennial of our teacher Professor Selim Krein, prominent mathematician and educator, who has played a major role in our lives, as well as in the lives of hundreds of other mathematicians.

Selim Krein has had a long and distinguished career as a researcher and teacher. Let us start by describing him as a teacher. Krein has always been surrounded by young mathematicians whom he taught and whose lives he influenced. He was a great lecturer. His lectures and talks were always well prepared, clearly structured, and targeted exactly the level that was appropriate for the audience. It had always been a distinct pleasure to hear him lecturing. However, the way he taught was only a small part of an explanation of his great success as a mentor of young researchers. The whole atmosphere in his school was filled with excitement and conducive to creative research.

The number of his PhD students has reached at least 83 (see the list in section 3). Quite a few of his former students have become prominent mathematicians and hold professorial positions in the USA, Bulgaria, Australia, France, Germany, Israel, Vietnam, as well as in Russia, Ukraine, and various other republics of the fSU. His first two PhD students, Professors Yu. Berezanskii[1] and Yu. Daletskii, have become members of the Ukrainian Academy of Sciences.

Although Krein's major achievements were in functional analysis, operator theory, partial differential equations, differential equations in Banach spaces, fluid dynamics, and numerical analysis, he was interested in mathematics as a whole, rather than just in a specific area. In order to emphasize freedom from rigid restrictions of topics, he called one of his seminars at the Voronezh Winter Math School (see [**183**] and [**KuLin, KuLin2**]) a "seminar on higher mathematics." In his seminars (even when they were designated to some specific areas) one could deliver talks on any subject one found interesting. Though in most cases he did follow the standard procedure of assigning a problem to a PhD student, he let the strongest of them

---

2010 *Mathematics Subject Classification.* Primary 01A70.

[1]The editors regret to report that as this volume was being prepared for the printer, Professor Yu. M. Berezanskii, prominent mathematician and the first PhD advisee of S. Krein, passed away at the age of 94.

"off the leash" to choose their own problems of interest. As the result, topics his students worked on, besides already listed functional analysis, operator theory and PDEs, included several complex variables, non-commutative integration, K-theory, commutative algebra, . . . , you name it. The amazing thing was that when a student would start to work on a problem seemingly very remote from Krein's current interests, he was able to get into a new field and help the student throughout the research and dissertation. It was also very common for his students to organize their own seminars in order to study new areas.

Another distinguishing feature of Krein as an advisor was that he helped his students not only to resolve their research problems, but also problems of their everyday life (which often were much harder). Helping former students in getting decent jobs and apartments and in overcoming unwritten restrictions imposed by the authoritarian regime were very common for him. He kept helping his former students long after their graduation. It was amazing to see that Krein remembers the names of thousands of his former (even undergraduate) students.

Krein's joyful character, readiness to joke, and ability to charm everyone added to the admiration we all felt toward him.

Selim Krein has always been eager to go an extra mile to strengthen the mathematics in Voronezh. He was the initiator of the Mathematics Research Institute, a series of lectures on contemporary topics in mathematics for professors and graduate students of numerous Voronezh institutions of higher education, and most importantly, the highly influential Voronezh Winter Mathematical School (see [**School, KuLin, KuLin2**], and [**183**]).

Let us now turn to a description of some of Krein's life and work in mathematics. One can find additional information in papers and books listed in the section 2. It is impossible to address (even without details) all the directions of Krein's research, so we will concentrate only on the major ones.

Selim Krein was born July 15, 1917 in Kiev. He got his undergraduate degree in mathematics at Kiev State University in 1935–1940 and after his graduate study in 1940–1941 defended in 1942 a PhD dissertation under advising of N. N. Bogolyubov. His first love in mathematics was functional analysis, which continued to be one of his major areas of research since then. His first works in this area were written jointly with his teacher Bogolyubov and his older brother, famous mathematician Mark Krein (1907–1989). One of the results of his dissertation was a theorem now known as the Kreins-Kakutani theorem.

When World War II started, Krein was not drafted, due to a childhood illness that damaged one of his legs and left him with a limp for the rest of his life. Being a research fellow at the Mathematics Institute of the Ukrainian Academy of Sciences (relocated to a region free of German occupation), he started to work on important war-related projects (e.g., theory of cumulative ordinances) in a group led by M. A. Lavrent'ev. This is also when his major research in fluid dynamics started. He was among the first to introduce methods of functional analysis into fluid dynamics. He studied, in particular, the motion of a vessel partially filled with fluid (one can imagine, for instance, a rocket with liquid fuel). Oscillations of a fluid in an open vessel was another topic to which he had devoted a lot of attention. All these considerations arose from important applied problems. Among the culminations of his research in this direction were his classified Doctor of Sciences dissertation, which he defended in 1950 in the Moscow Academy of Artillery Sciences, and the books [**18, 20, 21**].

In 1950–1951 Krein was the Head of the Numerical Mathematics Department of the Mathematics Institute of the Ukrainian Academy of Sciences. Here he developed with M. A. Krasnoselskii a minimal residual method for solving systems of linear equations. Numerical analysis has also become one of the areas to which Krein returned many times during his career. Several of his students have become numerical analysts.

It is remarkable (and little known) that at that time S. Krein was, together with S. A. Avramenko, actively involved into design of and programming for the first Soviet electronic computer MESM (Small Electronic Computing Machine), the predecessor of the much better known BESM (Big Electronic Computing Machine). In particular, they insisted on significant increase of the hardware precision of calculations. They have also been mentioned (see [Odin, pp. 53-54]) among the first Soviet computer programmers, writing MESM codes for numerical solutions of ODEs.

The end of 1940s and beginning of 1950s were the periods of a high wave of governmental anti-semitism, due to which both M. Krasnoselskii and S. Krein had to leave the Institute of Mathematics and Kiev. One can read about this in particular in [**Berez**].

After a short time spent in Krivoi Rog, Krein came to Voronezh in 1954, where he had worked since and together with M. Krasnoselski and V. Sobolev created the well known Voronezh school of functional analysis[2]. S. G. worked first as a Professor and Chairman of the Mathematics and Theoretical Mechanics Department at the Voronezh Forestry Institute (currently, Forestry Academy). He became Chair of the Department of Partial Differential Equations of Voronezh State University in 1964, and came back to his previous position at the Forestry Institute in 1971. Under his guidance, the Department of Partial Differential Equations of the Voronezh State University and the Mathematics and Theoretical Mechanics Department of the Forestry Institute had become places of very active mathematical research.

Among his first studies in functional analysis and function theory was the introduction with Yu. Daletskii the important double operator integral notion [**34, 38**], an area further developed by M. Birman and M. Solomyak and still very active currently. Together with Yu. Berezanskii, they introduced and studied hypercomplex systems with continual basis [**35–37, 59, 67**]. This work preceded much later studies of hypergroups, when various their results were re-discovered. One should also mention works on singular integrals [**27, 28, 30**].

Since then, his activity in functional analysis had mostly been concentrated on the theory of interpolation of linear operators, an area that has many applications, the most important ones in PDEs. Along with Calderon, Lions, Peetre and Gagliardo, Krein was one of the creators of the contemporary theory of interpolation of linear operators. He created the so called method of scales of Banach spaces (sometimes called "the Russian method" of operator interpolation), which has become one of the major methods in this theory. Several of his students have

[2]One can find a lot of historical information about mathematics at Voronezh State University in [**Adamova**].

become leading experts in interpolation theory. The milestones of this area of Krein's research were the surveys and books [**13, 16, 91, 95, 158**].

Research on fluid dynamics and PDEs had led S. Krein to another area of functional analysis: studying Fredholm and semi-Fredholm operators and continuous and analytic families of such operators, the behavior of spectra and eigenfunctions of such families being the main properties of interest. The techniques developed by Krein and his school in this area play an important role in many pure and applied studies (wave guides and photonic crystals among the most interesting ones). The survey books and papers [**7, 15, 136**] reflect the results of this research.

Krein also initiated, edited, and co-authored a popular reference book on functional analysis [**2, 4, 5, 8, 9**].

Problems of fluid dynamics have lead Krein to studying differential equations in Banach spaces and semigroups of operators. Here he has also become one of the world leaders. Books [**3, 6**] and survey [**151**] summarize results on the correctness of the Cauchy problem, analyticity of solutions, boundary value problems, asymptotic methods, and others, many of which were obtained by S. Krein and his students. In his last two decades, Krein also worked on singularly perturbed differential equations in Banach spaces.

A large part of Krein's scientific activity was devoted to various problems of partial differential equations. Among his best are, e.g. the results on homeomorphisms created by elliptic boundary value problems [**74, 76**], boundary value problems for overdetermined systems of PDEs, and boundary value problems in variable domains. Some of these studies are described in the books [**10, 11**].

For a quarter of century, his attention was attracted to differential equations on Lie groups and manifolds. For instance, papers [**111, 131, 132, 139**] treat relations between infinite dimensional representations of Lie groups and Cauchy problems for corresponding differential equations, differential equations of second order on Lie groups, representation of algebras of germs of differential operators with analytic coefficients by integro-differential operators, construction of infinitesimal operators for operators of generalized shift, and other topics. The book [**14**] develops analogs of Floquet theory for equations on manifolds, while [**19**] is devoted to function spaces related to representations of Lie groups and Lie algebras. The books [**10, 11, 14, 19**], regretfully, have never been translated to English.

Selim Krein was always eager to share his knowledge with others. He achieved this not only through his brilliant lectures, but also through the textbooks he wrote. The elementary calculus book [**1**] has been translated into several languages. A very nice analysis textbook [**12**] (also deserving translation) treats many important topics that are usually not discussed in standard textbooks. He has also written an introductory textbook [**17**] on linear programming for economics majors.

One of the S. Krein's great influences on the mathematical life in the fSU was through the remarkable Voronezh Winter Mathematical Schools (see [**School, KuLin, KuLin2**] and [**183**]).

The current volume and the preceding one [**Krein1**] contain mathematical works and memories of S. Krein's students, colleagues, and participants and lecturers of the Voronezh Winter Schools.

The editors express their sincere gratitude to all the contributors to the volume and to the AMS publishing staff, in particular to Ms. Christine Thivierge for great help provided. Special thanks go to Prof. R. S. Adamova for collecting and sharing with us information about the history of the department.

2. Some references

[**Adamova**] R. S. Adamova, *Derpt-Yuriev-Voronezh University. Mathematics and mathematicians*, Voronezh State University Publishing, Voronezh 2018. (In Russian)

[**Berez**] Yu. M. Berezanskii, Kiev, Fall of 1943 through 1946. The rebirth of Mathematics, in P. Kuchment and E. Semenov, *Functional Analysis and Geometry. Selim Grigorievich Krein Centennial*, AMS, Providence, R.I. 2019.

[**60**] Yu. M. Berezanskii, N. N. Bogolyubov, Yu. L. Daletskii, M. A. Lavrent'ev, E. M. Semenov, Selim Grigor'evich Krein (on the occasion of his sixtieth birthday), Uspehi Mat. Nauk **33** (1978), no. 2, 217–224, MR 57 #12102)

[**70**] Yu. M. Berezanskii, N. N. Bogolyubov, P. A. Kuchment, B. Ya. Levin, V. P. Maslov, S. P. Novikov, E. M. Semenov, Selim Grigor'evich Krein (on his seventieth birthday), Russian Math. Surveys **42**:5 (1987), 181–183, translation from the Russian original in Uspekhi Mat. Nauk **42**:5 (1987), 223–224.

[**Obit**] Yu. M. Berezanskii, S. G. Gindikin, S. S. Kutateladze, P. A. Kuchment, S. P. Novikov, Yu. G. Reshetnyak, E. M. Semenov, S. A. Sklyadnev, V. M. Tikhomirov, Selim Grigor'evich Krein (obituary), Russian Math. Surveys **55**:2 (2000), 327–328, translation of the Russian originalin Uspekhi Mat. Nauk **55**:2 (2000), 125–126. DOI: 10.1070/RM2000v055n02ABEH000270

[**100**] V.M. Buchstaber, P.A. Kuchment, S.P. Novikov, and E.M. Semenov, To the memory of Selim Grigor'evich Krein (1917–1998), Uspekhi Mat. Nauk **73**:1 (2018), 19–193 (in Rusian). English translation in Russian Math. Surveys **73**:1 (2018), 187–190. DOI: https://doi.org/10.1070/RM9816

[**School**] V. A. Kostin (Editor), *Proceedings of the international conference "S. G. Krein Voronezh Winter Mathematical School - 2018"*, Nauchnaya Kniga, Voronezh 2018. (In Russian)

[**2008**] V. A. Kostin, B. N. Sadovskii, and E. M. Semenov (Editors), *S. G. Krein. Recollections*, Voronezh State University 2008.

[**KuLin**] P. Kuchment and V. Lin (Ed.), *Voronezh Winter Mathematical Schools. Dedicated to Selim Krein*, AMS Translations, Series 2, v. 184, 1998.

[**KuLin2**] P. Kuchment and V. Lin, *Selim Krein and Voronezh Winter Mathematical Schools*, Contemporary Mathematics, vol. 734, American Mathematical Society, 2019

[**Krein1**] P. Kuchment and E. Semenov (Editors), *Functional Analysis and Geometry: Selim Grigorievich Krein Centennial*, Contemporary Mathematics, vol. 733, American Mathematical Society, 2019.

[**Comp**] V. P. Odinets, *Sketches of History of Computer Sciences*, Komi State Pedagogical Institute, Syktyvkar 2013.

3. List of 83 S. Krein's PhD students

Artemov, Georgii; Askerov, Nazim; Atlasov, Igor; Belyaeva, Elena; Berezansky, Yurii; Bryskin, Iliya; Chan, Thu; Chernyshov, Kornelii; Chuburin, Yurii; Daletsky, Yurii; Dement'eva, Olga; Denisov, Igor; Dmitriev, Vyacheslav; Duhovnyi, Mikhail; Fam, Ki; Fomin, Vasilii; Frolov, Nikolai; Furmenko, Aleksandr; Gasanov, Nasir; Glushko, Vladimir; Gohman, Aleksei; Gorohov, Evgenii; Grabovskaya, Revekka;

Gudovich, Nikolai; Ievleva, Oksana; Ivanov, Leonid; Kan, Ngo Zuy; Kolupanova, Galina; Kopachevskii, Nikolai; Kopaneva, Vera; Kostarchuk, Victor; Kostin, Vladimir; Kotko, Lyudmila; Kozlov, Ovsei; Ksendzenko, Lyudmila; Kuchment, Peter; Kulikov, Ivan; Kurina, Galina; Kvedaras, Bronyus; Laptev, Gennadii; Litvinkov, Stepan; Livchak, Aleksei; Nguen, Shi; Nikolov, Krasemir; Nikolova, Lyudmila; Osipov, Vasilii; Ovchinnikov, Vladimir; Pankov, Aleksandr; Pavlov, Evgenii; Pesenson, Isaac; Petunin, Yurii; Plushev, Yurii; Polichka, Nina; Prozorovskaya, Olga; Russman, Isaak; Rutitskaya, Alla; Salekhov, Dmitrii; Sapronov, Ivan; Savchenko, Galina; Savchenko, Yulia; Sedaev, Aleksandr; Semenov, Evgenii; Shablitzskaya, Liliya; Shihvatov, Aleksandr; Shmulev, Igor; Shneiberg, Iosif; Simonov, Aleksandr; Sklyadnev, Sergei Sobolevskii, Pavel; Solomatina, Lyuba; Susoev, Yurii; Tetievskaya, Irina; Tovbis, Aleksandr; Trofimov, Valerii; Venevitina, Svetlana; Yakut, Lidia; Yaroshenko, Nikolai; Yatzkin, Nikolai; Zaidenberg, Mikhail; Zarubin, Anatolii; Zobin, Nahum; Zubova, Svetlana; Zyukin, Pavel.

4. List of S. Krein's publications: books and papers

References

[1] Krein S. G. and Ushakova V. N., *Mathematical Analysis of Elementary Functions*, Fizmatgiz, Moscow 1963. (Translated to German in 1966)

[2] N. Ja. Vilenkin, E. A. Gorin, A. G. Kostjučenko, M. A. Krasnosel′skiĭ, S. G. Kreĭn, V. P. Maslov, B. S. Mitjagin, Ju. I. Petunin, Ja. B. Rutickiĭ, V. I. Sobolev, V. Ja. Stecenko, L. D. Faddeev, and È. S. Citlanadze, *Funktsionalyanyĭ analiz* (Russian), Edited by S. G. Kreĭn, Izdat. "Nauka", Moscow, 1964. MR0184056

[3] S. G. Kreĭn, *Lineĭ khye differentsial′nye uravneniya v Banakhovom prostranstve* (Russian), Izdat. "Nauka", Moscow, 1967. MR0247239

[4] S. G. Krein (ed.), *Analiza funkcjonalna* (Polish), A collection of articles edited by S. G. Kreĭn. Translated from the Russian by Ryszard Bittner, Państwowe Wydawnictwo Naukowe, Warsaw, 1967. MR0220034

[5] S. G. Krein (ed.), *Functional analysis*, Edited machine translation from the Russian. Foreign Technology Division MT-65-573, U.S. Department of Commerce, National Bureau of Standards, Washington, D.C., 1968. MR0234242

[6] S. G. Kreĭn, *Linear differential equations in Banach space*, American Mathematical Society, Providence, R.I., 1971. Translated from the Russian by J. M. Danskin; Translations of Mathematical Monographs, Vol. 29. MR0342804

[7] S. G. Kreĭn, *Lineĭ nye uravneniya v banakhovom prostranstve* (Russian), Izdat. "Nauka", Moscow, 1971. MR0374949

[8] M. Sh. Birman, N. Ya. Vilenkin, E. A. Gorin, P. P. Zabreĭko, I. S. Iokhvidov, M. Ĭ. Kadets′, A. G. Kostyuchenko, M. A. Krasnosel′skiĭ, S. G. Kreĭn, B. S. Mityagin, Yu. I. Petunin, Ya. B. Rutitskiĭ, E. M. Semenov, V. I. Sobolev, V. Ya. Stetsenko, L. D. Faddeev, and È. S. Tsitlanadze, *Funktsional′nyĭ analiz* (Russian), Izdat. "Nauka", Moscow, 1972. Edited by S. G. Kreĭn; Second edition, revised and augmented; Mathematical Reference Library. MR0352920

[9] N. Ya. Vilenkin, E. A. Gorin, A. G. Kostyuchenko, S. G. Krasnosel′skiĭ, S. G. Kreĭn, V. P. Maslov, B. S. Mityagin, Yu. I. Petunin, Ya. B. Rutitskii, V. I. Sobolev, V. Ya. Stetsenko, L. D. Faddeev, and E. S. Tsitlanadze, *Functional analysis*, Wolters-Noordhoff Publishing, Groningen, 1972. Translated from the Russian by Richard E. Flaherty; English edition edited by George F. Votruba with the collaboration of Leo F. Boron. MR0390693

[10] I. S. Gudovič and S. G. Kreĭn, *Boundary value problems for overdetermined systems of partial differential equations* (Russian, with Lithuanian and English summaries), Differencial′nye Uravnenija i Primenen.—Trudy Sem. Processy **Vyp. 9** (1974), 1–145. MR0481612

[11] L. Ivanov, L. Kotko, and S. Kreĭn, *Boundary value problems in variable domains* (Russian, with Lithuanian and English summaries), Differencial′nye Uravnenija i Primenen.—Trudy Sem. Processy **Vyp. 19** (1977), 161. MR0499710

[12] Krein S. G. and Zobin N. M., *Mathematical Analysis of Smooth Functions*, Voronezh State University, Voronezh 1978

[13] S. G. Kreĭn, Ju. I. Petunin, and E. M. Semënov, *Interpolyatsiya lineĭ nykh operatorov* (Russian), "Nauka", Moscow, 1978. MR506343

[14] S. G. Kreĭn and N. I. Yatskin, *Lineĭ nye differentsial′nye uravneniya na mnogoobraziyakh* (Russian), Voronezh. Gos. Univ., Voronezh, 1980. MR640266

[15] S. G. Kreĭn, *Linear equations in Banach spaces*, Birkhäuser, Boston, Mass., 1982. Translated from the Russian by A. Iacob; With an introduction by I. Gohberg. MR684836

[16] S. G. Kreĭn, Yu. Ī. Petunīn, and E. M. Semënov, *Interpolation of linear operators*, Translations of Mathematical Monographs, vol. 54, American Mathematical Society, Providence, R.I., 1982. Translated from the Russian by J. Szűcs. MR649411

[17] S. G. Kreĭn, *Matematicheskoe programmirovanie* (Russian), Voronezhskiĭ Gosudarstvennyĭ Universitet, Voronezh, 1983. MR768927

[18] N. D. Kopachevskiĭ, S. G. Kreĭn, and Ngo Huy Can, *Operatornye metody v lineĭ noĭ gidrodinamike* (Russian), "Nauka", Moscow, 1989. Èvolyutsionnye i spektral′nye zadachi. [Evolution and spectral problems]; With an English summary. MR1037258

[19] S. G. Kreĭn and I. Z. Pesenson, *Prostranstva gladkikh èlementov, porozhdennykh predstavleniem gruppy Li* (Russian), Voronezhskiĭ Gosudarstvennyĭ Universitet, Voronezh, 1990. Interpolyatsiya i priblizhenie. [Interpolation and approximation]; With an English summary. MR1071381

[20] N. D. Kopachevsky and S. G. Krein, *Operator approach to linear problems of hydrodynamics. Vol. 1*, Operator Theory: Advances and Applications, vol. 128, Birkhäuser Verlag, Basel, 2001. Self-adjoint problems for an ideal fluid. MR1860016

[21] N. D. Kopachevsky and S. G. Krein, *Operator approach to linear problems of hydrodynamics. Vol. 2*, Operator Theory: Advances and Applications, vol. 146, Birkhäuser Verlag, Basel, 2003. Nonself-adjoint problems for viscous fluids. MR2002951

Articles

1940

[22] (with Krein M. G.), An internal characterization of the space of all continuous functions defined on Hausdorff bicompact set, Dokl. Akad. Nauk SSSR 27(1940), no.5, 427-431. (In Russian)

1941

[23] (with Vershikov I. Kh. and Tovbin A. V.), on semi-ordered rings, Dokl. Akad. Nauk SSSR 30(1941), 778-780. (In Russian)

1943

[24] M. Krein and S. Krein, *Sur l'espace des fonctions continues définies sur un bicompact de Hausdorff et ses sousespaces semiordonnés* (French, with Russian summary), Rec. Math. [Mat. Sbornik] N.S. **13(55)** (1943), 1–38. MR0012209

1946

[25] (with Bogolyubov N. N.), On positive absolutely continuous operators, Trans. Inst. Matem. Ukarin. Akad. Nauk. 9(1946), 130-139. (In Ukrainian)

1947

[26] M. Krasnosel′skiĭ and S. Kreĭn, *On the center of a general dynamical system* (Russian), Doklady Akad. Nauk SSSR (N. S.) **58** (1947), 9–11. MR0022658

1948

[27] S. G. Kreĭn and B. Ya. Levin, *On the convergence of singular integrals* (Russian), Doklady Akad. Nauk SSSR (N.S.) **60** (1948), 13–16. MR0025606

[28] S. G. Kreĭn and B. Ya. Levin, *On the strong representation of functions by singular integrals* (Russian), Doklady Akad. Nauk SSSR (N.S.) **60** (1948), 195–198. MR0025607

[29] (with Levin B. Ya.), On one problem by I. P. Natanson, Uspehi Mat. Nauk 3(1948), no.3, 183-186. (In Russian)

[30] B. I. Korenblyum, S. G. Kreĭn, and B. Y. Levin, *On certain nonlinear questions of the theory of singular integrals* (Russian), Doklady Akad. Nauk SSSR (N.S.) **62** (1948), 17–20. MR0027439

[31] (with Kac G. I.), On the limit center of a dynamical system, Trans. Inst. Matem. Ukarin. Akad. Nauk. 1948, 121-134. (In Russian)

1949

[32] M. A. Krasnosel′skiĭ and S. G. Kreĭn, *On a proof of the theorem on category of a projective space* (Russian), Ukrain. Mat. Žurnal **1** (1949), no. 2, 99–102. MR0048809

1950

[33] Yu. L. Daleckiĭ and S. G. Kreĭn, *On differential equations in Hilbert space* (Russian), Ukrain. Mat. Žurnal **2** (1950), no. 4, 71–91. MR0047925

[34] (with Daleckii Yu. L.), Some properties of operators depending upon a parameter, Dokl. Akad. Nauk. Ukrain. SSR 6(1950), 433-436. (In Ukrainian)

[35] Yu. M. Berezanskiĭ and S. G. Kreĭn, *Some classes of continuous algebras* (Russian), Doklady Akad. Nauk SSSR (N.S.) **72** (1950), 237–240. MR0036946

[36] Yu. M. Berezanskiĭ and S. G. Kreĭn, *Continuous algebras* (Russian), Dokaldy Akad. Nauk SSSR (N.S.) **72** (1950), 5–8. MR0036945

1951

[37] Yu. M. Berezanskiĭ and S. G. Kreĭn, *Hypercomplex systems with a compact basis* (Russian), Ukrain. Mat. Žurnal **3** (1951), 184–204. MR0054585

[38] (with Daleckii Yu. L.), Formulas of differentiation with respect to a parameter of functions of Hermitian operators, Doklady Akad. Nauk SSSR (N.S.) 76(1951). 13–16. MR 12,617f.

1952

[39] M. A. Krasnosel′skiĭ and S. G. Kreĭn, *An iteration process with minimal residuals* (Russian), Mat. Sbornik N.S. **31(73)** (1952), 315–334. MR0052885

[40] M. A. Krasnosel′skiĭ and S. G. Kreĭn, *Remark on the distribution of errors in the solution of a system of linear equations by means of an iterative process* (Russian), Uspehi Matem. Nauk (N.S.) **7** (1952), no. 4(50), 157–161. MR0051582

1953

[41] S. G. Kreĭn, *On functional properties of operators of vector analysis and hydrodynamics* (Russian), Doklady Akad. Nauk SSSR (N.S.) **93** (1953), 969–972. MR0061954

[42] S. G. Kreĭn, *Uniform topology in the space of transformations* (Russian), Mat. Sbornik N.S. **33(75)** (1953), 627–638. MR0059542

[43] On fixed points of a conformal mapping, Uspehi Matem. Nauk (N.S.) 8(1953). no. 1(53), 155–159. MR 14,742f.

1954

[44] On an indeterminate equation in Hilbert space and its application in potential theory, Uspehi Matem. Nauk (N.S.) 9(1954). no. 3(61), 149–153. MR 16,262g.

1955

[45] M. A. Krasnosel′skiĭ and S. G. Kreĭn, *On the principle of averaging in nonlinear mechanics* (Russian), Uspehi Mat. Nauk (N.S.) **10** (1955), no. 3(65), 147–152. MR0071596

[46] M. A. Krasnosel′skiĭ and S. G. Kreĭn, *Nonlocal existence theorems and uniqueness theorems for systems of ordinary differential equations* (Russian), Dokl. Akad. Nauk SSSR (N.S.) **102** (1955), 13–16. MR0071588

1956

[47] Mathematical problems of the theory motion of a vessel filled with a fluid, Trans. of III All-union Math. Congress, v.1, p. 205, Akad. Nauk SSSR, Moscow 1956.

[48] (with Krasnoselskii M. A.), On differential equations in Banach spaces, Trans. of III All-union Math. Congress, v.2, p. 11, Akad. Nauk SSSR, Moscow 1956.

[49] M. A. Krasnosel′skiĭ, S. G. Kreĭn, and P. E. Sobolevskiĭ, *On differential equations with unbounded operators in Banach spaces* (Russian), Dokl. Akad. Nauk SSSR (N.S.) **111** (1956), 19–22. MR0088622

[50] M. A. Krasnosel′skiĭ and S. G. Kreĭn, *On the theory of ordinary differential equations in Banach spaces* (Russian), Voronoež. Gos. Univ. Trudy Sem. Funkcional. Anal. **1956** (1956), no. 2, 3–23. MR0086191

[51] Yu. L. Daleckiĭ and S. G. Kreĭn, *Integration and differentiation of functions of Hermitian operators and applications to the theory of perturbations* (Russian), Voronež. Gos. Univ. Trudy Sem. Funkcional. Anal. **1956** (1956), no. 1, 81–105. MR0084745

[52] M. A. Krasnosel′skiĭ and S. G. Kreĭn, *On a class of uniqueness theorems for the equation* $y' = f(x, y)$ (Russian), Uspehi Mat. Nauk (N.S.) **11** (1956), no. 1(67), 209–213. MR0079152

1957

[53] S. G. Krein, *On correctness classes for certain boundary problems* (Russian), Dokl. Akad. Nauk SSSR (N.S.) **114** (1957), 1162–1165. MR0089977

[54] S. G. Kreĭn, *Differential equations in a Banach space and their application in hydromechanics* (Russian), Uspehi Mat. Nauk (N.S.) **12** (1957), no. 1(73), 208–211. MR0085418
[55] M. A. Krasnosel′skiĭ and S. G. Kreĭn, *Continuity conditions for a linear operator in terms of properties of its square* (Russian), Voronež. Gos. Univ. Trudy Sem. Funkcional. Anal. **1957** (1957), no. 5, 98–101. MR0098323
[56] S. G. Kreĭn and O. I. Prozorovskaya, *An analogue of Seidel's method for operator equations* (Russian), Voronež. Gos. Univ. Trudy Sem. Funkcional. Anal. **5** (1957), 35–38. MR0095579
[57] M. A. Krasnoselsky, S. G. Krein, and P. E. Sobolevsky, *On differential equations with unbounded operators in Hilbert space* (Russian), Dokl. Akad. Nauk SSSR (N.S.) **112** (1957), 990–993. MR0089974
[58] S. G. Kreĭn and N. N. Moiseev, *On oscillations of a vessel containing a liquid with a free surface* (Russian), Prikl. Mat. Meh. **21** (1957), 169–174. MR0089572
[59] Yu. M. Berezanskiĭ and S. G. Kreĭn, *Hypercomplex systems with continual basis* (Russian), Uspehi Mat. Nauk (N.S.) **12** (1957), no. 1(73), 147–152. MR0086272
[60] (with Krasnosel'skii M. A. and Myshkis A. D.), An extended session in march 1957 of the Voronezh Seminar on Functional Analysis, Uspehi Mat. nauk 12(1957), no.4, 241-250.
1958
[61] V. P. Gluško and S. G. Kreĭn, *Fractional powers of differential operators and imbedding theorems.* (Russian), Dokl. Akad. Nauk SSSR **122** (1958), 963–966. MR0100144
[62] S. G. Kreĭn and P. E. Sobolevskiĭ, *A differential equation with an abstract elliptical operator in Hilbert space* (Russian), Dokl. Akad. Nauk SSSR (N.S.) **118** (1958), 233–236. MR0099604
[63] (with Krasnosel'skii M. A.), On differential equations in a Banach space,, Trans. of III All-union Math. Congress, v.3, Moscow (1956), p. 73-80, Akad. Nauk SSSR, Moscow 1958.
1960
[64] S. G. Kreĭn, *On an interpolation theorem in operator theory*, Soviet Math. Dokl. **1** (1960), 61–64. MR0119094
[65] S. G. Kreĭn, *On the concept of a normal scale of spaces*, Soviet Math. Dokl. **1** (1960), 586–589. MR0121627
[66] S. G. Kreĭn, *Differential equations in Banach space and their application in hydrodynamics*, Amer. Math. Soc. Transl. (2) **16** (1960), 423–426, DOI 10.1090/trans2/016/26. MR0117452
[67] Yu. M. Berezanski and S. G. Kreĭn, *Hypercomplex systems with continuous basis*, Amer. Math. Soc. Transl. (2) **16** (1960), 358–364. MR0117591
[68] S. G. Kreĭn and O. I. Prozorovskaja, *Analytic semi-groups and incorrect problems for evolutionary equations*, Soviet Math. Dokl. **1** (1960), 841–844. MR0151862
[69] V. P. Gluško and S. G. Kreĭn, *Inequalities for norms of derivatives in weighted L_p spaces* (Russian), Sibirsk. Mat. Ž. **1** (1960), 343–382. MR0133681
1961
[70] Ill-posed problems and estimates of solutions of parabolic equations, in "Ill-posed problems of mathematics and mechanics", 84-86, Siberian Branch of Acad. of Sci., Novosibirsk 1961
[71] S. G. Kreĭn and Ju. I. Petunin, *A relationship criterion for two Banach spaces* (Russian), Dokl. Akad. Nauk SSSR **139** (1961), 1295–1298. MR0141975
[72] S. G. Kreĭn and E. M. Semenov, *A scale of spaces* (Russian), Dokl. Akad. Nauk SSSR **138** (1961), 763–766. MR0140939
1962
[73] S. G. Kreĭn and G. I. Laptev, *Boundary-value problems for an equation in Hilbert space* (Russian), Dokl. Akad. Nauk SSSR **146** (1962), 535–538. MR0156068
1963
[74] Ju. M. Berezans′kiĭ, S. G. Kreĭn, and Ja. A. Roĭtberg, *A theorem on homeomorphisms and local increase of smoothness up to the boundary for solutions of elliptic equations*, Outlines Joint Sympos. Partial Differential Equations (Novosibirsk, 1963), Acad. Sci. USSR Siberian Branch, Moscow, 1963, pp. 33–38. MR0209665
[75] S. G. Kreĭn and O. I. Prozorovskaja, *Approximate methods of solving ill-posed problems* (Russian), Ž. Vyčisl. Mat. i Mat. Fiz. **3** (1963), 120–130. MR0153125
[76] Yu. M. Berezanski, S. G. Kreĭn, and Ja. A. Roĭtberg, *A theorem on homeomorphisms and local increase of smoothness up to the boundary for solutions of elliptic equations* (Russian), Dokl. Akad. Nauk SSSR **148** (1963), 745–748. MR0146508

1964

[77] S. G. Kreĭn, *Oscillations of a viscous fluid in a container* (Russian), Dokl. Akad. Nauk SSSR **159** (1964), 262–265. MR0182238

[78] S. G. Kreĭn, *Interpolation theorems in operator theory, and embedding theorems* (Russian), Proc. Fourth All-Union Math. Congr. (Leningrad, 1961), Izdat. "Nauka", Leningrad, 1964, pp. 504–510. MR0220077

[79] (with Krasnosel'skii M. A.), Operator equations in function spaces, in Proc. Fourth All-Union Math. Congr. (Leningrad, 1961), Vol. II, 292–299, "Nauka", Leningrad 1964. (In Russian) MR 36#2926.

[80] S. G. Kreĭn and Ju. I. Petunin, *On the concept of minimal scale of spaces* (Russian), Dokl. Akad. Nauk SSSR **154** (1964), 30–33. MR0161125

[81] N. G. Askerov, S. G. Kreĭn, and G. I. Laptev, *On a class of non-selfadjoint boundary-value problems* (Russian), Dokl. Akad. Nauk SSSR **155** (1964), 499–502. MR0160133

[82] (with Laptev G. I.), Boundary value problems with parameter in the boundary condition, Proc. 3rd All-Union Symposium on Diffraction and Waves, 39-41, Tbilisi 1964

[83] (with Krasnosel'skii M. A., Rutitskii Ya. B., and Sobolev V. I.), On mathematical life in Voronezh, Uspehi Mat. nauk 19(1964), no.3, 225-245. (In Russian)

1965

[84] (with Simonov A. S.), Theorem on homomorphisms and quai-linear equations, in "Abstracts of All-union Conf. on Appli. of Funct. Anal. and Non-linear problems", 62-63, Akad. Nauk Azerb. SSR, Baku 1965.

1966

[85] S. G. Kreĭn, *Correctness of the Cauchy problem and the analyticity of solutions of the evolution equation* (Russian), Dokl. Akad. Nauk SSSR **171** (1966), 1033–1036. MR0208115

[86] S. G. Kreĭn, Ju. I. Petunin, and E. M. Semenov, *Hyperscales of Banach lattices* (Russian), Dokl. Akad. Nauk SSSR **170** (1966), 265–267. MR0203448

[87] S. G. Kreĭn and G. I. Laptev, *Boundary value problems for second order differential equations in a Banach space. I* (Russian), Differencial'nye Uravnenija **2** (1966), 382–390. MR0199518

[88] S. G. Kreĭn and G. I. Laptev, *Correctness of boundary value problems for a differential equation of the second order in a Banach space. II* (Russian), Differencial'nye Uravnenija **2** (1966), 919–926. MR0203196

[89] S. G. Kreĭn and L. N. Šablickaja, *Stability of difference schemes for the Cauchy problem* (Russian), Ž. Vyčisl. Mat. i Mat. Fiz. **6** (1966), 648–664. MR0199978

[90] S. G. Kreĭn and A. S. Simonov, *A theorem of homeomorphisms and quasilinear equations* (Russian), Dokl. Akad. Nauk SSSR **167** (1966), 1226–1229. MR0197979

[91] S. G. Kreĭn and Ju. I. Petunin, *Scales of Banach spaces* (Russian), Uspehi Mat. Nauk **21** (1966), no. 2 (128), 89–168. MR0193499

[92] (with Glushko V. P. and Mukhin V. E.), On global and local optimal plans in problems of linear programming with two-sided bounds, in "Optimal programming in industrial problems", no.2, 89-90, Voronezh State Univ. 1966. (In Russian).

[93] (with Ievleva O. B.), On oscillations of a viscous fluid in a vessel, Abstracts of the Internat. Math. Congr., Sect. 12, p.37, Moscow 1966.

[94] (with Petunin Yu. I.), New results in the theory of scales of Banach spaces, Abstracts of the Internat. Math. Congr., Sect. 5, p.56, Moscow 1966.

1967

[95] S. G. Kreĭn, Ju. I. Petunin, and E. M. Semenov, *Scales of Banach lattices of measurable functions* (Russian), Trudy Moskov. Mat. Obšč. **17** (1967), 293–322. MR0223878

[96] Sixth joint conference in physics and mathematics of the Far East, Uspehi Mat. Nauk 22(1967), no.1, 197-198. (In Russian).

[97] First Voronezh Winter Math. School, Uspehi Mat. Nauk 22(1967), no.4, 189-190. (In Russian).

1968

[98] S. G. Kreĭn and G. I. Laptev, *On the problem of the motion of a viscous fluid in an open vessel* (Russian), Funkcional. Anal. i Priložen. **2** (1968), no. 1, 40–50. MR0248462

[99] S. G. Kreĭn, *The behavior of solutions of elliptic problems under variation of the domain* (Russian), Studia Math. **31** (1968), 411–424. MR0235276

[100] The behavior of solutions of elliptic problems under variation of the domain, Proc. of 7th Math. and Phys. Conf. of Far east, p. 18, Khabarovsk 1968. (In Russian)
[101] S. G. Kreĭn, *Lineĭ nye uravneniya v banakhovom prostranstve* (Russian), Voronež. Gosudarstv. Univ., Voronezh, 1968. MR0374950
[102] S. G. Kreĭn and G. I. Laptev, *On the problem of the motion of a viscous fluid in an open vessel* (Russian), Funkcional. Anal. i Priložen. **2** (1968), no. 1, 40–50. MR0248462
[103] N. K. Askerov, S. G. Kreĭn, and G. I. Laptev, *The problem of the oscillations of a viscous liquid and the operator equations connected with it* (Russian), Funkcional. Anal. i Priložen. **2** (1968), no. 2, 21–31. MR0232233
[104] V. P. Gluško and S. G. Kreĭn, *Degenerate linear differential equations in a Banach space* (Russian), Dokl. Akad. Nauk SSSR **181** (1968), 784–787. MR0232067
1969
[105] S. G. Kreĭn and I. M. Kulikov, *The Maxwell-Leontovič operator* (Russian), Differencial′nye Uravnenija **5** (1969), 1275–1282. MR0265991
[106] S. G. Kreĭn and V. P. Trofimov, *Holomorphic operator-valued functions of several complex variables.* (Russian), Funkcional. Anal. i Priložen. **3** (1969), no. 4, 85–86. MR0262861
[107] S. G. Kreĭn and Ngo Zui Kan, *The problem of small motions of a body with a cavity partially filled with a viscous fluid*, J. Appl. Math. Mech. **33** (1969), 110–117, DOI 10.1016/0021-8928(69)90118-X. MR0260269
[108] S. G. Kreĭn and Ngo Zui Kan, *Asymptotic method in the problem of oscillations of a strongly viscous fluid*, J. Appl. Math. Mech. **33** (1969), 442–450, DOI 10.1016/0021-8928(69)90059-8. MR0259257
[109] S. G. Kreĭn and G. I. Laptev, *An abstract scheme for the examination of parabolic problems in noncylindrical regions* (Russian), Differencial′nye Uravnenija **5** (1969), 1458–1469. MR0255989
[110] III Voronezh Winter Math. School, January 26 - February 6 1969, Uspehi Mat. Nauk 24(1969), no.4, 230-231. (In Russian)
1970
[111] S. G. Kreĭn and A. M. Šihvatov, *Linear differential equations on a Lie group* (Russian), Funkcional. Anal. i Priložen. **4** (1970), no. 1, 52–61. MR0439987
[112] S. G. Kreĭn and V. P. Trofimov, *Noetherian operators that depend holomorphically on parameters* (Russian), A collection of articles on function spaces and operator equations (Proc. Sem. Functional Anal., Math. Fac., Voronezh State Univ., Voronezh, 1970)(Russian), Voronež. Gos. Univ., Voronezh, 1970, pp. 63–85. MR0430834
[113] S. G. Kreĭn and V. P. Trofimov, *The multiplicity of a characteristic point of the holomorphic operator-function* (Russian), Mat. Issled. **5** (1970), no. 4(18), 105–114. MR0312305
[114] S. G. Kreĭn and Ju. B. Savčenko, *Exponential dichotomy for partial differential equations* (Russian), Differencial′nye Uravnenija **8** (1972), 835–844. MR0304809
[115] S. G. Kreĭn and V. B. Osipov, *Ljapunov functions and Cauchy problems for certain systems of partial differential equations* (Russian), Differencial′nye Uravnenija **6** (1970), 2053–2061. MR0299930
[116] S. G. Kreĭn, Ju. Ī. Petunīn, and E. M. Semenov, *Imbedding theorems and interpolation of linear operators* (Russian), Imbedding theorems and their applications (Proc. Sympos., Baku, 1966), Izdat. "Nauka", Moscow, 1970, pp. 127–131, 245. MR0313788
[117] S. G. Kreĭn, G. I. Laptev, and G. A. Cvetkova, *The Hadamard correctness of the Cauchy problem for an evolution equation* (Russian), Dokl. Akad. Nauk SSSR **192** (1970), 980–983. MR0265728
[118] Voronezh Winter Math. School, January 26 - February 7 1970, Uspehi Mat. Nauk 25(1970), no.5, 265-266. (In Russian)
1971
[119] S. G. Kreĭn and P. A. Kučment, *A certain approach to the interpolation problem for linear operators* (Russian), Voronež. Gos. Univ. Trudy Naučn.-Issled. Inst. Mat. VGU **3** (1971), 54–60. MR0355632
[120] S. G. Kreĭn, *Interpolation of linear operators, and properties of the solutions of elliptic equations* (Russian, with English summary), Elliptische Differentialgleichungen, Band II, Akademie-Verlag, Berlin, 1971, pp. 155–166. Schriftenreihe Inst. Math. Deutsch. Akad. Wissensch. Berlin, Reihe A, Heft 8. MR0344646

[121] I. S. Gudovič and S. G. Kreĭn, *Certain boundary value problems that are elliptic in a subspace* (Russian), Mat. Sb. (N.S.) **84 (126)** (1971), 595–606. MR0282050

[122] Fifth Voronezh Winter Math. School, Uspehi Mat. Nauk 26(1971), no.5, 270-272. (In Russian)

1972

[123] S. G. Kreĭn and N. I. Jackin, *Differential form equations* (Russian), Voronež. Gos. Univ. Trudy Naučn.-Issled. Inst. Mat. VGU **Vyp. 5 Sb. Stateĭ Teor. Operator. i Differencial. Uravneniĭ** (1972), 75–79. MR0448384

[124] I. S. Gudovič and S. G. Kreĭn, *Boundary value problems for operators of exterior differentiation* (Russian), Voronež. Gos. Univ. Trudy Naučn.-Issled. Inst. Mat. VGU **Vyp. 5 Sb. Stateĭ Teor. Operator. i Differencial. Uravneniĭ** (1972), 35–45. MR0448383

[125] S. G. Kreĭn and E. M. Semenov, *A certain property of equimeasurable functions* (Russian), Voronez. Gos. Univ. Trudy Naučn.-Issled. Inst. Mat. VGU **Vyp. 5 Sb. Stateĭ Teor. Operator. i Differencial. Uravneniĭ** (1972), 70–74. MR0447512

[126] I. S. Gudovič and S. G. Kreĭn, *Elliptic boundary value problems for the system* $\binom{d_0^*}{d_1}u = f$ (Russian), Funkcional. Anal. i Priložen. **6** (1972), no. 4, 75–76. MR0318666

[127] I. S. Gudovič, S. G. Kreĭn, and I. M. Kulikov, *Boundary value problems for the Maxwell equations* (Russian), Dokl. Akad. Nauk SSSR **207** (1972), 321–324. MR0316892

[128] (with Borisovich Yu. G.), VI Voronezh Winter Math. School, Uspehi Mat. Nauk 27(1972), no.5, 273-275. (In Russian)

[129] S. G. Kreĭn and Ju. B. Savčenko, *Exponential dichotomy for partial differential equations* (Russian), Differencial′nye Uravnenija **8** (1972), 835–844. MR0304809

1973

[130] S. G. Kreĭn and S. Ja. L′vin, *A general initial problem for a differential equation in a Banach space* (Russian), Dokl. Akad. Nauk SSSR **211** (1973), 530–533. MR0342805

[131] R. Ja. Grabovskaja and S. G. Kreĭn, *A certain representation of the algebra of differential operators, and the differential equations connected with it* (Russian), Dokl. Akad. Nauk SSSR **212** (1973), 280–283. MR0342889

[132] R. Ja. Grabovskaja and S. G. Kreĭn, *The formula for the permutation of functions of operators that represent a Lie algebra* (Russian), Funkcional. Anal. i Priložen. **7** (1973), no. 3, 81. MR0336405

[133] S. G. Kreĭn and L. N. Šablickaja, *Necessary conditions for the stability of difference schemes, and the eigenvalues of difference operators* (Russian), Ž. Vyčisl. Mat. i Mat. Fiz. **13** (1973), 647–657, 812. MR0331805

[134] S. G. Kreĭn and E. M. Semenov, *Interpolation of operators of weakened type* (Russian), Funkcional. Anal. i Priložen. **7** (1973), no. 2, 89–90. MR0315429

1974

[135] N. V. Efimov, L. V. Kantorovič, S. G. Kreĭn, I. S. Iohvidov, M. A. Krasnosel′s′kiĭ, and L. A. Ljusternik, *Vladimir Ivanovič Sobolev (on the occasion of his sixtieth birthday)* (Russian), Uspehi Mat. Nauk **29** (1974), no. 1(175), 247–250. (1 plate). MR0386949

1975

[136] M. G. Zaĭdenberg, S. G. Kreĭn, P. A. Kučment, and A. A. Pankov, *Banach bundles and linear operators* (Russian), Uspehi Mat. Nauk **30** (1975), no. 5(185), 101–157. MR0415661

[137] L. A. Kotko and S. G. Kreĭn, *The completeness of a system of eigen- and associated functions of boundary value problems with a parameter in the boundary conditions* (Russian), Collection of articles on applications of functional analysis (Russian), Voronež. Tehnolog. Inst., Voronezh, 1975, pp. 71–89. MR0481622

1976

[138] S. G. Kreĭn and E. M. Semenov, *Some interpolation theorems of the theory of linear operators and their applications* (Russian), Imbedding theorems and their applications (Proc. All-Union Sympos., Alma-Ata, 1973), Izdat. "Nauka" Kazah. SSR, Alma-Ata, 1976, pp. 64–68, 188. MR0482126

[139] R. Ya. Grabovskaya and S. G. Kreĭn, *Second order differential equations with operators generating a Lie algebra representation*, Math. Nachr. **75** (1976), 9–29, DOI 10.1002/mana.19760750103. MR0481673

[140] L. A. Kotko and S. G. Kreĭn, *The completeness of the system of eigen- and associated functions of boundary value problems with a parameter in the boundary conditions* (Russian), Dokl. Akad. Nauk SSSR **227** (1976), no. 2, 288–290. MR0402296

1977

[141] Interpolation of operators in spaces of smooth functions, in "Operator theory in functional spaces", 188-205, Nauka, Novosibirsk 1977.

[142] V. I. Dmitriev, S. G. Kreĭn, and V. I. Ovčinnikov, *Fundamentals of the theory of interpolation of linear operators* (Russian), Geometry of linear spaces and operator theory (Russian), Jaroslav. Gos. Univ., Yaroslavl, 1977, pp. 31–74. MR0634076

1978

[143] S. G. Kreĭn and A. I. Furmenko, *Dynamic equivalence of linear differential operators with nilpotent operators* (Russian), Approximate methods for investigating differential equations and their applications, No. 4 (Russian), Kuĭbyshev. Gos. Univ., Kuybyshev, 1978, pp. 47–56, 134. MR578334

[144] V. I. Dmitriev and S. G. Kreĭn, *Interpolation of operators of weak type* (English, with Russian summary), Anal. Math. **4** (1978), no. 2, 83–99, DOI 10.1007/BF02116975. MR505532

1980

[145] S. G. Kreĭn and N. P. Polička, *Behavior of the solutions of boundary value problems for the Lavrent′ev-Bicadze equation under variation of the domain* (Russian), Partial differential equations (Proc. Conf., Novosibirsk, 1978), "Nauka" Sibirsk. Otdel., Novosibirsk, 1980, pp. 39–41, 247. MR601401

[146] S. G. Kreĭn and L. Ĭ. Nikolova, *Holomorphic functions in a family of Banach spaces, interpolation* (Russian), Dokl. Akad. Nauk SSSR **250** (1980), no. 3, 547–550. MR557785

1981

[147] S. G. Kreĭn and K. I. Chernyshov, *Analogue of Tikhonov's theorem for the equation* $(A+\varepsilon B)\dot{x} = C(t)x$ (Russian), Approximate methods for investigating differential equations and their applications, Kuĭbyshev. Gos. Univ., Kuybyshev, 1981, pp. 103–115. MR708060

[148] S. G. Kreĭn and K. I. Chernyshov, *Behavior of solutions of general linear systems depending meromorphically on a small parameter* (Russian), Dokl. Akad. Nauk SSSR **260** (1981), no. 3, 530–535. MR631925

[149] S. G. Kreĭn and G. A. Kurina, *Singular perturbations in problems of optimal control* (Russian), Stability of motion. Analytical mechanics. Control of motion, "Nauka", Moscow, 1981, pp. 170–178, 303. MR663404

1982

[150] S. G. Kreĭn and L. Ĭ. Nikolova, *A complex interpolation method for a family of Banach spaces* (Russian), Ukrain. Mat. Zh. **34** (1982), no. 1, 31–42, 132. MR647928

1983

[151] S. G. Kreĭn and M. I. Khazan, *Differential equations in a Banach space* (Russian), Mathematical analysis, Vol. 21, Itogi Nauki i Tekhniki, Akad. Nauk SSSR, Vsesoyuz. Inst. Nauchn. i Tekhn. Inform., Moscow, 1983, pp. 130–264. MR736523

1984

[152] S. G. Kreĭn and L. Y. Nikolova, *On the method of complex interpolation*, Complex analysis and applications '81 (Varna, 1981), Publ. House Bulgar. Acad. Sci., Sofia, 1984, pp. 298–300. MR883250

[153] S. G. Kreĭn and N. P. Polichka, *A priori estimates for solutions of a problem with displacement for the Lavrent′ev-Bitsadze equation* (Russian), Differentsial′nye Uravneniya **20** (1984), no. 12, 2112–2120. MR772808

[154] S. G. Kreĭn, V. I. Levin, and A. S. Simonov, *L. M. Likhtarnikov (on the occasion of his sixtieth birthday)* (Russian), Mat. v Shkole **3** (1984), 77. MR755110

[155] S. G. Kreĭn and K. I. Chernyshov, *Singularly perturbed differential equations in a Banach space* (Russian), Ninth international conference on nonlinear oscillations, Vol. 1 (Kiev, 1981), "Naukova Dumka", Kiev, 1984, pp. 193–197, 443. MR800427

[156] (with Chernyshov K.I., Kuchment P.A., and Lvin S. L.) On the reconstruction of functions from empirical data with a priori information, preprint, VINITI, no. 1587–84, 53 pp., 1984. (in Russian).

1985

[157] S. G. Kreĭn, *Singularly perturbed linear differential equations in a Banach space* (Russian), Differentsial′nye Uravneniya **21** (1985), no. 10, 1814–1817, 1839. MR814583

1986

[158] Yu. A. Brudnyĭ, S. G. Kreĭn, and E. M. Semënov, *Interpolation of linear operators* (Russian), Mathematical analysis, Vol. 24 (Russian), Itogi Nauki i Tekhniki, Akad. Nauk SSSR, Vsesoyuz. Inst. Nauchn. i Tekhn. Inform., Moscow, 1986, pp. 3–163, 272. Translated in J. Soviet Math **42** (1988), no. 6, 2009–2112. MR887950

[159] S. G. Kreĭn, *Asymptotic decomposition of operator equations* (Russian), Lyapunov functions and their applications (Russian), "Nauka" Sibirsk. Otdel., Novosibirsk, 1986, pp. 206–214, 248. MR892764

1987

[160] S. G. Kreĭn and S. Ya. L′vin, *Overdetermined and underdetermined equations in Hilbert spaces* (Russian), Izv. Vyssh. Uchebn. Zaved. Mat. **9** (1987), 59–66, 83. MR923845

[161] Yu. M. Berezanskiĭ, I. M. Gel′fand, M. G. Kreĭn, S. G. Kreĭn, Yu. A. Mitropol′skiĭ, and A. V. Skorokhod, *Yuriĭ L′vovich Daletskiĭ (on the occasion of his sixtieth birthday)* (Russian), Uspekhi Mat. Nauk **42** (1987), no. 4(256), 213–214. MR912076

[162] S. G. Kreĭn and S. Ya. L′vin, *Overdetermined and underdetermined elliptic problems* (Russian), Functional analysis and mathematical physics (Russian), Akad. Nauk SSSR Sibirsk. Otdel., Inst. Mat., Novosibirsk, 1985, pp. 106–116, 135. MR895832

1988

[163] S. G. Kreĭn and N. P. Polichka, *Stability of solutions of a problem with shift for the Lavrent′ev-Bitsadze equation with variation of the domain* (Russian), Applied problems in statistical analysis (Russian), Akad. Nauk SSSR Sibirsk. Otdel., Dal′nevostochn. Filial, Vladivostok, 1988, pp. 16–29, 162. MR1082384

[164] S. G. Kreĭn and S. Ya. L′vin, *Partially overdetermined and underdetermined elliptic problems* (Russian), Izv. Vyssh. Uchebn. Zaved. Mat. **10** (1988), 15–23; English transl., Soviet Math. (Iz. VUZ) **32** (1988), no. 10, 19–30. MR982602

[165] S. G. Kreĭn and S. Ya. L′vin, *Underdetermined boundary value problems* (Russian), Function spaces and equations of mathematical physics (Russian), Voronezh. Gos. Univ., Voronezh, 1988, pp. 17–24, 85. MR958951

[166] S. G. Kreĭn and S. Ya. L′vin, *An approximation approach to overdetermined and underdetermined boundary value problems* (Russian), Functional and numerical methods in mathematical physics (Russian), "Naukova Dumka", Kiev, 1988, pp. 113–117, 268. MR1038560

1989

[167] N. D. Kopachevskiĭ and S. G. Kreĭn, *A problem of the flow of a viscous fluid* (Russian, with English and German summaries), Z. Anal. Anwendungen **8** (1989), no. 6, 557–561, DOI 10.4171/ZAA/374. MR1050094

[168] S. G. Kreĭn, *An abstract scheme for the consideration of boundary value problems* (Russian), Application of new methods of analysis to differential equations (Russian), Kach. Metody Kraev. Zadach, Voronezh. Gos. Univ., Voronezh, 1989, pp. 46–51, 108. MR1038813

[169] S. G. Kreĭn and S. Ya. L′vin, *Solution of overdetermined and underdetermined elliptic problems in the case of nonsmooth data* (Russian), Ukrain. Mat. Zh. **41** (1989), no. 9, 1222–1225, 1294, DOI 10.1007/BF01056278; English transl., Ukrainian Math. J. **41** (1989), no. 9, 1053–1056 (1990). MR1026421

[170] S. G. Kreĭn and Chan Tkhu Kha, *The problem of the flow of a nonuniformly heated viscous fluid* (Russian), Zh. Vychisl. Mat. i Mat. Fiz. **29** (1989), no. 8, 1153–1158, 1260, DOI 10.1016/0041-5553(89)90127-4; English transl., U.S.S.R. Comput. Math. and Math. Phys. **29** (1989), no. 4, 127–131 (1991). MR1032745

1990

[171] S. G. Kreĭn and V. I. Fomin, *Small perturbations of singular differential equations with unbounded operator coefficients* (Russian), Dokl. Akad. Nauk SSSR **314** (1990), no. 1, 77–79; English transl., Soviet Math. Dokl. **42** (1991), no. 2, 313–315. MR1118482

[172] S. G. Kreĭn and A. I. Tovbis, *Linear singular differential equations in finite-dimensional and Banach spaces* (Russian), Algebra i Analiz **2** (1990), no. 5, 1–62; English transl., Leningrad Math. J. **2** (1991), no. 5, 931–985. MR1086444

[173] S. G. Kreĭn and E. O. Utochkina, *An implicit canonical equation in a Hilbert space* (Russian, with Ukrainian summary), Ukrain. Mat. Zh. **42** (1990), no. 3, 388–390, DOI 10.1007/BF01057021; English transl., Ukrainian Math. J. **42** (1990), no. 3, 345–347. MR1054886

1991

[174] M. Sh. Birman, S. G. Kreĭn, O. A. Ladyzhenskaya, G. V. Rozenblyum, and Yu. G. Safarov, *Mikhail Zakharovich Solomyak (on the occasion of his sixtieth birthday)* (Russian), Uspekhi Mat. Nauk **46** (1991), no. 4(280), 183–184, DOI 10.1070/RM1991v046n04ABEH002828; English transl., Russian Math. Surveys **46** (1991), no. 4, 217–219. MR1138976

[175] V. V. Vasil′ev, S. G. Kreĭn, and S. I. Piskarëv, *Operator semigroups, cosine operator functions, and linear differential equations* (Russian), Mathematical analysis, Vol. 28 (Russian), Itogi Nauki i Tekhniki, Akad. Nauk SSSR, Vsesoyuz. Inst. Nauchn. i Tekhn. Inform., Moscow, 1990, pp. 87–202, 204. Translated in J. Soviet Math. **54** (1991), no. 4, 1042–1129. MR1060534

1994

[176] E. V. Belyaeva and S. G. Krein, *Homogeneous Volterra operator equations with a regular singularity*, Russian J. Math. Phys. **2** (1994), no. 1, 3–12. MR1297937

1995

[177] S. S. Venevitina and S. G. Kreĭn, *Small motions of an elastic medium in an open immobile container* (Russian, with Russian summary), Zh. Vychisl. Mat. i Mat. Fiz. **35** (1995), no. 7, 1095–1107; English transl., Comput. Math. Math. Phys. **35** (1995), no. 7, 875–884. MR1349101

1997

[178] S. G. Kreĭn and S. S. Venevitina, *Small motions of an elastic medium in a cavity of a rotating body* (Russian), Papers from the Conference on Functional Analysis and Equations of Mathematical Physics dedicated to the eightieth birthday of Selim Grigor′evich Kreĭn (Russian), Voronezh. Gos. Univ., Voronezh, 1997, pp. 49–54. MR1724191

[179] S. G. Kreĭn and I. V. Sapronov, *On the completeness of a system of solutions of a Volterra integral equation with a singularity* (Russian), Dokl. Akad. Nauk **355** (1997), no. 4, 450–452. MR1492018

[180] Yu. M. Berezans′kiĭ, V. S. Korolyuk, S. G. Kreĭn, Yu. O. Mitropol′s′kiĭ, A. M. Samoĭlenko, A. V. Skorokhod, and Ī. V. Skripnik, *Yuriĭ L′vovich Dalets′kiĭ (on the occasion of his seventieth birthday)* (Ukrainian), Ukraïn. Mat. Zh. **49** (1997), no. 3, 323–325, DOI 10.1007/BF02487238; English transl., Ukrainian Math. J. **49** (1997), no. 3, 357–359 (1998). MR1472222

1999

[181] S. Krein, P. Kuchment, V. Ovchinnikov, and E. Semenov, *Iosif Shneiberg July 26, 1950–July 21, 1992*, Function spaces, interpolation spaces, and related topics (Haifa, 1995), Israel Math. Conf. Proc., vol. 13, Bar-Ilan Univ., Ramat Gan, 1999, pp. 1–10. MR1707354

2004

[182] N. D. Kopachevskiĭ and S. G. Kreĭn, *An abstract Green formula for a triple of Hilbert spaces, and abstract boundary value and spectral problems* (Russian, with Russian summary), Ukr. Mat. Visn. **1** (2004), no. 1, 69–97, 146; English transl., Ukr. Math. Bull. **1** (2004), no. 1, 77–105. MR2180707

2008

[183] S. Krein, Voronezh Winter Mathematical Schools, in V. A. Kostin, B. N. Sadovskii, and E. M. Semenov (Editors), *S. G. Krein. Recollections*, Voronezh State University 2008. (In Russian)

Peter Kuchment, Mathematics Department, Texas A&M University, College Station, TX 77843-3368

Evgeny Semenov, Mathematics Department, Voronezh State University, Voronezh, Russia

Contemporary Mathematics
Volume **734**, 2019
https://doi.org/10.1090/conm/734/14758

Voronezh Winter Mathematical School

The presentation at the opening of the XXth School

Selim Krein

1. Pre-history

The 1966 ICM took place in Moscow. Alongside with vivid impressions of many wonderful lectures and conversations with many famous foreign mathematicians (Phillips, Iosida, Komatsu, Magenes, Calderon, and others), I could not help having a nagging feeling of dissatisfaction. My impression was that in several areas we were behind the current level. In a recent paper in "Itogi Nauki," V. Arnold quotes H. Poincaré, who said that "sometimes it is sufficient to invent a word, and this word becomes a creator."[1] So I got the impression that we did not know many new words.

In the Fall of 1966, I was told by B.S. Mityagin that that summer a school in topology took place in Gorky[2], with 15 strong young mathematicians participating. In that school there were lectures and practice sessions. This is how I got an idea of starting Voronezh Mathematical School. It seemed natural to me to conduct the school during universities' winter breaks. We also hoped that in winter it will be easier to get a place in resorts and camp sites.

In November, an university communist party conference took place. Giving a speech there as the secretary of the department party organization, I asked for help in organizing such winter schools. The city party secretary supported the idea and promised some assistance. Next morning I and my deputy[3] S. A. Sklyadnev went to his office. He called to local trade union office and asked to give us vouchers to the resort named after M. Gorky[4]. However, he was told that all vouchers for January had been already sent to professional organizations in the city. We thus went to the trade union office ourselves, and there two young men responsible for the distribution of those vouchers, confirmed that nothing could be done. Despite that, we decided to talk to then chairman V. A. Fetisov, in spite of being discouraged from doing so: we were told that he was a difficult and strong-willed man. In

---

2010 *Mathematics Subject Classification.* Primary 01A72, 01A70.
Translated by Dr. Mila Mogilevsky from the Russian version of "S. G. Krein, Recollections", Voronezh State University, 2008, with permission.

[1] Translator's remark: We could not find the exact quote in Poincare's work, so the translation is probably an approximation.

[2] Currently Nizhny Novgorod.

[3] S. Krein was the Dean of math faculty at that time.

[4] A prominent Russian writer.

our conversation Fetisov at first did not understand our goal, thinking that we were talking about high schools. However, when he learned that there professional mathematicians would teach each other, he became really excited. He immediately called to those two functionaires and told them to take back vouchers from other organizations and give them to the University. So, with Fetisov's blessing Voronezh Winter School was born in 1967.

2. The first school

The resort was situated on a right bank of river Voronezh, and lectures were conducted in Forestry Institute, uphill from it. So, after lectures the participants went down the (snowy and icy) hill on various parts of their bodies. There were hundred and nine participants from twenty two cities of the USSR, including fifty three Voronezh mathematicians. An unexpected frost of -35 degrees Celsius (-31 degrees Fahrenheit) at the beginning of the school made participants from southern parts of the USSR getting sick right away. After three days of school, fourteen people were ill. I thought that we would have to close the school, but by Sunday everyone got better and by the end of the school all were in full health.

We needed to decide on organization the school. We thought then and now that at this school, unlike at usual conferences, there must be lecture series on a state of research in various areas of mathematics. As a rule, we would choose areas less familiar to the local Voronezh mathematicians.

During the first school there was only one working seminar, on topology. Thus, in a school's poster newspaper[5] a child song verse was published:

Top-top hands, top-top feet[6]-
grandfathers and granddaughters learn topology.
Top-top feet, top-top hands-
Learned that in topology there are only spheres with handles.

During later Winter Schools there usually worked from five to eight seminars. Lecturing for the specific audience, where participants of various levels of preparation were mixed together, there was a real chance of active discussions between the lecturer and the experts present, leaving other, unprepared participants to wonder what's going on. To address this issue, it was decided that only one school participant E. A. Gorin was allowed to ask questions.

During the first School, the following lecture series were given:

- Topology of functional spaces and calculus of variations at large - by S. I. Al'ber
- Operator rings[7] - by G. I. Kats, and
- Multidimensional differential equations - by A. I. Perov.

3. The second school

At the opening of this School I was in pre-heart attack condition and indeed had heart attack during the same day. So, about this School I can only say that it took place in the same place as the first one. Among lecturers there were B. Ya. Levin, Yu. I. Lyubich, V. I. Matsaev, and S. V. Fomin.

[5]Such posters have become a staple feature of the School. The content consisted of making fun of the lectures and participants.

[6]to warm up.

[7]The Russian term of operator algebras.

4. The third school

One hundred seventy people from twenty six Soviet cities participated in this School, including forty six participants from Voronezh. It seems to me that the third school was in some sense pivotal. If in the first two schools the lecturers were mathematicians closely related to Voronezh researchers through personal friendship or math interests, the lecturers of the third school were personally not known to us.

I would like to specifically mention the first two lectures of S. P. Novikov with the title "Smooth manifolds, K-theory, elliptic operators," which were valuable to us not only due to their actual content, but also due to the new for us ideology that Novikov promoted with a great passion. (The effect of the second lecture was somewhat diminished by a banquet the preceding night.)

Great interest was aroused by series of lectures delivered by B. V. Shabat, S. G. Gindikin, G. M. Henkin, and E. M. Chirka on theory of functions of several complex variables. At that time, we did not know anything about this area.

G. I. Eskin gave a lecture on pseudodifferential equations. There was also a one hour lecture by G. A. Margulis (a Fields Prize winner).

5. The fourth school

The fourth School took place at the same place as the third, in Dzerzhinsky resort. 195 people from 22 cities, including 60 from Voronezh participated. The following series of lectures were delivered:

- Method of orbits in the theory of Lie Groups - by A.A. Kirillov,
- Axiomatic field theory - by Yu. M. Berezansky,
- Adiabatic and physical scattering matrix of quantum field theory - by A. S. Shwarts,
- Multidimensional manifolds - by Yu. L. Daletsky,
- Topology of Banach spaces- by B. S. Mityagin and A. S. Dynin.

There was also a discussion on problems of university mathematics education. Among the participants in the discussion were Yu. M. Berezansky, Yu. L. Daletsky, E. A. Gorin, B. A. Efremovitch, B. V. Shabat, V. P. Havin, Yu. I. Lyubich, and S. G. Krein.

6. A meeting at the Ministry of Higher Education of the Russian Federation

Once Russian Ministry of Higher Education gathered 27 professors to discuss problems of science development in universities. I devoted part of my presentation to the work of Voronezh Winter Mathematical School. At the conclusion of the gathering, Minister of Higher Education Stoletov praised our work. The very next day I submitted to him a project for organization of the 1-st All-Russian Mathematical School, based at Voronezh University. One of the goals was to get funding for creation of 0.5 position of a senior lab assistant.

Afterwards, a referent told me that the project set at the table of the minister, most likely ending up in a trash can. Soon I realised this was to the best: my impulsive action aimed at rising the rating of the Voronezh University could bring more harm than good. Indeed, if the School became an official institution under the helm of Ministry of Higher Education, faculty might be mandated to go there, even if they did not have any such desire. The strength of our School in fact is that it is not officially "registered" anywhere (though nowadays a report on School's activities has to be sent to the Regional Council). Everyone who even buys a voucher as a tourist may participate in School's work. Below I will illustrate this with an example.

7. The sixth school

Again, the threat loomed over the School. Vouchers to our previous resorts became unavailable ("our vouchers are for working class only"), and we were in despair. Unexpectedly, two of our participants Tanya Gareeva and Nadya Lapteva stumbled upon an advertizement that camp site "Kommunal'nik" invites tourists for a short term winter vacation. Professor Yu. G. Borisovich immediately went there and pretty quickly sealed the deal. The camp was located in a beautiful place, but consisted of small two-room wooden cabins without any amenities. As luck would have it, there was also a thermometer in every cabin. The only possible places to conduct lectures were the dining room and veranda.

Inviting B. Ya. Levin to School, I described the dire camp site conditions. He came anyway and said that the only reason was that if he did not come, I would have told everyone that Levin got scared.

One of the lecturers was Yu. I. Manin. He approached to me and said "I understand everything but ... it is -8 degrees in my room!"[8]. We immediately relocated him to another cabin, where the temperature was 13 degrees[9], and he was happy since.

Our organizing committee bought electric portable stoves, and they were somewhat saving us from cold. On top of that, Yu. G. Borisovich would wake up every night the camp site stoker and ask him to stoke the fire again (naturally, for an extra reward).

There were 115 participants from 19 cities, including 43 from Voronezh. The following series of lectures were delivered:

- Inner homologies and formal groups - by V. M. Buchshtaber and A. S. Mischenko,
- Theory of analytic J-stretching matrices - by I. V. Kovalishina and V. P. Potapov,
- Algebraic Geometry - by Yu. I. Manin,
- Functions of non-commutative operators and their application - by V. P. Maslov,
- Differential equations and Lie Groups - by A. L. Onishchik, and
- Topology of algebraic manifolds - by V. I. Arnold.

V. P. Maslov and V. V. Grushin told us that a cohort of undergraduate students with advanced math preparation was created at their Institute of Electronic Engineering. In particular, V.V. Grushin presents there the theory of differentiable functions as a limit of theory of polynomials. There was a discussion on that idea. V. I. Arnold, Yu. I. Lyubich, and others criticized it. I took a conciliatory position, claiming that no method of instruction (by a qualified lecturer) can ruin a group of strong students.

---

[8]-8 degrees Celsius is 17.6 degrees Fahrenheit.

[9]13 degrees Celsius is 55.4 degrees Fahrenheit.

Despite logistic difficulties, or may be even due to them, all participants of this school were left with warmest memories about sixth school as the most "romantic."

8. The seventh school

Here I would like to tell about one episode. The chairman of Organizing Committee this year was S. A. Sklyadnev. Meeting with me before this year School started, he mentioned that Voronezh University Rector allowed to invite to school from all Voronezh institutions professors, docents, and assistants[10]. I pretended that I did not understand why in this list there were no senior lecturers. And the thing was that the rector disliked one of the Voronezh University alumna who worked as a senior lecturer[11]. When vouchers to the resort were delivered, this person was refused one. Voronezh young mathematicians strongly resented the situation and were preparing for a rebellion, which could lead to the school liquidation. I found a simple solution, going to the director of the resort and asking him to sell us one more voucher, which he gladly did. So the disgraced senior lecturer became just a person having a vacation at the resort and could attend lectures and be an active participant in seminars.

We must explain the situation with the Organizing Committee, mentioned above. While I was working at the Voronezh University, the members of the "Committee" were G. G. Trofimova[12] and I. We needed only one memo from the university administration with the request for a certain number of vouchers. After I left the university to work at Forestry Institute, a friction started among the heads of various math departments at the university, who wanted to be in charge of the Winter School. Then everyone agreed on rotation. Members of Organising committee were chosen by the university administration. Each of the School organizers contributed to running the school. For instance, V. P. Glushko and P. E. Sobolevsky "discovered" the camp site called "Beryozka," which has became the permanent place for the Winter Schools. This place allowed us to increase the number of participants to 260-280. B. N. Sadovsky has introduced a very clear system of preparation to School work. As far as the scientific programs, I felt closest to the schools headed by Yu. G. Borisovich.

9. The impact of the school

Young mathematicians easily accept new ideas and boldly begin using them. We, the older ones, also have tried as much as we could to catch up with them.

I will only tell about the influence of the school on some of my students. Lectures of G. I. Kats have influenced the work of V. I. Ovchinnikov and his dissertation.

After lectures of A. I. Perov, we understood that he was working on theory of differential equations on a commutative Lie group, and so we developed this theory for a non-commutative case. This was the topic of the dissertations of A. M. Shihvatov, Yu. S. Sysoev, A. I. Furmenko, and I. Z. Pesenson (whom I met at one of the Winter Schools).

---

[10] In the West this would correspond to full professors, associate professors, and lecturers, with an intermediate position of a senor lecturer being suspiciously skipped over.

[11] This happened to be one of the editors of this volume, who had "bad political profile."

[12] A departmental office manager.

N. Zobin, M. Zaidenberg, and P. Kuchment struggle with lecture notes.
Picture provided by R. Grabovskaya and V. Kononenko.

Lectures on theory of functions of several complex variables influenced M. G. Zaidenberg, P. A. Kuchment, and A. A. Pankov (one of outcomes was our joint 1975 survey in Uspehi Matematicheskih Nauk [**2**]) . P.A. Kuchment widely used in his dissertation the knowledge he acquired from the lectures of B. S. Mityagin and A. S. Dynin.

The dissertation of N. I. Yatskin was closely related to lectures by A. P. Onishchik. Later on, we published a small monograph "Linear differential equations on manifolds" [**1**].

During the schools, many new collaborations developed. M. G. Zaidenberg and V. Ya. Lin worked together on problems of algebraic geometry, V. A. Kondratyev and S. D. Eidelman - on positive solutions of elliptic problems. Ya. A. Roitberg wrote his latest papers on theory of hyperbolic equations being influenced by lectures by R. L. Volevich.

It is impossible to overestimate the importance of mathematical conversations and discussions among school participants. Sometimes small "spontaneous" seminars were working till late at night.

Some of characteristic features of Voronezh Winter School are its goodwill, mutual aid, absence of priority arguments. The School does not have elections, prizes, or democracy, that's why everyone works in a calm, creative environment.

We did not forget about free time of participants. For sure, one of the most important activities during that time were ski trips. There were other activities, as well. Say, V. I. Arnold "infected" school participants with backgammon game, which they played with great enthusiasm. Traditionally, on January 28th there was a celebration dedicated to the birthday of the permanent participant E. A. Gorin.

Some of remarkable activities in school life were evening lectures of A. Ya. Helemsky on history of various countries and people of the world. We all were mesmerized by his extensive knowledge of dates, names of historic figures, their work and opinions, relationships, wars, etc. We remember also the lecture by A. T. Fomenko on his method of dating historic events, which lead to a heated discussion. During one of the Schools, there was an exhibit of Fomenko's art works. Laptev told us about his trip to England. There were so many of such cultural events, that it is hard to mention all of them.

Now, opening the 20th School, I would like to express confidence that Voronezh Winter Mathematical School will continue for many more years.

References

[1] S. G. Kreĭn and N. I. Yatskin, *Lineĭnye differentsial′nye uravneniya na mnogoobraziyakh* (Russian), Voronezh. Gos. Univ., Voronezh, 1980. MR640266
[2] M. G. Zaĭdenberg, S. G. Kreĭn, P. A. Kučment, and A. A. Pankov, *Banach bundles and linear operators* (Russian), Uspehi Mat. Nauk **30** (1975), no. 5(185), 101–157. MR0415661

Contemporary Mathematics
Volume **734**, 2019
https://doi.org/10.1090/conm/734/14759

Recollections of Mark Krein

Selim Krein

Mark Grigorievich Krein was born in April[1] 1907. There were seven children in our family. M.G. was the fourth one (I am the seventh). He managed to finish seven classes of public school and one class of some private school subsidized by parents. He had never completed the high school. (By the way, a number of our best mathematicians had not completed high school: Bogolyubov, Gelfand, Lavrentiev, Shnirelman, etc.). M. G. moonlighted sawing firewood (firewood was then the only source of heat). At the same time, M. G. became interested in mathematics and began to attend lectures and seminars at Kiev University. There he attracted attention of instructors. I remember how Professor Delone (a famous algebraist) came to our house and asked our parents to let M. G. go with him to Leningrad, where he (Delone) was moving from Kiev. However, he did not get our parents' consent.

One morning I woke up and learned that Mark had left. He left a note saying that he was already sitxeen years old and thus could no longer "sit on his parents' neck." His friend, who lived in our house, also ran away. The friend's parents somehow found out that the fugitives went to Odessa, and the elder brother went on a search. The search was unsuccessful and he, intending to leave soon, decided to go to the beach first. There he saw a crowd watching the gymnastic tricks performed by Mark and his companion. It transpired that they were going to get hired as cabin boys on a steamer and go abroad, trying to reach my uncle, who had lived in Berlin since 1908. That plan did not work out. Then they tried to get hired by a circus. I do not know why, but they were not accepted. Further there is a gap in my information. I only know that the brother was sheltered by the family of his future wife. To support himself, he gave private lessons. He soon contacted the university in Odessa and was able to get into the graduate school (without secondary and higher education) under advising of Nikolai Grigorievich Chebotarev (another famous algebraist). He graduated with PhD at the age of twenty two, presenting seven papers on algebra, geometry, and function theory. Simultaneously with his graduate studies, he lectured at the Donetsk Mining Institute (I think, in the rank of a professor).

Translated by Dr. Mila Mogilevsky from the Russian manuscript obtained from S. Krein's daughter T. Voronina.

The unpublished manuscript was provided by Dr. T. Voronina (nee Krein) and translated and published with her permission.

[1]April 3rd. - Ed.

Photograph provided by Dr. T. Voronina (nee Krein).

Brothers Krein, left to right: Mark, Selim, and Mikhail.

I lived in Kiev and thus rarely met my brother. In 1939, he was elected a corresponding member of the Academy of Sciences of Ukraine, and after that he would regularly visit Kiev. The first lectures he gave were on the theory of cones in Banach spaces. They fascinated many students. My adviser N. N. Bogolyubov suggested to use this technique to generalize one Fréchet's theorem on integral equations. With his help, I succeeded. During this research, I started thinking about a universal space with a minihedral cone. I took the first steps, and the final proof was obtained jointly with M. G. during my trip to Odessa. This result is known as the brothers Krein-Kakutani Theorem (Kakutani was a Japanese mathematician).

The war broke our cooperation. After the war, M. G., while still in Odessa, was appointed the Head of the Functional Analysis Department at the Institute of Mathematics of the Ukrainian Academy of Sciences. We formed a kind of a tandem: I was in charge of a permanent seminar, and M. G. would periodically come with talks and problems suggestions. I would start working with a young Kiev mathematician, and then M. G. would pick him up as the official adviser. One of the best students of M. G. was Mark Aleksandrovich Krasnoselsky, who went through such evolution. A similar thing happened with the famous Kiev mathematicians Yu.M. Berezansky and Yu.L. Daletsky.

In 1951, M. G. and I were forced to leave the Institute of Mathematics[2] and our creative paths diverged.

An article [**1**] in the journal "Uspekhi Matematicheskikh Nauk", devoted to the 70th anniversary of M. G., mentions several areas in which M. G. has obtained major results. Here they are: 1) The moments problem and the approximations theory. 2) The theory of stability of solutions of differential equations. 3) The geometry of Banach spaces. 4) The theory of extensions of Hermitian operators. 5) Spectral theory. 6) Operators in spaces with an indefinite metric. 7) The theory of non-self-adjoint operators. 8) The theory of spaces with a cone. 9) Oscillation matrices and operators. 10) The theory of Hamiltonian equations. 11) The theory of representations of compact groups.[3]

References

[1] V. M. Adamjan, Ju. M. Berezans′kiĭ, N. N. Bogoljubov, I. S. Iohvidov, A. N. Kolmogorov, M. A. Lavrent′ev, and Ju. A. Mitropol′skiĭ, *Mark Grigor′evič Kreĭn (on the occasion of his seventieth birthday)* (Russian), Uspehi Mat. Nauk **33** (1978), no. 3(201), 197–203. MR0490843

[2] D Z Arov, Yu M Berezanskii, N N Bogolyubov, V I Gorbachuk, M L Gorbachuk, Yu A Mitropol'skii and L D Faddeev, Mark Grigorievich Krein (on the occasion of his eightieth birthday) (Russian), Uspekhi Mat. Nauk 42 4(256) (1987), 201-206.

[3] Yu. M. Berezanskii, Kyiv, Fall of 1943 through 1946. The rebirth of Mathematics, in P. Kuchment and E. Semenov (Eds.), *Functional Analysis and Geometry. Selim Krein Centennial*, AMS 2019.

[4] I. Gohberg, *Mark Grigorievich Kreĭn, 1907–1989*, Notices Amer. Math. Soc. **37** (1990), no. 3, 284–285. MR1041733

[5] http://www.emomi.com/history/mechanics_odessa/university/krein.htm

[2]This was a period of a massive anti-semitic campaign in the USSR. One can find description of this particular event in [**3**].

[3]More information and references about M. G. Krein can be found, for instance, in [**1, 2, 4, 5**]. - Ed.

Contemporary Mathematics
Volume **734**, 2019
https://doi.org/10.1090/conm/734/14760

Selim Krein and Voronezh Winter Mathematical Schools

Peter Kuchment and Vladimir Lin

Dedicated to the centennial of Professor Selim G. Krein

We face a honorable, but formidable task to write about a unique mathematics gathering Voronezh Winter Mathematical School lead by Professor Selim Krein[1]. Let us start with some geography. Voronezh is a large city (with a population of around one million) in the middle of the European part of Russia, close to the Ukrainian border. It has a dozen institutions of higher education, including the large Voronezh State University. The mathematics department at the university has become well known since Professors Mark Krasnoselskii and Selim Krein came to it in the early '50s and started building their mathematical schools. The main areas of interest in Voronezh have been functional analysis, operator theory, partial differential equations, and topology. In the '50s and '60s the department became a center of active research in these areas and attracted many brilliant students. One of the greatest features of the Voronezh mathematical school was its diversity of interests and willingness to explore new venues. However, the possibilities for communication with leading researchers and the accessibility of literature were severely limited in a provincial city like Voronezh. Apparently, this observation led Professor Krein to the idea[2] of creating a yearly gathering that would attract leading researchers and let local mathematicians (especially students) learn about new trends and methods in mathematics. There were, though, some features planned that distinguished this event from most other similar mathematical gatherings. First of all, the main emphasis was on educating local mathematicians through lecture series devoted to interesting topics. This had several important implications:

- The name chosen was "school" rather than "conference": Voronezh Winter School in Mathematics.
- Voronezh mathematicians have practically never been (with very few exceptions) invited to give lectures, since they were available through the whole year anyway.
- Leading researchers were invited with lectures on major contemporary areas and recent developments in mathematics.

2010 *Mathematics Subject Classification.* Primary 01A72, 01A70.
Photo credits: The pictures in this article were provided by R. Grabovskaya and V. Kononenko.

[1]Partially based on the authors' introduction to [**2**]

[2]Read more about this story in S. Krein's article [**1**].

- Lecturers usually presented a series of lectures rather than a single one, which enabled them to provide a reasonably understandable introduction to, and/or survey of, a subject.
- Lecturers were asked to present topics in such a way that researchers from different areas of mathematics, and even undergraduate students, could comprehend them.
- Another unusual feature was that only S. Krein and E. Gorin were allowed to interrupt the lecturers with questions. One can be surprised by this, but those who know the "Russian seminar style," would imagine that otherwise instead of learning a new topic, one would see a heated math discussion between dozens of mathematicians.
- There was no restriction on the area of mathematics covered. It only had to be "hot" and important, which was guaranteed by the appropriate choice of speakers.
- A serious effort was made to attract as many students as possible (including undergraduates[3]).
- The School was planned to be at least a week long.

There was a problem, however, with attracting (especially on a yearly basis) leading researchers to a meeting. This problem was resolved by choosing a suitable place and time. The School was organized during winter breaks in a tourist resort in a beautiful forest, where all willing participants had the opportunity to ski. This has become, along with mathematics, one of the biggest attractions of the School.

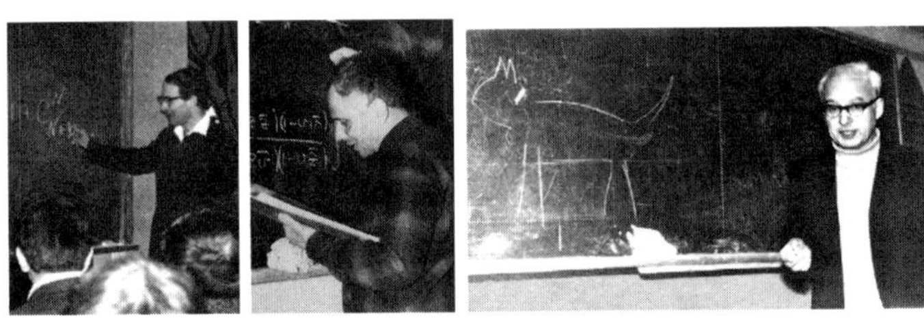

In the school. Left to right: V. Arnold, F. Berezin, S. Fomin

The first School was organized by S. Krein in January-February 1967 and proved to be a great success. Since then it had been organized on yearly basis for a quarter of a century. Since '90s, although the School would convene several times, and even produced several offsprings in summer and winter, its direction has changed, and we are not in a position to discuss its new status. The change was due to various factors, among them S. Krein's retirement from the leadership of the School and the emigration of many major, as well as junior, participants. So, when talking about the Voronezh Winter Mathematical School, we will mean its first 25 years after 1967.

Let us now briefly describe the organization of the School and its impact on the mathematical community of the former Soviet Union.

The location of the School varied sometimes, but it always was a tourist resort in a snowy forest close to Voronezh. The reader should beware, however, that the word "resort" had a meaning different from the one common to the "developed"

[3]One of the editors of this volume started attending the School as a sophomore, some others started as freshmen.

In the school. Left to right: V. Iskovskyh, Yu. Manin, A. Onishchik

world. One of the features was that normally the rooms were very poorly heated, which on the other hand helped to keep the audience at the lectures awake. The food was also of a mediocre quality. However, the quality of lectures, beautiful scenery, and the intensity of mathematical and social communication more than compensated for some "minor" inconveniences.

The School usually lasted for ten days, and normally worked according to the following schedule. An early breakfast was followed by three one-hour lectures. After lunch there was some free time that was used by many for skiing, by some for sleep, and by the most zealous mathematicians for discussions. Then one or two one hour lectures followed. Some free time before a supper was also spent in mathematical discussions and skiing. After supper, many parallel seminars with contributed talks continued well into the evening (till 9 or 10 p.m.).

Working hard. Left: lecture hall. One can notice S. Krein,
M. Solomyak, M. Birman. Center: V. Arnold and S. Krein.
Right: A. Markus and V. Lin have a problem.

And after that "the real life just started". What was it? Well, the list is long: lectures on history presented by A. Khelemsky and lately A. Fomenko, poetry readings, playing charades (in which V. Arnold was among the recognized leaders) and other games, singing songs, dancing through the night, and just attending warm gatherings of friends (sometimes twenty or thirty persons in a tiny hotel room) that could continue well into the next morning. Friends, normally separated by thousands of miles, when they met in the School, did not want to waste even an hour without communication. As was once said by a participant, "The School was the main benchmark of our lives. We lived from one School to the next."

The School accepted at different times from 100+ to almost 300 participants, and plenty of friendships, scientific careers, scientific cooperations, and even marriages started there.

The big success of the School was in attracting brilliant lecturers. We present a list of speakers (many of whom gave lectures more than once). Unfortunately, we

Skiing: Left - E. Gorin, center - V. Arnold, right - M. Mogilevsky.

Left to right: Singing. Guitarist in the middle - Yu. Samojlenko.
Lusya Lin is dancing. Playing charades:
left to right M. Zaidenberg, L. Gandel'sman, P. Kuchment; celebrating.

do not have complete data, so we apologize in advance for any incompleteness of the list. We ask lecturers whom we inadvertently left out to accept our apologies. So, here are some of the lecturers: S. Alber, V. Arnold, A. Belavin, Yu. Berezansky, F. Berezin, V. Buchstaber, E. Chirka, Yu. Daletsky, V. Drinfeld, B. Dubrovin, A. Dynin, B. Efremovich, G. Eskin, A. Fomenko, S. Fomin, D. Fuchs, S. Gindikin, E. Gorin, V. Havin, G. Henkin, M. Kadets, L. Kantorovich, G. Katz, V. Kharlamov, A. Khovanskii, A. Kirillov, S. Kislyakov, I. Kovalishina, M. Krasnoselskii, I. Krichever, P. Kuchment, B. Levin, V. Lin, Yu. Lyubich, Yu. Manin, V. Matsaev, G. Margulis, V. Maslov, V. Milman, A. Mishchenko, B. Mityagin, S. Novikov, G. Ol'shanskii, A. Onishchik, V. Palamodov, A. Perov, M. Postnikov, V. Potapov, A. Povzner, G. Rosenblum, S. Samborskii, A. Schwarz, B. Shabat, V. Shatalov, M. Shubin, M. Solomyak, B. Sternin, A. Varchenko, E. Vinberg, A. Vinogradov, O. Viro, L. Volevich, A. Vol'pert, M. Zaidenberg, and other distinguished mathematicians.

One can recognize in this list the names of many famous researchers, in particular of three Fields' Prize winners (V. Drinfeld, G. Margulis, and S. Novikov), one Nobel Prize winner (L. Kantorovich), and one Crafoort Prize winner (V. Arnold).

It is hard to overestimate the impact of the School on the mathematical life of the former SU. Many undergraduate and graduate students (including one of the authors) started their research careers influenced by lectures delivered at the School.

Many people still keep lecture notes taken there. The School provided a unique opportunity to learn from the leading experts about recent developments in various areas of mathematics. The list of theoretical and applied areas covered during the years of School's operation is so wide that it almost coincides with the Mathematics Subject Classification list! Topics ranged from quantum groups to geometry of

Left: G. Margulis, Yu. Lyubich, Z. Sheftel, V. Lin, L. Ronkin, V. Goldstein.
Center: Yu. Daletsky is lecturing, P. Sobolevsky takes notes.
Right: S. Krein and B. Levin

Working hard: "youngsters" are struggling
between the lectures (N. Zobin, M. Zaidenberg, P. Kuchment)

Banach spaces, fluid dynamics, random processes, integral geometry, K-theory, Lie algebras, boundary value problems for PDEs, inverse scattering method in nonlinear PDEs, classical harmonic analysis, mathematical methods in economics, math education, ... you name it! Many joint works and new areas of research have started in the School. Besides its purely scientific influence, the School also played an important role in the personal lives of many of its participants, providing a uniquely warm atmosphere that contrasted with the bleak "normal" way of life under the totalitarian Soviet regime.

The reader can find many other interesting stories about the School in [**1**].

References

[1] S. Krein, Voronezh Winter Mathematical School, in P. Kuchment and E. Semenov (Editors), *Differential Equations, Mathematical Physics and Applications. Selim Krein Centennial*, Contemporary Mathematics, vol. 734, Amer. Math. Soc., Providence, RI, 2019.

[2] P. Kuchment and V. Lin (eds.), *Voronezh Winter Mathematical Schools*, American Mathematical Society Translations, Series 2, vol. 184, American Mathematical Society, Providence, RI, 1998. Dedicated to Selim Krein; Papers from the school held in Voronezh; Advances in the Mathematical Sciences, 37. MR1729918

Mathematics Department, Texas A&M University, College Station, TX 77843-3368, USA
Email address: kuchment@math.tamu.edu

Department of Mathematics, Technion-Israel Institute of Technology, Haifa, Israel 32000
Email address: vlalin@gmail.com
Email address: vlin@technion.ac.il

Contemporary Mathematics
Volume **734**, 2019
https://doi.org/10.1090/conm/734/14761

A few words about our beloved teacher in mathematics and life

M. Grabovskaya and V. Kononenko

Years pass very quickly, it has already been eighteen years since Selim Grigorievich Krein passed away. We would like to share here a few little episodes we were fortunate to witness and memories that we treasure. We will use just his initials S.G., as most people (and all his pupils) did during his life[1].

In August of 1967, for the first time in the Far East, a Summer Math School was organized in Khabarovsk. S.G. was the prime organizer and to direct the final preparations, he flew from Voronezh to Khabarovsk two days before the opening of the School. The next day we with S. G. went to the airport to meet Professor V.I. Arnold, at that time a bright young star of Russian mathematics. After meeting V.I. Arnold, we hailed a taxi, to go about thirty miles away from the city, to a remote place on the bank of the Amur river, where the school was located. The driver was not eager at all to drive that far. One of us tried to convince the driver that it was extremely important to go there, since our guest was absolutely special. But S.G. solved this problem right away, having in his hand a "fan" of several 10-ruble bills (a high denomination at that time). The value of this offer was approximately equal to the price of a plane ticket from Moscow to Khabarovsk (five thousand miles). The taxi-driver understood this pantomime instantly and nodded a "yes". We were impressed by S.G.'s spectacular Charlie Chaplin- like performance and realized how highly he valued every minute of life and how little he did money.

Another event that impressed the attendees happened during the opening of the School the next day. The plenary session was given by two lecturers – V.I. Arnold and S.G. Krein. At that time, Far East, and in particular Khabarovsk, was very provincial and did not have many well educated mathematicians. Only guests from some European Russia universities were able to understand well the lecture that V.I.Arnold presented about his research on topology of algebraic manifolds. The second lecturer was S.G., who spoke about differential equations in Banach Spaces. He was one of the founders of this branch of Mathematics, but he started his talk with the words "What we have heard just now from V.I. Arnold, is a brilliant new discovery. What I am going to present, is some provincial math in

[1]In the mid 60s, when one of us lived in the dormitory of Voronezh University, all graduate students of S.G.Krein also called him "papa", which clearly expressed the feelings they had for him.

During the 1st Khabarovsk school. Left to right:
E. Gorin, S. Krein, Yu. Brudnyi, L. Lihtarnikov

comparison with that." For us, graduate students of S.G., this was surprising to hear, since we considered S.G. (correctly) as a mathematician of a highest rank. It was a lesson for us of how modest S.G. was in ranking his own research and how generous in emphasizing the results by Arnold.

Khabarovsk School

We have participated in many Math schools and conferences where often more senior people than S.G. took part: V. A. Efremovich, B.V. Shabat, B.Y.Levin, to name just a few. But somehow S.G. has always taken the lead, not only because he had completely grey hair since an early age, but because he eagerly took on his shoulders the (often heavy) responsibility for running the event. In any group meeting or celebration he was the leader in organizing, talking, joking and drinking. Even during our free time activities, we felt that everything was guided by S.G. He had served the math community not just simply as a professor, but also the Chair of various departments and important committees.

It has been a half of a century since S.G. organized the first Voronezh Winter Math School, which had influenced significantly the mathematics life in the USSR for decades. Now we see more clearly how democratic S.G. was and how much strain was put on him by keeping a school of more than 200 mathematicians from different regions of the country running smoothly. This involved keeping an uneasy balance between satisfying the often harsh demands of the communist authorities and maintaining a high ethical and scientific standards of the School.

As an example, one can mention the "information" spread about him by the "powers that be" in Voronezh. A friend of ours, a Philosophy Professor (and thus, mandatorily for such professors, a member of the Communist Party) once told us in a friendly dinner conversation, when we praised S.G.: " Be careful! S.G. is a leader of a secret Zionist organization in Voronezh. My advice is to keep your distance and avoid meetings at his house." (People not from the former USSR need to be told that such an accusation could easily lead to a job dismissal and even jailing of a "perpetrator".) It was a lie and a thus shock to hear at first. Our friend was an intelligent, highly educated and virtuous man. Only later we understood that he belonged to communist elite in the city, and that this was the "information" distributed by the communist authorities. S.G. had worked every day under the pressure of such and other "suspicions".

One of us (V. K.) had been fortunate sometimes to be a "personal driver" for S.G. I would drive him to the airport or bring him back home. Sometimes I would drive him to a local resort, or take him and his wife Evgenya Petrovna to the cemetery to visit the grave of his mother-in-law. I have always been happy to oblige. At those times cars in Russia were not so routine as in America, the roads were empty, people relaxed in the car and felt free to have a sincere and open talk. It was a great pleasure to listen to S.G., his thoughts, his points of view. It was a wonderful lesson for me about what is important in life and is of real value. I treasure the memory of these meetings and conversations.

Many of us were witnesses to the sad period of S.G.'s life when he broke his hip and was confined to his bed. The doctors were afraid to do a surgery, and S.G. suffered from pain and from a severe restriction in movement. During this time, we saw an enormous courage in this Man. His spirit was stronger than his serious illness. He continued to invite his former students to his home, was happy to communicate, showed lively interest in Math and life around him, smiled and joked.

After his death, we would always come to the cemetery on his birthday, the 15th of July. It was always strange to see the grave of S.G., in our memory he is still the most alive person that we have known.

We are happy and thankful to the fate that we were close witnesses, students, and friends of such great mathematician and human being.

Contemporary Mathematics
Volume **734**, 2019
https://doi.org/10.1090/conm/734/14762

Probabilistic interpretations of quasilinear parabolic systems

Ya. Belopolskaya

To the memory of Selim Krein and his deep and profound understanding of links between various fields of mathematics

ABSTRACT. We develop probabilistic interpretations for various types of the Cauchy problem solutions for nonlinear parabolic equations and systems. We extend the notion of nonlinear Markov processes and consider nonlinear evolution families which they generate in the spaces of functions and measures. We apply the results to construct probabilistic representations of classical or generalized solutions for forward and the backward Cauchy problem for various types of parabolic systems.

Introduction

It is well known that in the theory of linear second order parabolic equations stochastic differential equations (SDEs) play a part similar to one played by characteristics equations in the theory of linear hyperbolic equations. Similar connections exist both in nonlinear case and in the case of systems. One can say that stochastic equations correspond to a mycrodynamics of the phenomena described by PDEs while PDEs themselves describe the corresponding macrodynamics. It should be noted that in physical (biological, chemical and so on) applications one usually deals with the Cauchy problem for forward PDEs or systems and hence, initial data should be defined at the left hand side of a time interval, while in applications to financial mathematics one often deals with the Cauchy problem for a backward PDE which needs "initial" data fixed at the right hand side of the time interval. Even in a linear case second order backward parabolic equations called backward Kolmogorov equations are more natural objects from the SDE theory point of view, since a backward PDE can be immediately investigated provided one is given a solution of the correspondent SDE. On the other hand forward Kolmogorov equations present dual equations to the correspondent backward ones and need more sophisticated methods to be investigated as soon as we wish to obtain constructive probabilistic representations of their solutions.

A probabilistic approach to nonlinear parabolic equations based on the SDE theory was suggested in [**1**],[**2**] and developed in [**3**],[**4**]. Generally speaking this approach includes three steps.

1. At the first step we assume that there exists a (classical or regular enough generalized or viscosity) solution of the original Cauchy problem. Then keeping in

2010 *Mathematics Subject Classification.* Primary 60J60, 60H10.

mind this a priori assumption we can investigate the problem under consideration as if it was a linear one and construct the corresponding probabilistic representation of the required Cauchy problem solution for the original PDE.

2. At the second step we omit the above a priori assumption and attempt to construct a closed system of stochastic equations associated with the original PDE provided we have a probabilistic representation of its solution.

3.At the third step we have to investigate a closed stochastic system derived at the second step, prove the existence and uniqueness of its solution and verify that in this way we have constructed a required solution of the original problem.

Let us show how the above probabilistic approach works while constructing classical and viscosity solutions of the backward Cauchy problem for PDEs and systems as well as constructing generalized solutions of the forward Cauchy problem for them. In this paper we explain constructions essential for our purposes, state the results and give hints of their proof. We refer the reader to corresponding papers, where detailed proofs can be found.

Below the paper is organized as follows. In section 1 we recall some facts concerning probabilistic approaches to solution of the forward and backward Cauchy problem for scalar semilinear parabolic equations. In section 2 we extend some of these approaches to systems of semilinear parabolic equations with diagonal principal part. Finally in the concluding section we develop a probabilistic approach to construction of weak solution to nonlinear parabolic systems with cross-diffusion.

1. SDEs and nonlinear parabolic equations

Though the main target of this paper is to construct probabilistic representations of the Cauchy problem solutions for systems of nonlinear parabolic equations we start the exposition with a scalar case to make main ideas more transparent.

Consider the Cauchy problem for a nonlinear backward parabolic equation

$$(1.1)\qquad \frac{\partial u}{\partial s}+\frac{1}{2}A^u_{ik}(y)\frac{\partial^2 u}{\partial y_j\partial y_i}A^u_{kj}(y)+a^u_i(y)\frac{\partial u}{\partial y_i}=f(y,u)\quad u(T,y)=u_0(y)$$

$$i,k=1,\dots d,\quad x\in R^d,\quad 0\le s\le T, u_0:R^d\to R^1.$$

Here and below we assume a convention about summing over repeated indices if the contrary is not mentioned and use notations of the form $a^u(x)=a(x,u(t,x))$.

To derive a probabilistic representation of a classical solution to (1.1) we fix a probability space $(\Omega,\mathcal{F},P)$ and denote by $w(t)\in R^d$ a standard Wiener process defined on it.

A stochastic system associated with a classical solution of (1.1) includes a stochastic equation

$$(1.2)\qquad d\xi(\theta)=a^u(\xi(\theta))d\theta+A^u(\xi(\theta))dw(\theta),\quad \xi(s)=y\in R^d,0\le s\le\theta\le T,$$

and a relation

$$(1.3)\qquad u(s,y)=E\left[u_0(\xi_{s,y}(T))-\int_s^T f(\xi_{s,y}(\theta),u(\theta,\xi_{s,y}(\theta))d\theta\right].$$

Recall that when a^u, A^u and f^u do not depend on u, the last relation gives a probabilistic representation of a classical solution to the corresponding Cauchy problem. On the other hand, in the case under consideration (1.2), (1.3) make a closed system and have to be solved together.

Given a classical solution $u(s,y)$ to (1.1) we can easily verify by the Ito formula that (1.3) gives its probabilistic representation, provided $\xi_{s,y}(\theta)$ satisfies (1.2). In addition we can prove that the process $\xi_{s,y}(\theta)$ satisfying (1.2) possesses the Markov property.

To prove the existence and uniqueness of a solution to the system (1.2),(1.3) one needs certain conditions to be satisfied by a^u, A^u, f^u and u_0.

Below we denote by C, K, q positive constants which may vary from line to line.

We say condition **C1.1** holds if given nonrandom functions $a^u : R^d \to R^d$, $A^u : R^d \to R^d \otimes R^d$, $f^u : R^d \to R$ there exist positive constants C, K, q such that

$$\|a(x,u)\|^2 + \|A(x,u)\|^2 \le C[1 + \|x\|^2 + \|u\|^{2q}],$$
$$\|\nabla a(x,u)\|^2 + \|\nabla A(x,u)\|^2 \le C[1 + K|u|^q],$$

where $\|A^u\|^2 = TrA^u(A^u)^*$, $|f(x,u)| \le C[1 + K|u|^q]$, $\|\nabla f(x,u)\| < \infty$, and $u_0 : R^d \to R$ is differentiable and bounded.

We say condition **C1.2** holds when, provided **C1.1** holds, functions $a(x,u)$, $A(x,u)$, $u_0(x)$, $f(x,u)$ are twice differentiable in their arguments.

The correspondent results can be stated as follows.

THEOREM 1.1. *Let* **C1.1** *hold. Then there exists an interval* $[T_1, T]$ *with the length depending on constants in* **C1.1** *such that for all* $s \in [T_1, T]$ *there exists a unique solution* $\xi_{s,y}(\theta), u(s,y)$ *to (1.2) , (1.3). In addition the process* $\xi_{s,y}(\theta)$ *is a Markov process and* $u(s,y)$ *is a bounded Lipschitz continuous function.*

THEOREM 1.2. *Assume that* **C1.2** *holds. Then there exists an interval* $[T_2, T] \subset [T_1, T]$ *such that for all* $s \in [T_2, T]$ *the function* $u(s,x)$ *defined by (1.3) is a unique classical solution of (1.1).*

The proof of this assertions can be found in [**3**], [**4**].

Looking for a probabilistic representation of a classical solution $v(t,x)$ of the Cauchy problem for a nonlinear forward parabolic equation

$$\frac{\partial v}{\partial t} = a_i^v \frac{\partial v}{\partial x_i} + \frac{1}{2} A_{ik}^v \frac{\partial^2 v}{\partial x_i \partial x_j} A_{kj}^v, \quad v(0,x) = u_0(x) \tag{1.4}$$

one can reduce it to (1.1) setting

$$u(t,x) = v(T-t,x) \tag{1.5}$$

and proceeding in the above way.

On the other hand to construct a probabilistic representation of a generalized solution to (1.4) one needs more sophisticated considerations based on the theory of stochastic flows. The correspondent approach for linear parabolic equations was developed by Kunita [**5**]– [**7**]. It was extended to nonlinear parabolic equations and a certain class of systems in [**8**], [**9**].

The main idea of the Kunita approach that allows to construct a probabilistic representation of a generalized solution to (1.2) is the following. Consider a stochastic flow $\phi_{s,t} : y \to \xi_{s,y}(t)$ generated by a solution $\xi_{s,y}(t)$ of the SDE

$$d\xi(\theta) = a(\xi(\theta))d\theta + A(\xi(\theta))dw(\theta), \quad \xi(s) = y,$$

and given a generalized function u define a composition of u with $\phi_{s,t}$ as another generalized function satisfying

$$\int_{R^d} u \circ \phi_{s,t}(y) h(y) dy = \int_{R^d} u(x) h(\psi_{s,x}(t)) \hat{J}_{s,t}(x) dx.$$

Here $\psi_{s,t}$ is a flow time reversal to $\phi_{s,t}$ that is a map such that $\phi_{s,t}\circ\psi_{s,t}(x)=x$, and $\hat J_{s,t}$ is the Jacobian of the map $\psi_{s,t}: x\to\hat\xi(t)$. In a similar way we can define

$$\int_{R^d} u\circ\psi_{s,t}(x)h(x)dx=\int_{R^d} u(y)h(\phi_{s,y}(t))J_{s,t}(y)dy,$$

where $J_{s,t}(y)$ is the Jacobian of the map $\phi_{s,t}: y\to\xi_{s,y}(t)$.

It was shown by Kunita [**6**] that if $a^u(x)=a(x), A^u(x)=A(x)$ are smooth enough, then to construct a Markov process $\hat\xi(t)$ associated with (1.4) one needs to deal with a processes $\hat\xi(\theta)=\xi(t-\theta)$ which is time reversal to a solution $\xi(\theta)$ of an SDE

$$d\xi(\theta)=-a(\xi(\theta))d\theta-A(\xi(\theta))dw(\theta),\quad \xi(0)=y.$$

The corresponding process $\hat\xi_{0,x}(\theta)=\psi_{0,\theta}(x)$ satisfies an SDE

(1.6)
$$d\hat\xi(\theta)=a(\hat\xi_{0,x}(\theta))d\theta+A(\hat\xi_{0,x}(\theta))\nabla A(\hat\xi_{0,x}(\theta))d\theta+A(\hat\xi_{0,x}(\theta))d\tilde w(\theta),\quad \hat\xi(0)=x,$$

where $(A\nabla A)_i=\sum_{j,k}A_{jk}\nabla_j A_{ki}$ and $\tilde w(s)=w(t-s)-w(t)$. As a result in the case $a^u(x)=a(x), A^u(x)=A(x)$ one gets a probabilistic representation of a generalized solution to the Cauchy problem (1.4) in the form

$$u(t,x)=E\left[u_0(\hat\xi_{0,x}(t))\right], \tag{1.7}$$

where $\hat\xi_{0,x}(t))$ satisfies (1.6).

A probabilistic representation of a generalized solution to a nonlinear equation of the form (1.4) was derived in [**8**]. Unlike the Kunita case, to deal with nonlinear equation (1.4) one has to consider equations

$$d\hat\xi(\theta)=a^u(\hat\xi_{0,x}(\theta))d\theta+A^u(\hat\xi_{0,x}(\theta))\nabla A^u(\hat\xi_{0,x}(\theta))d\theta+A^u(\hat\xi_{0,x}(\theta))d\tilde w(\theta),\quad \hat\xi(0)=x,$$

$$u(t,x)=E\left[u_0(\hat\xi_{0,x}(t))\right]$$

as a closed system with respect to the couple $(\hat\xi_{0,x}(t),u(t,x))$. This system is similar to the system (1.2), (1.3) and can be studied by the above considerations [**8**].

2. Probabilistic approach to nonlinear systems of parabolic equations

Nonlinear systems of parabolic equations appear as mathematical models of various phenomena in a number of applications. The Cauchy problem for such systems was studied by many authors. Let us mention the book by Ladyzenskaya, Solonnikov, Uraltzeva [**10**] and pioneer papers by Amann [**11**] as well as more recent papers and references there [**12**], [**13**].

Probabilistic approaches to the investigation of systems of parabolic equations meet a number of obstacles. Nevertheless, being developed they allow to understand the specific features of different classes of such systems.

Note that among other things a probabilistic interpretation of parabolic systems justifies a certain classification of these systems. As a matter of fact, from probabilistic point of view we separate 4 classes among a variety of systems of second order parabolic equations, namely, 1) systems with equal diagonal principal parts and non diagonal entrance of the first and zero order terms, 2) systems with different diagonal principal parts and nondigaonal entrance of the zero order terms only, 3) systems with different diagonal principal parts and nondigaonal first order terms, 4) systems with nondiagonal principal parts, called systems with cross diffusion.

2.1. Probabilistic approach to semilinear systems of parabolic equations with diagonal principal parts. Let us start with investigation of systems from class 1). In a linear case a probabilistic representation of the classical Cauchy problem for a system of backward parabolic equations was derived in [**14**], [**15**] and it was extended to a semilinear case in [**3**].

Consider the backward Cauchy problem for a semilinear parabolic system

$$\frac{\partial u_k}{\partial s} + \mathcal{L}^u u_k + \sum_{l=1}^{d_1}\sum_{i=1}^{d} B^i_{lk}(x,u)\nabla_i u_l + \sum_{l=1}^{d_1} c_{lk}(x,u)u_l = f_k(u), \tag{2.1}$$

$$u_k(T,x) = u_{0k}(x),$$

where $\mathcal{L}^u u_k = \langle a(x,u), \nabla\rangle u_k + \frac{1}{2}TrA(x,u)\nabla^2 u_k A^*(x,u)$, $\langle a,b\rangle = \sum_{i=1}^d a_i b_i$ for $a, b \in R^d$ and $0 \le s \le T$. Semilinear (and quasilinear) systems of this kind were investigated by many authors (see [**10**]) from the PDE point of view.

A stochastic approach that allows to construct a classical solution for a system of the type (2.1) was developed in [**3**]. Consider a stochastic system associated with (2.1)

$$d\xi_{s,x}(\theta) = a^u(\xi_{s,x}(\theta))d\theta + A^u(\xi_{s,x}(\theta))dw(\theta), \quad \xi(s) = x \in R^d, \tag{2.2}$$

$$d\eta(\theta) = c^u(\xi_{s,x}(\theta))\eta(\theta)d\theta + C^u(\xi_{s,x}(\theta))(\eta(\theta), dw(\theta)), \quad \eta(s) = h \in R^{d_1}, \tag{2.3}$$

$$\langle h, u(s,x)\rangle = \mathbf{E}\left[\langle \eta_{s,h}(T), u_0(\xi_{s,x}(T))\rangle - \int_s^T \langle \eta(\theta), f(u(\theta,\xi(\theta)))\rangle d\theta\right], \tag{2.4}$$

where $B^i_{lk}(x,u) = \sum_j C^j_{kl}A_{ji}$ and $\langle h,u\rangle = \sum_{k=1}^{d_1} h_k u_k$.

One can state (see [**4**]) conditions to ensure that any bounded twice differentiable solution $u(s,x) = (u_1(s,x),\dots,u_{d_1}(s,x)$ of (2.1) admits a representation of the form (2.4) and on the contrary, if a function $u(s,x)$ defined by (2.4) is twice differentiable, then it satisfies (2.1) in the classical sense. The final step is to omit any a priori assumptions and study the closed system (2.2)– (2.4). We can state conditions to ensure the existence and uniqueness of a solution to (2.2)– (2.4) and conditions that allow to prove that $u(s,x)$ given by (2.4) is twice differentiable. This allows to verify that $u(s,x)$ is a unique classical solution to (2.1).

We say **C 2.1** holds if **C 1.1** holds for $a(x,u), A(x,u)$ and besides there exist positive constants C_1, q and a constant C_0 such that the estimates

$$\langle c(x,u)h, h\langle \le [C_0 + C_1\|u\|^{2q}]\|h\|^2, \quad \|C(x,v)h\|^2 \le C_1[1+\|u\|^{2q}]\|h\|^2,$$

hold, $c(x,u), C(x,u)$ are differentiable in x and u, derivatives $\nabla c, \nabla C$ satisfy the similar estimates with positive constants C^1_0, C^1_1 and there exist positive constants $K_{u_0}, K^1_{u_0}$ such that $\sup_x \|u_0(x)\|^2 \le K_{u_0}, \sup_x \|\nabla u_0(x)\|^2 \le K^1_0$.

We say **C 2.2** holds if **C 1.2** holds for $a^u(x), A^u(x), f^u(x)$, **C 2.1** holds and u_0, $c(x,u), C(x,u)$ are twice differentiable in x and u and satisfy similar estimates with positive constants C^2_0, C^2_1, K^2_0.

THEOREM 2.1. *Assume that* **C 2.1** *holds. Then there exists an interval* $[T_1, T]$ *with the length*

$$|T - T_1| \le \frac{1}{2C_0} log \left[1 + \frac{2C_0}{3C_1 K_0}\right]$$

such that for all $s \in [T_1, T]$ *there exists a unique solution to stochastic system (2.2)-(2.4). If in addition* $2C_0^1 + 3C_1^1 K_{u_0}^1 < 0$, *then a unique solution for (2.2)- (2.4) exits for all* $s \in [0, T]$.

THEOREM 2.2. *Assume that* **C 2.2** *holds. Then there exists an interval* $[T_2, T] \subset [T_1, T]$ *such that for all* $s \in [T_2, T]$ *there exists a unique solution to stochastic system (2.2)- (2.4) and the function* $u(s, x)$ *given by (2.4) is twice differentiable.*

As a consequence $u(s, x)$ given by (2.4) is a unique classical solution to (2.1). To be more precise the following assertion holds.

THEOREM 2.3. *Assume that* **C 2.2** *holds. Then there exists an interval* $[T_2, T] \subset [T_1, T]$ *such that for all* $s \in [T_2, T]$ *there exists a unique classical solution of the Cauchy problem (2.1) and this solution is given by (2.4).*

Detailed proofs of these results can be found in [**4**], [**3**].

2.2. Probabilistic approach to quasilinear systems of parabolic equations with diagonal principal parts. The probabilistic approach to systems of parabolic equations described above works when we consider semilinear systems which are nonlinear in u only. To apply it to investigation of systems of the same type as above which are nonlinear both in u and in ∇u we include a quasilinear system into a certain semilinear system with respect to a new vector function $V(s, x) = (u(s, x), \nabla u(s, x), \nabla^2 u(s, x))^*$. In other words we consider a second order differential prolongation of the system (2.1) with $a^u(x) = a(u(x), \nabla u(x))$, $A^u(x) = A(u(x), \nabla u(x)), c^u(x) = c(u(x), \nabla u(x)), C^u(x) = C(u(x), \nabla u(x))$ and $f^u(x) = f(u(x), \nabla u(x))$.

To illustrate the approach we consider the simplest version assuming that the only coefficient depending on $(u, \nabla u)$ is $a^u(x)$ while $A^u(x), C^u(x), c^u(x)$ and $f^u(x)$ depend only on $u(x)$. Besides, here and below we assume summation over all repeated indices if the contrary is not mentioned.

Let us derive an equation for $v_{mk}(s, x) = \nabla_m u_k(s, x)$ by formal differentiation of (2.1) with above chosen coefficients. As a result we get

$$\begin{aligned}(2.5)\qquad &\frac{\partial v_{mk}}{\partial s} + \mathcal{L}^u v_{mk} + B^i_{kq}(u)\nabla_m v_{qi} + c_{kl}(u) v_{lm} + \nabla_{u_q} a^u_j(x) v_{qm} v_{kj} \\ &+ \nabla_{v_{qp}} a^u_j(x) \nabla_m v_{qp} v_{kj} + \nabla_{u_q} A_{jq} v_{qm} \nabla_j v_{ki} A_{qi} - \nabla_{u_q} f_k v_{qm} + \nabla_{u_q} B^i_{lk}(u) v_{li} v_{qm} \\ &+ \nabla_{u_q} c_{lk}(u) u_l v_{qm} = 0.\end{aligned}$$

Analyzing a system (2.1), (2.5) one can see that this system has the same structure as the original one and can be written in the form

$$(2.6)\qquad \frac{\partial}{\partial s}\begin{pmatrix} u \\ v \end{pmatrix} + \mathcal{Q}\begin{pmatrix} u \\ v \end{pmatrix} - F(u, v) = 0,$$

where

$$\mathcal{Q}\begin{pmatrix} u \\ v \end{pmatrix} = \frac{1}{2}\begin{pmatrix} AA^* & 0 \\ 0 & AA^* \end{pmatrix}\nabla^2\begin{pmatrix} u \\ v \end{pmatrix} + \begin{pmatrix} a & 0 \\ 0 & a \end{pmatrix}\nabla\begin{pmatrix} u \\ v \end{pmatrix} + \begin{pmatrix} B & 0 \\ 0 & B + \bar{B} \end{pmatrix}\nabla\begin{pmatrix} u \\ v \end{pmatrix}$$

$$+\begin{pmatrix} c & 0 \\ 0 & c+\bar{c} \end{pmatrix}\begin{pmatrix} u \\ v \end{pmatrix}, \quad F\begin{pmatrix} u \\ v \end{pmatrix} = \begin{pmatrix} f^u(x) \\ \nabla_u f^u(x)v \end{pmatrix}$$

and

$$(\bar{B}v)_{mk} = \nabla_{v_{qp}} a_j^u(x)\nabla_m v_{qp} v_{kj} + \nabla_{u_q} A_{jq} v_{qm} \nabla_j v_{ki} A_{qi},$$

$$(\bar{c}v)_{mk} = \nabla_{u_q} B_{lk}^i(u) v_{li} v_{qm} + \nabla_{u_q} c_{lk}(u) u_l v_{qm}.$$

To construct a stochastic counterpart of the system (2.1),(2.5) we add to SDEs (2.2),(2.3) a system of SDEs for process $\alpha_{ij}(t) = \nabla_i \xi_j(t)$ and $\beta_{im}(t) = \nabla_i \eta_m(t)$. These equations have the form

$$(2.7) \qquad d\alpha_{ij}(\theta) = [\nabla_{u_q} a_i^u(\xi(\theta)) v_{jq} \alpha_{km}(\theta) + [\nabla_{v_{ql}} a_i^u(\xi(\theta))\nabla_m v_{ql} \alpha_{mj}(\theta)]]d\theta$$
$$+\nabla_{u_q} A_{ik}^u(\xi(\theta)) v_{kq} \alpha_{km}(\theta) dw_n(\theta),$$

$$(2.8) \qquad d\beta_{im}(\theta) = [\nabla_{u_q} c_{mk}^u(\xi(\theta)) v_{qr} \alpha_{ri}(\theta)\eta_k(\theta) + c_{mk}^u(\xi(\theta))\beta_{ik}(\theta)]d\theta$$
$$+[\nabla_{u_q} C_{mk}^j(u(\theta,\xi(\theta))) v_{jq} \alpha_{ji}(\theta)\eta_k(\theta) + C_{mk}^j(u(\theta,\xi(\theta)))\beta_{ki}(\theta)]dw_j(\theta).$$

Next, we consider a space $Y = R^d \oplus (R^d \otimes R^d) \otimes R^{d_1}$ and a tensor sum of operators $G \oplus \Gamma = G \otimes I + I \otimes \Gamma$ acting in Y as follows

$$G \oplus \Gamma(x,y) = Gx \otimes y + x \otimes \Gamma y.$$

Keeping in mind this notations we consider processes $\xi^k(\theta), \eta^k(\theta)$ satisfying (2.2), (2.3) and write the equations for $\beta_1(\theta) = \nabla \eta^k(\theta), \beta_2(\theta) = \alpha(\theta) \otimes \eta(\theta)$ in the form

$$\beta_1(\theta) = c(u(\theta,\xi(\theta))\beta_1(\theta)d\theta + C(u(\theta,\xi(\theta))\beta_1(\theta)dw(\theta)$$
$$+\nabla c(u(\theta,\xi(\theta)) \diamond \beta_2(\theta)d\theta + \nabla C(u(\theta,\xi(\theta)) \diamond \beta_2(\theta)dw(\theta),$$
$$\beta_2(\theta) = [\nabla a^u(\xi(\theta)) \oplus c^u(\xi(\theta)))]\beta_2(\theta)d\theta + [\nabla A^u(\xi(\theta)) \oplus C^u(\xi(\theta))]\beta_2(\theta)dw(\theta),$$
$$\beta_1(s) = 0, \beta_2(s) = y \otimes h.$$

Here $\nabla c^u \diamond (\alpha \otimes h) = c^u(\alpha, h)$, $\quad \nabla C^u \diamond (\alpha \otimes h) = \nabla C^u(\alpha, h)$. Finally, we can rewrite the system for $\beta(\theta) = (\beta_1(\theta), \beta_2(\theta))$ in the form

$$d\beta(\theta) = b(u(\theta,\xi^k(\theta)))\beta(\theta)d\theta + B(u(\theta,\xi(\theta)))\beta(\theta)dW(\theta),$$

where

$$W(t) = \begin{pmatrix} w(t) \\ w(t) \end{pmatrix}, \quad \beta(\theta) = \begin{pmatrix} \nabla\varsigma(\theta) \\ \alpha \otimes \eta(\theta) \end{pmatrix},$$

and coefficients $b(u), B(u)$ act as follows

$$b(u)\begin{pmatrix} \beta_1 \\ y \otimes \eta \end{pmatrix} = \begin{pmatrix} c^u\beta_1 + \nabla c^u(y,\eta) \\ 0 \cdot y \otimes \eta + y \otimes c^u \eta \end{pmatrix} = \begin{pmatrix} c^u & \nabla c^u \\ 0 & \nabla a^u \oplus c^u \end{pmatrix}\begin{pmatrix} \beta_1 \\ y \otimes \eta \end{pmatrix},$$

$$B(u)\begin{pmatrix} \beta_1 \\ y \otimes \eta \end{pmatrix} = \begin{pmatrix} C^u\beta_1 + \nabla C^u(y,h) \\ \nabla A^u y \otimes \eta + y \otimes C^u h \end{pmatrix}$$
$$= \begin{pmatrix} C^u & \nabla C^u \\ 0 & \nabla A^u y \oplus C^u \end{pmatrix}\begin{pmatrix} \beta_1 \\ y \otimes \eta \end{pmatrix}.$$

As a result we can rewrite the system (2.3) (2.7), (2.8) in the form

$$(2.9) \qquad d\beta(\theta) = b^u(\xi(\theta))\beta(\theta)d\theta + B^u(\xi(\theta))\beta(\theta)dW(\theta)$$

Let $G_0^k(y) = G^k(0,y) = \begin{pmatrix} n_k(y) \\ g_k(y) \end{pmatrix}$, where $g_k = \nabla n_k$ is a differentiable bounded function. Then we can deduce applying the Ito formula, that the function

$$(2.10) \qquad G_k(t,y) = \begin{pmatrix} n_k(t,x) \\ g_k(t,x) \end{pmatrix} = E\left[\begin{pmatrix} \eta^k(t) & 0 \\ \beta_1^k(t) & \beta_2(t) \end{pmatrix}\begin{pmatrix} n_k(\xi^k_{\theta,y}(t)) \\ \nabla n_k(\xi^k_{\theta,y}(t)) \end{pmatrix}\right]$$

$$= \begin{pmatrix} E[\eta^k(t) n_k(\xi^k_{\theta,y}(t))] \\ E[\beta_1^k(t) n_k(\xi^k_{\theta,y}(t)) + \beta_2^k(t)\nabla n_k(\xi^k_{\theta,y}(t)) \end{pmatrix}$$

is the required solution to (2.2) -(2.4), (2.9), (2.10). In other words the following statement holds [**16**].

THEOREM 2.4. *Assume that conditions* **C1.2**, **C 2.2** *hold. Then there exists a unique solution of the stochastic system (2.2) and (2.9), (2.10) and in addition $u(t,x)$ given by (2.4) is a unique classical solution to (2.1) defined for all $s \in [T_1, T]$, where $|T - T_1|$ depends on coefficients and initial data of (2.1).*

2. Systems with different diagonal diffusion coefficients.

The second class of systems still includes nonlinear parabolic systems with diagonal principal parts but this time diagonal terms may have different coefficients. At the same time there are no nondiagonal entrance of the first order terms. To be more precise we consider a system of the form

$$(2.11) \qquad \frac{\partial v_m}{\partial s} + \mathcal{L}^v_m v_m + [Q^v v]_m = 0, \quad v_m(t,x) = v_{0m}(x),$$

where

$$\mathcal{L}^v_m v_m = \langle a_m(x,v), \nabla\rangle v_m + \frac{1}{2} Tr A_m(x,v)\nabla^2 v_m A^*_m(x,v) \text{ and } [Q^v v]_m = \sum_{j=1}^{d_1} q_{lm} v_m.$$

Assume that for each $m \in V = \{1, \ldots, d_1\}$ $a_m(x,v), A_m(x,v)$ possess properties described above and the matrix $Q(x,v) = q_{lm}(x,v)$ possesses general properties of a generator of a Markov chain, namely,

1) for fixed $l, m \in V$ $q_{lm}(x,v)$ is a uniformly bounded in x and polylinear in v function, $x \in R^d, v \in R^M$;

2) $q_{lm}(x,v) \geq 0$ for all $x \in R^d, v \in R^{d_1}$ and $l \neq m$;

3) $q_{mm}(x,v) = -\sum_{l\neq m} q_{ml}(x,v)$ for all $x \in R^d, v \in R^{d_1}, m \in V$.

To derive a stochastic counterpart of (2.11) we consider a stochastic equation with coefficients depending on a Markov chain $\gamma(t) \in V$

$$(2.12) \qquad d\xi(t) = a^v(\xi(t),\gamma(t))dt + A^v(\xi(t),\gamma(t))dw(t), \xi(s) = x, \gamma(s) = m,$$

where $a : R^d \times V \times R^{d_1} \to R^d, A : R^d \times V \times R^{d_1} \to R^d \otimes R^d$ and

$$(2.13) \qquad P(\gamma(t+\Delta t) = l|\gamma(t) = j, (\xi(\theta),\gamma(\theta)), \theta \leq t\} = q^v_{jl}(\xi(t))\Delta t + o(\Delta t),$$

if $l \neq j$. Here and below to be brief we use notations of the form $a^v(x,l) \equiv a(x,l,v(x,l))$.

For our purposes it is convenient to change (2.13) for an SDE with respect to a jump process of the form

$$(2.14) \qquad d\gamma(\theta) = \int_R g^v(\xi(\theta),\gamma(\theta-),z)p(d\theta,dz), \quad \gamma(s) = l,$$

where $p(dt, dz)$ is the Poisson random measure with $Ep(dt, dz) = dtdz$,

$$g(x,l,z) = \begin{cases} m-l & if\ z \in \Delta_{lm}(x), \\ 0 & otherwise \end{cases}$$

and $\Delta_{lm}(x)$ are consecutive (with respect to the lexicographic ordering of $V \times V$) left closed, right open intervals of real line having length $q^v_{lm}(x)$.

To obtain a closed system playing a role of a stochastic counterpart to (2.11), we consider a system including (2.12), (2.14) and a closing relation

$$v(s,x,m) = E[v_0(\xi_{s,x}(T), \gamma_{s,m}(T))]. \tag{2.15}$$

We say that condition **C 2.3** holds if for each $m \in V$ functions $a(x,v,m)$, $A(x,v,m)$ satisfy condition **C 1.1** and a matrix $Q^v(x)$ acting as

$$Q^v(x)f(x,m) = \sum_{m\in V, m\neq l} q^v_{ml}(x)[f(x,m) - f(x,l)], \quad l \in V$$

satisfies inequalities

$$\|q^v_{lm}(x) - q^{v_1}_{lm}(y)\| \le L\|x-y\| + K_{v,v_1}|v - v_1|, \quad |q^v_{lm}(x)| \le K|v_l|^2.$$

THEOREM 2.5. [**17**] *Assume that* **C 2.3** *holds and* $v_0(x,m)$ *is a bounded twice differentiable function for each* $m \in V$. *Then the function* $v(s,x,m)$ *given by* (2.15) *is twice differentiable and presents a unique classical solution of the system* (2.11).

3. Probabilistic interpretation of the Cauchy problem viscosity solutions for parabolic systems with diagonal principal parts

In the previous section we have shown that under some conditions one can construct a classical solution to a parabolic system with a diagonal principal part while solving a certain stochastic problem. In this section we show that there exists a stochastic interpretation of the Cauchy problem viscosity solutions for systems of this type.

The notion of a viscosity solution of the Cauchy problem for a parabolic equation was defined in [**18**]. It was extend to systems of type (2.11) while analyzing stochastic representation of the Cauchy problem solution via the so called backward SDEs (or BSDEs). The BSDE theory was developed by Pardoux and Peng [**19**]- [**21**] and many other people and applied to many interesting nonlinear scalar equations. It was also applied to systems of the form (2.11) in [**22**], [**23**]. From the point of the PDE theory viscosity solutions to systems of the form (2.11)were studied later in [**24**]. A stochastic counterpart of the Cauchy problem (1.1) associated with a viscosity solution of this problem on the basis of forward- backward SDEs was recently developed in [**25**], [**26**].

Recall main ideas leading to BSDE theory associated with a parabolic equation of the form (1.1) with $f = f(u, A^*\nabla u)$. Assume that there exists a classical solution to (1.1) and derive an expression for a stochastic differential $dy(t)$ of the process $y(t) = u(t, \xi(t))$. By the Ito formula we know that

$$dy(\theta) = \left[\frac{\partial u}{\partial \theta} + a^u_i \frac{\partial u}{\partial x_i} + \frac{1}{2} Tr A^u_{ik} \frac{\partial^2 u}{\partial x_j \partial x_i} A^u_{kj}\right](\theta, \xi(\theta))d\theta \tag{3.1}$$
$$+ \frac{\partial u(\theta, \xi(\theta))}{\partial x_j} A^u_{kj}(\theta, \xi(\theta))dw_k(\theta).$$

In the approach used in previous sections we have integrated $dy(\theta)$ in θ from s to T and computed the mean value of the resulting expression to get a closed system (1.2), (1.3) associated with (1.1) assuming that $f = f(x, u)$. In the case when $f = f(u, \nabla u)$ in (1.1) within a framework of section 2 we have to construct a differential prolongation of the PDE system in order to include (1.1) into a new semilinear system.

In the framework of the BSDE approach to (1.1) with $f = f(u, A^*\nabla u)$ we choose an alternative way to obtain a closing relation. Namely, we define a stochastic process $z(t) = \frac{\partial u(t,\xi(t))}{\partial x_j} A^u_{kj}(\xi(t))$ and rewrite (3.1) in the form

$$dy(\theta) = f(y(\theta), z(\theta))d\theta + \langle z(\theta), dw(\theta)\rangle, y(T) = u_0(\xi(T)). \tag{3.2}$$

The system (1.2), (3.2) seems not to be closed but a necessary closing relation can be obtained due to the Ito theorem about martingale representations (see e.g.[**27**]).

Note that the BSDE approach gives a way to find both processes $y(t)$ and $z(t)$ simultaneously and hence gives immediately the probabilistic representation both for a solution to (1.1) and its gradient. The resulting probabilistic counterpart appears to be well suited to construct a viscosity solution of the backward Cauchy problem (1.1).

Recall that the notion of a viscosity solution was defined in [**18**] for scalar parabolic equations. A remarkable feature of a probabilistic interpretation of systems (2.11) and (2.1) is the possibility to reveal that both systems are equivalent to certain scalar equations.

As one can see from the probabilistic representation (2.15), the system (2.11) is equivalent to a scalar equation

$$\frac{\partial u(s,x,m)}{\partial s} + \mathcal{L}^u_m u(s,x,m) + [Q^u u](s,x,m) = 0, \quad u(t,x,m) = v_0(x,m),$$

with respect to the scalar function $u(s,x,m)$ defined on $[0,T] \times R^d \times V$.

Using this fact and results from the BSDE theory it was proved in [**22**],[**23**] that one can construct a probabilistic representation of a viscosity solution to the system (2.11).

On the other hand the above probabilistic approach to a system of the form (2.1) allows to reveal that it is equivalent to a scalar equation as well. To verify it we consider a scalar equation

$$\frac{\partial \Phi}{\partial s} + \frac{1}{2} TrG(\kappa, u)\nabla^2\Phi G^* z, u) + \langle g(z,u), \nabla\Phi\rangle = 0, \tag{3.3}$$

with respect to a scalar function $\Phi(s,z) = \langle h, u(s,x)\rangle, \quad z = (x,h)$. Here

$$TrG\nabla^2\Phi(s,x,h)G^* = A_{ik}\frac{\partial^2\Phi(s,x,h)}{\partial x_j \partial x_i}A_{kj} + 2C^{lm}_k h_l \frac{\partial^2\Phi(s,x,h)}{\partial x_j \partial h_m}A_{jk}$$

$$+C^{qm}_k h_m \frac{\partial^2\Phi(s,x,h)}{\partial h_q \partial h_p} C^{pn}_k h_n = A_{ik}\frac{\partial^2\Phi(s,x,h)}{\partial x_i \partial x_j}A_{kj} + 2C^{lm}_k h_l \frac{\partial^2\Phi(s,x,h)}{\partial x_j \partial h_m}A_{jk}$$

and

$$\langle q, \nabla\Phi(s,x,h)\rangle = a_j \frac{\partial\Phi(s,x,h)}{\partial x_j} + c_{lm}h_m \frac{\partial\Phi(s,x,h)}{\partial h_l}.$$

This gives a way to define a viscosity solution of the system (2.1) as a viscosity solution of the scalar equation (3.3).

In addition this allows to extend the notion of a viscosity solution to a quasilinear system of type (2.1) and to apply the BSDE approach to construct its probabilistic representation [**25**],[**26**].

Consider a parabolic system of the form

$$\frac{\partial u_k}{\partial s} + \mathcal{L}^u u_k + \sum_{l=1}^{d_1}\sum_{i=1}^{d} B^i_{lk}(x,u)\nabla_i u_l + \sum_{l=1}^{d_1} c_{lk}(x,u)u_l = f(u,\nabla u), \tag{3.4}$$

$$u_k(s,x) = u_{0k}(x).$$

To construct its probabilistic counterpart based on the BSDE approach we consider a stochastic system of the form (2.2), (2.3) and choose a BSDE of the form

$$dy(\theta) = g(y(\theta), z(\theta))d\theta + z(\theta)dw(\theta), \quad y(T) = \Gamma^*(s,T)u_0(\xi(T)), \tag{3.5}$$

for a closing relation. Here $\Gamma(s,t)$ is defined by $\eta(t) = \Gamma(s,t)h$,

$$y(t) = \Gamma^*(s,t)u(t,\xi(t)), \quad \langle \eta(T), u_0(\xi_{s,x}(T))\rangle = \langle h, \Gamma^*(s,T)u_0(\xi_{s,x}(T))\rangle,$$

$$g(y(\theta), z(\theta)) = \Gamma^*(s,\theta) f(y(\theta), [\Gamma^*(s,\theta)]^{-1} z(\theta)),$$

and

$$z(\theta) = \Gamma^*(s,t)[[C^u] * (\xi(\theta), u(\theta,\xi(\theta)) + [A^u]^*(\xi(\theta))\nabla u(\theta,\xi(\theta))].$$

A remarkable feature of this approach is that due to the Ito theorem about a martingale representation one can define immediately a stochastic representation of both the function $u(t,x)$ satisfying (2.5) and its gradient ∇u, namely, we get

$$u(s,x) = y(s), \quad [C^u]^*(x)u(s,x) + [A^u]^*(x)\nabla u(s,x) = z(s).$$

We say that condition **C 3.1** holds when condition **C 2.1** holds, $f : R^{d_1} \times R^{d\times d_1} \to R^{d_1}$, and there exist constants L, L_3, such that

$$\|g(y^1,z^1) - g(y^2,z^2)\| \le L[\,\|y^1 - y^2\| + \|z^1 - z^2\|\,],$$

or

$$\langle y - y_1, g(y^1,z) - g(y^2,z)\rangle \le \mu\|y^1 - y^2\|^2.$$

A combination of the BSDE theory results and the approach of section 2.1 allows to deduce the following statement [**25**].

THEOREM 3.1. *Assume conditions* **C 2.1** *and* **C 3.1** *hold. Then there exists a unique solution* $((\xi(t),\eta(t)), y(t), z(t))$ *of the system (2.2), (2.3), (3.5). In other words there exist processes* $(\xi(t),\eta(t),y(t),z(t))$ *such that with probability 1 the following relations hold*

$$\xi(t) = x + \int_s^t a^u(\xi_{s,x}(\theta))d\theta + \int_s^t A^u(\xi_{s,x}(\theta))dw(\theta), \tag{3.6}$$

$$\eta(t) = h + \int_s^t c^u(\xi_{s,x}(\theta))\eta(\theta)d\theta + \int_s^t C^u(\xi_{s,x}(\theta))(\eta(\theta), dw(\theta)), \tag{3.7}$$

$$y(\theta) = \Gamma^*(\theta,T)u_0(\xi_{s,x}(T)) - \int_\theta^T g(y(\tau), z(\tau))d\tau - \int_\theta^T z(\tau)dw(\tau). \tag{3.8}$$

In addition the function $u(s,x) = y(s)$ *is a viscosity solution of the Cauchy problem (3.4).*

4. Stochastic representation of the Cauchy problem solution for a system with cross-diffusion

Consider a class of fully nondiagonal systems of parabolic equations of the form

$$\frac{\partial g_m}{\partial t}+\sum_{j=1}^{d}\frac{\partial f_m^j(g)}{\partial x_j}=\sum_{ij=1}^{d}\frac{\partial}{\partial x_i}\left[F_{ml}^{ij}(g)\frac{\partial g_l}{\partial x_j}\right],\quad g_m(0,x)=g_{m0}(x), \tag{4.1}$$

where $m=1,2,\dots d_1$.

A probabilistic approach to construct a solution of the Cauchy problem (4.1) meets two obstacles. First we have to keep in mind that systems of this class are systems of the forward Kolmogorov equation type and hence to construct Markov processes associated with such a system we have to consider the corresponding dual system. Actually this is compatible with the fact that we are looking for a generalized solution of the system. The second problem is the following. The main tool in developing of a probabilistic approach to parabolic equations and systems is the Ito formula which allows to find out the form of a generator of a required Markov process. Unfortunately, there is no chance to obtain in this way a generator having the form of the operator in the right hand side of (4.1). This shows that there are no chance to apply the formal trick mentioned in section 1 (see(1.5)).

We present a probabilistic approach to interpretation of the Cauchy problem for the system (4.1) taking as an example a particular case of the SKT system suggested by Shigesada-Kawasaki-Teramoto in [**29**] for modeling spatial segregation phenomena of competing species in population dynamics. We consider a version studied in [**30**], namely,

$$\frac{\partial u_1}{\partial t}=\frac{1}{2}\Delta[(d_1+\alpha_1u_2)u_1]+u_1(a_1-b_1u_1-c_1u_2),\quad u_1(0,x)=u_{01}(x), \tag{4.2}$$

$$\frac{\partial u_2}{\partial t}=\frac{1}{2}\Delta[(d_2+\alpha_2u_1)u_2]+u_2(a_2-b_2u_1-c_2u_2),\quad u_2(0,x)=u_{02}(x). \tag{4.3}$$

Here $u_1(t,x)$ is the density of a population moving under chemotaxis and $u_2(t,x)$ is the density of the chemical substance, $d_k,a_k,b_k,c_k, k=1,2$ are positive constants, u_{k0} are positive bounded functions square integrable together with they first derivatives.

Denote by $H^1(R^d)$ the Sobolev space of functions defined on R and twice integrable along with their derivatives.

We say that $u_k(t,x)$ satisfy (4.2),(4.3) in a generalized sense if

$$u_k\in L_\infty([0,T];H^1(R^d))$$

and for any $h_k\in C_0^\infty(R^d), k=1,2$ the following integral identities hold

$$\frac{\partial}{\partial\theta}\int_{R^d}u_k(\theta,y)h_k(y)dy=\int_{R^d}u_k(\theta,y)\left[\frac{1}{2}M_k^2(u)\Delta h_k(y)+m_k(u)h_k(y)\right]dy, \tag{4.4}$$

where

$$M_1^2(u)=d_1+\alpha_1u_2,\ M_2^2(u)=d_2+\alpha_2u_1,\ m_k(u)=a_k-b_ku_1-c_ku_2,\ k=1,2. \tag{4.5}$$

To give a probabilistic interpretation to the system under consideration we use an alternative definition of a (very) weak solution [**13**] to (4.2),(4.3). Namely, we say that $u_k(t,x)$ satisfy (4.2),(4.3) in a very weak sense if $u_k \in L_\infty([0,T]; H^1(R^d))$ for any $h_k \in C_0^\infty([0,T) \times R^d)$ the following integral identities hold

$$\int_0^T \int_{R^d} u_k(\theta,y) \left[\frac{\partial h_k(\theta,y)}{\partial\theta} + \frac{1}{2} M_k^2(u)\Delta h_k(\theta,y) + m_k(u)h_k(\theta,y)\right] dyd\theta \tag{4.6}$$

$$= - \int_{R^d} u_k(0,y)h_k(0,y)dy, \quad k = 1,2.$$

It should be mentioned that relations (4.6) help to understand what is the form of generators of required Markov processes associated with (4.2),(4.3), while actually we will construct stochastic representation of a generalized solution to (4.2),(4.3) in the sense of (4.4). Note that connections between these two types of generalized solutions in the case of scalar parabolic equations are discussed in [**23**].

Assume that $u_k(t,x), k = 1,2$ satisfying (4.2),(4.3) are strictly positive bounded functions differentiable in spatial variable and derive a probabilistic representation for the solution of the Cauchy problem for this system.

The relations (4.6) prompt that one can consider a dual system

$$\frac{\partial h_1(s,y)}{\partial\theta} + \frac{1}{2} M_1^2(u)\Delta h_1(s,y) + m_1(u)h_1(s,y) = 0, \quad h_1(T,y) = h_{10}(y), \tag{4.7}$$

$$\frac{\partial h_2(s,y)}{\partial\theta} + \frac{1}{2} M_2^2(u)\Delta h_2(s,y) + m_2(u)h_2(s,y) = 0, \quad h_2(T,y) = h_{20}(y), \tag{4.8}$$

with coefficients defined by (4.5).

Let us start with an a priori assumption that there exists a smooth solution $u = (u_1, u_2)$ of (4.2), (4.3). Under this assumption we can construct stochastic processes $\xi^k(\theta), \eta^k(\theta), k = 1,2$ associated with (4.7), (4.8). To this end we consider SDEs

$$d\xi^k(\theta) = M_k(u(\theta,\xi^k(\theta))dw(\theta), \quad \xi^k(s) = y, \tag{4.9}$$

and linear SDEs

$$d\eta^k(\theta) = n_k(u(\theta,\xi^1(\theta))\eta^k(\theta)d\theta + \eta_k(\theta)\langle N_k(u(\theta,\xi^1(\theta)), dw(\theta)\rangle, \quad \eta^k(s) = 1. \tag{4.10}$$

with coefficients to be specified below.

Note that if $u_1(t,x), u_2(t,x)$ are smooth bounded functions then from classical results of the SDE theory we deduce that there exist processes $\xi^k(t)$ satisfying (4.9). Moreover the corresponding flows $\phi^1_{s,\theta}(x) = \xi^1(\theta)$ and $\phi^2_{s,\theta}(x) = \xi^2(\theta)$ are smooth in x and processes $\mathbf{J}^k(\theta) = \nabla\xi^k(\theta)$ satisfy SDEs

$$d\mathbf{J}^1_{ij}(\theta) = \frac{\alpha_1 \sum_{q=1}^d \nabla_q u_2(\theta,\xi^1(\theta))\mathbf{J}^1_{iq}(\theta)}{2\sqrt{d_1 + \alpha_1 u_2(\theta,\xi^1(\theta))}} dw_j(\theta), \quad \mathbf{J}^1_{ij}(s) = \delta_{ij} \tag{4.11}$$

$$d\mathbf{J}^2_{ij}(\theta) = \frac{\alpha_2 \sum_{q=1}^d \nabla_q u_1(\theta,\xi^2(\theta))\mathbf{J}^2_{iq}(\theta)}{2\sqrt{d_2 + \alpha_2 u_1(\theta,\xi^2(\theta))}} dw_j(\theta), \quad \mathbf{J}^2_{ij}(s) = \delta_{ij} \tag{4.12}$$

where δ_{ij} is the Kronecker symbol. Along with Jacobian matrices $\mathbf{J}^k(\theta)$ we derive an SDE for their determinants $J^k(\theta) = det\mathbf{J}^k(\theta)$. The corresponding equations have the form

$$dJ^1(\theta) = J^1(\theta)\frac{\alpha_1\langle\nabla u_2(\theta,\xi^1(\theta)), dw(\theta)\rangle}{2\sqrt{d_1 + \alpha_1 u_2(\theta,\xi^1(\theta))}}, \quad J^1(s) = 1. \tag{4.13}$$

$$dJ^2(\theta) = J^2(\theta)\frac{\alpha_2\langle\nabla u_1(\theta,\xi^2(\theta)), dw(\theta)\rangle}{2\sqrt{d_2+\alpha_2 u_1(\theta,\xi^2(\theta))}}, \quad J^2(s)=1. \tag{4.14}$$

Consider a stochastic processes $\gamma^k(\theta) = \eta^k(\theta)h(\xi^k(\theta))J^k(\theta)$ which will be used as a random test function and evaluate its stochastic differential assuming that h is twice differentiable.

LEMMA 4.1. *Let coefficients n_u^k and N_u^k have the form*

$$n_u^1 = m_1(u) - \frac{\alpha_1^2\|\nabla u_2\|^2}{4[d_1+\alpha_1 u_2]}, \quad N_u^1 = -\frac{\alpha_1\nabla u_2}{2\sqrt{d_1+\alpha_1 u_2}}. \tag{4.15}$$

$$n_u^2 = m_2(u) - \frac{\alpha_2^2\|\nabla u_1\|^2}{4[d_2+\alpha_2 u_1]}, \quad N_u^2 = -\frac{\alpha_2\nabla u_1}{2\sqrt{d_2+\alpha_2 u_1}}. \tag{4.16}$$

Then the processes $\Gamma^k(\theta) = \eta^k(\theta)h_k(\xi_{0,y}^k(\theta))J^k(\theta), k=1,2,$ have stochastic differentials of the form

$$d\Gamma^k(\theta) = \left[\frac{1}{2}M_k^2(u)\Delta h_k + m_k(u)h\right](\theta,\xi_{0,y}^k(\theta))\eta^k(\theta)J^k(\theta)d\theta + \langle\Lambda^k(\theta), dw(\theta)\rangle, \tag{4.17}$$

where $\Lambda^k = [1-M_k(u)]\nabla M_k(u)h_k + M_k(u)\nabla h_k$.

PROOF. We apply the Ito formula to evaluate $d\gamma^k(t)$

$$\begin{aligned} d\Gamma^k(\theta) = d[\eta^k(\theta)h_k(\xi^k(\theta))J^k(\theta)] &= d[\eta^k(\theta)]h_k(\xi^k(\theta))J^k(\theta) + \eta^k(\theta)d[h_k(\xi^k(\theta))]J^k(\theta)\\ &+\eta^k(\theta)h_k(\xi^k(\theta))dJ^k(\theta) + d[\eta^k(\theta)]d[h_k(\xi^k(\theta))]J^k(\theta)\\ &+\eta^k(\theta)d[h_k(\xi^k(\theta))]dJ^k(\theta) + d[\eta^k(\theta)]h_k(\xi^k(\theta))dJ^k(\theta). \end{aligned}$$

Taking into account the expressions for $d\xi^k(t), dJ^k(t)$ and $d\eta^k(t)$ from (4.9)–(4.10) and (4.13)–(4.14) we deduce

$$\begin{aligned} d\Gamma^k(\theta) &= \eta^k(\theta)\{n_k(u)h_k + \frac{1}{2}M_u^2\Delta h_k + \langle\nabla M_k(u), N_k\rangle h_k + \langle N_k, \nabla h_k\rangle\}J^k(\theta)d\theta\\ &+\eta^k(\theta)\{\langle N_k(u)+\nabla M_u, dw(\theta)\rangle h_k + \langle M_k(u)\nabla h_k, dw(\theta)\rangle\}J^k(\theta) \end{aligned}$$

and setting

$$N_k(u) = -\nabla M_k(u)$$

and

$$n_k(u) = m_k(u) + \|\nabla M_k(u)\|^2$$

we get

$$\begin{aligned} d\Gamma^k(\theta) &= \left[m_k(u(\theta,\xi^k(\theta))h_k(\xi^k(\theta)) + \frac{1}{2}M_k^2(u(\theta,\xi^k(\theta))\Delta h_k(\xi^k(\theta))\right]\eta^k(\theta)J^k(\theta)d\theta\\ &+M_k(\xi^k(\theta))\langle\nabla h_k(\xi^k(\theta)), dw(\theta)\rangle\eta^k(\theta)J^k(\theta). \end{aligned}$$

□

To construct a stochastic representation of a generalized solution $u(t,x) = (u_1(,x), u_2(t,x))$ to (4.2),(4.3) we need as well time reversal processes $\hat{\xi}^k(\theta)$. Given processes $\xi^k(\theta)$ satisfying (4.9) we denote by $\psi_{s,t}^k(x)$ the corresponding time reversal flows such that $\phi_{s,t}^k(\psi_{s,t}^k(x) = x$. Let $\hat{\xi}_{t,x}^k(\theta)$ denote a process defined by $\hat{\xi}_{t,x}^k(\theta) = \psi_{s,t}^k(x)$. Now we can state the following assertion.

THEOREM 4.2. *Let a differentiable function $u = (u_1, u_2)$ be a unique generalized solution of (4.2),(4.3). Then there exist stochastic processes $\hat{\xi}^k(t), \tilde{\eta}^k(t)$ such that functions $u_k, k = 1,2$ admit a probabilistic representation of the form*

$$u_k(t) = E[\tilde{\eta}^k(t)u_{0k}(\hat{\xi}^k(t))], \tag{4.18}$$

where $\tilde{\eta}_k(t)$ have the form

$$\tilde{\eta}^k(t) = exp\left\{\int_0^t [n_k(u(\theta, \hat{\psi}^k_{\theta,t}(x)) - \frac{1}{2}\|N_k(u(\theta, \hat{\psi}^k_{\theta,t}(x))\|^2]d\theta \right. \tag{4.19}$$
$$\left. + \int_0^t \langle N_k(u(\theta, \hat{\psi}^k_{\theta,t}(x)), dw(\theta)\rangle\right\}$$

with coefficients n_k, N_k given by (4.15)–(4.16).

PROOF. Provided $u_k(t)$ are bounded differentiable functions, the existence and uniqueness of $\phi^k_{0,t}(y) = \xi^k_{0,y}(t)$ satisfying (4.9) is justified by classical results of SDE theory as well as smooth dependence of $\xi^k_{0,y}(t)$ on the initial value y. Hence the time reversal process $\hat{\xi}^k_{0,x}(t)$ does exist as well. Consider a random process $\zeta^k(t)$ defined by

(4.20)
$$\int_{R^d} \zeta^k \circ \psi^k_{0,t}(x)h_k(x)dx = \int_{R^d} u^k_0(y)\eta^k(t)h_k \circ \phi_{0,t}(y)J^k(t)dy = \int_{R^d} u_{0k}(y)\gamma^k(t,y)dy.$$

One can easily check that

$$\int_{R^d}\int_0^t u_{0k}(y)d\gamma^k(\theta)dy = \int_{R^d} u_{0k}(y)\gamma^k(t)dy - \int_{R^d} u_{0k}(y)h_k(y)dy \tag{4.21}$$
$$= \int_{R^d} \tilde{\eta}(t)u_{0k}(\xi^k(t))h_k(x)dx - \int_{R^d} u_{0k}(x)h(x)dx.$$

Set $\mathcal{M}_k(u)h_k = \frac{1}{2}M_k^2(u)\Delta h_k + m_k(u)h_k$ and deduce from (4.21) a relation

(4.22)
$$E\left[\int_{R^d}\int_0^t u_{0k}(y)d\gamma^k(\theta)dy\right] = E\left[\int_0^t\int_{R^d} u_{0k}(y)d\left[\eta^k(\theta)h_k(\xi^k_{0,y}(\theta))J^k(\theta))\right]dy\right]$$
$$= E\left[\int_0^t\int_{R^d} u_{0k}(y)[\mathcal{M}_k(u)h_k(\xi^k_{0,y}(\theta))]\eta^k(\theta)J^k(\theta))dyd\theta\right].$$

As a result we get the equality

$$E\left[\int_{R^d} u_{0k}(y)\gamma^k(t)dy\right] - \int_{R^d} u_{0k}(y)h_k(y)dy \tag{4.23}$$
$$= E\left[\int_0^t\int_{R^d} u_{0k}(y)[\mathcal{M}_k(u)h_k(\xi^k_{0,y}(\theta))]\eta^k(\theta)J^k(\theta)dyd\theta\right].$$

By the change of variables $\xi^k_{0,y}(\theta) = x$ due to the stochastic Fubini theorem we deduce from (4.22)

(4.24)
$$E\left[\int_{R^d}\int_0^t u_{0k}(y)d\gamma^k(\theta)dy\right] = E\left[\int_0^t\int_{R^d}[\tilde{\eta}^k(\theta)u_{0k}(\hat{\xi}^k_{0,x}(\theta))]\mathcal{M}_k(u)h_k(x)dxd\theta\right]$$
$$= \int_0^t\int_{R^d} E[\tilde{\eta}^k(\theta)u_{0k}(\hat{\xi}^k_{0,x}(\theta))]\mathcal{M}_k(u)h_k(x)dxd\theta$$

$$= \int_0^t \int_{R^d} \mathcal{L}_k(u) E[\tilde{\eta}^k(\theta) u_{0k}(\hat{\xi}^k_{0,x}(\theta))] h_k(x) dx d\theta.$$

Hence, from (4.21) - (4.24) we obtain that functions

$$v_k(t,x) = E[\tilde{\eta}^k(\theta) u_{0k}(\hat{\xi}^k_{0,x}(\theta))], k = 1,2$$

satisfy equality

$$\int_{R^d} v_k(t,x) h_k(x) dx - \int_{R^d} u_{0k}(x) h_k(x) dx = \int_0^t \int_{R^d} \mathcal{L}_k(u) v_k(\theta, x) h_k(x) dx d\theta$$

and due to assumed above uniqueness of a solution to (4.7), (4.8) that $v_k(t,x) = u_k(t,x)$. Thus, we obtain the required probabilistic representation (4.18) of solutions to (4.7), (4.8). □

As one can see from (4.18) we need stochastic equations for $\hat{\xi}^k(\theta)$. From general theory of stochastic flows (see [**6**]) we deduce that processes $\hat{\xi}^k(\theta)$ satisfy SDEs

$$(4.25) \quad d\hat{\xi}^k(\theta) = [M_k(u)\nabla M_k(u)](\hat{\xi}^k(\theta))d\theta - M_k(u)(\hat{\xi}^k(\theta))d\tilde{w}(\theta), \quad \hat{\xi}^k(0) = x,$$

where $\tilde{w}(\theta) = w(t-\theta) - w(t)$.

As a result we have proved the following assertion.

THEOREM 4.3. *Assume that there exists a function $u(t,x) = (u_1(t,x), u_2(t,x))$ which belongs to $C([0,T]; H^1)$ and satisfies (4.7), (4.8). Then there exists a process $\hat{\xi}^k(t)$ satisfying (4.25) and a process $\tilde{\eta}^k(t)$ of the form (4.20) such that the representation (4.18) is valid.*

REMARK 4.4. Obviously, the system (4.18), (4.20), (4.25) is not closed since

$$\nabla M^1(u(\theta, \hat{\xi}^1(\theta)) = \frac{\alpha_1 \nabla u_2(\theta, \hat{\xi}^1(\theta)) \hat{\mathbf{J}}^1(\theta)}{2\sqrt{d_1 + \alpha_1 u_2(\theta, \xi^1(\theta))}},$$

$$\nabla M^2(u(\theta, \hat{\xi}^2(\theta)) = \frac{\alpha_2 \nabla u_1(\theta, \hat{\xi}_2(\theta)) \hat{\mathbf{J}}^2(\theta)}{2\sqrt{d_1 + \alpha_2 u_1(\theta, \xi_2(\theta))}}.$$

To make this system closed we need additionally stochastic representations for $\nabla u_k(t,x)$. Hence, for systems with cross-diffusion we have realized only the first step of the probabilistic approach mentioned in the introduction. In general for this class of systems development of this approach is in progress though in some simple cases all three steps are fulfilled. In particular it concerns the Cauchy problem for the MHD-Burgers system [**28**] which will be mentioned below.

Consider one more class of nonlinear PDE systems which is number 3 in the list presented at the beginning of section 2, namely, the class of systems having diagonal entrance of second order terms with different coefficients and non diagonal entrance of the first order terms. In other words consider the Cauchy problem for a system of the form

$$(4.26) \qquad \frac{\partial u_m}{\partial s} + \mathcal{L}^u_m u_m + B^i_{lm} \nabla_i u_l = 0, \quad u_m(t,x) = u_{0m}(x),$$

An example of a system of this type is the MHD-Burgers system [**28**]

$$(4.27) \qquad \frac{\partial v}{\partial t} + v \cdot \nabla v = \mu^2 \nabla^2 v + (\nabla \times B) \times B, \quad v(0,x) = v_0(x),$$

$$\frac{\partial B}{\partial t} = \nabla \times (v \times B) + \sigma^2 \nabla^2 B, \quad B(0,x) = B_0(x), \tag{4.28}$$

where B is a magnetic field, v is a flow velocity.

Though systems of the form (4.26) seem to be intermediate between first two classes of parabolic systems in the above mentioned list but this impression is deceiving. In fact, this class of systems should be unified with a class of fully non diagonal systems and a stochastic approach to construct a weak solution of the Cauchy problem for such a system is similar to one developed in this section. It will be presented in a forthcoming paper. This system as an example of a system which allows to fulfill all three steps of the probabilistic approach described in this section. The corresponding results in the one dimensional framework are presented in [**33**].

Acknowledgement

Financial support of RSF grant No 17-11-01136 is gratefully acknowledged.

References

[1] H. P. McKean Jr., *A class of Markov processes associated with nonlinear parabolic equations*, Proc. Nat. Acad. Sci. U.S.A. **56** (1966), 1907–1911, DOI 10.1073/pnas.56.6.1907. MR0221595

[2] M. I. Freĭdlin, *Quasilinear parabolic equations, and measures on a function space* (Russian), Funkcional. Anal. i Priložen. **1** (1967), no. 3, 74–82. MR0224985

[3] Ya. I. Belopolskaya, Yu. L. Dalecky, *Investigation of the Cauchy problem for systems of quasilinear equations via Markov processes*, Izv. VUZ Matematika, 12 (1978), 6–17.

[4] Ya. I. Belopol′skaya and Yu. L. Dalecky, *Stochastic equations and differential geometry*, Mathematics and its Applications (Soviet Series), vol. 30, Kluwer Academic Publishers Group, Dordrecht, 1990. Translated from the Russian. MR1050097

[5] H. Kunita, *Stochastic flows and stochastic differential equations*, Cambridge Studies in Advanced Mathematics, vol. 24, Cambridge University Press, Cambridge, 1990. MR1070361

[6] H. Kunita, *Stochastic flows acting on Schwartz distributions*, J. Theoret. Probab. **7** (1994), no. 2, 247–278, DOI 10.1007/BF02214270. MR1270603

[7] H. Kunita, *Generalized solutions of a stochastic partial differential equation*, J. Theoret. Probab. **7** (1994), no. 2, 279–308, DOI 10.1007/BF02214271. MR1270604

[8] Ya. Belopolskaya and W. A. Woyczynski, *Generalized solutions of nonlinear parabolic equations and diffusion processes*, Acta Appl. Math. **96** (2007), no. 1-3, 55–69, DOI 10.1007/s10440-007-9095-0. MR2327525

[9] Y. Belopolskaya and W. A. Woyczynski, *Generalized solutions of the Cauchy problem for systems of nonlinear parabolic equations and diffusion processes*, Stoch. Dyn. **12** (2012), no. 1, 1150001, 31, DOI 10.1142/S0219493712003523. MR2887913

[10] O. A. Ladyženskaja, V. A. Solonnikov, and N. N. Ural′ceva, *Linear and quasilinear equations of parabolic type* (Russian), Translated from the Russian by S. Smith. Translations of Mathematical Monographs, Vol. 23, American Mathematical Society, Providence, R.I., 1968. MR0241822

[11] H. Amann, *Dynamic theory of quasilinear parabolic systems. III. Global existence*, Math. Z. **202** (1989), no. 2, 219–250, DOI 10.1007/BF01215256. MR1013086

[12] A. Jüngel, *The boundedness-by-entropy method for cross-diffusion systems*, Nonlinearity **28** (2015), no. 6, 1963–2001, DOI 10.1088/0951-7715/28/6/1963. MR3350617

[13] L. Desvillettes, Th. Lepoutre, and A. Moussa, *Entropy, duality, and cross diffusion*, SIAM J. Math. Anal. **46** (2014), no. 1, 820–853, DOI 10.1137/130908701. MR3165911

[14] Ju. L. Daleckiĭ, *Representability of solutions of operator equations in the form of continual integrals* (Russian), Dokl. Akad. Nauk SSSR **134** (1960), 1013–1016. MR0171180

[15] D. W. Stroock, *On certain systems of parabolic equations*, Comm. Pure Appl. Math. **23** (1970), 447–457, DOI 10.1002/cpa.3160230313. MR0272075

[16] S. Albeverio and Ya. Belopolskaya, *Probabilistic approach to systems of nonlinear PDEs and vanishing viscosity method*, Markov Process. Related Fields **12** (2006), no. 1, 59–94. MR2223421

[17] Ya. I. Belopol′skaya, *Probabilistic models of conservation and balance laws in switching regimes* (Russian, with English summary), Zap. Nauchn. Sem. S.-Peterburg. Otdel. Mat. Inst. Steklov. (POMI) **454** (2016), no. Veroyatnost′i Statistika. 24, 5–42, DOI 10.1007/s10958-018-3701-8; English transl., J. Math. Sci. (N.Y.) **229** (2018), no. 6, 601–625. MR3602399

[18] M. G. Crandall, H. Ishii, and P.-L. Lions, *User's guide to viscosity solutions of second order partial differential equations*, Bull. Amer. Math. Soc. (N.S.) **27** (1992), no. 1, 1–67, DOI 10.1090/S0273-0979-1992-00266-5. MR1118699

[19] É. Pardoux and S. G. Peng, *Adapted solution of a backward stochastic differential equation*, Systems Control Lett. **14** (1990), no. 1, 55–61, DOI 10.1016/0167-6911(90)90082-6. MR1037747

[20] É. Pardoux and S. Peng, *Backward stochastic differential equations and quasilinear parabolic partial differential equations*, Stochastic partial differential equations and their applications (Charlotte, NC, 1991), Lect. Notes Control Inf. Sci., vol. 176, Springer, Berlin, 1992, pp. 200–217, DOI 10.1007/BFb0007334. MR1176785

[21] É. Pardoux, *Backward stochastic differential equations and viscosity solutions of systems of semilinear parabolic and elliptic PDEs of second order*, Stochastic analysis and related topics, VI (Geilo, 1996), Progr. Probab., vol. 42, Birkhäuser Boston, Boston, MA, 1998, pp. 79–127. MR1652339

[22] E. Pardoux, F. Pradeilles, and Z. Rao, *Probabilistic interpretation of a system of semilinear parabolic partial differential equations* (English, with English and French summaries), Ann. Inst. H. Poincaré Probab. Statist. **33** (1997), no. 4, 467–490, DOI 10.1016/S0246-0203(97)80101-X. MR1465798

[23] E. Pardoux and A. Răşcanu, *Stochastic differential equations, backward SDEs, partial differential equations*, Stochastic Modelling and Applied Probability, vol. 69, Springer, Cham, 2014. MR3308895

[24] W. Liu, Y. Yang, and G. Lu, *Viscosity solutions of fully nonlinear parabolic systems*, J. Math. Anal. Appl. **281** (2003), no. 1, 362–381. MR1980097

[25] Y. I. Belopolskaya, *Probabilistic counterparts of nonlinear parabolic partial differential equation systems*, Modern stochastics and applications, Springer Optim. Appl., vol. 90, Springer, Cham, 2014, pp. 71–94, DOI 10.1007/978-3-319-03512-3_5. MR3236069

[26] Ya. Belopolskaya, W.Woyczynski, *Probabilistic approach to viscosity solutions of the Cauchy problem for systems of fully nonlinear parabolic equations,* J. of Math. Sci. **188**, N 6 (2013), 655–672.

[27] B. Øksendal, *Stochastic differential equations*, 6th ed., Universitext, Springer-Verlag, Berlin, 2003. An introduction with applications. MR2001996

[28] H.-Y. Jin, Z.-A. Wang, and L. Xiong, *Cauchy problem of the magnetohydrodynamic Burgers system*, Commun. Math. Sci. **13** (2015), no. 1, 127–151, DOI 10.4310/CMS.2015.v13.n1.a7. MR3238142

[29] N. Shigesada, K. Kawasaki, and E. Teramoto, *Spatial segregation of interacting species*, J. Theoret. Biol. **79** (1979), no. 1, 83–99, DOI 10.1016/0022-5193(79)90258-3. MR540951

[30] S. Shim, *On the properties of solutions to predator-prey models with cross-diffusions*, Proc. National Inst. for Math. Sci. **1**, N 3 (2006), 77–87.

[31] Ya. Belopolskaya, *Probabilistic counterparts for strongly coupled parabolic systems*, Springer Proceedings in Mathematics & Statistics. Topics in Statistical Simulation. **114** (20114), 33–42.

[32] H.-Y. Jin, Z.-A. Wang, and L. Xiong, *Cauchy problem of the magnetohydrodynamic Burgers system*, Commun. Math. Sci. **13** (2015), no. 1, 127–151, DOI 10.4310/CMS.2015.v13.n1.a7. MR3238142

[33] Ya. I. Belopol′skaya and A. O. Stepanova, *Stochastic interpretation of a Burgers-MHD system* (Russian, with English summary), Zap. Nauchn. Sem. S.-Peterburg. Otdel. Mat. Inst. Steklov. (POMI) **466** (2017), no. Veroyatnost′i Statistika. 26, 7–29. MR3760039

Department of Mathematics, St Petersburg University of Architecture and Civil Engineering, 2-ja Krasnoarmejskaja str. 4 and POMI RAN, nab. Fontanka 27, St Petersburg, Russia

Email address: yana.belopolskaya@gmail.com

Contemporary Mathematics
Volume **734**, 2019
https://doi.org/10.1090/conm/734/14763

On sectorial L-systems with Shrödinger operator

S. Belyi and E. Tsekanovskiĭ

In respectful memory of Selim Grigor'evich Krein

ABSTRACT. We study L-systems with sectorial main operator and connections of their impedance functions with sectorial Stieltjes and inverse Stieltjes functions. Conditions when the main and state space operators (the main and associated state space operators) of a given L-system have the same or not angle of sectoriality are presented in terms of their impedance functions with discussion provided. Detailed analysis of L-systems with one-dimensional sectorial Shroödinger operator on half-line is given as well as connections with the Kato problem on sectorial extensions of sectorial forms. Examples that illustrate the obtained results are presented.

1. Introduction

In the current paper we focus on sectorial L-systems and, in particular, on L-systems with Shrödinger operators whose impedance functions are sectorial Stieltjes or sectorial inverse Stieltjes functions. The formal definition, exposition and discussions of *sectorial* classes S^{α} and S^{α_1,α_2} of Stieltjes functions and *sectorial* classes $S^{-1,\alpha}$ and S^{-1,α_1,α_2} of inverse Stieltjes functions are presented in Sections 2, 3 and 4 (see [**1, 5, 9**]). Theorems for these sectorial classes allow us to observe the geometric properties of the corresponding L-systems. Moreover, the knowledge of the limit values at zero and infinity of the impedance function allows to find angle of sectoriality of the main, state space or associated state space operators of a given L-system that leads to the connection of Kato's problem about sectorial extension of sectorial forms. Section 5 is devoted to L-systems with Schrödinger operator in $L_2[a,+\infty)$ and non-self-adjoint boundary conditions. A complete description of such L-systems as well as the formulas for their transfer and impedance functions are presented. Section 6 contains the main results of the present paper. Utilizing theorems covered in Section 4, we obtain some new properties of L-systems with Schrödinger operator whose impedance function falls into a particular class. Most of the results are given in terms of the real parameter μ that appears in the construction of L-system. Finally, it is worth to mention that a Stieltjes function can be a coefficient of dynamic pliability of a string (this was established by M. Krein, see [**16**]), and that maximal sectorial operators have profound connections with

---

2010 *Mathematics Subject Classification.* Primary 47A10; Secondary 47N50, 81Q10.
Key words and phrases. L-system, transfer function, impedance function, Herglotz-Nevanlinna function, inverse Stieltjes function, sectorial operators, Shrödinger operator.

holomorphic contraction semigroups [**15, 18**]. The present paper is a further development of the theory of open physical systems conceived by M. Livšic in [**19**].

This paper is written on the occasion of the centenary of Selim Grigor'evich Krein, a remarkable human being and great mathematician. A couple of generations of mathematicians (including the authors) from the former USSR are in debt to him for an opportunity to learn a lot as participants of the famous Voronezh Winter Mathematical School created and conducted by S.G. for many years despite certain singularities of life at that time. His apostolic devotion to mathematics and mathematical community, his help and support for those who needed it will always be in our hearts.

2. Preliminaries

For a pair of Hilbert spaces $\mathcal{H}_1$, $\mathcal{H}_2$ we denote by $[\mathcal{H}_1,\mathcal{H}_2]$ the set of all bounded linear operators from $\mathcal{H}_1$ to $\mathcal{H}_2$. Let $\dot{A}$ be a closed, densely defined, symmetric operator in a Hilbert space $\mathcal{H}$ with inner product $(f,g), f,g\in\mathcal{H}$. Any non-symmetric operator T in $\mathcal{H}$ such that

$$\dot{A}\subset T\subset \dot{A}^*$$

is called a *quasi-self-adjoint extension* of $\dot{A}$.

Consider the rigged Hilbert space (see [**2, 10**]) $\mathcal{H}_+\subset\mathcal{H}\subset\mathcal{H}_-$, where $\mathcal{H}_+=\operatorname{Dom}(\dot{A}^*)$ and

$$(f,g)_+=(f,g)+(\dot{A}^*f,\dot{A}^*g),\quad f,g\in\operatorname{Dom}(A^*). \tag{1}$$

Let $\mathcal{R}$ be the *Riesz-Berezansky operator* $\mathcal{R}$ (see [**2, 10**]) which maps $\mathcal{H}_-$ onto $\mathcal{H}_+$ such that $(f,g)=(f,\mathcal{R}g)_+$ ($\forall f\in\mathcal{H}_+$, $g\in\mathcal{H}_-$) and $\|\mathcal{R}g\|_+=\|g\|_-$. Note that identifying the space conjugate to $\mathcal{H}_\pm$ with $\mathcal{H}_\mp$, we get that if $\mathbb{A}\in[\mathcal{H}_+,\mathcal{H}_-]$, then $\mathbb{A}^*\in[\mathcal{H}_+,\mathcal{H}_-]$. An operator $\mathbb{A}\in[\mathcal{H}_+,\mathcal{H}_-]$ is called a *self-adjoint bi-extension* of a symmetric operator $\dot{A}$ if $\mathbb{A}=\mathbb{A}^*$ and $\mathbb{A}\supset\dot{A}$. Let $\mathbb{A}$ be a self-adjoint bi-extension of $\dot{A}$ and let the operator $\hat{A}$ in $\mathcal{H}$ be defined as follows:

$$\operatorname{Dom}(\hat{A})=\{f\in\mathcal{H}_+:\mathbb{A}f\in\mathcal{H}\},\quad \hat{A}=\mathbb{A}\restriction\operatorname{Dom}(\hat{A}).$$

The operator $\hat{A}$ is called a *quasi-kernel* of a self-adjoint bi-extension $\mathbb{A}$ (see [**25**] and [**2**, Section 2.1]). A self-adjoint bi-extension $\mathbb{A}$ of a symmetric operator $\dot{A}$ is called *t-self-adjoint* (see [**2**, Definition 4.3.1]) if its quasi-kernel $\hat{A}$ is self-adjoint operator in $\mathcal{H}$. An operator $\mathbb{A}\in[\mathcal{H}_+,\mathcal{H}_-]$ is called a *quasi-self-adjoint bi-extension* of an operator T if $\mathbb{A}\supset T\supset\dot{A}$ and $\mathbb{A}^*\supset T^*\supset\dot{A}$. We will be mostly interested in the following type of quasi-self-adjoint bi-extensions. Let T be a quasi-self-adjoint extension of $\dot{A}$ with nonempty resolvent set $\rho(T)$. A quasi-self-adjoint bi-extension $\mathbb{A}$ of an operator T is called (see [**2**, Definition 3.3.5]) a *(*)-extension* of T if $\operatorname{Re}\mathbb{A}$ is a t-self-adjoint bi-extension of $\dot{A}$. In what follows we assume that $\dot{A}$ has deficiency indices $(1,1)$. In this case it is known [**2**] that every quasi-self-adjoint extension T of $\dot{A}$ admits $(*)$-extensions. The description of all $(*)$-extensions via Riesz-Berezansky operator $\mathcal{R}$ can be found in [**2**, Section 4.3].

Recall that a linear operator T in a Hilbert space $\mathcal{H}$ is called **accretive** [**17**] if $\operatorname{Re}(Tf,f)\geq 0$ for all $f\in\operatorname{Dom}(T)$. We call an accretive operator T **α-sectorial** [**17**] if there exists a value of $\alpha\in(0,\pi/2)$ such that

$$(\cot\alpha)|\operatorname{Im}(Tf,f)|\leq\operatorname{Re}(Tf,f),\qquad f\in\operatorname{Dom}(T). \tag{2}$$

We say that the angle of sectoriality α is **exact** for an α-sectorial operator T if

$$\tan\alpha = \sup_{f\in \mathrm{Dom}(T)} \frac{|\,\mathrm{Im}(Tf,f)|}{\mathrm{Re}\,(Tf,f)}.$$

A $(*)$-extension $\mathbb{A}$ of T is called **accretive** if $\mathrm{Re}\,(\mathbb{A}f,f)\ge 0$ for all $f\in\mathcal{H}_+$. This is equivalent to that the real part $\mathrm{Re}\,\mathbb{A}=(\mathbb{A}+\mathbb{A}^*)/2$ is a nonnegative self-adjoint bi-extension of $\dot{A}$. A $(*)$-extensions $\mathbb{A}$ of an operator T is called **accumulative** if

$$(\mathrm{Re}\,\mathbb{A}f,f)\le(\dot{A}^*f,f)+(f,\dot{A}^*f),\quad f\in\mathcal{H}_+. \tag{3}$$

The following definition is a "lite" version of the definition of L-system given for a scattering L-system with one-dimensional input-output space. It is tailored for the case when the symmetric operator of an L-system has deficiency indices $(1,1)$. The general definition of an L-system can be found in [**2**, Definition 6.3.4].

DEFINITION 1. *An array*

$$\Theta=\begin{pmatrix} & \mathbb{A} & & K & 1\\ \mathcal{H}_+\subset & \mathcal{H} & \subset\mathcal{H}_- & & \mathbb{C}\end{pmatrix} \tag{4}$$

is called an ***L-system*** *if:*

(1) *T is a dissipative ($\mathrm{Im}(Tf,f)\ge 0$, $f\in\mathrm{Dom}(T)$) quasi-self-adjoint extension of a symmetric operator $\dot{A}$ with deficiency indices* $(1,1)$*;*
(2) *$\mathbb{A}$ is a $(*)$-extension of T;*
(3) $\mathrm{Im}\,\mathbb{A}=KK^*$*, where* $K\in[\mathbb{C},\mathcal{H}_-]$ *and* $K^*\in[\mathcal{H}_+,\mathbb{C}]$.

Operators T and $\mathbb{A}$ are called a *main and state-space operators respectively* of the system Θ, and K is a *channel operator*. It is easy to see that the operator $\mathbb{A}$ of the system (4) can be chosen such that $\mathrm{Im}\,\mathbb{A}=(\cdot,\chi)\chi$, $\chi\in\mathcal{H}_-$ and $Kc=c\cdot\chi$, $c\in\mathbb{C}$. A system Θ in (4) is called *minimal* if the operator $\dot{A}$ is a prime operator in $\mathcal{H}$, i.e., there exists no non-trivial reducing invariant subspace of $\mathcal{H}$ on which it induces a self-adjoint operator. Minimal L-systems of the form (4) with one-dimensional input-output space were also considered in [**6**].

We associate with an L-system Θ the function

$$W_\Theta(z)=I-2iK^*(\mathbb{A}-zI)^{-1}K,\quad z\in\rho(T), \tag{5}$$

which is called the **transfer function** of the L-system Θ. We also consider the function

$$V_\Theta(z)=K^*(\mathrm{Re}\,\mathbb{A}-zI)^{-1}K, \tag{6}$$

that is called the **impedance function** of an L-system Θ of the form (4). The transfer function $W_\Theta(z)$ of the L-system Θ and function $V_\Theta(z)$ of the form (6) are connected by the following relations valid for $\mathrm{Im}\,z\neq 0$, $z\in\rho(T)$,

$$\begin{aligned} V_\Theta(z)&=i[W_\Theta(z)+I]^{-1}[W_\Theta(z)-I],\\ W_\Theta(z)&=(I+iV_\Theta(z))^{-1}(I-iV_\Theta(z)).\end{aligned}$$

The class of all Herglotz-Nevanlinna functions that can be realized as impedance functions of L-systems and connections with Weyl-Titchmarsh functions can be found in [**2, 6, 11, 12, 14**] and references therein.

An L-system Θ of the form (4) is called an **accretive system** ([**8, 13**]) if its state-space operator $\mathbb{A}$ is accretive and **accumulative** ([**7**]) if its state-space operator $\mathbb{A}$ is accumulative, i.e., satisfies (3). It is easy to see that if an L-system is accumulative, then (3) implies that the operator $\dot{A}$ of the system is non-negative and both operators T and T^* are accretive. We also associate another operator $\tilde{\mathbb{A}}$ to an accumulative L-system Θ. It is given by

$$\tilde{\mathbb{A}} = 2\ \mathrm{Re}\ \dot{A}^* - \mathbb{A}, \tag{7}$$

where $\dot{A}^*$ is in $[\mathcal{H}_+, \mathcal{H}_-]$. Obviously, Re $\dot{A}^* \in [\mathcal{H}_+, \mathcal{H}_-]$ and $\tilde{\mathbb{A}} \in [\mathcal{H}_+, \mathcal{H}_-]$. Clearly, $\tilde{\mathbb{A}}$ is a bi-extension of $\dot{A}$ and is accretive if and only if $\mathbb{A}$ is accumulative. It is also not hard to see that even though $\tilde{\mathbb{A}}$ is not a $(*)$-extensions of the operator T but the form $(\tilde{\mathbb{A}}f, f)$, $f \in \mathcal{H}_+$ extends the form (f, Tf), $f \in \mathrm{Dom}(T)$. An accretive L-system is called **sectorial** if its state-space operator $\mathbb{A}$ is sectorial, i.e., satisfies (2) for some $\alpha \in (0, \pi/2)$. Similarly, an accumulative L-system is **sectorial** if its operator $\tilde{\mathbb{A}}$ of the form (7) is sectorial.

3. Realization of Stieltjes and inverse Stieltjes functions

A scalar function $V(z)$ is called the Herglotz-Nevanlinna function if it is holomorphic on $\mathbb{C} \setminus \mathbb{R}$, symmetric with respect to the real axis, i.e., $V(z)^* = V(\bar{z})$, $z \in \mathbb{C} \setminus \mathbb{R}$, and if it satisfies the positivity condition $\mathrm{Im}\, V(z) \geq 0$, $z \in \mathbb{C}_+$. The following definition can be found in [**16**]. A scalar Herglotz-Nevanlinna function $V(z)$ is a *Stieltjes function* if it is holomorphic in $\mathrm{Ext}[0, +\infty)$ and

$$\frac{\mathrm{Im}[zV(z)]}{\mathrm{Im}\, z} \geq 0. \tag{8}$$

It is known [**16**] that a Stieltjes function $V(z)$ admits the following integral representation

$$V(z) = \gamma + \int_0^\infty \frac{dG(t)}{t - z}, \tag{9}$$

where $\gamma \geq 0$ and $G(t)$ is a non-decreasing on $[0, +\infty)$ function such that $\int_0^\infty \frac{dG(t)}{1+t} < \infty$. We are going to focus on the class $S_0(R)$ (see [**8**], [**13**], [**2**]), whose definition is the following. A scalar Stieltjes function $V(z)$ is said to be a member of the **class** $S_0(R)$ if the measure $G(t)$ in representation (9) is unbounded. It was shown in [**2**] (see also [**8**]) that such a function $V(z)$ can be realized as the impedance function of an accretive L-system Θ of the form (4) with a densely defined symmetric operator if and only if it belongs to the class $S_0(R)$.

Now we turn to inverse Stieltjes functions. A scalar Herglotz-Nevanlinna function $V(z)$ is called **inverse Stieltjes** if $V(z)$ it is holomorphic in $\mathrm{Ext}[0, +\infty)$ and

$$\frac{\mathrm{Im}[V(z)/z]}{\mathrm{Im}\ z} \geq 0. \tag{10}$$

It can be shown (see [**16**]) that every inverse Stieltjes function $V(z)$ admits the following integral representation

$$V(z) = \gamma + z\beta + \int_0^\infty \left(\frac{1}{t - z} - \frac{1}{t}\right) dG(t), \tag{11}$$

where $\gamma \leq 0$, $\beta \geq 0$, and $G(t)$ is a non-decreasing on $[0, +\infty)$ function such that $\int_0^\infty \frac{dG(t)}{t+t^2} < \infty$. The following definition provides the description of a realizable subclass of inverse Stieltjes functions. A scalar inverse Stieltjes function $V(z)$ is a member of the **class** $S_0^{-1}(R)$ if the measure $G(t)$ in representation (11) is unbounded and $\beta = 0$. It was shown in [**2**] that a function $V(z)$ belongs to the class $S_0^{-1}(R)$ if and only if it can be realized as impedance function of an accumulative L-system Θ of the form (4) with a non-negative densely defined symmetric operator $\dot{A}$.

4. Sectorial classes and their realizations

In this section we are going to introduce sectorial subclasses of scalar Stieltjes and inverse Stieltjes functions. Let $\alpha \in (0, \frac{\pi}{2})$. First, we introduce **sectorial subclasses** S^α of Stieltjes functions as follows. A scalar Stieltjes function $V(z)$ belongs to S^α if

$$(12) \qquad K_\alpha = \sum_{k,l=1}^{n} \left[\frac{z_k V(z_k) - \bar{z}_l V(\bar{z}_l)}{z_k - \bar{z}_l} - (\cot\alpha)\ V(\bar{z}_l)V(z_k)\right] h_k \bar{h}_l \geq 0,$$

for arbitrary sequences of complex numbers $\{z_k\}$, $(\operatorname{Im} z_k > 0)$ and $\{h_k\}$, $(k = 1, ..., n)$. For $0 < \alpha_1 < \alpha_2 < \frac{\pi}{2}$, we have

$$S^{\alpha_1} \subset S^{\alpha_2} \subset S,$$

where S denotes the class of all Stieltjes functions (which corresponds to the case $\alpha = \frac{\pi}{2}$). Let Θ be a minimal L-system of the form (4) with a densely defined non-negative symmetric operator $\dot{A}$. Then (see [**2**]) the impedance function $V_\Theta(z)$ defined by (6) belongs to the class S^α if and only if the operator $\mathbb{A}$ of the L-system Θ is α-sectorial.

Let $0 \leq \alpha_1 \leq \alpha_2 \leq \frac{\pi}{2}$. We say that a scalar Stieltjes function $V(z)$ belongs to the **class** S^{α_1,α_2} if

$$(13) \qquad \tan\alpha_1 = \lim_{x\to-\infty} V(x), \qquad \tan\alpha_2 = \lim_{x\to-0} V(x).$$

The following connection between the classes S^α and S^{α_1,α_2} can be found in [**2**]. Let Θ be an L-system of the form (4) with a densely defined non-negative symmetric operator $\dot{A}$ with deficiency numbers $(1, 1)$. Let also $\mathbb{A}$ be an α-sectorial $(*)$-extension of T. Then the impedance function $V_\Theta(z)$ defined by (6) belongs to the class S^{α_1,α_2}, $\tan\alpha_2 \leq \tan\alpha$. Moreover, the main operator T is $(\alpha_2-\alpha_1)$-sectorial with the exact angle of sectoriality $(\alpha_2-\alpha_1)$. In the case when α is the exact angle of sectoriality of the operator T we have that $V_\Theta(z) \in S^{0,\alpha}$ (see [**2**]). It also follows that under this set of assumptions, the impedance function $V_\Theta(z)$ is such that $\gamma = 0$ in representation (9).

Now let Θ be an L-system of the form (4), where $\mathbb{A}$ is a $(*)$-extension of T and $\dot{A}$ is a closed densely defined non-negative symmetric operator with deficiency numbers $(1, 1)$. It was proved in [**2**] that if the impedance function $V_\Theta(z)$ belongs to the class S^{α_1,α_2}, then $\mathbb{A}$ is α-sectorial, where

$$(14) \qquad \tan\alpha = \tan\alpha_2 + 2\sqrt{\tan\alpha_1(\tan\alpha_2 - \tan\alpha_1)}.$$

Under the above set of conditions on L-system Θ it is shown in [**2**] that $\mathbb{A}$ is α-sectorial $(*)$-extension of an α-sectorial operator T with the exact angle $\alpha \in (0, \pi/2)$ if and only if $V_\Theta(z) \in S^{0,\alpha}$. Moreover, the angle α can be found via the formula

$$(15) \qquad \tan\alpha = \int_0^\infty \frac{dG(t)}{t},$$

where $G(t)$ is the measure from integral representation (9) of $V_\Theta(z)$.

Now we introduce **sectorial subclasses** $S^{-1,\alpha}$ of scalar inverse Stieltjes functions as follows. An inverse Stieltjes function $V(z)$ belongs to $S^{-1,\alpha}$ if

$$K_\alpha=\sum_{k,l=1}^{n}\left[\frac{V(z_k)/z_k-V(\bar z_l)/\bar z_l}{z_k-\bar z_l}-(\cot\alpha)\,\frac{V(\bar z_l)}{\bar z_l}\frac{V(z_k)}{z_k}\right]h_k\bar h_l\geq 0, \tag{16}$$

for an arbitrary sequences of complex numbers $\{z_k\}$, $(\operatorname{Im} z_k>0)$ and $\{h_k\}$, $(k=1,...,n)$. For $0<\alpha_1<\alpha_2<\frac{\pi}{2}$, we have

$$S^{-1,\alpha_1}\subset S^{-1,\alpha_2}\subset S^{-1},$$

where S^{-1} denotes the class of all inverse Stieltjes functions (which corresponds to the case $\alpha=\frac{\pi}{2}$).

Let Θ be an accumulative minimal L-system of the form (4). It was shown in [**9**] that the impedance function $V_\Theta(z)$ defined by (6) belongs to the class $S^{-1,\alpha}$ if and only if the operator $\tilde{\mathbb{A}}$ of the form (7) associated to the L-system Θ is α-sectorial.

Let $0\leq\alpha_1<\alpha_2\leq\frac{\pi}{2}$. We say that a scalar inverse Stieltjes function $V(z)$ of the class $S_0^{-1}(R)$ belongs to the **class** S^{-1,α_1,α_2} if

$$\tan(\pi-\alpha_1)=\lim_{x\to 0}V(x),\qquad \tan(\pi-\alpha_2)=\lim_{x\to-\infty}V(x). \tag{17}$$

The following connection between the classes $S^{-1,\alpha}$ and S^{-1,α_1,α_2} was established in [**9**]. Let Θ be an accumulative L-system of the form (4) with a densely defined non-negative symmetric operator $\dot A$. Let also $\tilde{\mathbb{A}}$ of the form (7) be α-sectorial. Then the impedance function $V_\Theta(z)$ defined by (6) belongs to the class S^{-1,α_1,α_2}. Moreover, the operator T of Θ is $(\alpha_2-\alpha_1)$-sectorial with the exact angle of sectoriality $(\alpha_2-\alpha_1)$, and $\tan\alpha_2\leq\tan\alpha$. Note, that this also remains valid for the case when the operator $\tilde{\mathbb{A}}$ is accretive but not α-sectorial for any $\alpha\in(0,\pi/2)$. It also follows that under the same set of assumptions, if α is the exact angle of sectoriality of the operator T, then $V_\Theta(z)\in S^{-1,0,\alpha}$ and is such that $\gamma=0$ and $\beta=0$ in (11).

Let Θ be a minimal accumulative L-system of the form (4) as above. Let also $\tilde{\mathbb{A}}$ be defined via (7). It was shown in [**9**] that if the impedance function $V_\Theta(z)$ belongs to the class S^{-1,α_1,α_2}, then $\tilde{\mathbb{A}}$ is α-sectorial, where $\tan\alpha$ is defined via (15) where $G(t)$ is the measure from integral representation (11) of $V_\Theta(z)$. Moreover, both $\tilde{\mathbb{A}}$ and T are α-sectorial operators with the exact angle $\alpha\in(0,\pi/2)$ if and only if $V_\Theta(z)\in S^{-1,0,\alpha}$ (see [**9**, Theorem 13]).

5. L-systems with Schrödinger operator

Let $\mathcal{H}=L_2[a,+\infty)$ and $l(y)=-y''+q(x)y$, where q is a real locally summable function. Suppose that the symmetric operator

$$\begin{cases}\dot A y=-y''+q(x)y\\ y(a)=y'(a)=0\end{cases} \tag{18}$$

has deficiency indices (1,1). Let D^* be the set of functions locally absolutely continuous together with their first derivatives such that $l(y)\in L_2[a,+\infty)$. Consider $\mathcal{H}_+=\operatorname{Dom}(\dot A^*)=D^*$ with the scalar product

$$(y,z)_+=\int_a^\infty\Big(y(x)\overline{z(x)}+l(y)\overline{l(z)}\Big)\,dx,\quad y,\ z\in D^*.$$

Let $\mathcal{H}_+ \subset L_2[a,+\infty) \subset \mathcal{H}_-$ be the corresponding triplet of Hilbert spaces. Consider the operators

$$\begin{cases} T_h y = l(y) = -y'' + q(x)y \\ hy(a) - y'(a) = 0 \end{cases}, \quad \begin{cases} T_h^* y = l(y) = -y'' + q(x)y \\ \bar{h}y(a) - y'(a) = 0 \end{cases}, \tag{19}$$

Let $\dot{A}$ be a symmetric operator of the form (18) with deficiency indices (1,1), generated by the differential operation $l(y) = -y'' + q(x)y$. Let also $\varphi_k(x,\lambda) (k = 1,2)$ be the solutions of the following Cauchy problems:

$$\begin{cases} l(\varphi_1) = \lambda\varphi_1 \\ \varphi_1(a,\lambda) = 0 \\ \varphi_1'(a,\lambda) = 1 \end{cases}, \quad \begin{cases} l(\varphi_2) = \lambda\varphi_2 \\ \varphi_2(a,\lambda) = -1 \\ \varphi_2'(a,\lambda) = 0 \end{cases}.$$

It is well known [**20**] that there exists a function $m_\infty(\lambda)$ for which

$$\varphi(x,\lambda) = \varphi_2(x,\lambda) + m_\infty(\lambda)\varphi_1(x,\lambda)$$

belongs to $L_2[a,+\infty)$.

Suppose that the symmetric operator $\dot{A}$ of the form (18) with deficiency indices (1,1) is nonnegative, i.e., $(\dot{A}f,f) \geq 0$ for all $f \in \mathrm{Dom}(\dot{A})$). For one-dimensional Shrödinger operator on the semi-axis the Phillips-Kato extension problem in restricted sense has the following form.

THEOREM 2 ([**3**, **21**, **22**]). *Let $\dot{A}$ be a nonnegative symmetric Schrödinger operator of the form* (18) *with deficiency indices* $(1,1)$ *and locally summable potential in* $\mathcal{H} = L^2[a,\infty)$. *Consider operator T_h of the form* (19). *Then*

(1) *operator $\dot{A}$ has more than one non-negative self-adjoint extension, i.e., the Friedrichs extension A_F and the Kreĭn-von Neumann extension A_K do not coincide, if and only if $m_\infty(-0) < \infty$;*
(2) *operator T_h coincides with the Kreĭn-von Neumann extension if and only if $h = -m_\infty(-0)$;*
(3) *operator T_h is accretive if and only if*

$$\mathrm{Re}\; h \geq -m_\infty(-0); \tag{20}$$

(4) *operator T_h, $(h \neq \bar{h})$ is α-sectorial if and only if* $\mathrm{Re}\; h > -m_\infty(-0)$ *holds;*
(5) *operator T_h, $(h \neq \bar{h})$ is accretive but not α-sectorial for any $\alpha \in (0, \frac{\pi}{2})$ if and only if* $\mathrm{Re}\; h = -m_\infty(-0)$;
(6) *If T_h, $(\mathrm{Im}\, h > 0)$ is α-sectorial, then the angle α can be calculated via*

$$\tan\alpha = \frac{\mathrm{Im}\, h}{\mathrm{Re}\; h + m_\infty(-0)}. \tag{21}$$

For the remainder of this paper we assume that $m_\infty(-0) < \infty$. Then according to Theorem 2 above (see also [**23**, **24**]) we have the existence of the operator T_h, $(\mathrm{Im}\, h > 0)$ that is accretive and/or sectorial. The following was shown in [**2**]. Let T_h $(\mathrm{Im}\, h > 0)$ be an accretive Schrödinger operator of the form (19). Then for all real μ satisfying the following inequality

$$\mu \geq \frac{(\mathrm{Im}\, h)^2}{m_\infty(-0) + \mathrm{Re}\; h} + \mathrm{Re}\; h, \tag{22}$$

the operators

$$\text{(23)}\quad \begin{aligned} \mathbb{A}y &= -y'' + q(x)y - \frac{1}{\mu - h}\,[y'(a) - hy(a)]\,[\mu\delta(x-a) + \delta'(x-a)], \\ \mathbb{A}^*y &= -y'' + q(x)y - \frac{1}{\mu - \overline{h}}\,[y'(a) - \overline{h}y(a)]\,[\mu\delta(x-a) + \delta'(x-a)], \end{aligned}$$

define the set of all accretive $(*)$-extensions $\mathbb{A}$ of the operator T_h. The accretive operator T_h has a unique accretive $(*)$-extension $\mathbb{A}$ if and only if

$$\operatorname{Re} h = -m_\infty(-0).$$

In this case this unique $(*)$-extension has the form

$$\text{(24)}\quad \begin{aligned} \mathbb{A}y &= -y'' + q(x)y + [hy(a) - y'(a)]\,\delta(x-a), \\ \mathbb{A}^*y &= -y'' + q(x)y + [\overline{h}y(a) - y'(a)]\,\delta(x-a). \end{aligned}$$

Now we shall construct an L-system based on a non-self-adjoint Schrödinger operator T_h. It was shown in [**2**, **4**] that the set of all $(*)$-extensions of a non-self-adjoint Schrödinger operator T_h of the form (19) in $L_2[a,+\infty)$ can be represented in the form (23). Moreover, the formulas (23) establish a one-to-one correspondence between the set of all $(*)$-extensions of a Schrödinger operator T_h of the form (19) and all real numbers $\mu \in [-\infty,+\infty]$. One can easily check that the $(*)$-extension $\mathbb{A}$ in (23) of the non-self-adjoint dissipative Schrödinger operator T_h, $(\operatorname{Im} h > 0)$ of the form (19) satisfies the condition

$$\operatorname{Im} \mathbb{A} = \frac{\mathbb{A} - \mathbb{A}^*}{2i} = (.,g)g,$$

where

$$\text{(25)}\quad g = \frac{(\operatorname{Im} h)^{\frac{1}{2}}}{|\mu - h|}\,[\mu\delta(x-a) + \delta'(x-a)]$$

and $\delta(x-a), \delta'(x-a)$ are the delta-function and its derivative at the point a, respectively. Moreover,

$$(y,g) = \frac{(\operatorname{Im} h)^{\frac{1}{2}}}{|\mu - h|}\,[\mu y(a) - y'(a)],$$

where $y \in \mathcal{H}_+$, $g \in \mathcal{H}_-$, $\mathcal{H}_+ \subset L_2(a,+\infty) \subset \mathcal{H}_-$ and the triplet of Hilbert spaces discussed above. Let $E = \mathbb{C}$, $Kc = cg$ $(c \in \mathbb{C})$. It is clear that

$$\text{(26)}\quad K^*y = (y,g), \quad y \in \mathcal{H}_+,$$

and $\operatorname{Im} \mathbb{A} = KK^*$. Therefore, the array

$$\text{(27)}\quad \Theta = \begin{pmatrix} \mathbb{A} & K & 1 \\ \mathcal{H}_+ \subset L_2[a,+\infty) \subset \mathcal{H}_- & & \mathbb{C} \end{pmatrix},$$

is an L-system with the main operator $\mathbb{A}$ of the form (4) with the channel operator K of the form (26). It was shown in [**2**, **4**] that the transfer and impedance functions of Θ are

$$W_\Theta(\lambda) = \frac{\mu - h}{\mu - \overline{h}}\,\frac{m_\infty(\lambda) + \overline{h}}{m_\infty(\lambda) + h},$$

and

$$\text{(28)}\quad V_\Theta(\lambda) = \frac{(m_\infty(\lambda) + \mu)\operatorname{Im} h}{(\mu - \operatorname{Re} h)\,m_\infty(\lambda) + \mu \operatorname{Re} h - |h|^2}.$$

It was proved in [**2**] that if Θ is an L-system of the form (27), where $\mathbb{A}$ is a $(*)$-extension of the form (23) of an accretive Schrödinger operator T_h of the form (19), then its impedance function $V_\Theta(z)$ is Stieltjes function if and only if (22) holds and inverse Stieltjes function if and only if

$$-m_\infty(-0) \le \mu \le \operatorname{Re} h. \tag{29}$$

Using formulas (23) and direct calculations one can obtain the formula for operator $\tilde{\mathbb{A}}$ of the form (7) as follows

$$\begin{aligned}\tilde{\mathbb{A}}y &= -y'' + q(x)y - y'(a)\delta(x-a) - y(a)\delta'(x-a)\\ &\quad + \frac{1}{\mu - h}\left[y'(a) - hy(a)\right]\left[\mu\delta(x-a) + \delta'(x-a)\right].\end{aligned} \tag{30}$$

6. Sectorial L-systems with Schrödinger operator

Let Θ be an L-system of the form (27), where $\mathbb{A}$ is a $(*)$-extension (23) of the accretive Schrödinger operator T_h. According to Theorem 2 we have that if an accretive Schrödinger operator T_h, $(\operatorname{Im} h > 0)$ is α-sectorial, then (21) holds. Conversely, if h, $(\operatorname{Im} h > 0)$ is such that $\operatorname{Re} h > -m_\infty(-0)$, then operator T_h of the form (19) is α-sectorial and α is determined by (21). Moreover, T_h is accretive but not α-sectorial for any $\alpha \in (0, \pi/2)$ if and only if $\operatorname{Re} h = -m_\infty(-0)$. Also (see [**2**]) the operator $\mathbb{A}$ of Θ is accretive if and only if (22) holds. Consider our system Θ with $\mu = +\infty$. It was shown in [**5**] that in this case $V_\Theta(z)$ belongs to the class $S^{0,\alpha}$. In the case when $\mu \ne +\infty$ we have $V_\Theta(z) \in S^{\alpha_1,\alpha_2}$ (see [**5**]).

THEOREM 3 ([**2**]). *Let Θ be an L-system of the form* (27), *where $\mathbb{A}$ is a $(*)$-extension of an α-sectorial operator T_h with the exact angle of sectoriality $\alpha \in (0, \pi/2)$. Then $\mathbb{A}$ is an α-sectorial $(*)$-extension of T_h (with the same angle of sectoriality) if and only if $\mu = +\infty$ in* (23).

We note that if T_h is α-sectorial with the exact angle of sectoriality α, then it admits only one α-sectorial $(*)$-extension $\mathbb{A}$ with the same angle of sectoriality α. Consequently, $\mu = +\infty$ and $\mathbb{A}$ has the form (24).

THEOREM 4 ([**2**]). *Let Θ be an L-system of the form* (27), *where $\mathbb{A}$ is a $(*)$-extension of an α-sectorial operator T_h with the exact angle of sectoriality $\alpha \in (0, \pi/2)$. Then $\mathbb{A}$ is accretive but not α-sectorial for any $\alpha \in (0, \pi/2)$ $(*)$-extension of T_h if and only if in* (23)

$$\mu = \mu_0 = \frac{(\operatorname{Im} h)^2}{m_\infty(-0) + \operatorname{Re} h} + \operatorname{Re} h. \tag{31}$$

Note that it follows from the above theorem that any α-sectorial operator T_h with the exact angle of sectoriality $\alpha \in (0, \pi/2)$ admits only one accretive $(*)$-extension $\mathbb{A}$ that is not α-sectorial for any $\alpha \in (0, \pi/2)$. This extension takes form (23) with $\mu = \mu_0$ where μ_0 is given by (31).

THEOREM 5. *Let Θ be an accretive L-system of the form* (27), *where $\mathbb{A}$ is a $(*)$-extension of a θ-sectorial operator T_h. Let also $\mu_* \in (\mu_0, +\infty)$ be a fixed value that determines $\mathbb{A}$ via* (23), *μ_0 be defined by* (31), *and $V_\Theta(z) \in S^{\alpha_1,\alpha_2}$. Then a $(*)$-extension $\mathbb{A}_\mu$ of T_h is β-sectorial for any $\mu \in [\mu_*, +\infty)$ with*

$$\tan\beta = \tan\alpha_1 + 2\sqrt{\tan\alpha_1\,\tan\alpha_2}. \tag{32}$$

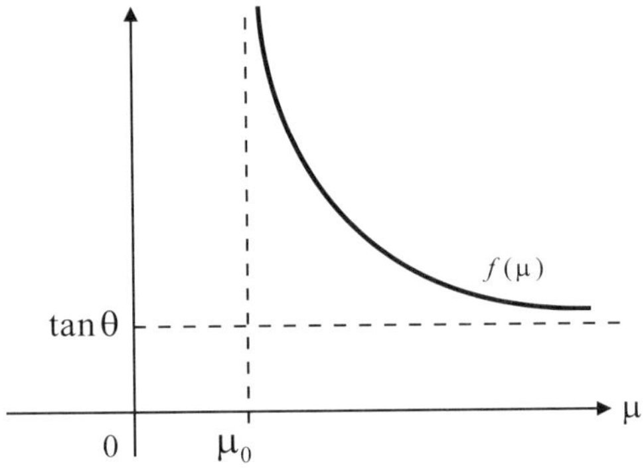

FIGURE 1. Function $f(\mu)$.

PROOF. According to [**2**], a φ-sectorial operator $\mathbb{A}$ of an L-system of the form (27) with the impedance function of the class S^{α_1,α_2} is also α-sectorial with $\tan\alpha$ described by (14). But then, clearly

$$\tan\alpha < \tan\beta = \tan\alpha_1 + 2\sqrt{\tan\alpha_1\,\tan\alpha_2}, \tag{33}$$

and hence this $\mathbb{A}$ is also β-sectorial.

Now suppose $\mu \in (\mu_0, +\infty)$. Then it follows from Theorem 4 that the operator $\mathbb{A}$ in L-system Θ of the form (27) is φ-sectorial (with some angle φ) for any such μ in parametrization (23). Using (37) and (36) on the impedance function $V_\Theta(z)$ of this L-system we can define a function

$$\begin{aligned} f(\mu) = \tan\alpha_1 + 2\sqrt{\tan\alpha_1\,\tan\alpha_2} &= \frac{(m_\infty(-0)+\mu)\operatorname{Im} h}{(\mu - \operatorname{Re} h)\,(m_\infty(-0)+\operatorname{Re} h) - (\operatorname{Im} h)^2} \\ &+ 2\sqrt{\frac{\operatorname{Im} h}{\mu - \operatorname{Re} h}\cdot\frac{(m_\infty(-0)+\mu)\operatorname{Im} h}{(\mu - \operatorname{Re} h)\,(m_\infty(-0)+\operatorname{Re} h) - (\operatorname{Im} h)^2}}. \end{aligned} \tag{34}$$

Recall that $\operatorname{Im} h > 0$ and (20) together with (22) imply $\mu > \operatorname{Re} h$. It also follows from (20) and (22) that the first fraction in the right side of (34) is positive for every $\mu \in (\mu_0, +\infty)$. Moreover, direct check reveals that the derivative of this fraction is negative and hence it is a decreasing function on $\mu \in (\mu_0, +\infty)$. Consequently, the expression under the square root in the second term has a negative derivative and hence is a decreasing. This can be seen by applying the product rule and taking into account that $\frac{\operatorname{Im} h}{\mu - \operatorname{Re} h}$ is a positive term with a negative derivative. Thus, one confirms that $f(\mu)$ is a decreasing function defined on $(\mu_0, +\infty)$ with the range $[\tan\theta, +\infty)$, where θ is the angle of sectoriality of the operator T_h and $\tan\theta$ is given by (21). The graph of this functions is schematically given on the Figure 1.

Next we take the $(*)$-extension $\mathbb{A}$ that is determined via (23) by the fixed value $\mu_* \in (\mu_0, +\infty)$ from the premise of our theorem. According to our derivations above this $\mathbb{A}$ is β-sectorial with β given by (32). But then for every $\mu \in (\mu_*, +\infty)$ the values of $f(\mu)$ are going to be smaller than $\tan\beta$ (see Figure 2). Consequently, for a $(*)$-extension $\mathbb{A}_\mu$ that is parameterized by the value of $\mu \in [\mu_*, +\infty)$ the following obvious inequalities take place

$$|\operatorname{Im}(\mathbb{A}_\mu f, f)| \le f(\mu)\operatorname{Re}(\mathbb{A}_\mu f, f) \le (\tan\beta)\operatorname{Re}(\mathbb{A}_\mu f, f), \qquad f \in \mathcal{H}_+.$$

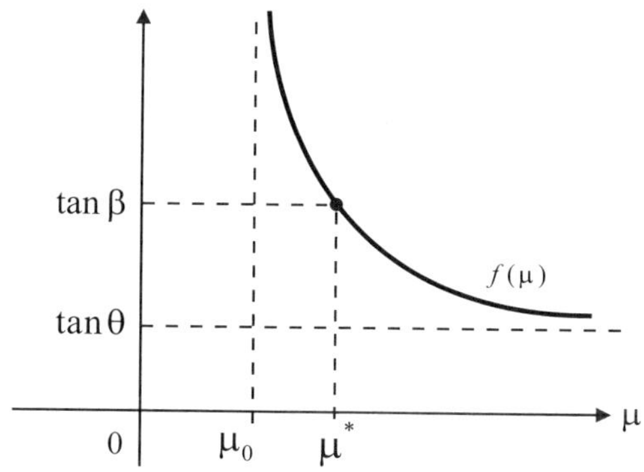

FIGURE 2. Angle of sectoriality β.

Hence, any $(*)$-extension $\mathbb{A}_\mu$ parameterized by a $\mu \in [\mu_*, +\infty)$ is β-sectorial. □

Note that Theorem 5 provides us with a value β which serves as a universal angle of sectoriality for the entire family of $(*)$-extensions $\mathbb{A}$ of the form (23).

It was shown in [**2**] that the operator $\mathbb{A}$ of Θ is accumulative if and only if (29) holds. Using (28) we can write the impedance function $V_\Theta(z)$ in the form

$$V_\Theta(z) = \frac{(m_\infty(z) + \mu)\,\mathrm{Im}\, h}{(\mu - \mathrm{Re}\ h)\,(m_\infty(z) + \mathrm{Re}\ h) - (\mathrm{Im}\, h)^2}. \tag{35}$$

Let μ satisfy the inequality (29). Then

$$\lim_{x \to -0} V_\Theta(x) = \frac{(m_\infty(-0) + \mu)\,\mathrm{Im}\, h}{(\mu - \mathrm{Re}\ h)\,(m_\infty(-0) + \mathrm{Re}\ h) - (\mathrm{Im}\, h)^2} = \tan(\pi - \alpha_1) = -\tan\alpha_1 \tag{36}$$

and

$$\begin{aligned}\lim_{x \to -\infty} V_\Theta(x) &= \lim_{x \to -\infty} \frac{(m_\infty(x) + \mu)\,\mathrm{Im}\, h}{(\mu - \mathrm{Re}\ h)\,(m_\infty(x) + \mathrm{Re}\ h) - (\mathrm{Im}\, h)^2} = \frac{\mathrm{Im}\, h}{\mu - \mathrm{Re}\ h} \\ &= \tan(\pi - \alpha_2).\end{aligned} \tag{37}$$

Therefore, $V_\Theta(z) \in S^{-1,\alpha_1,\alpha_2}$, where α_1 and α_2 are defined by (36) and (37).

THEOREM 6. *Let Θ be an L-system of the form (27), where $\mathbb{A}$ is an accumulative $(*)$-extension of an α-sectorial operator T_h with the exact angle of sectoriality $\alpha \in (0, \pi/2)$. Then the associated operator $\tilde{\mathbb{A}}$ is α-sectorial (with the same angle of sectoriality as T_h) if and only if $\mu = -m_\infty(-0)$ in (30).*

PROOF. It follows from (36)-(37) that in this case $V_\Theta(z) \in S^{-1,0,\alpha}$ if and only if $\mu = -m_\infty(-0)$. Thus, using [**9**, Theorem 13] for the function $V_\Theta(z)$ we obtain that $\tilde{\mathbb{A}}$ is α-sectorial. □

THEOREM 7. *Let Θ be an L-system of the form (27), where $\mathbb{A}$ is a $(*)$-extension of an α-sectorial operator T_h with the exact angle of sectoriality $\alpha \in (0, \pi/2)$. Then the associated operator $\tilde{\mathbb{A}}$ is accretive but not α-sectorial for any $\alpha \in (0, \pi/2)$ $(*)$-extension of T_h if and only if in (23)*

$$\mu = \mu_0 = \mathrm{Re}\ h. \tag{38}$$

PROOF. Let $V_\Theta(z)$ be the impedance function of our system Θ. If in (36) we set $\mu = \mu_0 = \text{Re } h$, then

$$\begin{aligned}\lim_{x\to 0} V_\Theta(x) &= -\frac{m_\infty(-0)+\text{Re } h}{\text{Im } h} = -\frac{1}{\tan\alpha} = -\tan\left(\frac{\pi}{2}-\alpha\right) = -\tan\alpha_1 \\ &= \tan(\pi-\alpha_1),\end{aligned} \tag{39}$$

where $\alpha_1 = \frac{\pi}{2}-\alpha$. On the other hand, using (37) with $\mu=\mu_0=\text{Re } h$ we obtain

$$\lim_{x\to-\infty} V_\Theta(x) = \frac{\text{Im } h}{\mu_0 - \text{Re } h} = -\infty = -\tan\frac{\pi}{2} = \tan(\pi-\alpha_2). \tag{40}$$

Hence, (39) and (40) yield that $V_\Theta(z)\in S^{-1,\frac{\pi}{2}-\alpha,\frac{\pi}{2}}$. Now, if we assume the α-sectoriality of $\mathbb{A}$, then by [**9**, Theorem 11]

$$\tan\alpha > \tan\alpha_2 = \infty.$$

Therefore, $\tilde{\mathbb{A}}$ is accretive but not α-sectorial for any $\alpha\in(0,\pi/2)$.

Conversely, suppose $\tilde{\mathbb{A}}$ is an α-sectorial $(*)$-extension for some $\alpha\in(0,\pi/2)$. Then, according to Theorem [**9**, Theorem 14], $\tilde{\mathbb{A}}$ is also β-sectorial and

$$\tan\beta = \tan\alpha_2 + 2\sqrt{\tan\alpha_1(\tan\alpha_2-\tan\alpha_1)} < \infty.$$

Hence, $\tan\alpha_2\neq\infty$ and it follows from (40) that $\mu\neq\mu_0$. The theorem is proved. □

THEOREM 8. *Let Θ be an accretive L-system of the form* (27), *where $\mathbb{A}$ is a $(*)$-extension of a θ-sectorial operator T_h. Let also $\mu_*\in[-m_\infty(-0),\mu_0)$ be a fixed value that parameterizes the associated operator $\tilde{\mathbb{A}}$ via* (30), *$\mu_0=\text{Re } h$, and $V_\Theta(z)\in S^{-1,\alpha_1,\alpha_2}$. Then operator $\tilde{\mathbb{A}}_\mu$ of T_h is β-sectorial for any $\mu\in[-m_\infty(-0),\mu_*)$ with*

$$\tan\beta = \tan\alpha_1 + 2\sqrt{\tan\alpha_1\,\tan\alpha_2}. \tag{41}$$

PROOF. According to [**9**, Theorem 11 and Theorem 14], a φ-sectorial associated operator $\tilde{\mathbb{A}}$ of an L-system of the form (27) with the impedance function of the class S^{-1,α_1,α_2} is also α-sectorial with

$$\tan\alpha = \tan\alpha_2 + 2\sqrt{\tan\alpha_1(\tan\alpha_2-\tan\alpha_1)}.$$

But then, clearly

$$\tan\alpha < \tan\beta = \tan\alpha_1 + 2\sqrt{\tan\alpha_1\,\tan\alpha_2}, \tag{42}$$

and hence this $\tilde{\mathbb{A}}$ is also β-sectorial.

Now suppose $\mu\in[-m_\infty(-0),\mu_*)$. Then it follows from Theorem 7 that the operator $\tilde{\mathbb{A}}$ associated with L-system Θ of the form (27) is φ-sectorial (with some angle φ) for any such μ in parametrization (30). Using (37) and (36) on the impedance function $V_\Theta(z)$ of this L-system we can define a function

$$\begin{aligned} f(\mu) = \tan\alpha_1 + 2\sqrt{\tan\alpha_1\,\tan\alpha_2} &= \frac{-(m_\infty(-0)+\mu)\,\text{Im } h}{(\mu-\text{Re } h)\,(m_\infty(-0)+\text{Re } h)-(\text{Im } h)^2} \\ &+ 2\sqrt{\frac{\text{Im } h}{\mu-\text{Re } h}\cdot\frac{(m_\infty(-0)+\mu)\,\text{Im } h}{(\mu-\text{Re } h)\,(m_\infty(-0)+\text{Re } h)-(\text{Im } h)^2}}.\end{aligned} \tag{43}$$

Recall that $\text{Im } h>0$ and (29) implies $\mu<\text{Re } h$ and $m_\infty(-0)+\mu\geq 0$. Consequently, the first fraction in the right side of (34) is positive. Furthermore, direct check reveals that the derivative of this fraction is also positive and hence it is an increasing function on $[-m_\infty(-0),\mu_*)$. Also, since $\mu<\text{Re } h$ and $\text{Im } h>0$, the first fraction under the square root is negative and has a negative derivative. Taking the

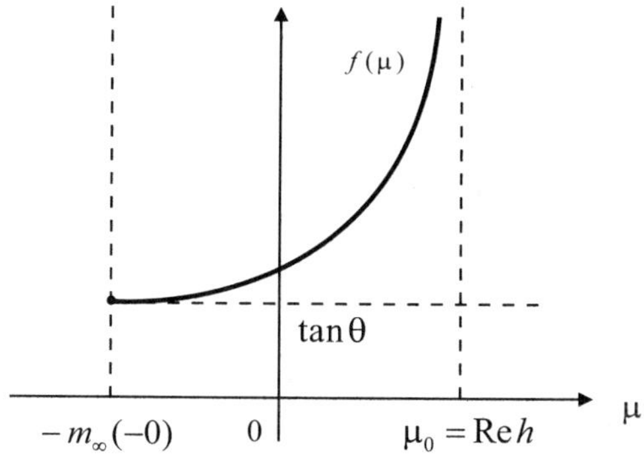

FIGURE 3. Function $f(\mu)$.

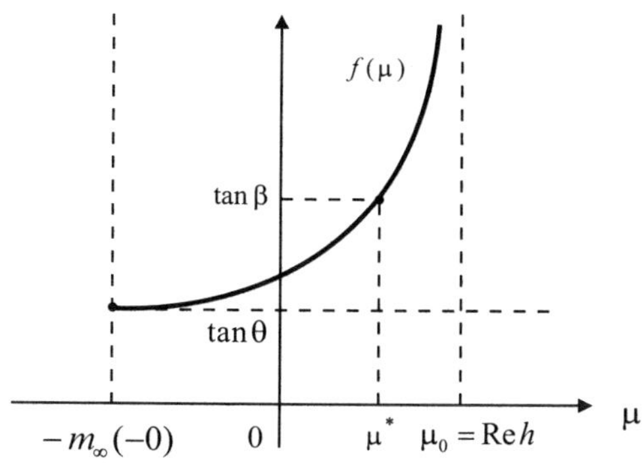

FIGURE 4. Angle of sectoriality β.

above into account and applying the product rule to the product under the square root we obtain that the derivative of the product is positive. Therefore, the entire second term in (43) an increasing function on $[-m_\infty(-0), \mu_*)$. Consequently, $f(\mu)$ is an increasing function defined on $[-m_\infty(-0), \mu_0)$ with the range $[\tan\theta, +\infty)$, where θ is the angle of sectoriality of the operator T_h and $\tan\theta$ is given by (21). The graph of this functions is schematically given on the Figure 3.

Next we take the associated operator $\tilde{\mathbb{A}}$ that is determined via (30) by the fixed value $\mu_* \in [-m_\infty(-0), \mu_0)$ from the premise of our theorem. According to our derivations above, this $\tilde{\mathbb{A}}$ is β-sectorial with β given by (41). But then for every $\mu \in [-m_\infty(-0), \mu_0)$ the values of $f(\mu)$ are going to be smaller than $\tan\beta$ (see Figure 4). Consequently, for an operator $\tilde{\mathbb{A}}_\mu$ that is parameterized by the value of $\mu \in [\mu_*, +\infty)$ the following obvious inequalities take place

$$|\operatorname{Im}(\tilde{\mathbb{A}}_\mu f, f)| \leq f(\mu)\operatorname{Re}(\tilde{\mathbb{A}}_\mu f, f) \leq (\tan\beta)\operatorname{Re}(\tilde{\mathbb{A}}_\mu f, f), \qquad f \in \mathcal{H}_+.$$

Hence, any associated operator $\tilde{\mathbb{A}}_\mu$ parameterized by a $\mu \in [\mu_*, +\infty)$ is β-sectorial. □

Note that Theorem 8 provides us with a value β which serves as a universal angle of sectoriality for the entire family of associated operators $\tilde{\mathbb{A}}$ of the form (30).

We conclude this paper with a couple of simple illustrations.

Example 1. Consider a function

$$V(z) = 1 + \frac{i}{\sqrt{z}}. \tag{44}$$

By direct check (see also [**2**]) one can see that $V(z)$ is a Stieltjes function of the sectorial class $S^{\frac{\pi}{4},\frac{\pi}{2}}$. Setting the values of parameters $h = \frac{1}{2}(1+i)$ and $\mu = 1$ into (28) and taking into account that $m_\infty(z) = -i\sqrt{z}$ (see [**2**]), we see that $V_\Theta(z)$ in (28) matches $V(z)$ in (44). Thus, this set of parameters corresponds to the L-system Θ whose impedance function is $V(z)$. Applying (23) yields

$$\begin{aligned} \mathbb{A}\, y &= -y'' - \frac{1}{1-i}[2y'(0) - (1+i)y(0)](\delta(x) + \delta'(x)), \\ \mathbb{A}^*\, y &= -y'' - \frac{1}{1+i}[2y'(0) - (1-i)y(0)](\delta(x) + \delta'(x)). \end{aligned} \tag{45}$$

The operator T_h in this case is

$$\begin{cases} T_h y = -y'' \\ 2y'(0) = (1+i)y(0). \end{cases}$$

The channel vector g of the form (25) then equals $g = \delta(x) + \delta'(x)$, satisfying

$$\operatorname{Im} \mathbb{A} = \frac{\mathbb{A} - \mathbb{A}^*}{2i} = KK^* = (.,g)g,$$

and operator $Kc = cg$, $(c \in \mathbb{C})$ with $K^*y = (y,g) = y(0) - y'(0)$. The real part

$$\operatorname{Re} \mathbb{A}\, y = -y'' - y'(0)(\delta(x) + \delta'(x))$$

contains the self-adjoint quasi-kernel

$$\begin{cases} \widehat{A} y = -y'' \\ y'(0) = 0. \end{cases}$$

An L-system with Schrödinger operator of the form (4) that realizes $V(z)$ can now be written as

$$\Theta = \begin{pmatrix} \mathbb{A} & K & 1 \\ \mathcal{H}_+ \subset L_2[a,+\infty) \subset \mathcal{H}_- & & \mathbb{C} \end{pmatrix}, \tag{46}$$

where $\mathbb{A}$ and K are defined above. By direct calculations we obtain that

$$(\operatorname{Re} \mathbb{A}y, y) = \int_0^\infty |y'(x)|dx + |y'(0)|^2 \geq 0,$$

and hence $\mathbb{A}$ is accretive. Applying Theorem 4 and (31) for $h = \frac{1}{2}(1+i)$ and $\mu = 1$ we get that $\mathbb{A}$ is not α-sectorial for any α even though T_h is α-sectorial for $\alpha = \pi/4$.

Example 2. Consider a function

$$V(z) = -\frac{\sqrt{z}}{\sqrt{z}+2i}. \tag{47}$$

By direct check (see also [**9**]) $V(z)$ is an inverse Stieltjes function of the class $S^{-1,0,\pi/4}$. Setting the values of parameters $h = 1+i$ and $\mu = 0$ into (28) and taking into account that $m_\infty(z) = -i\sqrt{z}$, we see that $V_\Theta(z)$ in (28) matches $V(z)$ in (47). Thus, this set of parameters corresponds to the L-system Θ whose impedance function is $V(z)$. Now we assemble an L-system Θ of the form (46) with this set of parameters. We have

$$\begin{cases} T_h y = -y'', \\ y'(0) = (1+i)y(0). \end{cases} \tag{48}$$

It was discussed in [**9**] that T of the form (48) is α-sectorial with the exact angle $\alpha = \pi/4$. Furthermore, the state-space operator is

$$\begin{aligned} \mathbb{A}\, y &= -y'' + \frac{1}{1+i}[y'(0) - (1+i)y(0)]\delta'(x), \\ \mathbb{A}^*\, y &= -y'' + \frac{1}{1-i}[y'(0) - (1-i)y(0)]\delta'(x), \end{aligned} \tag{49}$$

where $\operatorname{Im}\mathbb{A} = KK^*$ and $Kc = c \cdot g$ with $g = \frac{1}{\sqrt{2}}\delta'(x)$, $c \in \mathbb{C}$. The associated operator $\tilde{\mathbb{A}}$ of the form (7) (see also (30)) is

$$\tilde{\mathbb{A}}y = -y'' - y'(0)\delta(x) - y(0)\delta'(x) + \frac{1}{1+i}[y'(0) - (1+i)y(0)]\delta'(x).$$

By direct calculations we obtain that

$$(\operatorname{Re}\tilde{\mathbb{A}}y, y) = \|y'(x)\|^2_{L^2} + \frac{1}{2}|y'(0)|^2, \quad (\operatorname{Im}\tilde{\mathbb{A}}y, y) = -\frac{1}{2}|y'(0)|^2,$$

and hence $(\operatorname{Re}\tilde{\mathbb{A}}y, y) \geq |(\operatorname{Im}\tilde{\mathbb{A}}y, y)|$. Thus, $\tilde{\mathbb{A}}$ is α-sectorial with $\alpha = \pi/4$. According to [**9**, Theorem 13] this angle of sectoriality is exact. Consequently, we have shown that the α-sectorial sesquilinear form $(y, T_h y)$ defined on a subspace $\operatorname{Dom}(T_h)$ of $\mathcal{H}_+$ can be extended to the α-sectorial form $(\tilde{\mathbb{A}}y, y)$ defined on $\mathcal{H}_+$ having the exact (for both forms) angle of sectoriality $\alpha = \pi/4$. A general problem of extending sectorial sesquilinear forms to sectorial ones was mentioned by T. Kato in [**17**].

References

[1] D. Alpay and E. Tsekanovskii, *Interpolation theory in sectorial Stieltjes classes and explicit system solutions*, Linear Algebra Appl. **314** (2000), no. 1-3, 91–136, DOI 10.1016/S0024-3795(00)00113-0. MR1769016

[2] Y. Arlinskii, S. Belyi, and E. Tsekanovskii, *Conservative realizations of Herglotz-Nevanlinna functions*, Operator Theory: Advances and Applications, vol. 217, Birkhäuser/Springer Basel AG, Basel, 2011. MR2828331

[3] Yu. Arlinskiĭ and E. Tsekanovskiĭ, *M. Kreĭn's research on semi-bounded operators, its contemporary developments, and applications*, Modern analysis and applications. The Mark Krein Centenary Conference. Vol. 1: Operator theory and related topics, Oper. Theory Adv. Appl., vol. 190, Birkhäuser Verlag, Basel, 2009, pp. 65–112, DOI 10.1007/978-3-7643-9919-1_5. MR2568624

[4] Yu. Arlinskiĭ and E. Tsekanovskiĭ, *Linear systems with Schrödinger operators and their transfer functions*, Current trends in operator theory and its applications, Oper. Theory Adv. Appl., vol. 149, Birkhäuser, Basel, 2004, pp. 47–77. MR2063747

[5] S. Belyi, *Sectorial Stieltjes functions and their realizations by L-systems with a Schrödinger operator*, Math. Nachr. **285** (2012), no. 14-15, 1729–1740, DOI 10.1002/mana.201100226. MR2988003

[6] S. Belyi, K. A. Makarov, and E. Tsekanovskiĭ, *Conservative L-systems and the Livšic function*, Methods Funct. Anal. Topology **21** (2015), no. 2, 104–133. MR3407905

[7] S. V. Belyi and E. R. Tsekanovskii, *Inverse Stieltjes-like functions and inverse problems for systems with Schrödinger operator*, Characteristic functions, scattering functions and transfer functions, Oper. Theory Adv. Appl., vol. 197, Birkhäuser Verlag, Basel, 2010, pp. 21–49. MR2642707

[8] S. Belyi and E. Tsekanovskiĭ, *Stieltjes like functions and inverse problems for systems with Schrödinger operator*, Oper. Matrices **2** (2008), no. 2, 265–296, DOI 10.7153/oam-02-17. MR2420727

[9] S. Belyi and E. Tsekanovskiĭ, *Sectorial classes of inverse Stieltjes functions and L-systems*, Methods Funct. Anal. Topology **18** (2012), no. 3, 201–213. MR3051790

[10] Ju. M. Berezans′kiĭ, *Expansions in eigenfunctions of selfadjoint operators*, Translated from the Russian by R. Bolstein, J. M. Danskin, J. Rovnyak and L. Shulman. Translations of Mathematical Monographs, Vol. 17, American Mathematical Society, Providence, R.I., 1968. MR0222718

[11] V. Derkach, M.M. Malamud, *Extension theory of symmetric operators and boundary value problems.* Proceedings of Institute of Mathematics of NAS of Ukraine., Vol. 104., Institute of Mathematics of NAS of Ukraine, 2017.

[12] V. A. Derkach, M. M. Malamud, and È. R. Tsekanovskiĭ, *Sectorial extensions of a positive operator, and the characteristic function* (Russian), Ukrain. Mat. Zh. **41** (1989), no. 2, 151–158, 286, DOI 10.1007/BF01060376; English transl., Ukrainian Math. J. **41** (1989), no. 2, 136–142. MR992814

[13] I. N. Dovzhenko and È. R. Tsekanovskiĭ, *Classes of Stieltjes operator functions and their conservative realizations* (Russian), Dokl. Akad. Nauk SSSR **311** (1990), no. 1, 18–22; English transl., Soviet Math. Dokl. **41** (1990), no. 2, 201–204 (1991). MR1050823

[14] F. Gesztesy and E. Tsekanovskii, *On matrix-valued Herglotz functions*, Math. Nachr. **218** (2000), 61–138, DOI 10.1002/1522-2616(200010)218:1⟨61::AID-MANA61⟩3.3.CO;2-4. MR1784638

[15] J. A. Goldstein, *Semigroups of linear operators and applications*, Oxford Mathematical Monographs, The Clarendon Press, Oxford University Press, New York, 1985. MR790497

[16] I.S. Kac, M.G. Krein, *R-functions – analytic functions mapping the upper halfplane into itself*, Amer. Math. Soc. Transl., Vol. 2, **103**, 1-18, 1974.

[17] T. Kato, *Perturbation theory for linear operators*, Die Grundlehren der mathematischen Wissenschaften, Band 132, Springer-Verlag New York, Inc., New York, 1966. MR0203473

[18] S. G. Kreĭn, *Lineĭ nye uravneniya v banakhovom prostranstve* (Russian), Izdat. "Nauka", Moscow, 1971. MR0374949

[19] M.S. Livšic, *Operators, oscillations, waves.* Moscow, Nauka, (1966)

[20] M. A. Naĭmark, *Lineĭ nye differentsial′nye operatory* (Russian), Izdat. "Nauka", Moscow, 1969. Second edition, revised and augmented; With an appendix by V. È. Ljance. MR0353061

[21] È. R. Tsekanovskiĭ, *Accretive extensions and problems on the Stieltjes operator-valued functions relations*, Operator theory and complex analysis (Sapporo, 1991), Oper. Theory Adv. Appl., vol. 59, Birkhäuser, Basel, 1992, pp. 328–347. MR1246823

[22] È. R. Tsekanovskiĭ, *The characteristic function and sectorial boundary value problems* (Russian), Trudy Inst. Mat. (Novosibirsk) **7** (1987), no. Issled. Geom. Mat. Anal., 180–194, 200. MR905266

[23] E. R. Tsekanovskiĭ, *The Friedrichs and Kreĭn extensions of positive operators, and holomorphic semigroups of contractions* (Russian), Funktsional. Anal. i Prilozhen. **15** (1981), no. 4, 91–92. MR639213

[24] E. Tsekanovskiĭ, *Non-self-adjoint accretive extensions of positive operators and theorems of Friedrichs-Krein-Phillips.* Funct. Anal. Appl. **14**, 156–157 (1980)

[25] È. R. Cekanovskiĭ and Ju. L. Šmul′jan, *The theory of biextensions of operators in rigged Hilbert spaces. Unbounded operator colligations and characteristic functions* (Russian), Uspehi Mat. Nauk **32** (1977), no. 5(197), 69–124, 239. MR0463955

Department of Mathematics, Troy State University, Troy, Alabama 36082
Email address: sbelyi@troy.edu

Department of Mathematics, Niagara University, New York 14109
Email address: tsekanov@niagara.edu

Contemporary Mathematics
Volume **734**, 2019
https://doi.org/10.1090/conm/734/14764

The L^p-dissipativity of certain differential and integral operators

Alberto Cialdea and Vladimir Maz'ya

In Memory of Selim G. Krein

ABSTRACT. The first part of the paper is a survey of some of the results previously obtained by the authors concerning the L^p-dissipativity of scalar and matrix partial differential operators. In the second part we give new necessary and, separately, sufficient conditions for the L^p-dissipativity of the "complex oblique derivative" operator. In the case of real coefficients we provide a necessary and sufficient condition. We prove also the L^p-positivity for a certain class of integral operators.

1. Introduction

In a series of papers [**2**–**4**, **6**] we have studied the problem of characterizing the L^p-dissipativity of scalar and matrix partial differential operators. The main result we have obtained is that the algebraic condition

$$|p-2|\,|\langle \mathscr{I}m\, \mathscr{A}\, \xi, \xi\rangle| \leq 2\sqrt{p-1}\,\langle \mathscr{R}e\, \mathscr{A}\, \xi, \xi\rangle \tag{1.1}$$

for any $\xi \in \mathbb{R}^n$, is necessary and sufficient for the L^p-dissipativity of the Dirichlet problem for the scalar differential operator $\nabla^t(\mathscr{A}\,\nabla)$, where $\mathscr{A}$ is a matrix whose entries are complex measures, not necessarily absolutely continuous, and whose imaginary parts is symmetric. Specifically we have proved that condition (1.1) is necessary and sufficient for the L^p-dissipativity of the related sesquilinear form

$$\mathscr{L}(u,v) = \int_\Omega \langle \mathscr{A}\,\nabla u, \nabla v\rangle\,.$$

Such condition characterizes the L^p-dissipativity individually, for each p, while in the literature previous results dealt with the L^p-dissipativity for any $p \in [1,+\infty)$. Later on we have considered more general operators. Our results are described in the monograph [**5**].

We remark that, if $\mathscr{I}m\,\mathscr{A}$ is symmetric, (1.1) is equivalent to the the condition

$$\frac{4}{p\,p'}\langle \mathscr{R}e\,\mathscr{A}\,\xi,\xi\rangle + \langle \mathscr{R}e\,\mathscr{A}\,\eta,\eta\rangle + 2\langle (p^{-1}\,\mathscr{I}m\,\mathscr{A} + p'^{-1}\,\mathscr{I}m\,\mathscr{A}^*)\xi,\eta\rangle \geq 0 \tag{1.2}$$

for any $\xi, \eta \in \mathbb{R}^n$. If the matrix $\mathscr{I}m\,\mathscr{A}$ is not symmetric, condition (1.2) is only sufficient for the L^p-dissipativity of the corresponding form.

2010 *Mathematics Subject Classification.* Primary 47B44; Secondary 47F05, 74B05.
The publication was prepared with the support of the "RUDN University Program 5–100".

Let us consider the class of partial differential operators of the second order whose principal part is such that the form (1.2) is not merely non-negative, but strictly positive. This class of operators, which could be called *p-strongly elliptic*, was recently considered by Carbonaro and Dragičević [**1**], and Dindoš and Pipher [**7**].

In what follows, saying the L^p-dissipativity of an operator A, we mean the L^p-dissipativity of the corresponding form $\mathscr{L}$, just to simplify the terminology.

In the present paper, after surveying some of our more recent results for systems which are not contained in [**5**], we give some new theorems.

The first ones concern the "complex oblique derivative" operator

$$\lambda \cdot \nabla u = \frac{\partial u}{\partial x_n} + \sum_{j=1}^{n-1} a_j \frac{\partial u}{\partial x_j}$$

where $\lambda = (1, a_1, \dots, a_{n-1})$ and a_j are complex valued functions. We give necessary and, separately, sufficient conditions under which such boundary operator is L^p-dissipative on $\mathbb{R}^{n-1}$. If the coefficients a_j are real valued, we provide a necessary and sufficient condition.

The last result concerns a class of integral operators which can be written as

$$\int_{\mathbb{R}^n}^{*} [u(x) - u(y)]\, K(dx, dy) \tag{1.3}$$

where the integral has to be understood as a principal value in the sense of Cauchy and the kernel $K(dx, dy)$ is a Borel positive measure defined on $\mathbb{R}^n \times \mathbb{R}^n$ satisfying certain conditions. The class of operators we consider includes the fractional powers of Laplacian $(-\Delta)^s$, with $0 < s < 1$. We establish the L^p-positivity of operator (1.3), extending in this way a result we obtained in [**5**, p.230–231].

The paper is organized as follows. Section 2 presents a review of our main results concerning scalar differential operators of the second order.

Section 3 is dedicated to systems. In particular we describe some necessary and sufficient conditions for the L^p-dissipativity of systems of partial differential operators of the first order, recently obtained in [**6**].

The topic of Section 4 is elasticity system. After recalling the necessary and sufficient conditions we previously obtained in the planar case, we describe sufficient conditions holding in any dimension and proved in [**4**]. Finding necessary and sufficient conditions for the L^p-dissipativity of elasticity system in the three-dimensional case is still an open problem.

In Sections 5 and 6 we prove the above mentioned results concerning the "complex oblique derivative" operator and the operator (1.3) respectively.

2. The scalar operators

Let Ω be an open set in $\mathbb{R}^n$. Consider the sesquilinear form

$$\mathscr{L}(u, v) = \int_\Omega (\langle \mathscr{A} \nabla u, \nabla v\rangle - \langle \mathbf{b}\nabla u, v\rangle + \langle u, \overline{\mathbf{c}}\nabla v\rangle - a\langle u, v\rangle)$$

defined on $C_0^1(\Omega) \times C_0^1(\Omega)$. Here $\mathscr{A}$ is a $n \times n$ matrix function with complex valued entries $a^{hk} \in (C_0(\Omega))^*$, $\mathbf{b} = (b_1, \dots, b_n)$ and $\mathbf{c} = (c_1, \dots, c_n)$ are complex valued vectors with $b_j, c_j \in (C_0(\Omega))^*$ and a is a complex valued scalar distribution in $(C_0^1(\Omega))^*$. The symbol $\langle \cdot, \cdot \rangle$ denotes the inner product either in $\mathbb{C}^n$ or in $\mathbb{C}$.

In what follows, if $p \in (1, \infty)$, p' denotes its conjugate exponent $p/(p-1)$.

The integrals appearing in this definition have to be understood in a proper way. The entries a^{hk} being measures, the meaning of the first term is

$$\int_\Omega \langle \mathscr{A}\, \nabla u, \nabla v\rangle = \int_\Omega \partial_k u\, \partial_h \overline{v}\, da^{hk}\,.$$

Similar meanings have the terms involving $\mathbf{b}$ and $\mathbf{c}$. Finally, the last term is the action of the distribution $a \in (C_0^1(\Omega))^*$ on the functions $u\,\overline{v}$ belonging to $C_0^1(\Omega)$.

We say that the form $\mathscr{L}$ is *L^p-dissipative* $(1 < p < \infty)$ if for all $u \in C_0^1(\Omega)$

$$\mathscr{R}e\, \mathscr{L}(u, |u|^{p-2}u) \geq 0 \qquad \text{if } p \geq 2; \tag{2.1}$$

$$\mathscr{R}e\, \mathscr{L}(|u|^{p'-2}u, u) \geq 0 \qquad \text{if } 1 < p < 2 \tag{2.2}$$

(we use here that $|u|^{q-2}u \in C_0^1(\Omega)$ for $q \geq 2$ and $u \in C_0^1(\Omega)$).

The form $\mathscr{L}$ is related to the operator

$$Au = \operatorname{div}(\mathscr{A}\, \nabla u) + \mathbf{b}\nabla u + \operatorname{div}(\mathbf{c}u) + au. \tag{2.3}$$

where div denotes the divergence operator. The operator A acts from $C_0^1(\Omega)$ to $(C_0^1(\Omega))^*$ through the relation

$$\mathscr{L}(u,v) = -\int_\Omega \langle Au, v\rangle$$

for any $u, v \in C_0^1(\Omega)$. The integration is understood in the sense of distributions.

As we already remarked, saying the L^p-dissipativity of the operator A, we mean the L^p-dissipativity of the form $\mathscr{L}$.

The following Lemma provides a necessary and sufficient condition for the L^p-dissipativity of the form $\mathscr{L}$.

LEMMA 2.1 ([**2**]). *The operator A is L^p-dissipative if and only if for all $v \in C_0^1(\Omega)$*

$$\begin{aligned} \mathscr{R}e \int_\Omega \Big[\langle \mathscr{A}\, \nabla v, \nabla v\rangle - (1-2/p)\langle (\mathscr{A} - \mathscr{A}^*)\nabla(|v|), |v|^{-1}\overline{v}\nabla v\rangle - \\ (1-2/p)^2 \langle \mathscr{A}\, \nabla(|v|), \nabla(|v|)\rangle \Big] + \int_\Omega \langle \mathscr{I}m(\mathbf{b}+\mathbf{c}), \mathscr{I}m(\overline{v}\nabla v)\rangle + \\ \int_\Omega \mathscr{R}e(\operatorname{div}(\mathbf{b}/p - \mathbf{c}/p') - a)|v|^2 \geq 0. \end{aligned}$$

Here and in what follows the integrand is extended by zero on the set where v vanishes.

This result has several consequences. The first one is a necessary condition for the L^p-dissipativity.

COROLLARY 2.2 ([**2**]). *If the operator A is L^p-dissipative, we have*

$$\langle \mathscr{R}e\, \mathscr{A}\, \xi, \xi\rangle \geq 0 \tag{2.4}$$

(in the sense of measures) for any $\xi \in \mathbb{R}^n$.

Obviously condition (2.4) is not sufficient for the L^p-dissipativity of the form $\mathscr{L}$.

COROLLARY 2.3 ([**2**]). *Let α, β two real constants. If*

$$\begin{aligned}\frac{4}{pp'}\langle \mathscr{R}e\, \mathscr{A}\, \xi,\xi\rangle + \langle \mathscr{R}e\, \mathscr{A}\, \eta,\eta\rangle + 2\langle (p^{-1}\, \mathscr{I}m\, \mathscr{A} + p'^{-1}\, \mathscr{I}m\, \mathscr{A}^*)\xi,\eta\rangle + \\ \langle \mathscr{I}m(\mathbf{b}+\mathbf{c}),\eta\rangle - 2\langle \mathscr{R}e(\alpha\mathbf{b}/p - \beta\mathbf{c}/p'),\xi\rangle + \\ \mathscr{R}e\,[\operatorname{div}((1-\alpha)\mathbf{b}/p - (1-\beta)\mathbf{c}/p') - a] \geq 0\end{aligned} \tag{2.5}$$

for any $\xi,\eta \in \mathbb{R}^n$, the operator A is L^p-dissipative.

Generally speaking, conditions (2.5) are not necessary, as the following example shows.

EXAMPLE 2.4. Let $n = 2$ and

$$\mathscr{A} = \begin{pmatrix} 1 & i\gamma \\ -i\gamma & 1 \end{pmatrix}$$

where γ is a real constant, $\mathbf{b} = \mathbf{c} = a = 0$. In this case polynomial (2.5) is given by

$$(\eta_1 - \gamma\xi_2)^2 + (\eta_2 - \gamma\xi_1)^2 - (\gamma^2 - 4/(pp'))|\xi|^2.$$

Taking $\gamma^2 > 4/(pp')$, condition (2.5) is not satisfied, while we have the L^p-dissipativity, because the corresponding operator A is the Laplacian.

Note that in this example the matrix $\mathscr{I}m\, \mathscr{A}$ is not symmetric. Later we give another example showing that, even for symmetric matrices $\mathscr{I}m\, \mathscr{A}$, conditions (2.5) are not necessary for L^p-dissipativity (see Example 2.11). Nevertheless in the next section we show that the conditions are necessary for the L^p-dissipativity, provided the operator A has no lower order terms and the matrix $\mathscr{I}m\, \mathscr{A}$ is symmetric (see Theorem 2.5 and Remark 2.6).

In the case of an operator (2.3) without lower order terms:

$$Au = \operatorname{div}(\mathscr{A}\, \nabla u) \tag{2.6}$$

with the coefficients $a^{hk} \in (C_0(\Omega))^*$, we can give an algebraic necessary and sufficient condition for the L^p-dissipativity.

THEOREM 2.5 ([**2**]). *Let the matrix $\mathscr{I}m\, \mathscr{A}$ be symmetric, i.e. $\mathscr{I}m\, \mathscr{A}^t = \mathscr{I}m\, \mathscr{A}$. The form*

$$\mathscr{L}(u,v) = \int_\Omega \langle \mathscr{A}\, \nabla u, \nabla v\rangle$$

is L^p-dissipative if and only if

$$|p-2|\,|\langle \mathscr{I}m\, \mathscr{A}\, \xi,\xi\rangle| \leq 2\sqrt{p-1}\,\langle \mathscr{R}e\, \mathscr{A}\, \xi,\xi\rangle \tag{2.7}$$

for any $\xi \in \mathbb{R}^n$, where $|\cdot|$ denotes the total variation.

REMARK 2.6. One can prove that condition (2.7) holds if and only if

$$\frac{4}{pp'}\langle \mathscr{R}e\, \mathscr{A}\, \xi,\xi\rangle + \langle \mathscr{R}e\, \mathscr{A}\, \eta,\eta\rangle - 2(1-2/p)\langle \mathscr{I}m\, \mathscr{A}\, \xi,\eta\rangle \geq 0 \tag{2.8}$$

for any $\xi,\eta \in \mathbb{R}^n$. This means that conditions (2.5) are necessary and sufficient for the operators considered in Theorem 2.5.

REMARK 2.7. Let us assume that either A has lower order terms or they are absent and $\mathscr{I}m\,\mathscr{A}$ is not symmetric. Using the same arguments as in Theorem 2.5, one could prove that (2.7) (or, equivalently, (2.8)) is still a necessary condition for A to be L^p-dissipative. However, in general, it is not sufficient. This is shown by the next example (see also Theorem 2.9 below for the particular case of constant coefficients).

EXAMPLE 2.8. Let $n = 2$ and let Ω be a bounded domain. Denote by σ a not identically vanishing real function in $C_0^2(\Omega)$ and let $\lambda \in \mathbb{R}$. Consider operator (2.6) with

$$\mathscr{A} = \begin{pmatrix} 1 & i\lambda\partial_1(\sigma^2) \\ -i\lambda\partial_1(\sigma^2) & 1 \end{pmatrix}$$

i.e.

$$Au = \partial_1(\partial_1 u + i\lambda\partial_1(\sigma^2)\,\partial_2 u) + \partial_2(-i\lambda\partial_1(\sigma^2)\,\partial_1 u + \partial_2 u),$$

where $\partial_i = \partial/\partial x_i$ $(i = 1, 2)$.

By definition, we have L^2-dissipativity if and only if

$$\mathscr{R}e \int_\Omega ((\partial_1 u + i\lambda\partial_1(\sigma^2)\,\partial_2 u)\partial_1\overline{u} + (-i\lambda\partial_1(\sigma^2)\,\partial_1 u + \partial_2 u)\partial_2\overline{u})\,dx \geq 0$$

for any $u \in C_0^1(\Omega)$, i.e. if and only if

$$\int_\Omega |\nabla u|^2 dx - 2\lambda \int_\Omega \partial_1(\sigma^2)\,\mathscr{I}m(\partial_1\overline{u}\,\partial_2 u)\,dx \geq 0$$

for any $u \in C_0^1(\Omega)$. Taking $u = \sigma\,\exp(itx_2)$ $(t \in \mathbb{R})$, we obtain, in particular,

$$t^2 \int_\Omega \sigma^2 dx - t\lambda \int_\Omega (\partial_1(\sigma^2))^2 dx + \int_\Omega |\nabla\sigma|^2 dx \geq 0. \tag{2.9}$$

Since

$$\int_\Omega (\partial_1(\sigma^2))^2 dx > 0,$$

we can choose $\lambda \in \mathbb{R}$ so that (2.9) is impossible for all $t \in \mathbb{R}$. Thus A is not L^2-dissipative, although (2.7) is satisfied.

Since A can be written as

$$Au = \Delta u - i\lambda(\partial_{21}(\sigma^2)\,\partial_1 u - \partial_{11}(\sigma^2)\,\partial_2 u),$$

the same example shows that (2.7) is not sufficient for the L^2-dissipativity in the presence of lower order terms, even if $\mathscr{I}m\,\mathscr{A}$ is symmetric.

Generally speaking, it is impossible to obtain an algebraic characterization for an operator with lower order terms. Indeed, let us consider, for example, the operator

$$Au = \Delta u + a(x)u$$

in a bounded domain $\Omega \subset \mathbb{R}^n$ with zero Dirichlet boundary data. Denote by λ_1 the first eigenvalue of the Dirichlet problem for the Laplace equation in Ω. A sufficient condition for the L^2-dissipativity of A has the form $\operatorname{Re} a \leq \lambda_1$, and we cannot give an algebraic characterization of λ_1.

Consider, as another example, the operator

$$A = \Delta + \mu$$

where μ is a nonnegative Radon measure on Ω. The operator A is L^p-dissipative if and only if

$$\int_\Omega |w|^2 d\mu \leq \frac{4}{pp'} \int_\Omega |\nabla w|^2 dx \tag{2.10}$$

for any $w \in C_0{}^\infty(\Omega)$ (cf. Lemma 2.1). Maz'ya [**9**–**11**] proved that the following condition is sufficient for (2.10):

$$\frac{\mu(F)}{\mathrm{cap}_\Omega(F)} \leq \frac{1}{pp'} \tag{2.11}$$

for all compact set $F \subset \Omega$ and the following condition is necessary:

$$\frac{\mu(F)}{\mathrm{cap}_\Omega(F)} \leq \frac{4}{pp'} \tag{2.12}$$

for all compact set $F \subset \Omega$. Here, $\mathrm{cap}_\Omega(F)$ is the capacity of F with respect to Ω, i.e.,

$$\mathrm{cap}_\Omega(F) = \inf\left\{ \int_\Omega |\nabla u|^2 dx \ :\ u \in C_0{}^\infty(\Omega),\ u \geq 1 \text{ on } F\right\}. \tag{2.13}$$

The condition (2.11) is not necessary and the condition (2.12) is not sufficient.

However, it is possible to find necessary and sufficient conditions in the case of constant coefficients. Namely, let A be the differential operator

$$Au = \nabla^t(\mathscr{A}\,\nabla u) + \mathbf{b}\nabla u + au$$

with constant complex coefficients. Without loss of generality we assume that the matrix $\mathscr{A}$ is symmetric.

THEOREM 2.9 ([**2**]). *Let Ω be an open set in $\mathbb{R}^n$ which contains balls of arbitrarily large radius. The operator A is L^p-dissipative if and only if there exists a real constant vector V such that*

$$2\,\mathscr{R}e\,\mathscr{A}\,V + \mathscr{I}m\,\mathbf{b} = 0$$

$$\mathscr{R}e\,a + \langle \mathscr{R}e\,\mathscr{A}\,V, V\rangle \leq 0$$

and the inequality

$$|p-2|\,|\langle \mathscr{I}m\,\mathscr{A}\,\xi,\xi\rangle| \leq 2\sqrt{p-1}\,\langle \mathscr{R}e\,\mathscr{A}\,\xi,\xi\rangle \tag{2.14}$$

holds for any $\xi \in \mathbb{R}^n$.

COROLLARY 2.10 ([**2**]). *Let Ω be an open set in $\mathbb{R}^n$ which contains balls of arbitrarily large radius. Let us suppose that the matrix $\mathscr{R}e\,\mathscr{A}$ is not singular. The operator A is L^p-dissipative if and only if* (2.14) *holds and*

$$4\,\mathscr{R}e\,a \leq -\langle (\mathscr{R}e\,\mathscr{A})^{-1}\,\mathscr{I}m\,\mathbf{b}, \mathscr{I}m\,\mathbf{b}\rangle\,. \tag{2.15}$$

EXAMPLE 2.11. Let $n = 1$ and $\Omega = \mathbb{R}^1$. Consider the operator

$$\left(1 + 2\,\frac{\sqrt{p-1}}{p-2}\,i\right) u'' + 2iu' - u,$$

where $p \neq 2$ is fixed. Conditions (2.14) and (2.15) are satisfied and this operator is L^p-dissipative, in view of Corollary 2.10.

On the other hand, the polynomial considered in Corollary 2.3 is

$$Q(\xi,\eta) = \left(2\,\frac{\sqrt{p-1}}{p}\,\xi - \eta\right)^2 + 2\eta + 1$$

which is not nonnegative for any $\xi, \eta \in \mathbb{R}$. This shows that, in general, condition (2.5) is not necessary for the L^p-dissipativity, even if the matrix $\mathscr{I}m\,\mathscr{A}$ is symmetric.

3. L^p-dissipativity for systems

In this Section we describe criteria we have obtained for some systems of partial differential equations.

3.1. Systems of the first order. Let $\mathscr{B}^h$ and $\mathscr{C}^h$ $(h = 1, \ldots, n)$ be $m \times m$ matrices with complex-valued entries $b^h_{ij}, c^h_{ij} \in (C_0(\Omega))^*$ $(1 \leq i, j \leq m)$. Let $\mathscr{D}$ stand for a matrix whose elements d_{ij} are complex-valued distributions in $(C^1_0(\Omega))^*$.

Let $\mathscr{L}(u, v)$ be the sesquilinear form

$$\mathscr{L}(u,v) = \int_\Omega \langle \mathscr{B}^h\, \partial_h u, v\rangle - \langle \mathscr{C}^h\, u, \partial_h v\rangle + \langle \mathscr{D}\, u, v\rangle \tag{3.1}$$

defined in $(C^1_0(\Omega))^m \times (C^1_0(\Omega))^m$, where $\partial_h = \partial/\partial x_h$.

The form $\mathscr{L}$ is related to the system of partial differential operators of the first order:

$$Eu = \mathscr{B}^h\, \partial_h u + \partial_h(\mathscr{C}^h\, u) + \mathscr{D}\, u \tag{3.2}$$

As in the scalar case, we say that the form $\mathscr{L}$ is L^p-dissipative if (2.1)-(2.2) hold for all $u \in (C^1_0(\Omega))^m$.

We have found necessary and sufficient conditions for the L^p-dissipativity when

$$Eu = \mathscr{B}^h\, \partial_h u + \mathscr{D}\, u$$

and the entries of the matrices $\mathscr{B}^h, \mathscr{D}$ are locally integrable functions. Moreover we suppose that also $\partial_h \mathscr{B}^h$ (where the derivatives are in the sense of distributions) is a matrix with locally integrable entries.

THEOREM 3.1 ([**6**]). *The form*

$$\mathscr{L}(u,v) = \int_\Omega \langle \mathscr{B}^h\, \partial_h u, v\rangle + \langle \mathscr{D}\, u, v\rangle$$

is L^p – dissipative if, and only if, the following conditions are satisfied:

(1)

$$\mathscr{B}^h(x) = b_h(x)\, I, \qquad \text{if } p \neq 2, \tag{3.3}$$

$$\mathscr{B}^h(x) = (\mathscr{B}^h)^*(x), \qquad \text{if } p = 2, \tag{3.4}$$

for almost any $x \in \Omega$ and $h = 1, \ldots, n$. Here b_h are real locally integrable functions $(1 \leq h \leq n)$.

(2)

$$\mathscr{R}e\langle (p^{-1}\partial_h\, \mathscr{B}^h(x) - \mathscr{D}(x))\zeta, \zeta\rangle \geq 0 \tag{3.5}$$

for any $\zeta \in \mathbb{C}^m$, $|\zeta| = 1$ and for almost any $x \in \Omega$.

As far as the more general operator (3.2) is concerned, we have the following result, under the assumption that $\mathscr{B}^h$, $\mathscr{C}^h$, $\mathscr{D}$, $\partial_h \mathscr{B}^h$ and $\partial_h \mathscr{C}^h$ are matrices with complex locally integrable entries.

THEOREM 3.2 ([**6**]). *The form (3.1) is L^p-dissipative if, and only if, the following conditions are satisfied*

(1)
$$\mathscr{B}^h(x)+\mathscr{C}^h(x)=b_h(x)\,I, \qquad \text{if } p\neq 2,$$
$$\mathscr{B}^h(x)+\mathscr{C}^h(x)=(\mathscr{B}^h)^*(x)+(\mathscr{C}^h)^*(x), \qquad \text{if } p=2,$$
for almost any $x\in\Omega$ and $h=1,\ldots,n$. Here b_h are real locally integrable functions ($1\leq h\leq n$).

(2)
$$\mathscr{R}e\langle(p^{-1}\partial_h\,\mathscr{B}^h(x)-p'^{-1}\partial_h\,\mathscr{C}^h(x)-\mathscr{D}(x))\zeta,\zeta\rangle\geq 0$$
for any $\zeta\in\mathbb{C}^m$, $|\zeta|=1$ and for almost any $x\in\Omega$.

3.2. Systems of the second order. In this section we consider the class of systems of partial differential equations of the form

$$Eu=\partial_h(\mathscr{A}^h(x)\partial_h u)+\mathscr{B}^h(x)\partial_h u+\mathscr{D}(x)u, \tag{3.6}$$

where $\mathscr{A}^h$, $\mathscr{B}^h$ and $\mathscr{D}$ are $m\times m$ matrices with complex locally integrable entries.

If the operator (3.6) has no lower order terms, we have necessary and sufficient conditions:

THEOREM 3.3 ([**3**]). *The operator*
$$\partial_h(\mathscr{A}^h(x)\partial_h u)$$
is L^p-dissipative if and only if

$$\begin{aligned}&\mathscr{R}e\langle\mathscr{A}^h(x)\lambda,\lambda\rangle-(1-2/p)^2\,\mathscr{R}e\langle\mathscr{A}^h(x)\omega,\omega\rangle(\mathscr{R}e\langle\lambda,\omega\rangle)^2\\&-(1-2/p)\,\mathscr{R}e(\langle\mathscr{A}^h(x)\omega,\lambda\rangle-\langle\mathscr{A}^h(x)\lambda,\omega\rangle)\,\mathscr{R}e\langle\lambda,\omega\rangle\geq 0\end{aligned}\tag{3.7}$$

for almost every $x\in\Omega$ and for every $\lambda,\omega\in\mathbb{C}^m$, $|\omega|=1$, $h=1,\ldots,n$.

Combining this result with Theorem 3.1 we find

THEOREM 3.4 ([**6**]). *Let E be the operator (3.6), where $\mathscr{A}^h$ are $m\times m$ matrices with complex locally integrable entries and the matrices $\mathscr{B}^h(x)$, $\mathscr{D}(x)$ satisfy the hypothesis of Theorem 3.1. If (3.7) holds for almost every $x\in\Omega$ and for every $\lambda,\omega\in\mathbb{C}^m$, $|\omega|=1$, $h=1,\ldots,n$, and if conditions (3.3)-(3.4) and (3.5) are satisfied, the operator E is L^p-dissipative.*

Consider now the operator (3.6) in the scalar case (i.e. $m=1$)
$$\partial_h(a^h(x)\partial_h u)+b^h(x)\partial_h u+d(x)u$$
(a^h, b^h and d being scalar functions). In this case such an operator can be written in the form

$$Eu=\operatorname{div}(\mathscr{A}(x)\nabla u)+\mathscr{B}(x)\nabla u+d(x)\,u \tag{3.8}$$

where $\mathscr{A}=\{c_{hk}\}$, $c_{hh}=a^h$, $c_{hk}=0$ if $h\neq k$ and $\mathscr{B}=\{b^h\}$. For such an operator one can show that (3.7) is equivalent to

$$\frac{4}{pp'}\langle\mathscr{R}e\,\mathscr{A}(x)\xi,\xi\rangle+\langle\mathscr{R}e\,\mathscr{A}(x)\eta,\eta\rangle-2(1-2/p)\langle\mathscr{I}m\,\mathscr{A}(x)\xi,\eta\rangle\geq 0 \tag{3.9}$$

for almost any $x\in\Omega$ and for any $\xi,\eta\in\mathbb{R}^n$ (see [**5**, Remark 4.21, p.115]). Condition (3.9) is in turn equivalent to the inequality:

$$|p-2|\,|\langle\mathscr{I}m\,\mathscr{A}(x)\xi,\xi\rangle|\leq 2\sqrt{p-1}\,\langle\mathscr{R}e\,\mathscr{A}(x)\xi,\xi\rangle \tag{3.10}$$

for almost any $x\in\Omega$ and for any $\xi\in\mathbb{R}^n$ (see [**5**, Remark 2.8, p.42]). We have then

THEOREM 3.5 ([**6**]). *Let E be the scalar operator* (3.8) *where $\mathscr{A}$ is a diagonal matrix. If inequality* (3.10) *and conditions* (3.3)-(3.4) *and* (3.5) *are satisfied, the operator E is L^p-dissipative.*

More generally, consider the scalar operator (3.8) with a matrix $\mathscr{A} = \{a_{hk}\}$ not necessarily diagonal. The following result holds true.

THEOREM 3.6 ([**6**]). *Let the matrix $\mathscr{I}m\,\mathscr{A}$ be symmetric. If inequality* (3.10) *and conditions* (3.3)-(3.4) *and* (3.5) *are satisfied, the operator* (3.8) *is L^p-dissipative.*

4. The L^p-dissipativity of the Lamé operator

Let us consider the classical operator of linear elasticity

$$Eu = \Delta u + (1-2\nu)^{-1}\nabla \operatorname{div} u \tag{4.1}$$

where ν is the Poisson ratio. We assume that either $\nu > 1$ or $\nu < 1/2$. It is well known that E is strongly elliptic if and only if this condition is satisfied.

We remark that the elasticity system is not of the form considered in the subsection 3.2.

Let $\mathscr{L}$ be the bilinear form associated with operator (4.1), i.e.

$$\mathscr{L}(u,v) = -\int_\Omega (\langle \nabla u, \nabla v\rangle + (1-2\nu)^{-1} \operatorname{div} u \operatorname{div} v)\, dx\,,$$

The following lemma holds in any dimensions:

LEMMA 4.1 ([**3**]). *Let Ω be a domain of $\mathbb{R}^n$. The operator* (4.1) *is L^p-dissipative if and only if*

$$\int_\Omega [C_p|\nabla|v||^2 - \sum_{j=1}^{2} |\nabla v_j|^2 + \gamma\, C_p\, |v|^{-2}|v_h\partial_h|v||^2 - \gamma\, |\operatorname{div} v|^2]\, dx \leq 0$$

for any $v \in (C_0^1(\Omega))^2$, where

$$C_p = (1-2/p)^2, \qquad \gamma = (1-2\nu)^{-1}. \tag{4.2}$$

More precise results are known in the case of planar elasticity. At first we have an algebraic necessary condition:

LEMMA 4.2 ([**3**]). *Let Ω be a domain of $\mathbb{R}^2$. If the operator* (4.1) *is L^p-dissipative, we have*

$$C_p[|\xi|^2 + \gamma\,\langle\xi,\omega\rangle^2]\langle\lambda,\omega\rangle^2 - |\xi|^2|\lambda|^2 - \gamma\,\langle\xi,\lambda\rangle^2 \leq 0$$

for any ξ, λ, $\omega \in \mathbb{R}^2$, $|\omega| = 1$ (the constants C_p and γ being given by (4.2)*).*

Hinging on Lemmas 4.1 and 4.2, we proved

THEOREM 4.3 ([**3**]). *Let Ω be a domain of $\mathbb{R}^2$. The operator* (4.1) *is L^p-dissipative if and only if*

$$\left(\frac{1}{2}-\frac{1}{p}\right)^2 \leq \frac{2(\nu-1)(2\nu-1)}{(3-4\nu)^2}\,. \tag{4.3}$$

Concerning the elasticity system in any dimension, the next Theorem shows that condition (4.3) is necessary, even in the case of a non constant Poisson ratio. Here Ω is a bounded domain in $\mathbb{R}^n$ whose boundary is in the class C^2.

THEOREM 4.4 ([**4**]). *Suppose $\nu = \nu(x)$ is a continuous function defined in Ω such that*

$$\inf_{x\in\Omega} |2\nu(x) - 1| > 0.$$

If the operator (4.1) *is L^p-dissipative in Ω, then*

$$\left(\frac{1}{2} - \frac{1}{p}\right)^2 \leq \inf_{x\in\Omega} \frac{2(\nu(x) - 1)(2\nu(x) - 1)}{(3 - 4\nu(x))^2}.$$

We do not know if condition (4.3) is sufficient for the L^p-dissipativity of the n-dimensional elasticity. The next Theorem provides a more strict sufficient condition.

THEOREM 4.5 ([**4**]). *Let Ω be a domain in $\mathbb{R}^n$. If*

$$(1 - 2/p)^2 \leq \begin{cases} \dfrac{1 - 2\nu}{2(1 - \nu)} & \text{if } \nu < 1/2 \\[2ex] \dfrac{2(1 - \nu)}{1 - 2\nu} & \text{if } \nu > 1 \end{cases}$$

the operator (4.1) *is L^p-dissipative.*

We have also a kind of weighted L^p-negativity of elasticity system defined on rotationally symmetric vector functions.

Let Φ be a point on the $(n-2)-$dimensional unit sphere S^{n-2} with spherical coordinates $\{\vartheta_j\}_{j=1,\dots,n-3}$ and φ, where $\vartheta_j \in (0, \pi)$ and $\varphi \in [0, 2\pi)$. A point $x \in \mathbb{R}^n$ is represented as a triple $(\varrho, \vartheta, \Phi)$, where $\varrho > 0$ and $\vartheta \in [0, \pi]$. Correspondingly, a vector u can be written as $u = (u_\varrho, u_\vartheta, u_\Phi)$ with $u_\Phi = (u_{\vartheta_{n-3}}, \dots, u_{\vartheta_1}, u_\varphi)$. We call $u_\varrho, u_\vartheta, u_\Phi$ the spherical components of the vector u.

THEOREM 4.6 ([**4**]). *Let the spherical components u_ϑ and u_Φ of the vector u vanish, i.e. $u = (u_\varrho, 0, 0)$, and let u_ϱ depend only on the variable ϱ. Then, if $\alpha \geq n - 2$, we have*

$$\int_{\mathbb{R}^n} \left(\Delta u + (1 - 2\nu)^{-1}\nabla \operatorname{div} u\right) |u|^{p-2} u \frac{dx}{|x|^\alpha} \leq 0$$

for any $u \in (C_0^\infty(\mathbb{R}^n \setminus \{0\}))^n$ satisfying the aforesaid symmetric conditions, if and only if

$$-(p - 1)(n + p' - 2) \leq \alpha \leq n + p - 2.$$

If $\alpha < n - 2$ the same result holds replacing $(C_0^\infty(\mathbb{R}^n \setminus \{0\}))^n$ by $(C_0^\infty(\mathbb{R}^n))^n$.

5. The L^p-dissipativity of the "complex oblique derivative" operator

In this section we consider the L^p-dissipativity of the "complex oblique derivative" operator, i.e. of the boundary operator

$$\lambda \cdot \nabla u = \frac{\partial u}{\partial x_n} + \sum_{j=1}^{n-1} a_j \frac{\partial u}{\partial x_j}, \tag{5.1}$$

the coefficients a_j being L^∞ complex valued functions defined on $\mathbb{R}^{n-1}$.

We start with a Lemma, in which we use the concept of multiplier (see [**12**]). In particular we consider the space - we denote by $\mathscr{M}$ - of the multipliers acting from $H^{1/2}(\mathbb{R}^{n-1})$ into itself. Necessary and sufficient conditions for a function to be a multiplier and equivalent expressions for the relevant norms are given in [**12**] (see, in particular, Theorem 4.1.1, p.134).

LEMMA 5.1. *Let $a = (a_1, \dots, a_{n-1})$ be a vector multiplier belonging to $\mathscr{M}$. We have*

$$\left| \int_{\mathbb{R}^{n-1}} a_j\, f\, \partial_j g\, dx' \right| \leq \|a\|_{\mathscr{M}} \|\nabla f\|_{L^2(\mathbb{R}^n_+)} \|\nabla g\|_{L^2(\mathbb{R}^n_+)}$$

for any $f, g \in H^1(\mathbb{R}^n_+)$, $j = 1, \dots, n-1$. Here the derivatives are understood in the sense of distributions.

PROOF. Let us denote by Λ the operator $\sqrt{-\Delta}$ and write

$$\int_{\mathbb{R}^{n-1}} a_j\, f\, \partial_j g\, dx' = \int_{\mathbb{R}^{n-1}} \Lambda^{1/2}(a_j\, f)\, \Lambda^{-1/2}(\partial_j g)\, dx' .$$

We have

$$\left| \int_{\mathbb{R}^{n-1}} a_j\, f\, \partial_j g\, dx' \right| \leq \|\Lambda^{1/2}(a_j\, f)\|_{L^2(\mathbb{R}^{n-1})} \|\Lambda^{-1/2}(\partial_j g)\|_{L^2(\mathbb{R}^{n-1})} \leq$$

$$\|a\|_{\mathscr{M}} \|\Lambda^{1/2} f\|_{L^2(\mathbb{R}^{n-1})} \|\Lambda^{1/2} g\|_{L^2(\mathbb{R}^{n-1})} \leq \|a\|_{\mathscr{M}} \|\nabla f\|_{L^2(\mathbb{R}^n_+)} \|\nabla g\|_{L^2(\mathbb{R}^n_+)} .$$

□

The next Theorem provides a sufficient condition for the L^p-dissipativity of operator (5.1) under the assumption that

$$\| \mathscr{I}m\, a\|_{\mathscr{M}} < \frac{4}{p\,p'} . \tag{5.2}$$

THEOREM 5.2. *Suppose condition (5.2) is satisfied. If there exists a real vector $\Gamma \in L^2_{loc}(\mathbb{R}^n)$ such that*

$$-\partial_j(\mathscr{R}e\, a_j)\, \delta(x_n) \leq \frac{p}{2} \left(\frac{4}{p\,p'} - \| \mathscr{I}m\, a\|_{\mathscr{M}} \right) (\operatorname{div} \Gamma - |\Gamma|^2) \tag{5.3}$$

in $\mathbb{R}^n$, in the sense of distributions, then the operator (5.1) is L^p-dissipative.

PROOF. It is well known that $-\partial/\partial_{x_n} u(x', 0) = \Lambda(u)$ where, as in the previous Lemma, $\Lambda = \sqrt{-\Delta}$. This permits us to introduce the sesquilinear form

$$\mathscr{L}(u, v) = -\int_{\mathbb{R}^{n-1}} \langle \Lambda^{1/2} u, \Lambda^{1/2} v \rangle\, dx' + \int_{\mathbb{R}^{n-1}} \langle \sum_{j=1}^{n-1} a_j \partial_j u, v \rangle\, dx' . \tag{5.4}$$

We say that the operator $\lambda \cdot \nabla$ is L^p-dissipative if conditions (2.1) and (2.2) are satisfied for any $u \in C^1_0(\mathbb{R}^{n-1})$, the form $\mathscr{L}$ being given by (5.4).

Suppose $p \geq 2$. Denote by U the harmonic extension of u, i.e.

$$U(x', x_n) = \frac{2}{\omega_n} \int_{\mathbb{R}^{n-1}} u(y')\, \frac{x_n}{(|x' - y'|^2 + x_n^2)^{n/2}}\, dy'$$

ω_n being the measure of the unit sphere in $\mathbb{R}^n$.

Integrating by parts we get

$$\mathscr{R}e\, \mathscr{L}(u, |u|^{p-2} u) =$$

$$\mathscr{R}e \int_{\mathbb{R}^{n-1}} \frac{\partial U}{\partial x_n}\, |u|^{p-2} \overline{u}\, dx' + \mathscr{R}e \int_{\mathbb{R}^{n-1}} \sum_{j=1}^{n-1} a_j \frac{\partial u}{\partial x_j}\, |u|^{p-2} \overline{u}\, dx' =$$

$$-\mathscr{R}e \int_{\mathbb{R}^n_+} \nabla U \cdot \nabla (|U|^{p-2} \overline{U})\, dx + \mathscr{R}e \int_{\mathbb{R}^{n-1}} \sum_{j=1}^{n-1} a_j \frac{\partial u}{\partial x_j}\, |u|^{p-2} \overline{u}\, dx' .$$

Therefore, for $p \geq 2$, the operator $\lambda \cdot \nabla$ is L^p-dissipative if and only if

$$\mathscr{R}e \int_{\mathbb{R}^{n-1}} \sum_{j=1}^{n-1} a_j \frac{\partial u}{\partial x_j} |u|^{p-2}\overline{u}\, dx' \leq \mathscr{R}e \int_{\mathbb{R}^n_+} \nabla U \cdot \nabla(|U|^{p-2}\overline{U})\, dx \tag{5.5}$$

for any $u \in C_0^1(\mathbb{R}^{n-1})$, U being the harmonic extension of u to $\mathbb{R}^n_+$.

Setting $V = |U|^{(p-2)/2}U$ and $v = |u|^{(p-2)/2}u$, we get

$$|u|^{p-2}\overline{u}\, \partial_j u = -(1-2/p)\, |v|\, \partial_j |v| + \overline{v}\, \partial_j v$$

and then

$$\mathscr{R}e(|u|^{p-2}\overline{u}\, \partial_j u) = -(1-2/p)\, |v|\, \partial_j |v| + \mathscr{R}e(\overline{v}\, \partial_j v) = \frac{1}{p}\partial_j(|v|^2).$$

With similar computations we find

$$\mathscr{R}e(\nabla U \cdot \nabla(|U|^{p-2}\overline{U})) = |\nabla V|^2 - (1-2/p)^2 |\, \nabla |V|\, |^2.$$

Inequality (5.5) becomes

$$\begin{aligned} -\frac{1}{p}\int_{\mathbb{R}^{n-1}} \partial_j(\mathscr{R}e\, a_j)\, |v|^2\, dx' - \int_{\mathbb{R}^{n-1}} \mathscr{I}m\, a_j\; \mathscr{I}m(\overline{v}\, \partial_j v)\, dx' \leq \\ \int_{\mathbb{R}^n_+} (|\nabla V|^2 - (1-2/p)^2 |\, \nabla |V|\, |^2)\, dx\, . \end{aligned} \tag{5.6}$$

Lemma 5.1 implies that

$$\left| \int_{\mathbb{R}^{n-1}} \mathscr{I}m\, a_j\; \mathscr{I}m(\overline{v}\, \partial_j v)\, dx' \right| \leq \| \mathscr{I}m\, a\|_{\mathscr{M}} \int_{\mathbb{R}^n_+} |\nabla V|^2 dx \tag{5.7}$$

On the other hand, inequality (5.3) is the necessary and sufficient condition for the validity of the inequality

$$-\int_{\mathbb{R}^{n-1}} \partial_j(\mathscr{R}e\, a_j)\, |v|^2\, dx' \leq \frac{p}{2}\left(\frac{4}{p\,p'} - \| \mathscr{I}m\, a\|_{\mathscr{M}}\right) \int_{\mathbb{R}^n} |\nabla V|^2\, dx \tag{5.8}$$

for any $V \in C_0^\infty(\mathbb{R}^n)$, v being the restriction of V on $\mathbb{R}^{n-1}$ (see [**8**, Th. 5.1]). In particular, we find

$$-\int_{\mathbb{R}^{n-1}} \partial_j(\mathscr{R}e\, a_j)\, |v|^2\, dx' \leq p\left(\frac{4}{p\,p'} - \| \mathscr{I}m\, a\|_{\mathscr{M}}\right) \int_{\mathbb{R}^n_+} |\nabla V|^2\, dx \tag{5.9}$$

for any function $V \in C_0^\infty(\mathbb{R}^n)$ which is even with respect to x_n. Since $|\, \nabla |V|\, | \leq |\nabla V|$ and $1-(1-2/p)^2 = 4/(p\,p')$, we have also

$$\frac{4}{p\,p'}\int_{\mathbb{R}^n_+} |\nabla V|^2\, dx \leq \int_{\mathbb{R}^n_+} (|\nabla V|^2 - (1-2/p)^2 |\, \nabla |V|\, |^2)\, dx\, .$$

This inequality, together with (5.7) and (5.9), show that (5.6) holds and the operator $\lambda \cdot \nabla$ is L^p-dissipative.

If $1 < p < 2$ we have to show that

$$\mathscr{R}e\, \mathscr{L}(|u|^{p'-2}u, u) \leq 0$$

for any $u \in C_0^1(\mathbb{R}^{n-1})$.

Arguing as for (5.5) we find that the operator $\lambda \cdot \nabla$ is L^p-dissipative if and only if

$$\mathscr{R}e \int_{\mathbb{R}^{n-1}} \sum_{j=1}^{n-1} a_j \frac{\partial(|u|^{p'-2}u)}{\partial x_j}\, \overline{u}\, dx' \leq \mathscr{R}e \int_{\mathbb{R}^n_+} \nabla(|U|^{p'-2}U) \cdot \nabla\overline{U}\, dx\, .$$

Setting $V = |V|^{p'-2}V$ and $v = |u|^{p'-2}v$, we have

$$\mathscr{R}e(|u|^{p'-2}u\,\partial_j\overline{u}) = (1-2/p')\,|v|\,\partial_j|v| + \mathscr{R}e(v\,\partial_j\overline{v}) = \frac{1}{p}\partial_j(|v|^2),$$

$$\mathscr{R}e(\nabla(|U|^{p'-2}U)\cdot\nabla\overline{U}) = |\nabla V|^2 - (1-2/p)^2|\,\nabla|V|\,|^2.$$

Therefore the operator $\lambda\cdot\nabla$ is L^p-dissipative if and only if (5.6) holds and the proof proceeds as in the case $p\geq 2$. □

The next Theorem provides a necessary condition, similar to the previous one, but which contains a different constant. We remark that in the next Theorem we do not require the smallness of $\|\mathscr{I}m\,a\|_{\mathscr{M}}$.

THEOREM 5.3. *If the operator $\lambda\cdot\nabla$ is L^p-dissipative, then there exists a real vector $\Gamma\in L^2_{loc}(\mathbb{R}^n)$ such that we have the inequality*

$$-\partial_j(\mathscr{R}e\,a_j)\,\delta(x_n) \leq \frac{p}{2}\,(1+\|\mathscr{I}m\,a\|_{\mathscr{M}})\,(\operatorname{div}\Gamma - |\Gamma|^2)$$

in the sense of distributions.

PROOF. If $\lambda\cdot\nabla$ is L^p-dissipative, we have the inequality (5.5). Keeping in mind (5.7) we find

$$-\frac{1}{p}\int_{\mathbb{R}^{n-1}}\partial_j(\mathscr{R}e\,a_j)\,|v|^2\,dx' \leq \int_{\mathbb{R}^n_+}(|\nabla V|^2 - (1-2/p)^2|\,\nabla|V|\,|^2)\,dx +$$

$$\int_{\mathbb{R}^{n-1}}\mathscr{I}m\,a_j\;\mathscr{I}m(\overline{v}\,\partial_j v)\,dx' \leq (1+\|\mathscr{I}m\,a\|_{\mathscr{M}})\int_{\mathbb{R}^n_+}|\nabla V|^2\,dx\,.$$

Let us now consider $V\in C^\infty_0(\mathbb{R}^n)$ and write $V = V_o + V_e$, where V_o and V_e are odd and even respectively. The last inequality we have written leads to

$$-\frac{1}{p}\int_{\mathbb{R}^{n-1}}\partial_j(\mathscr{R}e\,a_j)\,|v|^2\,dx' \leq \frac{1+\|\mathscr{I}m\,a\|_{\mathscr{M}}}{2}\int_{\mathbb{R}^n}|\nabla V_o|^2\,dx \leq$$

$$\frac{1+\|\mathscr{I}m\,a\|_{\mathscr{M}}}{2}\int_{\mathbb{R}^n}|\nabla V|^2\,dx$$

and the thesis follows from the already quoted result [**8**, Th. 5.1].

□

A different sufficient condition can be obtained by using the concept of capacity (see (2.13)).

THEOREM 5.4. *Suppose condition* (5.2) *is satisfied. If*

$$-\frac{1}{\operatorname{cap}_\Omega(F)}\int_{F\cap\mathbb{R}^{n-1}}\partial_j(\mathscr{R}e\,a_j)\,dx' \leq \frac{p}{8}\left(\frac{4}{p\,p'} - \|\mathscr{I}m\,a\|_{\mathscr{M}}\right) \tag{5.10}$$

for all compact sets $F\subset\mathbb{R}^n$, then the operator $\lambda\cdot\nabla$ is L^p-dissipative.

PROOF. We know that inequality (2.10) holds for any test function w if condition (2.11) is satisfied for all compact sets $F\subset\Omega$. As remarked in [**11**, Remark 5.2], (2.11) implies (2.10) even without the requirement that $\mu\geq 0$, i.e. μ can be an arbitrary locally finite real valued charge.

Therefore, condition (5.10) implies inequality (5.8) and, as in Theorem 5.2, the result follows. □

In the case of the real oblique derivative problem we have a necessary and sufficient condition.

THEOREM 5.5. *Let us suppose that the coefficients a_j in (5.1) are real valued $(j = 1, \dots, n-1)$. The operator $\lambda \cdot \nabla$ is L^p-dissipative if and only if there exists a real vector $\Gamma \in L^2_{loc}(\mathbb{R}^n)$ such that*

$$-\partial_j(\mathscr{R}e\, a_j)\, \delta(x_n) \leq \frac{2}{p'}(\operatorname{div} \Gamma - |\Gamma|^2) \tag{5.11}$$

in the sense of distributions.

PROOF. The operator $\lambda \cdot \nabla$ being real, we consider the conditions (2.1) and (2.2) for real valued functions. We remark that, if V is a real valued function, we have $|\nabla V| = |\nabla |V||$ and then

$$\int_{\mathbb{R}^n_+} (|\nabla V|^2 - (1-2/p)^2 |\nabla |V||^2)\, dx = \frac{4}{p\,p'} \int_{\mathbb{R}^n_+} |\nabla V|^2 dx\,.$$

Therefore the L^p-dissipativity of $\lambda \cdot \nabla$ occurs if and only if

$$-\frac{1}{p} \int_{\mathbb{R}^{n-1}} \partial_j(\mathscr{R}e\, a_j)\, |v|^2\, dx' \leq \frac{4}{p\,p'} \int_{\mathbb{R}^n_+} |\nabla V|^2 dx\,. \tag{5.12}$$

Arguing as in the proof of Theorem 5.2, we find that (5.12) holds if and only if

$$-\int_{\mathbb{R}^{n-1}} \partial_j(\mathscr{R}e\, a_j)\, |v|^2\, dx' \leq \frac{2}{p'} \int_{\mathbb{R}^n} |\nabla V|^2 dx$$

for any $V \in C^1_0(\mathbb{R}^n)$. Appealing again to [**8**, Th. 5.1], we see that this inequality holds if and only if condition (5.11) is satisfied. □

6. L^p-positivity of certain integral operators

The aim of this section is to prove the L^p-positivity of the operator

$$Tu(E) = \int_E \int^*_{\mathbb{R}^n} [u(x) - u(y)]\, K(dx, dy)\,. \tag{6.1}$$

Here E is a Borel set in $\mathbb{R}^n$, $K(dx, dy)$ is a nonnegative Borel measure on $\mathbb{R}^n \times \mathbb{R}^n$, locally finite outside the diagonal $\{(x, y) \in \mathbb{R}^n \times \mathbb{R}^n \mid x = y\}$ and the function u belongs to $C^1_0(\mathbb{R}^n)$.

The integral in (6.1) has to be understood as a principal value in the sense of Cauchy

$$\lim_{\varepsilon \to 0^+} \int_E \int_{\mathbb{R}^n \setminus B_\varepsilon(x)} [u(x) - u(y)]\, K(dx, dy)$$

and we assume that such singular integral does exist for any $u \in C^1_0(\mathbb{R}^n)$.

In what follows we shall make also the following assumptions on the kernel $K(dx, dy)$:

(i) $K(E, F) = K(F, E)$ for any Borel sets $E, F \subset \mathbb{R}^n$;

(ii) for any compact set $E \subset \mathbb{R}^n$ we have

$$\iint_{E \times E} |x - y|^2 K(dx, dy) < \infty\,;$$

(iii) for any $R > 0$ we have

$$\int_{|x|<R} \int_{|y-x|>2R} K(dx, dy) < \infty\,.$$

As an example, consider the measure $K(dx,dy) = |x-y|^{-n-2s}dxdy$ $(0 < s < 1)$; it satisfies all the previous conditions and in this case the operator T coincides, up to a constant factor, to the fractional power of Laplacian $(-\Delta)^s$ (see, e.g., [**5**, p.230]).

As in (2.1)-(2.2), we say that T is L^p-positive if

$$\begin{aligned} &\int_{\mathbb{R}^n} \langle Tu, |u|^{p-2}u\rangle \geq 0, \quad \text{if } p \geq 2, \\ &\int_{\mathbb{R}^n} \langle T(|u|^{p'-2}u), u\rangle \geq 0, \quad \text{if } 1 < p < 2, \end{aligned} \tag{6.2}$$

for any $u \in C_0^1(\mathbb{R}^n)$.

THEOREM 6.1. *Let $K(dx,dy)$ be a kernel satisfying the previous conditions. Then the operator* (6.1) *is L^p-positive. More precisely we have the inequalities*

(6.3)

$$\int_{\mathbb{R}^n} \langle Tu, |u|^{p-2}u\rangle \geq \frac{4}{pp'} \iint_{\mathbb{R}^n\times\mathbb{R}^n} (|u(y)|^{p/2} - |u(x)|^{p/2})^2 K(dx,dy)\,, \textit{ if } p \geq 2;$$

$$\int_{\mathbb{R}^n} \langle T(|u|^{p'-2}u), u\rangle \geq \frac{4}{pp'} \iint_{\mathbb{R}^n\times\mathbb{R}^n} (|u(y)|^{p'/2} - |u(x)|^{p'/2})^2 K(dx,dy)\,, \textit{ if } 1 < p < 2.$$

PROOF. Let us observe that, in view of (i), we may write

$$\int_{\mathbb{R}^n} \langle Tu, v\rangle = \frac{1}{2} \iint_{\mathbb{R}^n\times\mathbb{R}^n} [u(x) - u(y)][v(x) - v(y)]\, K(dx,dy) \tag{6.4}$$

for any $u, v \in C_0^1(\mathbb{R}^n)$. In fact, since u and v have compact support and thanks to conditions (ii) and (iii), the integral in the right hand side of (6.4) is absolutely convergent. Now we may appeal to the dominated convergence Theorem to obtain (6.4).

Let $p \geq 2$ and consider

$$\int_{\mathbb{R}^n} \langle Tu, |u|^{p-2}u\rangle =$$

$$\frac{1}{2} \iint_{\mathbb{R}^n\times\mathbb{R}^n} [u(x) - u(y)][|u(x)|^{p-2}u(x) - |u(y)|^{p-2}u(y)]\, K(dx,dy)$$

for any $u \in C_0^1(\mathbb{R}^n)$. Note that in this case $|u|^{p-2}u \in C_0^1(\mathbb{R}^n)$. Since

$$(x-y)(|x|^{p-2}x - |y|^{p-2}y) \geq \frac{4}{pp'}\,(|x|^{p/2} - |y|^{p/2})^2$$

for any $x, y \in \mathbb{R}$ (see [**5**, p.231]), we have that (6.2) holds for any $u \in C_0^1(\mathbb{R}^n)$.

If $1 < p < 2$, the second condition in (6.2) can be written as

$$\int_{\mathbb{R}^n} \langle Tv, |v|^{p'-2}v\rangle \geq 0$$

for any $v \in C_0^1(\mathbb{R}^n)$ and the result follows as in the case $p \geq 2$ already considered. □

As a Corollary, we have that under an additional condition, we have a lower estimate involving a Besov semi-norm.

COROLLARY 6.2. *Let the kernel $K(dx,dy)$ satisfy the conditions of Theorem* 6.1. *Moreover suppose that there exist $C > 0$ and $s \in (0,1)$ such that*

$$K(dx,dy) \geq C\frac{dxdy}{|x-y|^{n+2s}} \quad \textit{on } \mathbb{R}^n\times\mathbb{R}^n\,. \tag{6.5}$$

Then we have

$$\int_{\mathbb{R}^n} \langle Tu, u\rangle\, |u|^{p-2} \geq \frac{2C}{p\,p'}\, \|\, |u|^{p/2}\|^2_{\mathcal{L}^{s,2}(\mathbb{R}^n)}, \quad \textit{if } p \geq 2;$$

$$\int_{\mathbb{R}^n} \langle T(|u|^{p'-2}u), u\rangle \geq \frac{2C}{p\,p'}\, \|\, |u|^{p'/2}\|^2_{\mathcal{L}^{s,2}(\mathbb{R}^n)}, \quad \textit{if } 1 < p < 2,$$

where

$$\|v\|_{\mathcal{L}^{s,2}(\mathbb{R}^n)} = \left(\iint_{\mathbb{R}^n\times\mathbb{R}^n} |v(y) - v(x)|^2 \frac{dxdy}{|y-x|^{n+2s}} \right)^{1/2}.$$

PROOF. The result follows immediately from (6.3) and (6.5). □

We conclude with a remark. In Sections 5 and 6, the space $C_0^1(\mathbb{R}^n)$ was the class of admissible functions. Actually, in these cases, we could extend this class and consider more general functions like, for example, compactly supported Lipschitz functions or even bounded functions in proper Sobolev spaces.

References

[1] A. Carbonaro and O. Dragičević, *Convexity of power functions and bilinear embedding for divergence-form operators with complex coefficients*, arXiv:1611.00653.

[2] A. Cialdea and V. Maz'ya, *Criterion for the L^p-dissipativity of second order differential operators with complex coefficients* (English, with English and French summaries), J. Math. Pures Appl. (9) **84** (2005), no. 8, 1067–1100, DOI 10.1016/j.matpur.2005.02.003. MR2155899

[3] A. Cialdea and V. Maz'ya, *Criteria for the L^p-dissipativity of systems of second order differential equations*, Ric. Mat. **55** (2006), no. 2, 233–265, DOI 10.1007/s11587-006-0014-x. MR2279424

[4] A. Cialdea and V. Maz'ya, *L^p-dissipativity of the Lamé operator* (English, with English and Georgian summaries), Mem. Differ. Equ. Math. Phys. **60** (2013), 111–133. MR3288172

[5] A. Cialdea and V. Maz'ya, *Semi-bounded differential operators, contractive semigroups and beyond*, Operator Theory: Advances and Applications, vol. 243, Birkhäuser/Springer, Cham, 2014. MR3235527

[6] A. Cialdea and V. Maz'ya, *The L^p-dissipativity of first order partial differential operators*, Complex Var. Elliptic Equ. **63** (2018), no. 7-8, 945–960, DOI 10.1080/17476933.2017.1321638. MR3802808

[7] M. Dindoš and J. Pipher, *Regularity theory for solutions to second order elliptic operators with complex coefficients and the L^p Dirichlet problem*, Adv. Math. **341** (2019), 255–298, DOI 10.1016/j.aim.2018.07.035. MR3872848

[8] B. J. Jaye, V. Maz'ya, and I. E. Verbitsky, *Existence and regularity of positive solutions of elliptic equations of Schrödinger type*, J. Anal. Math. **118** (2012), no. 2, 577–621, DOI 10.1007/s11854-012-0045-z. MR3000692

[9] V. Maz'ya, *The negative spectrum of the higher-dimensional Schrödinger operator* (Russian), Dokl. Akad. Nauk SSSR **144** (1962), 721–722. MR0138880

[10] V. Maz'ya, *On the theory of the higher-dimensional Schrödinger operator* (Russian), Izv. Akad. Nauk SSSR Ser. Mat. **28** (1964), 1145–1172. MR0174879

[11] V. Maz'ya, *Analytic criteria in the qualitative spectral analysis of the Schrödinger operator*, Spectral theory and mathematical physics: A Festschrift in honor of Barry Simon's 60th birthday, Proc. Sympos. Pure Math., vol. 76, Amer. Math. Soc., Providence, RI, 2007, pp. 257–288, DOI 10.1090/pspum/076.1/2310207. MR2310207

[12] V. Maz'ya and T. Shaposhnikova, *Theory of Sobolev multipliers*: *With applications to differential and integral operators*, Grundlehren der Mathematischen Wissenschaften [Fundamental Principles of Mathematical Sciences], vol. 337, Springer-Verlag, Berlin, 2009. MR2457601

DIPARTIMENTO DI MATEMATICA, ECONOMIA ED INFORMATICA, UNIVERSITY OF BASILICATA, V.LE DELL'ATENEO LUCANO 10, 85100, POTENZA, ITALY

Email address: cialdea@email.it

DEPARTMENT OF MATHEMATICAL SCIENCES, M& O BUILDING, UNIVERSITY OF LIVERPOOL, LIVERPOOL L69 7ZL, UK –AND– DEPARTMENT OF MATHEMATICS, LINKÖPING UNIVERSITY, SE-581 83, LINKÖPING, SWEDEN –AND– RUDN UNIVERSITY, 6 MIKLUKHO-MAKLAY ST, MOSCOW, 117198, RUSSIA.

Email address: vladimir.mazya@liu.se

Contemporary Mathematics
Volume **734**, 2019
https://doi.org/10.1090/conm/734/14765

Probabilistic approach to a cell growth model

Gregory Derfel, Yaqin Feng, and Stanislav Molchanov

Dedicated to 100th anniversary of the birth of Selim Grigorievich Krein.

ABSTRACT. We consider the time evolution of the supercritical Galton-Watson model of branching particles with extra parameter (mass). In the moment of the division the mass of the particle (which is growing linearly after the birth) is divided in random proportion between two offsprings (mitosis). Using the technique of moment equations we study asymptotics of the mass distribution of the particles. Mass distribution of the particles is the solution of the equation with linearly transformed argument: functional, functional-differential or integral. We derive several limit theorems describing the fluctuations of the density of the particles, first two moments of the total masses etc.

1. Introduction

A model for the simultaneous growth and division of a cell population, structured by size, was introduced and studied by Hall and Wake [**18**] (cf. [**23**]). The original model deals with symmetrical cell-division, where each cell divides into κ equally sized daughter cells. Under this assumption Hall and Wake proved that the steady-size mass distribution exists and satisfies the celebrated pantograph functional-differential equation

$$y'(x) = ay(\kappa x) + by(x) \tag{1}$$

where $\kappa > 1$. Since then, different variations and extensions of the original model have been studied and used to describe plant cells, diatoms ([**2**], [**4**], [**11**], [**8**]) and also tumor growth [**3**].

In the present paper, in order to describe cell growth model we use the supercritical Galton-Watson model of branching particles with extra parameter (mass). Similar approach was applied earlier in [**11**].

Namely, we assume that the mass of the particle is growing linearly between the exponentially distributed splitting moments and that in the moment of the division the mass of the particle is divided in a random proportion between two offspring (see Figure 1 later in the paper). Notice that under these assumptions splitting moments depend on mass.

The third author was partially supported by Russian Science Foundation, Project No 17-11-01098.

The model described above gives only rather schematic description of the cell growth process, but it is interesting from the mathematical point of view, and hopefully in some cases may reflect an important qualitative features of real biological systems ([**2**], [**4**], [**11**], [**8**], [**3**]).

We start from the study of the total number of the particles $N(t)$, their distribution with respect to the mass, and first two moments of the total mass distribution. We study the asymptotic of the mass distribution of the particles and prove several limit theorems. In particular, we derive asymptotics of the steady-size mass distribution $\Pi(m)$, as $m \to \infty$.

Let $N(t)$ be the supercritical Galton-Watson process [**16**] with mortality rate μ and splitting rate $\beta > \mu \geq 0$. Assume that at initial time 0, $N(0) = 1$. For the generating function $u_z(t) = Ez^{N(t)}$, we have the well known equation [**19**]:

$$\begin{aligned} \frac{\partial u_z(t)}{\partial t} &= \beta u_z(t)^2 - (\beta+\mu)u_z(t) + \mu \\ u_z(0) &= 1. \end{aligned} \tag{2}$$

Let $\delta = \beta - \mu$ and $\gamma = \frac{\mu}{\beta}$, elementary calculations give for $N(t)$ the geometric distribution

$$P(N(t) = k) = \frac{(1-\gamma)^2 \left(1-e^{\delta t}\right)^{k-1} e^{\delta t}}{\left(\gamma - e^{\delta t}\right)^{k+1}}, k \geq 1, \tag{3}$$

$$P(N(t) = 0) = \frac{1 - e^{\delta t}}{1 - \frac{1}{\gamma} e^{\delta t}}, \tag{4}$$

$$EN(t) = e^{\delta t},$$

and for any $a \geq 0$, as $t \to \infty$,

$$P\left\{\frac{N(t)}{e^{\delta t}} \in (a, a+da)\right\} = (1-\gamma)e^{-a} + \gamma\delta_0(a).$$

Assume now that the initial particles have mass $m > 0$. The evolution of the mass $m(t)$ includes two features. First of all we assume the linear growth:

$$m(t) = m + vt,$$

where $v > 0$ until the splitting. Probability of the splitting in each time interval $[t, t+dt]$ equals βdt, i.e. the moment τ_1 of splitting of the initial particle has exponential law with parameter β and $P\{\tau > t\} = e^{-\beta t}$. At the moment $\tau_1 + 0$, the particle with mass $m + v\tau_1$ is divided into two particles with the random masses:$m' = \theta(m + v\tau_1)$, $m'' = (1-\theta)(m + v\tau_1)$. Here $\theta \in [0,1]$ is symmetrically distributed (with respect to the center $0.5 \in [0,1]$) random variable which has the density $q(x) = q(1-x)$, $x \in [0,1]$. As usually we assume that the random variable θ_i, $i = 1, 2, \cdots$ for different splittings are independent. The dynamics of the subpopulations generated by different offspring are also independent.

Let us introduce the main object of our study: the moment generating function of the two random variables: $N(t) :=$ total numbers of particles at the moment $t > 0$, $M(t) :=$ total mass of the particles at the moment t. We introduce the generating function of the form:

$$u(t, m; z, k) = E_m z^{N(t)} e^{-kM(t)}, |z| \leq 1, k \geq 0. \tag{5}$$

The following results are the basis for the further analysis:

THEOREM 1. *Let $u(t,m;z,k) = Ez^{N(t)}e^{-kM(t)}$, then $u(t,m;z,k)$ satisfy the following functional-differential equation:*

$$\begin{aligned} \frac{\partial u(t,m;z,k)}{\partial t} &= \frac{\partial u(t,m;z,k)}{\partial m}v + \beta \int_0^1 u(t,\theta m;z,k)\cdot u(t,(1-\theta)m;z,k)q(\theta)d\theta \\ &\quad -(\beta+\mu)u(t,m;z,k)+\mu \\ u(0,m;z,k) &= ze^{-km}. \end{aligned} \tag{6}$$

PROOF. The formal derivation of this equation is based on the standard technique: balance of the probabilities in the infinitesimal initial time interval $[0, dt]$. Namely, let's consider

$$u(t+dt,m;z,k) = E_m z^{N(t+dt)}e^{-kM(t+dt)}$$

and then let's split the interval $[0, t+dt]$ into two parts $[0, dt] \cup [dt, t+dt]$. At the moment $t = 0$, we have one particle in the point x with mass m and during $[0, dt]$, we observe one of the following :

- splitting of the initial particle into two particles with probability βdt;
- annihilation of the initial particle with probability μdt;
- nothing happen, no annihilation and no splitting with probability $1 - \beta dt - \mu dt$.

Now one can apply the full expectation formula:

$$\begin{aligned} u(t+dt,m;z,k) &= u(t,m+vdt;z,k)(1-\beta dt-\mu dt)+ \\ &\quad \beta dt \int_0^1 u(t,\theta m;z,k)\cdot u(t,(1-\theta)m;z,k)q(\theta)d\theta + \mu dt \end{aligned}$$

Theorem 1 is obtained by letting $dt \to 0$. □

For $k = 0$ in equation (5), it will lead to the equation for $Ez^{N(t)}$, which we already discussed. Let's put $z = 1$ and study the equation (6) as a function of k. Denote $L_1(t,m) := E_m(M(t)) = \frac{-\partial u}{\partial k}|_{z=1,k=0}$, then from equation (6), we have

$$\begin{cases} \frac{\partial L_1(t,m)}{\partial t} = \frac{\partial L_1(t,m)}{\partial m} v + 2\beta \int_0^1 (L_1(t,\theta m) - L_1(t,m))q(\theta)d\theta + (\beta-\mu)L_1(t,m) \\ \\ L_1(0,m) = m. \end{cases} \tag{7}$$

Differentiate the equation (6) twice over k and substituting $z = 1$ and $k = 0$, we will get the equation for the second moment $L_2(t,m) := E_m(M^2(t)) = \frac{\partial^2 u}{\partial k^2}|_{z=1,k=0}$

$$\begin{cases} \frac{\partial L_2(t,m)}{\partial t} = \frac{\partial L_2(t,m)}{\partial m} v + 2\beta \int_0^1 (L_2(t,\theta m) - L_2(t,m))q(\theta)d\theta + (\beta-\mu)L_2(t,m) \\ +2\beta \int_0^1 (L_1(t,\theta m)L_1(t,(1-\theta)m))q(\theta)d\theta \\ \\ L_2(0,m) = m^2. \end{cases} \tag{8}$$

The rest of the paper is organized as follows. In section 2, we discuss the mass process. In section 3, we study the analytic properties of the limiting mass distribution density. Following this, we discuss the moment of total mass of the population in section 4.

2. Mass process

The equation (7) contains the constant potential $\beta - \mu$ and the operator

$$\mathcal{L}_m f = v\frac{\partial f}{\partial m} + 2\beta \int_0^1 (f(\theta m) - f(m))\, q(\theta) d\theta, \tag{9}$$

which is the generator of the one dimension Markov process $m(t)$. This mass process $m(t)$ has the following description: it starts at $t = 0$ with the initial mass m and grows linearly $m(t) = m + vt$, $t \leq \tau_1$, where τ_1 is exponential distributed random variable with parameter 2β. At the moment $\tau_1 + 0$, this particle splits into two particles with corresponding masses

$$m' = (m + v\tau_1)\theta_1,$$
$$m'' = (m + v\tau_1)(1 - \theta_1),$$

where θ_1 and $1 - \theta_1$ has the same density $q(\theta)$. By definition,

$$m(\tau_1 + 0) = (m + v\tau_1)\theta_1.$$

The graph of $m(t)$ is presented in the following Figure 1.

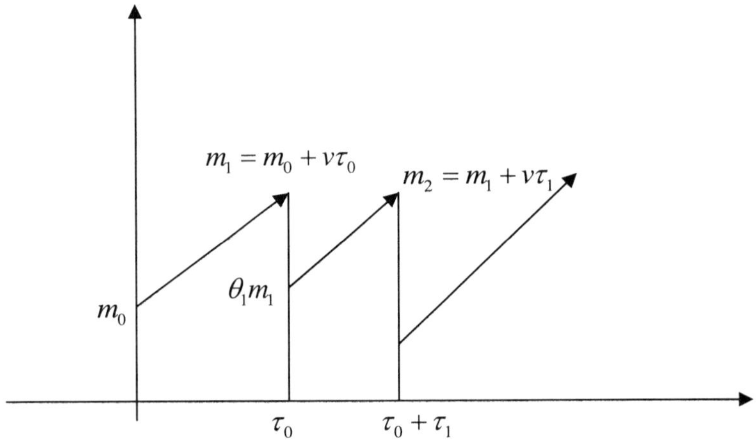

FIGURE 1. Mass process

Remark: Factor 2β instead of β appears due to the fact that after splitting, we have two identical particles.

Let us consider the embedded chain $m_n = (m_{n-1} + v\tau_n)\theta_n$, $n \geq 1$, i.e. the mass process $m(t)$ at the Poisson moments $T_1 = \tau_1$, $T_2 = \tau_1 + \tau_2$, $\cdots$, $T_n = \tau_1 + \cdots + \tau_n$, we will get recursively

$$m(T_1) = \theta_1(m + v\tau_1)$$

Similarly, at the moment of the second splitting,

$$\begin{aligned} m(T_2) &= \theta_2(m_1 + v\tau_2) \\ &= \theta_1\theta_2 m + v\tau_2\theta_2 + v\tau_1\theta_1\theta_2 \\ &\overset{law}{=} \theta_1\theta_2 m + v\tau_1\theta_1 + v\tau_2\theta_1\theta_2 \end{aligned}$$

In general,

$$m(T_n) \overset{law}{=} \theta_1 \cdots \theta_n m + v\tau_1\theta_1 + \cdots + v\tau_n\theta_1 \cdots \theta_n$$

so as $n \to \infty$, the limit will have the the form

$$m(T_n) \xrightarrow[n\to\infty]{law} m_\infty = v\tau_1\theta_1 + \cdots + v\tau_n\theta_1\cdots\theta_n + \cdots$$

The last random series has all moments since τ_i is exponential distributed with parameter 2β and θ_i, $i = 1, 2, \cdots$ are bounded. This chain describes the distribution of the mass of new born particles at the moments of splitting. Unfortunately, the law of m_∞ is the invariant distribution for the chain $m(T_n) = m(\tau_1 + \cdots + \tau_n)$, but not for $m(t)$. Let us find the invariant density $\Pi(m)$ for the process $m(t)$.

Denote $\nu(t)$ the number of the Poisson point T_i, $i = 1, 2, \cdots$ on the time interval $[0, t]$, i.e, $\nu(t) \sim Poisson(2\beta t)$. Then for $\nu(t) = n$,

$$m(t) = m_n + v(t - T_n).$$

The points $T_1, \cdots, T_n$ divide $[0, t]$ onto $n+1$ sub-interval (spacing) $\Delta_1, \cdots, \Delta_{n+1}$ with the same distribution. They are not independent of course since $\Delta_1 + \cdots + \Delta_{n+1} = t$. But the points $T_1, T_2, \cdots, T_n$ are the ordered statistics for the set of n independent and uniformly distributed on $[0, t]$ random variable. It is well known [**12**] that the spacing can be presented in the form

$$\Delta_i = \frac{Z_i t}{Z_1 + \cdots + Z_{n+1}} \qquad i \ = 1, \cdots, n+1$$

where Z_i are i.i.d random variable with exponential law $Exp(1)$, then for $\nu(t) = n$

$$\begin{aligned}
m(t) &= (\cdots(((m + \Delta_1 v)\,\theta_1 + \Delta_2 v)\,\theta_2 + \Delta_3 v)\,\theta_3 + \cdots + \Delta_n v)\,\theta_n + \Delta_{n+1} v \\
&= \theta_1 \cdots \theta_n m + \frac{t}{Z_1 + \cdots + Z_{n+1}} ((\cdots(((Z_1 v)\theta_1 + Z_2 v)\theta_2 + Z_3 v)\theta_3 + \\
&\quad \cdots + \xi_n v)\theta_n + \xi_{n+1} v) \\
&= \theta_1 \cdots \theta_n m + \frac{t}{Z_1 + \cdots + Z_{n+1}} ((\cdots(((Z_1 v)\theta_1 + Z_2 v)\theta_2 + Z_3 v)\theta_3 + \\
&\quad \cdots + Z_{\nu(t)} v)\theta_{\nu(t)} + Z_{\nu(t)+1} v) \\
&\overset{law}{=} \theta_1 \cdots \theta_{\nu(t)} m + \frac{t}{\nu(t)} \frac{\nu(t)}{Z_1 + \cdots + Z_{\nu(t)+1}} (\xi_0 v + \xi_1 v\theta_1 + \xi_2 v\theta_1\theta_2 + \cdots) \\
&\xrightarrow[t\to\infty]{law} \frac{v}{2\beta} (\xi_0 + \xi_1\theta_1 + \xi_2\theta_1\theta_2 + \cdots)
\end{aligned}$$

where ξ_i are standard independent exponentially distributed random variables with parameter 1: $Exp(1)$. Note that in the last step we use the following facts :

(1) Z_i are i.i.d $Exp(1)$ random variable and $\frac{\nu(t)}{Z_1 + \cdots + Z_{\nu(t)+1}} \xrightarrow[t\to\infty]{law} \frac{1}{E(Z_i)} = 1$;
(2) $\nu(t) \sim Poisson(2\beta t)$ and $E(\nu(t)) = 2\beta t$;
(3) θ_i are i.i.d random variable.

We proved the following result: Markov mass process $m(t)$ has the limiting distribution $\Pi(m)$ which is the law of the random variable

$$\begin{aligned}
m_\infty &= \frac{v}{2\beta} (\xi_0 + \xi_1\theta_1 + \xi_2\theta_1\theta_2 + \cdots) \\
&= v\,(\tau_0 + \tau_1\theta_1 + \tau_2\theta_1\theta_2 + \cdots)
\end{aligned}$$

where $\tau_i \sim Exp(2\beta)$ and ξ_i are i.i.d $Exp(1)$ random variable and θ_i are also i.i.d random variable with the symmetric density $q(x) = q(1-x)$ for $x \in [0, 1]$. We will assume that $q(x) = 0$ if $|x - \frac{1}{2}| \geq \delta$, $0 < \delta < \frac{1}{2}$, $i.e.$ $0 < \delta \leq \theta_i \leq 1 - \delta$, $i = 1, 2, \cdots$

The transition density of the mass process $\rho(t,m,m')$, i.e. the fundamental solution of

$$\begin{cases} \frac{\partial\rho(t,m,m')}{\partial t} = \mathcal{L}_m\rho(t,m,m') \\ \rho(0,m,m') = \delta_{m'}(m) \end{cases}$$

have a limit $\Pi(m') = \lim_{t\to\infty}\rho(t,m,m')$.

THEOREM 2. *Process $m(t)$ has the invariant density $\Pi(m)$, this density equals to the distribution density of the random geometric series*

$$\xi = v\tau_0 + v\tau_1\theta_1 + \cdots + v\tau_n\theta_1\cdots\theta_n + \cdots \tag{10}$$

Where $\tau_i, i \geq 0$ are i.i.d $Exp(2\beta)$ random variable and $\theta_i, i \geq 0$ are i.i.d random variable with the probability density $q(\theta)$, τ_i and θ_i are independent.

Remark:

- The operator $\mathcal{L}_m$ is the unusual Markov generator. It belongs to the class functional-differential operator with linearly transformed argument which appear in many applications. See Derfel et al. [**11**]. All such Markov processes are directly or indirectly related to the solvable group $Aff(R^1)$ of the transformations $x \to ax + b$ of $R^1 \to R^1$. This group has the standard matrix representation $g = \begin{bmatrix} a & b \\ 0 & 1 \end{bmatrix}$, $a > 0$.The simplest symmetric random walks on this group have the form

$$g_n = \begin{bmatrix} e^{X_1+\cdots+X_n} & \sum Y_i e^{X_1+\cdots+X_{i-1}} \\ 0 & 1 \end{bmatrix} \tag{11}$$

  $\{X_i\}$ and $\{Y_i\}$, $i \geq 1$ are symmetric i.i.d random vector. The upper of diagonal term has the same structure like m_∞.
- One can check that the law m_∞ is invariant density for the mass process directly. It is not difficult to verify that $\Pi(m)$ is the solution of the conjugate equation

$$\mathcal{L}^\star g = -v\frac{\partial g}{\partial m} + 2\beta\int_0^1 [g(\frac{m}{\theta}) - g(m)]q(\theta)d\theta = 0 \tag{12}$$

  This functional-differential equation with rescaling is similar to the archetypal equation which was studied in [**10**], [**5**] and [**6**].

3. Analytic properties of the limiting mass distribution density

Now we'll calculate the moment for the invariant limiting distribution $\Pi(m)$, the calculation will be based on the following fact:

$$\xi = v\tau + \theta_1\tilde{\xi}$$

Here $\tilde{\xi} \overset{law}{=} \xi$ and τ, θ_1 are independent on $\tilde{\xi}$. Since τ is exponential random variable with parameter 2β, so

$$E\tau = \frac{1}{2\beta}, E\tau^2 = \frac{1}{2\beta^2}$$

As a result,

$$E\xi = Ev\tau + E\theta_1\tilde{\xi}$$

so

$$E\xi = \frac{v}{2\beta} + \frac{1}{2}E\xi$$

thus,

$$E\xi = \frac{v}{\beta}$$

The second moment

$$E\xi^2 = v^2 E(\tau^2) + 2vE(\tau\theta_1\tilde{\xi}) + E(\theta_1^2\tilde{\xi})^2$$

from the independence, then

$$E\xi^2 = \frac{v^2}{\beta^2(1 - E\theta_1^2)}$$

so

$$var\xi = \frac{v^2}{\beta^2}(\frac{1}{1 - E\theta_1^2} - 1).$$

Similarly, the third moments

$$E\xi^3 = \frac{3v^3}{2\beta^3(1 - E\theta_1^2)(1 - E\theta_1^3)}.$$

In general,

$$E\xi^k = E(v\tau + \theta_1\tilde{\xi})^k$$

i.e.

$$E\xi^k(1 - E\theta_1^k) = \sum_{i=0}^{k-1} \left(\begin{array}{c} k \\ i \end{array} \right) (vE\tau)^i E(\theta\xi)^{k-i}$$

Let's find the asymptotic of $\Pi(m)$ for large m and small m. Since

$$\xi = v\tau_0 + v\tau_1\theta_1 + \cdots + v\tau_n\theta_1 \cdots \theta_n + \cdots$$

Therefore,

$$\begin{aligned}
E_\theta[e^{-\lambda\xi}] &= E_\theta[e^{-\lambda(v\tau_0 + v\tau_1\theta_1 + v\tau_2\theta_1\theta_2\cdots)}] \\
&= E_\theta[e^{-\lambda v\tau_0}]E_\theta[e^{-\lambda v\tau_1\theta_1}]\cdots[e^{-\lambda v\tau_2\theta_1\cdots\theta_n}]\cdots \\
&= \frac{1}{(1+\frac{\lambda v}{2\beta})(1+\frac{\lambda v\theta_1}{2\beta})\cdots(1+\frac{\lambda v\theta_1\cdots\theta_n}{\xi 2\beta})\cdots} \\
&= \frac{c_0}{1+\frac{\lambda v}{2\beta}} + \frac{c_1}{1+\frac{\lambda v\theta_1}{2\beta}} + \cdots + \frac{c_n}{1+\frac{\lambda v\theta_1\cdots\theta_n}{2\beta}} + \cdots \\
&= \frac{c_0}{1+\frac{\lambda}{\frac{2\beta}{v}}} + \frac{c_1}{1+\frac{\lambda}{\frac{2\beta}{v\theta_1}}} + \cdots + \frac{c_n}{1+\frac{\lambda}{\frac{2\beta}{v\theta_1\cdots\theta_n}}} + \cdots
\end{aligned}$$

Here,

$$c_0 = \frac{1}{(1-\theta_1)(1-\theta_1\theta_2)\cdots}$$

$$c_1 = \frac{1}{(1-\frac{1}{\theta_1})(1-\theta_2)(1-\theta_2\theta_3)\cdots}$$

$$c_n = \frac{1}{(1-\frac{1}{\theta_1\cdots\theta_n})(1-\frac{1}{\theta_2\cdots\theta_n})\cdots(1-\frac{1}{\theta_n})(1-\theta_{n+1})(1-\theta_{n+1}\theta_{n+2})\cdots}$$

Now one can find conditional density $p_\xi(m)$ of random variable if $\vec{\theta} = (\theta_1, \theta_2, \cdots)$ are known,

$$
\begin{aligned}
p_\xi(m) &= E_{\vec{\theta}}[\frac{2\beta}{v} e^{-\frac{2\beta m}{v}} c_0] + E_{\vec{\theta}}[\frac{2\beta}{v\theta_1} e^{-\frac{2\beta m}{v\theta_1}} c_1] + \cdots \\
&= E_{\vec{\theta}}(\frac{2\beta}{v} e^{-\frac{2\beta m}{v}} \frac{1}{(1-\theta_1)(1-\theta_1\theta_2)\cdots}) \\
&+ E_{\vec{\theta}}(\frac{2\beta}{v\theta_1} e^{-\frac{2\beta m}{v\theta_1}} \frac{1}{(1-\frac{1}{\theta_1})(1-\theta_2)(1-\theta_2\theta_3)\cdots}) + \cdots \\
&= \frac{2\alpha\beta}{v} e^{-\frac{2\beta m}{v}} - E_{\theta_1} \frac{2\alpha\beta}{1-\theta_1} e^{-\frac{2\beta m}{v\theta_1}} + \cdots \qquad (13)
\end{aligned}
$$

where

$$\alpha = E \frac{1}{(1-\theta_1)(1-\theta_1\theta_2)\cdots} \qquad (14)$$

Let's formulate several analytic results about the invariant density .

THEOREM 3. *Assume that* $Supp\theta = [a, 1-a]$, $0 < a \leq \frac{1}{2}$, *then for large* m,

$$\Pi(m) \xrightarrow[m\to\infty]{} \frac{2\alpha\beta}{v} e^{-\frac{2\beta m}{v}} + R(m)$$

The remainder term with the maximum on the boundary satisfies,

$$R(m) \sim \frac{2\alpha\beta}{a} e^{-\frac{2\beta m}{v(1-a)}} L(m),$$

where $L(m) \xrightarrow[m\to\infty]{} 0$ *and* $L(m)$ *depends on the structure of the distribution* $q(d\theta)$ *near the maximum point* $\theta_{critical} = 1 - a$.

PROOF. From (13), due to the Laplace method, it is trivial to get the result. □

The behavior of $p_\xi(m)$ as $m \to 0$ is much more interesting. Here we will use the Exponential Chebyshev's inequality. More detailed analysis in the case when $q(d\theta)$ is a discrete (atomic) measure, has been done in Derfel [**9**], Cooke & Derfel [**7**].

For instance, the following result is true for the pantograph equation (1).

(i) Steady- state solution of (1) satisfies the following estimate

$$|y(x)| < D\exp\{-b\ln^2|x|\}; \qquad D > 0, \qquad b = \frac{1}{2\ln\alpha} \qquad (15)$$

in some neighborhood of zero.

(ii) On the other hand, every solution of (1) which satisfies estimate $y(x)| < D\exp\{-a\ln^2|x|\}$ for with some $a > b$ is identically equal zero.

Similar results are valid also for more general equation

$$y(x) = \sum_{j=0}^{l}\sum_{k=0}^{n} a_{jk} y^{(k)}(\lambda_j x), \qquad (16)$$

where $\lambda_j \neq 0$ under the assumption that $\Lambda = \max|\lambda_j| < 1$. Namely, statements (i) and (ii) are fulfilled with $b = \frac{1}{2|\ln\lambda|}$ and $a > \frac{m|\ln\lambda|}{2\ln^2\Lambda}$, where $\lambda = \min|\lambda_j|$.

We conjecture that similar asymptotic behavior occurs also for our model, but currently can prove the following weaker result, only.

THEOREM 4. *Assume that* $Supp\theta = [a, 1-a]$, $0 < a \le \frac{1}{2}$, *then if* $m \to 0$, *then*

$$P\{\xi \le m\} \le e^{-c_1 \ln^2(\frac{1}{m})}$$

where $c_1 > 0$ *is some constant (see Fig. 2)*

PROOF. Let's start from the standard calculations, for $\lambda > 0$ and fix $a \le \theta_i \le 1-a$, $i = 1, 2, \cdots$

$$\begin{aligned} P\left\{\xi \le m|\vec{\theta}\right\} &= P\{e^{-\lambda\xi} > e^{-\lambda m}|\vec{\theta}\} \le \min_{\lambda>0} \frac{Ee^{-\lambda\xi}}{e^{-\lambda m}} \\ &= \min_{\lambda>0} e^{\lambda m - \ln(1+\frac{\lambda v}{2\beta}) - \ln(1+\frac{\lambda v\theta_1}{2\beta}) - \ln(1+\frac{\lambda v\theta_1\theta_2}{2\beta}) - \cdots} \end{aligned} \tag{17}$$

Equation for the critical point $\lambda_0 = \lambda_0(m)$ has a form:

$$m = \frac{\frac{v}{2\beta}}{1+\frac{\lambda v}{2\beta}} + \frac{\frac{v\theta_1}{2\beta}}{1+\frac{\lambda v\theta_1}{2\beta}} + \cdots + \frac{\frac{v\theta_1\cdots\theta_k}{2\beta}}{1+\frac{\lambda v\theta_1\theta_2}{2\beta}} + \cdots$$

i.e.

$$m = \frac{1}{\frac{2\beta}{v}+\lambda} + \frac{1}{\frac{2\beta}{v\theta_1}+\lambda} + \cdots + \frac{1}{\frac{2\beta}{v\theta_1\cdots\theta_k}+\lambda} + \cdots$$

Define $k(\lambda) = \min\{k : \frac{2\beta}{v\theta_1\cdots\theta_k} \sim \lambda\}$, then $m \sim \frac{k(\lambda)}{\lambda}$. From $\frac{2\beta}{v\theta_1\cdots\theta_k} \sim k$, we then have $k(\lambda) \sim \frac{\ln\lambda}{E\ln(\frac{1}{\theta})}$. Hence, the critical point

$$\lambda \sim \frac{\ln(\frac{1}{m})}{mE\ln(\frac{1}{\theta})} \tag{18}$$

Substitute (18) into Chebyshev's inequality (17) gives

$$\begin{aligned} P\left\{\xi \le m|\vec{\theta}\right\} &\le e^{\frac{\ln(\frac{1}{m})}{E\ln(\frac{1}{\theta})} - \ln\left(1+\frac{v\ln(\frac{1}{m})}{2m\beta E\ln(\frac{1}{\theta})}\right) - \cdots - \ln\left(1+\frac{v\ln(\frac{1}{m})\theta_1\cdots\theta_k}{2m\beta E\ln(\frac{1}{\theta})}\right)} \\ &\le e^{\frac{\ln(\frac{1}{m})}{E\ln(\frac{1}{\theta})} - \ln\left(1+\frac{v\ln(\frac{1}{m})}{2m\beta E\ln(\frac{1}{\theta})}\right) - \cdots - \ln\left(1+\frac{v\ln(\frac{1}{m})a^k}{2m\beta E\ln(\frac{1}{\theta})}\right)} \\ &\le e^{\frac{\ln(\frac{1}{m})}{E\ln(\frac{1}{\theta})} - \sum_{i=0}^{k}\ln\left(\frac{v\ln(\frac{1}{m})a^i}{2m\beta E\ln(\frac{1}{\theta})}\right)} \\ &\le e^{-c_1\ln^2\left(\frac{1}{m}\right)}. \end{aligned}$$

□

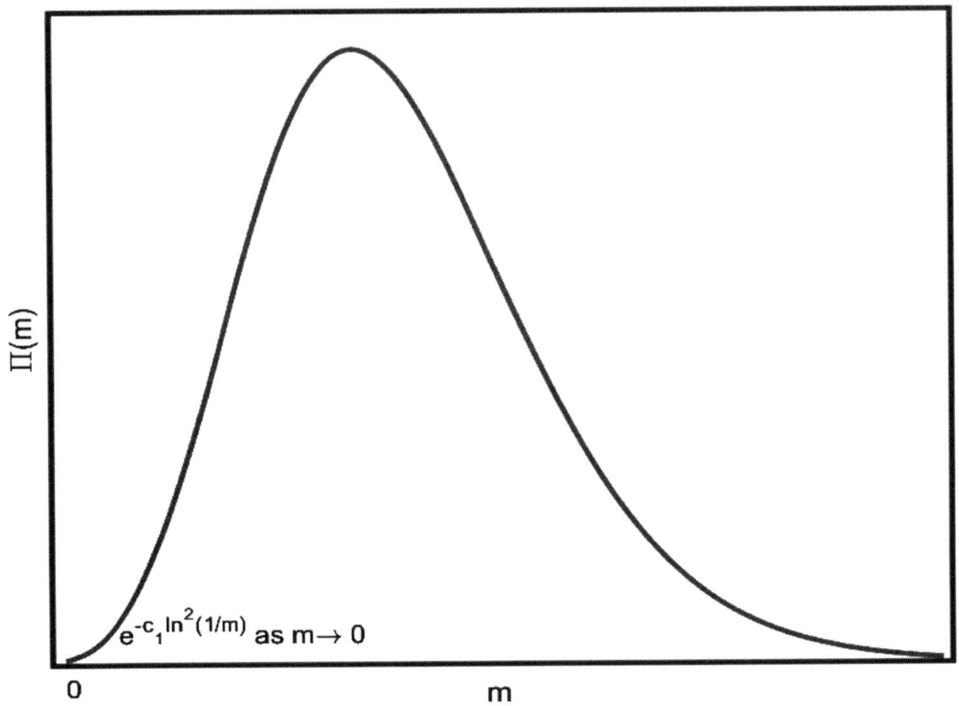

FIGURE 2. Asymptotic behavior of $\Pi(m)$

Remark: Asymptotic behaviour of $\Pi(m)$ for large m under the assumption that $q(x)$ is a discrete (atomic) measure, was studied in a number of papers (see, for example [**18**] ,[**11**]) and it was shown that $\Pi(m)$ decays exponentially, as $m \to \infty$ (cf. [**18**]. formula (45), Fig.1)

From Theorem 3 given above, it follows that for continuous distribution function $q(x)$ same asymptotic behaviour (exponential decay) holds valid.

On the other hand, to the best of our knowledge, asymptotics of $\Pi(m)$ for small m has not been discussed up till now, and the result of Theorem 4 (see Fig.2) is new.

4. Moments of total mass of population $M(t)$

In this section, we will study the first moment and second moment of the total mass of population $M(t)$. As discussed in section 1, the first moment $L_1(t,m)$ is given by equation

(19)
$$\begin{cases} \frac{\partial L_1(t,m)}{\partial t} = \frac{\partial L_1(t,m)}{\partial m} v + 2\beta \int_0^1 (L_1(t,\theta m) - L_1(t,m))q(\theta)d\theta + (\beta-\mu)L_1(t,m) \\ L_1(0,m) = m \end{cases}$$

COROLLARY 5. *Let* $L_1(t,m) = E_m(M(t))$ *then for* $t \to \infty$,

$$L_1(t,m) \to e^{(\beta-\mu)t}\frac{v}{\beta}$$

PROOF. From equation (19), Duhamel's formula gives us

$$L_1(t,m) = e^{(\beta-\mu)t}\int_0^\infty \rho(t,m,m')m'dm'$$

as $t \to \infty$,

$$L_1(t,m) \to e^{(\beta-\mu)t}\int_0^\infty \Pi(m')m'dm' = e^{(\beta-\mu)t}\frac{v}{\beta}.$$

The last equality use both Theorem 2 $\rho(t,m,m') \to \Pi(m')$ and the fact that $E\xi = \frac{v}{\beta}$. □

The second moment $L_2(t,m) = E_m(M(t)^2)$ is given by

(20)

$$\begin{cases} \frac{\partial L_2(t,m)}{\partial t} = \frac{\partial L_2(t,m)}{\partial m} v + 2\beta\int_0^1 (L_2(t,\theta m) - L_2(t,m))q(\theta)d\theta + (\beta-\mu)L_2(t,m) \\ +2\beta\int_0^1 (L_1(t,\theta m)L_1(t,(1-\theta)m))q(\theta)d\theta \\ \\ L_2(0,m) = m^2 \end{cases}$$

From equation (20), we have $\frac{\partial L_2}{\partial t} = \mathcal{L}_m L_2 + f(t,m)$, here $\mathcal{L}_m$ is the operator of the mass process. By applying Duhamel's principle one can find that

$$L_2(t,m) = 2\left(e^{(\beta-\mu)t}\frac{v}{\beta}\right)^2 + O(e^{(\beta-\mu)t})$$

Acknowledgements

We are grateful to the anonymous referee for the careful reading of the manuscript and for the helpful remarks.

S.A.M. thankfully acknowledges hospitality and support provided by the Center for Advanced Studies in Mathematics (Ben-Gurion University) during his research visits in Beer-Sheva.

References

[1] S. Albeverio, L. V. Bogachev, and E. B. Yarovaya, *Asymptotics of branching symmetric random walk on the lattice with a single source* (English, with English and French summaries), C. R. Acad. Sci. Paris Sér. I Math. **326** (1998), no. 8, 975–980, DOI 10.1016/S0764-4442(98)80125-0. MR1649878

[2] Basse, B., Wake, G. C., Wall, D. J. N., and Van Brunt, B. (2004). On a cell-growth model for plankton. *Mathematical Medicine and Biology*, 21, 49-61.

[3] B. Basse, B. C. Baguley, E. S. Marshall, W. R. Joseph, B. van Brunt, G. Wake, and D. J. N. Wall, *A mathematical model for analysis of the cell cycle in cell lines derived from human tumors*, J. Math. Biol. **47** (2003), no. 4, 295–312, DOI 10.1007/s00285-003-0203-0. MR2024498

[4] R. E. Begg, D. J. N. Wall, and G. C. Wake, *The steady-states of a multi-compartment, age-size distribution model of cell-growth*, European J. Appl. Math. **19** (2008), no. 4, 435–458, DOI 10.1017/S0956792508007535. MR2431699

[5] L. V. Bogachev, G. Derfel, and S. A. Molchanov, *On bounded continuous solutions of the archetypal equation with rescaling*, Proc. A. **471** (2015), no. 2180, 20150351, 19, DOI 10.1098/rspa.2015.0351. MR3394793

[6] L. V. Bogachev, G. Derfel, and S. A. Molchanov, *Analysis of the archetypal functional equation in the non-critical case*, Discrete Contin. Dyn. Syst. **Dynamical systems, differential equations and applications. 10th AIMS Conference. Suppl.** (2015), 132–141, DOI 10.3934/proc.2015.0132. MR3462442

[7] K. L. Cooke and G. Derfel, *On the sharpness of a theorem by Cooke and Verduyn Lunel*, J. Math. Anal. Appl. **197** (1996), no. 2, 379–391, DOI 10.1006/jmaa.1996.0026. MR1372185

[8] L. Daukste, B. Basse, B. C. Baguley, and D. J. N. Wall, *Mathematical determination of cell population doubling times for multiple cell lines*, Bull. Math. Biol. **74** (2012), no. 10, 2510–2534, DOI 10.1007/s11538-012-9764-7. MR2978790

[9] Derfel, G. (1978). On the asymptotics of the solution of some linear functional-differential equations. *Reports of the I.N. Vekua Institute of Applied Mathematics, Tbilisi* , N12-13, 21-23 (in Russian).

[10] G. A. Derfel, *A probabilistic method for studying a class of functional-differential equations* (Russian), Ukrain. Mat. Zh. **41** (1989), no. 10, 1322–1327, 1436, DOI 10.1007/BF01057249; English transl., Ukrainian Math. J. **41** (1989), no. 10, 1137–1141 (1990). MR1034672

[11] G. Derfel, B. van Brunt, and G. Wake, *A cell growth model revisited*, Funct. Differ. Equ. **19** (2012), no. 1-2, 75–85. MR3307313

[12] W. Feller, *An introduction to probability theory and its applications. Vol. II.*, Second edition, John Wiley & Sons, Inc., New York-London-Sydney, 1971. MR0270403

[13] Y. Feng, S. Molchanov, and J. Whitmeyer, *Random walks with heavy tails and limit theorems for branching processes with migration and immigration*, Stoch. Dyn. **12** (2012), no. 1, 1150007, 23, DOI 10.1142/S0219493712003626. MR2887919

[14] R. A. Fisher (1937). The wave of advance of advantageous genes, *Ann Eugenics*, 7, 355-369.

[15] F. Galton (1873). Problem 4001: On the extinction of surnames. *Educational Times*, 26, 1-17.

[16] F. Galton and H. Watson (1875). On the probability of the extinction of families. *The Journal of the Anthropological Institute of Great Britain and Ireland*, 4, 138-144.

[17] I. I. Gikhman and A. V. Skorokhod, *The theory of stochastic processes. II*, Classics in Mathematics, Springer-Verlag, Berlin, 2004. Translated from the Russian by S. Kotz; Reprint of the 1975 edition. MR2058260

[18] A. J. Hall and G. C. Wake, *A functional-differential equation arising in modelling of cell growth*, J. Austral. Math. Soc. Ser. B **30** (1989), no. 4, 424–435, DOI 10.1017/S0334270000006366. MR982622

[19] T. E. Harris, *The theory of branching processes*, Die Grundlehren der Mathematischen Wissenschaften, Bd. 119, Springer-Verlag, Berlin; Prentice-Hall, Inc., Englewood Cliffs, N.J., 1963. MR0163361

[20] T. E. Harris, *Contact interactions on a lattice*, Ann. Probability **2** (1974), 969–988, DOI 10.1214/aop/1176996493. MR0356292

[21] A. N. Kolmogorov, I. G. Petrovskii, and N. S. Piskunov (1937). A study of the diffusion equation with increase in the quantity of matter, and its application to a biological problem. *Moscow University Mathematics Bulletin Ser.* A(1), 1-25.

[22] L. Koralov and S. Molchanov, *Structure of population inside propagating front*, J. Math. Sci. (N.Y.) **189** (2013), no. 4, 637–658, DOI 10.1007/s10958-013-1212-1. Problems in mathematical analysis. No. 69. MR3098334

[23] F. E. Round, R. M. Crawford, and D. G. Mann (1990). The Diatoms. Cambridge University Press, Cambridge.

Department of Mathematics, Ben-Gurion University of the Negev Beersheba, Israel
Email address: derfel@math.bgu.ac.il

Department of Mathematics, Ohio University, Athens, Ohio 45701
Email address: fengy@ohio.edu

Department of Mathematics and Statistics, University of North Carolina at Charlotte, Charlotte, NC 28223,USA and National Research University, Higher School of Economics, Russian Federation
Email address: smolchan@uncc.edu

Contemporary Mathematics
Volume **734**, 2019
https://doi.org/10.1090/conm/734/14766

New examples of Hawking radiation from acoustic black holes

Gregory Eskin

In memory of Selim Grigorievich Krein on occasion of his 100th birthday

ABSTRACT. Rotating acoustic metrics may have black holes inside the ergosphere. Simple cases of acoustic black holes were studied in the references [21], [4]. In the present paper we study the Hawking radiation for more complicated cases of acoustic black holes including black holes that have corner points.

1. Introduction

In classical general relativity black hole is the region such that no signal or disturbance can escape it. It was a remarkable discovery by S. Hawking [10] that once the quantum effects are added the black holes emit quantum particles. This effect is called the Hawking radiation and it was studied subsequently in many papers (cf. [1], [2], [3], [8], [9], [12], [13], [14] and others). In an astrophysical experiment the Hawking effect is too weak to be detected on the earth.

In quest to find the experimental confirmation of the Hawking radiation effect W. Unruh considered the Hawking radiation for the acoustic black holes [18]. Acoustic black holes are one of the examples of analogue black holes (cf. [21], see also [15], [16]). The Hawking radiation for analogue black hole have been studied in [19], [20], [22] and others.

In all previous works on Hawking radiation the case of spherically symmetric metric or the case of one space dimension was treated.

In present paper, as in [6], we study the case of two space dimensional (rotating) acoustic metric. It was shown in [4] and [5] that there is a rich class of acoustic metrics having black holes. In [6] we consider the case of smooth black holes but as it was shown in [5] there exist black holes that are not smooth and that event horizons may have corners.

The Hawking radiation from such black holes will be considered in this paper.

The plan of the paper is the following:

In §2 we set up the framework of the quantum field theory on curved spacetimes following mostly T. Jacobson [12], [13] (see also [8], [11]). In departure from [6], where the Unruh type vacuum [17] was considered, we use in §2 a more common vacuum state as in S. Hawking [10] and T. Jacobson [12].

2010 *Mathematics Subject Classification.* Primary 35L05, Secondary 83C57.

In §3 we briefly consider the case of a simple acoustic black hole similar to the one in §3 of [6]. There is a difference in computations with §3 of [6] since we consider a different vacuum state in this paper.

In §4 we study the Hawking radiation for the acoustic metrics when the ergosphere has a finite number of characteristic points. This case is different from the case considered in section 4.1 of [6] where the characteristic points were not allowed. In §5 we consider the Hawking radiation from a black holes having corners. Note that constructions in §4 and §5 requires a localization in the angular coordinate.

2. The number of particles operator

Consider a fluid flow in a vortex with the velocity field

$$v = (v^1, v^2) = \frac{A}{r}\hat{x} + \frac{B}{r}\hat{\theta}, \tag{2.1}$$

where $r = |x|$, $\hat{x} = \frac{(x_1,x_2)}{|x|}$, $\hat{\theta} = \frac{(-x_2,x_1)}{|x|}$, A, B are functions of $x = (x_1, x_2)$.

The acoustic waves in the moving fluid are described by the wave equation (cf. [21])

(2.2)

$$\Box_g u(x_0, x_1, x_2) = \frac{1}{\sqrt{g(x_1,x_2)}} \sum_{j,k=0}^{2} \frac{\partial}{\partial x_k}\left(\sqrt{g(x_1,x_2)}g^{jk}(x_1,x_2)\frac{\partial u(x_0,x_1,x_2)}{\partial x_j}\right) = 0,$$

x_0 is the time variable, $g^{00} = 1$, $g^{0j} = g^{j0} = v^j$, $g^{jk} = (-\delta_{jk} + v^j v^k)$, $1 \leq j, k \leq 2$, $g(x_1, x_2) = 1$ and we, for the simplicity, assume the sound speed and density are equal to 1.

The Hamiltonian corresponding to (2.2) has the following form in polar coordinates (ρ, φ):

$$H(\rho, \varphi, \eta_0, \eta_\rho, \eta_\varphi) = \left(\eta_0 + \frac{A}{\rho}\eta_\rho + \frac{B}{\rho^2}\eta_\varphi\right)^2 - \eta_\rho^2 - \frac{1}{\rho^2}\eta_\varphi^2, \tag{2.3}$$

where $(\eta_0, \eta_\rho, \eta_\varphi)$ are dual to (x_0, ρ, φ).

Note that (2.3) can be factored

$$H = H^+ H^-,$$

where

$$H^\pm(\rho, \varphi, \eta_0, \eta_\rho, \eta_\varphi) = \eta_0 + \frac{A}{\rho}\eta_\rho + \frac{B}{\rho^2}\eta_\varphi \pm \sqrt{\eta_\rho^2 + \frac{1}{\rho^2}\eta_\varphi^2}. \tag{2.4}$$

We refer to [11], [12] to introduce the main facts of quantum field theory on curved spacetimes.

We define $f_k^+(x_0, x)$ as the solution of (2.2) in $\mathbb{R}^2 \times \mathbb{R}$ having the following initial conditions in polar coordinates (ρ, φ)

$$f_k^+(x_0, x_1, x_2)\Big|_{x_0=0} = \gamma_k e^{ik\cdot x}, \tag{2.5}$$

$$\frac{\partial}{\partial x_0} f_k^+(x_0, x_1, x_2)\Big|_{x_0=0} = i\lambda_0^-(k)\gamma_k e^{ik\cdot x},$$

where $x = (\rho, \varphi)$, $k = (\eta_\rho, m)$, $m \in \mathbb{Z}$, $k \cdot x = \rho\eta_\rho + m\varphi$, $\gamma_k = \frac{1}{2\pi\sqrt{2}\sqrt{\rho}\left(\eta_\rho^2+a^2\right)^{\frac{1}{4}}}$, $a > 0$ is arbitrary,

$$\lambda_0^-(k) = -\frac{A}{\rho}\eta_\rho - \frac{Bm}{\rho^2} - \sqrt{\eta_\rho^2 + a^2}. \tag{2.6}$$

Let

$$f_{-k}^-(x_0, x) = \overline{f_k^+(x_0, x)}. \tag{2.7}$$

Let u, v be the solutions of (2.2) and $< u, v >$ be their Klein-Gordon inner product

$$< u, v >= i \int\limits_{x_0=t} |g|^{\frac{1}{2}}\{u, v\} dx_1 dx_2, \tag{2.8}$$

where $\{u, v\} = \sum_{j=0}^{2} g^{0j}\left(\overline{u}\frac{\partial v}{\partial x_j} - \frac{\partial \overline{u}}{\partial x_j} v\right)$.

Note that (2.8) is independent of t (cf. [12]).

It is easy to see (cf. [6]) that

(2.9)
$$< f_k^+, f_{k'}^+ >= \delta(\eta_\rho - \eta_\rho')\delta_{mm'}, \quad < f_k^-, f_{k'}^- >= -\delta(\eta_\rho - \eta_\rho')\delta_{mm'}, \quad < f_k^+, f_{k'}^- >= 0,$$

where $k = (\eta_\rho, m), k' = (\eta_\rho', m')$.

Thus f_k^+, f_{-k}^- form a basis of solutions of (2.2).

Now we introduce the selfadjoint field operator Φ

$$\Phi = \sum_{m=-\infty}^{\infty} \int\limits_{-\infty}^{\infty} \left(\alpha_k^+ f_k^+(x_0, x_1, x_2) + \alpha_{-k}^- f_{-k}^-(x_0, x_1, x_2)\right) d\eta_\rho, \tag{2.10}$$

where $k = (\eta_\rho, m), \eta_\rho \in \mathbb{R}, m \in \mathbb{Z}$,

$$\alpha_k^+ =< f_k^+, \Phi >, \quad \alpha_k^- = - < f_k^-, \Phi > . \tag{2.11}$$

Since $\overline{f_k^+} = f_{-k}^-$ we have that $(\alpha_{-k}^-)^* = \alpha_k^+$.

Operators $\alpha_k^+, \alpha_{-k}^-$ are called the annihilation and creation operators, respectively, and they satisfy the following commutation relations (cf. [12]):

$$[\alpha_k^+, \alpha_{-k'}^- = \delta(\eta_\rho - \eta_\rho')\delta_{mm'} I, \quad [\alpha_k^+, \alpha_{k'}^+] = 0, \tag{2.12}$$

$[\alpha_k^-, \alpha_{-k'}^-] = 0$, I is the identity operator.

Let $C(x_0, x_1, x_2)$ be a solution of (2.2) with some initial conditions at $x_0 = 0$ that will be specified later.

Expanding $C(x_0, x_1, x_2)$ in the basis f_k^+, f_{-k}^- we get

$$C = \sum_{m=-\infty}^{\infty} \int\limits_{-\infty}^{\infty} \left(C^+(k) f_k^+(x_0, x_1, x_2) + C^-(k) f_{-k}^-(x_0, x_1, x_2)\right) d\eta_\rho, \tag{2.13}$$

where $k = (\eta_\rho, m)$, $C^+(k) =< f_k^+, C >$, $C^-(k) = - < f_{-k}^-, C >$.

Let

$$C^\pm = \sum_{m=-\infty}^{\infty} \int\limits_{-\infty}^{\infty} C^\pm(k) f_{\pm k}^\pm(x_0, x) d\eta_\rho, \tag{2.14}$$

Thus

$$C = C^+ + C^-. \tag{2.15}$$

It follows from (2.10), (2.13) that

$$< C, \Phi >= \sum_{m=-\infty}^{\infty} \int_{-\infty}^{\infty} \left(\overline{C^+}(k)\alpha_k^+ - \overline{C^-}(k)\alpha_{-k}^-\right) d\eta_\rho. \tag{2.16}$$

The vacuum state $|0\rangle$ is defined by the conditions

$$\alpha_k^+ |0\rangle = 0 \quad \text{for all } k. \tag{2.17}$$

Let $N(C)$ be the number of particles operator created by the wave packet C (cf. [12])

$$N(C) =< C, \Phi >^* < C, \Phi >, \tag{2.18}$$

and let $\langle 0|N(C)|0\rangle$ be the average number of particles.

As in [6] one have the following theorem

THEOREM 2.1. *The average number of particles created by the wave packet C is given by the formula*

$$\langle 0|N(C)|0\rangle = \sum_{m=-\infty}^{\infty} \int_{-\infty}^{\infty} |C^-(k)|^2 d\eta_\rho, \tag{2.19}$$

where C^- is the same as in (2.14).

Note that (cf. (2.14))

$$< C^-, C^- >= - \sum_{m=-\infty}^{\infty} \int_{-\infty}^{\infty} |C^-(k)|^2 d\eta_\rho. \tag{2.20}$$

Therefore

$$\langle 0|N(C)|0\rangle = - < C^-, C^- >, \tag{2.21}$$

where C^- is the same as in (2.14).

Proof of Theorem 2.1. We have

$$< C, \Phi > |0\rangle = - \sum_{m=-\infty}^{\infty} \int_{-\infty}^{\infty} \overline{C^-(k)}\alpha_{-k}^- |0\rangle d\eta_\rho. \tag{2.22}$$

Analogously,

$$\langle 0| < C, \Phi >^* = - \sum_{m=-\infty}^{\infty} \int_{-\infty}^{\infty} \langle 0|C^-(k)\alpha_k^+ d\eta_\rho, \tag{2.23}$$

since $\langle 0|\alpha_{-k}^- = 0$. Therefore

$$\begin{aligned} &\langle 0| < C, \Phi >^* < C, \Phi > |0\rangle \\ &= \sum_{m=-\infty}^{\infty} \int_{-\infty}^{\infty} \langle 0|C^-(k)\alpha_k^+ d\eta_\rho \cdot \sum_{m'=-\infty}^{\infty} \int_{-\infty}^{\infty} \overline{C^-(k')}\alpha_{-k'}^- |0\rangle d\eta_\rho' \\ &= \sum_{m=-\infty}^{\infty} \int_{-\infty}^{\infty} |C^-(k)|^2 d\eta_\rho, \end{aligned} \tag{2.24}$$

since $\alpha_k^+\alpha_{-k'}^- = \alpha_{-k}^-\alpha_k^+ + I\delta(k-k')$ and $\alpha_k^+|0\rangle = 0,\ \forall k$.

3. Hawking radiation from the simple rotating acoustic black hole

Consider the simplest case when $A<0, B\neq 0$ are constants. This case was studied in [6], §3, and we briefly repeat the computations taking into account that the vacuum state in this paper is different from [6], §3.

Let $S(x_0,\rho,\varphi)$ be the eikonal satisfying the equation

$$\Big(-\eta_0+\frac{A}{\rho}S_\rho+\frac{B}{\rho^2}S_\varphi\Big)^2 - S_\rho^2 - \frac{1}{\rho^2}S_\varphi^2 = 0.$$

We are looking for the eikonal of the form $S = -\eta_0 x_0 + s(\rho) + m\varphi$ where $s(\rho)\to+\infty$ when $\rho>|A|$ and $\rho\to|A|$.

It can be shown (cf. [6]) that when $\rho\to|A|$

$$S_\rho = \frac{|A|\Big(\eta_0-\frac{Bm}{|A|^2}\Big)}{\rho-|A|} + O(\rho-|A|).$$

We will not try to solve the eikonal equation exactly. Instead we use the approximation

$$S_1 = -\eta_0 x_0 + |A|\Big(\eta_0 - \frac{Bm}{|A|^2}\Big)\ln(\rho-|A|) + m\varphi$$

of $S(x_0,\rho,\varphi)$ to construct the exact solution of (2.2) having the following initial conditions

$$\text{(3.1)}\qquad C(x_0,\rho,\varphi)\Big|_{x_0=0} = \theta(\rho-|A|)\frac{(\rho-|A|)^\varepsilon}{\sqrt{\rho}}\,e^{-a(\rho-|A|)}\,e^{i\xi_0|A|\ln(\rho-|A|)+im\varphi},$$
$$\frac{\partial C(x_0,\rho,\varphi)}{\partial x_0}\Big|_{x_0=0} = i\beta C(x_0,\rho,\varphi_0)\Big|_{x_0=0},$$

where $a>0,\varepsilon>0$ are arbitrary, $\theta(\rho-|A|)=1$ for $\rho>|A|$ and $\theta(\rho-|A|)=0$ for $\rho<|A|$, $\xi_0=\eta_0-\frac{Bm}{|A|^2}$,

$$\text{(3.2)}\qquad \beta = -\frac{A}{\rho}\frac{\xi_0|A|}{\rho-|A|} - \frac{Bm}{\rho^2} - \frac{\xi_0|A|}{\rho-|A|}.$$

For the convenience we take $a>0$ in (3.2) equal to a in (2.6). It is easy to compute the KG norm of $C(x_0,\rho,\varphi)$ (cf. (3.8) in [6]):

$$\text{(3.3)}\qquad <C,C> = \frac{4\pi\Gamma(2\varepsilon)\xi_0|A|}{(2a)^\varepsilon}.$$

We shall call $C(x_0,\rho,\varphi)$ a wave packet.

By Theorem 2.1 to compute the average number of created particles we need to find $C^-(k) = -<f_{-k}^-, C>$. Computing the KG norm we have

$$\text{(3.4)}\qquad C^- = C_1^-(k) + C_2^-(k),$$

where (cf. (3.11), (3.12) in [6])

(3.5)

$$C_1^-(k) = \int_0^\infty \int_0^{2\pi} \frac{e^{i\rho\eta_\rho + im'\varphi}\theta(\rho - |A|)}{\sqrt{\rho}(\eta_\rho^2 + a^2)^{\frac{1}{4}}\sqrt{2}\ 2\pi} \frac{(\rho - |A|)^\varepsilon}{\sqrt{\rho}} \Big(\frac{\xi_0|A|}{\rho - |A|} - \frac{A}{\rho}\Big(\frac{-i\varepsilon}{\rho - |A|} + ia\Big)\Big) \cdot e^{-a(\rho - |A|) + i\xi_0|A|\ln(\rho - |A|)}\, e^{im\varphi}\rho\, d\rho d\varphi,$$

$$(3.6) \quad C_2^-(k) = -\int_0^\infty \int_0^{2\pi} \frac{e^{i\rho\eta_\rho + im'\varphi}(\eta_\rho^2 + a^2)^{\frac{1}{4}}}{\sqrt{\rho}\sqrt{2}\ 2\pi} \frac{\theta(\rho - |A|)(\rho - |A|)^\varepsilon}{\sqrt{\rho}} \cdot e^{-a(\rho - |A|) + i\xi_0|A|\ln(\rho - |A|) + im\varphi}\rho d\rho d\varphi, \quad k = (\eta_\rho, m').$$

Integrating in φ and using the formula (cf. [6])

$$(3.7) \qquad \int_0^\infty e^{it\eta_\rho}t^\lambda e^{-at}dt = \frac{e^{i\frac{\pi}{2}(\lambda+1)}\Gamma(\lambda+1)}{(\eta_\rho + ia)^{\lambda+1}},$$

we get

$$(3.8) \quad C_1^-(k) = \frac{\delta_{m',-m}}{\sqrt{2}} e^{i|A|\eta_\rho} \frac{e^{i\frac{\pi}{2}(i\xi_0|A| + \varepsilon)}\Gamma(i\xi_0|A| + \varepsilon)(\xi_0|A| - i\varepsilon)}{(\eta_\rho^2 + a^2)^{\frac{1}{4}}(\eta_\rho + ia)^{i\xi_0|A| + \varepsilon}}\Big(1 - \frac{ia}{\eta_\rho + ia}\Big) + \delta_{m',-m}O\Big(\frac{1}{|\eta_\rho + ia|^{1+\varepsilon}}\Big),$$

$$(3.9) \qquad C_2^-(k) = -\frac{\delta_{m',-m}}{\sqrt{2}} e^{i|A|\eta_\rho} \frac{e^{i\frac{\pi}{2}(i\xi_0|A| + \varepsilon + 1)}\Gamma(i\xi_0|A| + \varepsilon + 1)(\eta_\rho^2 + a^2)^{\frac{1}{4}}}{(\eta_\rho + ia)^{i\xi_0|A| + \varepsilon + 1}},$$

where $\delta_{m_1,m_2} = 1$ when $m_1 = m_2$ and $\delta_{m_1,m_2} = 0$ when $m_1 \neq m_2$. Since $\big|e^{i\frac{\pi}{2}(i\xi_0|A| + \varepsilon)}\big| = e^{-\frac{\pi}{2}\xi_0|A|}$, $\Gamma(i\xi_0|A| + \varepsilon + 1) = (i\xi_0|A| + \varepsilon)e^{-\frac{\pi}{2}\xi_0|A|}\Gamma_1(i\xi_0|A| + \varepsilon)$, Γ_1 is bounded, we have

(3.10)

$$|C^-|^2 = |C_1^- + C_2^-|^2 = \frac{\delta_{m',m}}{2}\Big|\frac{\xi_0|A| - i\varepsilon}{(\eta_\rho^2 + a^2)^{\frac{1}{4}}} \cdot \frac{\eta_\rho}{\eta_\rho + ia} - \frac{i(i\xi_0|A| + \varepsilon)(\eta_\rho^2 + a^2)^{\frac{1}{4}}}{\eta_\rho + ia}\Big|^2 \cdot e^{-2\pi\xi_0|A|}|\Gamma_1(i\xi_0|A| + \varepsilon)|^2 e^{2\xi_0|A|\arg(\eta_\rho + ia)}(\eta_\rho^2 + a^2)^{-\varepsilon} + \delta_{m',-m}O\big(|\eta_\rho + ia|^{-2-2\varepsilon}\big).$$

Taking the sum in m', integrating in η_ρ and making change of variables $\eta_\rho \to a\eta_\rho$, we get

$$(3.11) \qquad \langle 0|N(C)|0\rangle = a^{-2\varepsilon}\int_{-\infty}^\infty C_3(\eta_\rho)d\eta_\rho + O\big(a^{-2\varepsilon-1}\big),$$

where

$$(3.12) \quad C_3(\eta_\rho) = \frac{1}{2}e^{-2\pi\xi_0|A|}|\Gamma_1(i\xi_0|A| + \varepsilon)|^2\, |\xi_0|A| + i\varepsilon|^2\Big|\frac{\eta_\rho}{(\eta_\rho^2 + 1)^{\frac{1}{4}}} + (\eta_\rho^2 + 1)^{\frac{1}{4}}\Big|^2 \cdot e^{2\xi_0|A|\arg(\eta_\rho + i)}(\eta_\rho^2 + 1)^{-\varepsilon-1}.$$

Denote by C_n the normalized wave packet, i.e. $C_n = \frac{C}{<C,C>}$. Thus $N(C_n) = \frac{N(C)}{<C,C>}$. Dividing (3.11) by $<C,C>$ and taking the limit as $a \to \infty$ we get

$$\lim_{a\to\infty} \langle 0|N(C_n)|0\rangle = \frac{2^{\varepsilon}}{4\pi\Gamma(2\varepsilon)} \int_{-\infty}^{\infty} \frac{1}{\xi_0|A|} C_3(\eta_\rho) d\eta_\rho. \tag{3.13}$$

Therefore we proved the following theorem:

THEOREM 3.1. *The average number of particles created by the normalized wave packet C_n is given by (3.13) where $C_3(\eta_\rho)$ is the same as in (3.12).*

Now we will analyze the decay of the right size of (3.13) when $\xi_0|A| \to \infty$. Note

$$\arg(\eta_\rho + i) = \sin^{-1} \frac{1}{\sqrt{\eta_\rho^2 + 1}} \quad \text{when} \quad \eta_\rho > 0, \tag{3.14}$$

$$\arg(\eta_\rho + i) = \pi - \sin^{-1} \frac{1}{\sqrt{\eta_\rho^2 + 1}} \quad \text{when} \quad \eta_\rho < 0. \tag{3.15}$$

We have

$$\Big| \frac{1}{(\eta_\rho^2 + 1)^{\frac{1}{4}}} + \frac{(\eta_\rho^2 + 1)^{\frac{1}{4}}}{\eta_\rho + i} \Big|^2 \tag{3.16}$$

$$\text{is} \quad O\Big(\frac{1}{(\eta_\rho^2 + 1)^{\frac{1}{2}}}\Big) \quad \text{when} \quad \eta_\rho > 0, \tag{3.17}$$

$$\text{and} \quad O\Big(\frac{1}{\eta_\rho^2 + 1}\Big) \quad \text{when} \quad \eta_\rho < 0, \tag{3.18}$$

Note that $\sin^{-1} \frac{1}{\sqrt{\eta_\rho^2+1}} \le \frac{\pi}{2}$. Therefore for $\eta_\rho > 0$ we have

$$C_3 \le C \frac{e^{-2\pi\xi_0|A| + \pi\xi_0|A|}((\xi_0|A|)^2 + \varepsilon^2)}{(\eta_\rho^2 + 1)^{\varepsilon + \frac{1}{2}}}, \quad \eta_\rho > 0, \tag{3.19}$$

i.e. C_3 is exponentially decaying on $(0, +\infty)$.

When $\eta_\rho < 0$ we have

$$C_3 \le C \, \frac{(\xi_0|A|)^2 e^{-2\xi_0|A| \sin^{-1} \frac{1}{\sqrt{\eta_\rho^2+1}}}}{(\eta_\rho^2 + 1)^{\varepsilon + 1}}, \quad \eta_\rho < 0. \tag{3.20}$$

If $|\eta_\rho| \le (\xi_0|A|)^\delta, \ 0 < \delta < 1$, then

$$e^{-2\xi_0|A| \sin^{-1} \frac{1}{\sqrt{\eta_\rho^2+1}}} \le e^{-C(\xi_0|A|)^{1-\delta}}. \tag{3.21}$$

Thus C_3 is exponentially decaying on $(-(\xi_0|A|)^\delta, 0)$. On $(-\infty, -(\xi_0|A|)^\delta)$ we get

$$\int_{-\infty}^{-(\xi_0|A|)^\delta} C_3 \ d\eta_\rho \ \le \int_{-\infty}^{-(\xi_0|A|)^\delta} C \frac{(\xi_0|A|)^2}{(\eta_\rho^2 + 1)^{\varepsilon+1}} d\eta_\rho \ \le \ C \frac{(\xi_0|A|)^2}{((\xi_0|A|)^{2\delta} + 1)^{\varepsilon + \frac{1}{2}}}. \tag{3.22}$$

Hence

(3.23)
$$\int_{-\infty}^{0} C_3\, d\eta_\rho \leq C(\xi_0|A|)^{2-\delta(1+2\varepsilon)}.$$

Therefore normalizing C to make $< C_n, C_n >= 1$, we get (cf. (3.3))

(3.24)
$$\int_{-\infty}^{0} \frac{1}{\xi_0|A|} C_3 d\eta_\rho \leq c_1(\xi_0|A|)^{1-\delta(1+2\varepsilon)}.$$

Remark 3.1 It follows from (3.19) that $\int_0^\infty C_3\, d\eta_\rho$ is exponentially decaying in $\xi_0|A|$ (cf. [10]) and so it contributes to the Hawking radiation. However, the integral $\int_{-\infty}^0 C_3\, d\eta_\rho$ has only decay of order (3.24) and it dos not contribute to the Hawking radiation. Note that for $f_k^+(x_0, x_1, x_2)$ and $f_k^-(x_0, x_1, x_2) = \overline{f_k^*}$ to form a basis of solutions of (2.2) one needs to use all $\eta_\rho \in (-\infty, +\infty)$. Therefore $\langle 0|N(C)|0\rangle$ must include also $\int_{-\infty}^0 C_3\, d\eta_\rho$. Only when we replace $|0\rangle$ by the Unruh type vacuum $|\psi\rangle$ (cf. [6]), we get that all term in $\langle\Psi|N(C)|\Psi\rangle$ contribute to the Hawking radiation.

4. Hawking radiation from rotating acoustic black holes

In this and the next sections we shall continue to study the Hawking radiation from rotating acoustic black holes started in [6]. We shall consider the case of the fluid flow (2.1) where $A(\rho, \varphi), B(\rho, \varphi)$ are functions of (ρ, φ), $A < 0$. The ergosphere of the acoustic metric is $\frac{A^2(\rho,\varphi)+B^2(\rho,\varphi))}{\rho^2} = 1$ and we assume that it is a smooth Jordan curve. It was proven in [5] that there are always black holes for the acoustic matrics. In [6] we consider a particular case when the normals to the ergosphere are not characteristic at any point of the ergosphere. In this case there is a smooth black hole inside the ergosphere. When the ergosphere has characteristic points the black hole, in general, may have corners points. We shall study two typical examples. In this section we consider the case of finite number of characteristic points on the ergosphere where the black hole is tangent to the ergosphere, and in the next section we shall consider an example of acoustic black hole having a corner.

Let the velocity field be of the form

(4.1)
$$v = \frac{A}{\rho}\hat{x} + \frac{B(\varphi)}{\rho}\hat{\theta},$$

where $A < 0$ is a constant and $B(\varphi)$ has finite number of zeros

(4.2)
$$B(\theta_k) = 0, \quad 0 \leq \theta_1 < \theta_2 < ... < \theta_p < 2\pi.$$

Here $\rho = |A|$ is the boundary of the black hole and it touches the ergosphere at points $\theta_k, 1 \leq k \leq p$.

Consider the eikonal $S(\rho, \varphi)$:

(4.3)
$$-\eta_0 + \frac{A}{\rho}S_\rho + \frac{B(\varphi)}{\rho^2}S_\varphi + \sqrt{S_\rho^2 + \frac{1}{\rho^2}S_\varphi^2} = 0$$

such that $S_\rho \to +\infty$ when $\rho \to |A|$. As in [6] (cf. also §3), for small $\rho - |A|$ it is convenient to use an approximation $S_1(\rho, \varphi)$ of eikonal $S(\rho, \varphi)$ that satisfies the equation

$$-\eta_0 + \frac{\rho - |A|}{|A|} S_{1\rho} + \frac{B(\varphi)}{|A|^2} S_{1\varphi} = 0. \tag{4.4}$$

We look for a solution of (4.4) in a form

$$S_1(\rho, \varphi) = \eta_0 |A| \ln(\rho - |A|) + S_2(\rho, \varphi), \tag{4.5}$$

where $S_2(\rho, \varphi)$ satisfies

$$(\rho - |A|) S_{2\rho} + \frac{B(\varphi)}{|A|} S_{2\varphi} = 0, \quad \rho > |A|.$$

Let $[\varphi_r, \varphi_{r1}]$ be a closed interval where $B(\varphi) \neq 0$ and $\varphi_{r0} \in (\varphi_r, \varphi_{r1})$. For $\varphi \in [\varphi_r, \varphi_{r1}]$ we have

$$S_2(\rho, \varphi) = \alpha_r \ln |\rho - A| - \alpha_r \int_{\varphi_{r0}}^{\varphi} |A| B^{-1}(\varphi') d\varphi' + d_r, \tag{4.6}$$

where α_k and d_k are arbitrary. Therefore

$$S_1(\rho, \varphi) = (\eta_0 |A| + \alpha_r) \ln |\rho - |A| + S_{3r}, \tag{4.7}$$

where

$$S_{3r}(\varphi) = -\alpha_r \int_{\varphi_{r0}}^{\varphi} |A| B^{-1}(\varphi') d\varphi + d_r, \tag{4.8}$$

for $\varphi \in [\varphi_r, \varphi_{r1}]$.

Let $\hat{C}_r(x_0, \rho, \varphi)$ be a solution of (2.2) (wave packet) having the following initial conditions

$$\hat{C}_r \Big|_{x_0 = 0} = \theta(\rho - |A|) c_r(\varphi) f(\rho) e^{i(\eta_0 |A| + \alpha_r) | \ln |\rho - |A| \, | + S_{3r}} \tag{4.9}$$

where

$$f(\rho) = \frac{(\rho - |A|)^{\varepsilon} e^{-a(\rho - |A|)}}{\sqrt{\rho}}, \tag{4.10}$$

$a > 0, \varepsilon > 0$ (cf. §3), $\operatorname{supp} c_r(\varphi) \subset (\varphi_r, \varphi_{r_1})$,

$$\frac{\partial \hat{C}_r}{\partial x_0} \Big|_{x_0 = 0} = i \beta_r \theta(\rho - |A|) c_r(\varphi) f(\rho) e^{i(\eta_0 |A| + \alpha_r) | \ln |\rho - |A| \, | + S_{3r}(\varphi)}, \tag{4.11}$$

$$\begin{aligned} \beta_r &= -\frac{A}{\rho} S_{1\rho} - \frac{B(\varphi)}{\rho^2} S_{1\varphi} - \frac{\eta_0 |A| + \alpha_r}{\rho - |A|} \\ &= -\frac{A}{\rho} \frac{(\eta_0 |A| + \alpha_r)}{\rho - |A|} - \frac{(\eta_0 |A| + \alpha_r)}{\rho - |A|} + \frac{\alpha_k |A|}{\rho^2} \\ &= -\frac{\eta_0 |A| + \alpha_r}{\rho} + \frac{\alpha_r |A|}{\rho^2} = -\eta_0 + O(\rho - |A|). \end{aligned} \tag{4.12}$$

Note that although (4.7) is an approximation of the eikonal, the solution $C_r(x_0, \rho, \varphi)$ is an exact solution of the wave equation (2.2).

Finally we define wave packet $\hat{C}(x_0,\rho,\varphi)$ as the sum

$$\hat{C}(x_0,\rho,\varphi)=\sum_{r=1}^{p}\hat{C}_r(x_0,\rho,\varphi). \tag{4.13}$$

Similarly to (3.3) the KG norm of $\hat{C}$ is

$$<\hat{C},\hat{C}>=\sum_{r=1}^{p}\int_0^{2\pi}|c_r(\varphi)|^2\frac{\Gamma(2\varepsilon)(\eta_0|A|+\alpha_r)}{(2a)^{\varepsilon}}. \tag{4.14}$$

To compute the average number of particles created by $\hat{C}(x_0,\rho,\varphi)$ we use Theorem 2.1. Note that

$$\hat{C}^-(k)=-<\bar{f}_k^+,\hat{C}>=\sum_{r=1}^{p}-<\bar{f}_k^+,\hat{C}_r>=\sum_{r=1}^{p}\hat{C}_r^-(k).$$

We have

$$\begin{aligned}<f_k^+,\hat{C}_r>=i\int_0^{\infty}\int_0^{2\pi}\overline{f_k^+}\Big(\frac{\partial\hat{C}_r}{\partial x_0}+\frac{A}{\rho}\frac{\partial\hat{C}_r}{\partial\rho}+\frac{B(\varphi)}{\rho^2}\frac{\partial\hat{C}_r}{\partial\varphi}\Big)\\ -\hat{C}_r\Big(\frac{\partial\overline{f_k^+}}{\partial x_0}+\frac{A}{\rho}\frac{\partial\overline{f_k^+}}{\partial\rho}+\frac{B(\varphi)}{\rho^2}\frac{\partial\overline{f_k^+}}{\partial\varphi}\Big)\rho d\rho d\varphi.\end{aligned} \tag{4.15}$$

Note that

$$\frac{\partial\hat{C}_r}{\partial\rho}=\Big(\frac{i(\eta_0|A|+\alpha_r)}{\rho-|A|}+\frac{\varepsilon}{\rho-|A|}-a\Big)\hat{C}_r, \tag{4.16}$$

$$\frac{\partial\hat{C}_r}{\partial\varphi}=i\frac{\partial S_{3r}}{\partial\varphi}\hat{C}_r+\frac{\partial c_r}{\partial\varphi}e^{i(\eta_0|A|+\alpha_r)\ln(\rho-|A|)+iS_{3r}(\varphi)}f(\rho). \tag{4.17}$$

Therefore we need to take care of the extra turns when we take derivatives in ρ and in φ. These extra terms will dissapear when we will take the limit when the parameter $a\to\infty$.

Computing $\hat{C}_r^-(k)$ as in (3.4)-(3.9) in §3 we obtain

$$\hat{C}_r^-(k)=\hat{C}_{r1}^-+\hat{C}_{r2}^-, \tag{4.18}$$

where

$$\hat{C}_{r1}^-=\frac{\gamma_{m'}^{(r)}}{\sqrt{2}}\frac{(\eta_\rho^2+a^2)^{\frac14}e^{i|A|\eta_\rho}\Gamma(i(\eta_0|A|+\alpha_r)+\varepsilon+1)e^{i(i(\eta_0|A|+a_r)+\varepsilon+1)\frac{\pi}{2}}}{(\eta_\rho+ia)^{i(\eta_r|A|+\alpha_r)+\varepsilon+1}}, \tag{4.19}$$

$$\begin{aligned}\hat{C}_{r2}^{+-}=&\frac{-\gamma_{m'}^{(r)}(\eta_0|A|+\alpha_r)}{\sqrt{2}}\\ &\cdot\frac{(\eta_\rho^2+a^2)^{-\frac14}e^{i|A|\eta_\rho}\Gamma(i(\eta_0|A|+\alpha_r)+\varepsilon)e^{i(i(\eta_0|A|+\alpha_r)+\varepsilon)\frac{\pi}{2}}}{(\eta_\rho+ia)^{i(\eta_0|A|+\alpha_r)+\varepsilon}}\\ &+\gamma_{m'}^{(r)}O(|\eta_\rho+ia|^{-\varepsilon-1})+\hat{\gamma}_{m'}^{(r)}O(|\eta_\rho+ia|^{-\varepsilon-1}),\end{aligned} \tag{4.20}$$

where

$$\gamma_{m'}^{(r)} = \frac{1}{2\pi}\int_0^{2\pi} c_r(\varphi)e^{im'\varphi+iS_{3r}(\varphi)}\,d\varphi. \tag{4.21}$$

$$\hat{\gamma}_{m'}^{(r)} = \frac{1}{2\pi}\int_0^{2\pi} iB(\varphi)\frac{\partial c_r}{\partial\varphi}e^{im'\varphi+iS_{3r}(\varphi)}\,d\varphi, \tag{4.22}$$

Note that

$$\sum_{m'=-\infty}^{\infty}\gamma_{m'}^{(r_1)}\overline{\gamma}_{m'}^{(r_2)} = \frac{1}{2\pi}\int_0^{2\pi} c_{r_1}(\varphi)e^{iS_{3r_1}(\varphi)}\overline{c_r(\varphi)}e^{-iS_{3r_2}(\varphi)}d\varphi = 0, \tag{4.23}$$

since $c_{r_1}(\varphi)c_{r_2}(\varphi)\equiv 0$. Therefore taking into account (4.23) we get, as in §3:

$$\begin{aligned}\langle 0|N(\hat{C})|0\rangle &= \sum_{m=-\infty}^{\infty}\int_{-\infty}^{\infty}|\hat{C}^-(k)|^2d\eta_\rho = \sum_{r=1}^{p}\sum_{m'=-\infty}^{\infty}\int_{-\infty}^{\infty}|\hat{C}_r^-|^2d\eta_\rho \\ &= \sum_{r=1}^{p}\frac{1}{2\pi}\int_0^{2\pi}|c_r(\varphi)|^2d\varphi\,\frac{1}{2}\int_{-\infty}^{\infty}\big|\eta_0|A|+\alpha_r-i\varepsilon\big|^2\Big|\frac{\eta_\rho}{(\eta_\rho^2+a^2)^{\frac14}}+(\eta_\rho^2+a^2)^{\frac14}\Big|^2 \\ &\cdot e^{-2\pi(\eta_0|A|+\alpha_r)}|\Gamma_1(i(\eta_0|A|+\alpha_r)+\varepsilon)|^2e^{2(\eta_0|A|+\alpha_r)\arg(\eta_\rho+ia)}(\eta_\rho^2+a^2)^{-\varepsilon-1}d\eta_\rho. \\ &+\int_{-\infty}^{\infty}O(|\eta_\rho+ia|^{-2\varepsilon-2})d\eta_\rho,\end{aligned} \tag{4.24}$$

where

$$\Gamma_1((\eta_0|A|+\alpha_r)+\varepsilon) = i\int_0^{\infty}e^{i(i(\eta_0|A|+\alpha_r)+\varepsilon-1)\ln y+i(\varepsilon-1)\frac{\pi}{2}-iy}dy, \tag{4.25}$$

and we used that

$$\sum_{m'=-\infty}^{\infty}|\gamma_{m'}^{(r)}|^2 = \frac{1}{2\pi}\int_0^{2\pi}|c_r(\varphi)|^2d\varphi, \tag{4.26}$$

and that (cf. (4.22)) $|\hat{\gamma}_{m'}^{(r)}|\le\frac{C}{|m'|^2}$.

Therefore we proved the following theorem.

THEOREM 4.1. *The average number of particles created by the wave packet (4.9), (4.11) is given by the formula (4.24).*

Finally, making the change of variables $\eta_\rho \to a\eta_\rho$, replacing $\hat{C}$ by the normalized wave packet $\hat{C}_n = \frac{\hat{C}}{<\hat{C},\hat{C}>}$ and taking the limit where $a \to \infty$ we get

$$(4.27)\quad \lim_{a\to\infty} \langle 0|N(\hat{C}_n)|0\rangle = \Big(\sum_{r=1}^{p}\int_0^{2\pi} |c_r(\varphi)|^2 d\varphi \frac{\Gamma(2\varepsilon)}{2^\varepsilon}(\eta_0|A|+\alpha_r)\Big)^{-1}$$
$$\cdot \sum_{r=1}^{p}\frac{1}{2\pi}\int_0^{2\pi}|c_r(\varphi)|^2 d\varphi \int_{-\infty}^{\infty} C_{3r}(\eta_\rho)d\eta_\rho,$$

where $\hat{C}_{3r}(\eta_\rho)$ is the same as C_3 in (3.12) with $|\xi_0|A|$ replaced by $\eta_0|A|+\alpha_r$.

Note that when α_r is the same for all $1 \le r \le p$ then the sum $\sum_{r=1}^{p}\int_0^{2\pi}|c_k(\varphi)|^2 d\varphi$ in (4.27) cancels.

Remark 4.1 (cf. Remark 3.1). As in the end of §3 we have that $\int_0^\infty \hat{C}_{3r}(\eta_\rho)d\eta_\rho$ is exponentially decaying when $(\eta_0|A|+\alpha_r)\to\infty$ and

$$\int_{-\infty}^{0}\hat{C}_{3r}(\eta_\rho)d\eta_\rho = O((\eta_0|A|+\alpha_r)^{2-\delta(1+2\varepsilon)}$$

(cf. (3.24)).

5. The Hawking radiation in the case of black hole with corners

In this section we study the Hawking radiation from rotating acoustic black holes having corners. It was proven in [4], [5] and [7] that the zero energy null geodesics form two family of smooth curves inside the ergosphere and the boundary of the black hole (the event horizon) consists of segments of zero energy null geodesics belonging to one or another family. In the case when the normal to the ergosphere is not characteristic at any point, the black hole is formed by one family of null geodesics and it is smooth closed curve. When the ergosphere contains the characteristic points then the black hole (or black holes) consists of segments belonging to the different families and therefore when adjacent segments belong to different families they intersect and form a corner.

Let, as in [5], Example 4.2,

$$(5.1)\qquad A = A_0 + \varepsilon r\sin\varphi,\quad B = \varepsilon r\cos\varphi,\quad 0<\varepsilon<1,\ A_0<-1,$$

A and B define the velocity field by formula (2.1).

The equation of the ergosphere is

$$(5.2)\qquad r = r_0(\varphi) = -\frac{A_0}{1-\varepsilon^2}\Big(-\varepsilon\sin\varphi + \sqrt{1-\varepsilon^2\cos^2\varphi}\Big).$$

Points $\alpha_1 = (r_0(\frac{\pi}{2}),\frac{\pi}{2})$ and $\alpha_2 = (r_0(-\frac{\pi}{2}),-\frac{\pi}{2})$ on the ergosphere are characteristic points. Note that both families of zero-energy null-geodesics are tangent to the ergosphere at the characteristic point. For the ergosphere (5.2) we have zero energy null-geodesics γ_1 and γ_2 that starts at $\alpha_1 = (r_0(\frac{\pi}{2}),\frac{\pi}{2})$ and intersected at some point α_3 forming an angle (see Fig. 1).

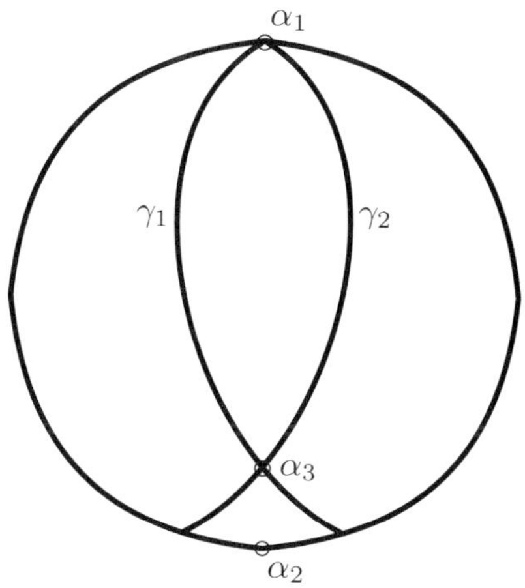

Fig,. 1. Null-geodesics γ_1 and γ_2 intersect at point α_3 and bound a black hole having corner point α_3.

Denote by γ_{10} the part of γ_1 where $\varphi_1 < \varphi < \varphi_{11}$ on γ_1. Analogously let γ_{20} be the part of γ_2 such that $\varphi_2 < \varphi < \varphi_{20}$ on γ_2. Let $\rho = \rho_i(\varphi)$ be the equation of γ_{i0}, $\varphi_i < \varphi < \varphi_{i1}, i = 1, 2$.

Connect $\rho_1(\varphi_1)$ and $\rho_2(\varphi_{21})$ and connect $\rho_1(\varphi_{11})$ and $\rho_2(\varphi_2)$ by smooth curves to get a smooth periodic curve $\rho = \rho_0(\varphi)$, $0 < \varphi < 2\pi$ such that $\rho_0(\varphi) = \rho_i(\varphi)$ for $\varphi_1 \leq \varphi \leq \varphi_{11}$ and $\rho_0(\varphi) = \rho_2(\varphi)$ for φ between φ_{21} and φ_2.

The following arguments are not restricted to Example 4.2 and apply to any situation when we have two smooth segments γ_{10} and γ_{20} of the boundary of the black hole.

As in [6], §4.2, make change of variable

$$\tilde{\rho} = \rho - \rho_0(\varphi), \qquad \varphi = \varphi. \tag{5.3}$$

Denote by $\tilde{f}_k^+, k = (\eta_\rho, m)$, the solutions of the wave equation in $(x_0, \tilde{\rho}, \varphi)$ coordinates such that

$$\tilde{f}_k^+(x_0, x)\Big|_{x_0=0} = \tilde{\gamma}_k e^{i\eta_{\tilde{\rho}}\tilde{\rho} + im'\varphi}, \tag{5.4}$$

$$\frac{\partial f_k^+}{\partial x_0}\Big|_{x_0=0} = i\tilde{\lambda}_0^-(k)\tilde{\gamma}_k e^{i\eta_{\tilde{\rho}}\tilde{\rho} + im'\varphi}, \tag{5.5}$$

where

$$\tilde{\gamma}_k = \frac{1}{\sqrt{\rho_0(\varphi) + \tilde{\rho}}\ (\eta_{\tilde{\rho}}^2 + a^2)^{\frac{1}{4}}\sqrt{2(2\pi)^2}}, \quad \rho = \rho_0(\varphi) + \tilde{\rho}. \tag{5.6}$$

$$\tilde{\lambda}_0^-(k) = -\Big(\frac{A}{\rho} - \frac{B}{\rho^2}\rho_0'(\varphi)\Big)\eta_{\tilde{\rho}} - \frac{B}{\rho^2}m' - \sqrt{\eta_{\tilde{\rho}}^2 + a^2}. \tag{5.7}$$

The eikonal equation in $(\tilde{\rho}, \varphi)$ coordinates takes the form (cf. (4.19) in [6])

$$\tilde{A}(\tilde{\rho}, \varphi)\tilde{S}_{\tilde{\rho}}^2 + 2\tilde{B}(\eta_0, \tilde{\rho}, \varphi, \tilde{S}_\varphi)\tilde{S}_{\tilde{\rho}} + \tilde{C}(\eta_0, \tilde{\rho}, \varphi, \tilde{S}_\varphi) = 0,$$

where $\tilde{A}, \tilde{B}, \tilde{C}$ are the same as in (4.20) in [6]. Since $\tilde{\rho} = 0$ is characteristic when $\varphi_1 < \varphi < \varphi_{11}$ and when $\varphi_2 < \varphi < \varphi_{21}$ we have that $\tilde{A}(\tilde{\rho}, \varphi) = A_{01}(\tilde{\rho}, \varphi)\tilde{\rho}$ when $\varphi_1 < \varphi < \varphi_{11}$ and when $\varphi_2 < \varphi < \varphi_{21}$.

As in [6] we have that

$$\tilde{\rho} S_{\tilde{\rho}}^{(j)}(\tilde{\rho}, \varphi) + B_1^{(j)}(\varphi)\eta_0 + B_2^{(j)} S_\varphi^{(j)} = 0, \tag{5.8}$$

where equation (5.8) for $j = 1$ holds on $(\varphi_1, \varphi_{11})$ and (5.8) for $j = 2$ holds on $(\varphi_2, \varphi_{21})$.

Similarly to §4 the solution of (5.8) has the form

$$S^{(j)}(\tilde{\rho}, \varphi) = \tilde{\alpha}_j \ln \tilde{\rho} - \int_{\varphi_{j0}}^{\varphi} \frac{\tilde{\alpha}_j + B_1^{(j)}(\varphi')\eta_0}{B_2^{(j)}} d\varphi' + \tilde{d}_j, \tag{5.9}$$

where $\tilde{\alpha}_j > 0$ and d_j are arbitrary, $\varphi_{j0} \in (\varphi_j, \varphi_{j1})$.

Note that (5.9) for $j = 1, 2$ holds when $\varphi_j < \varphi < \varphi_{j1}$.

Now we shall construct the wave packet $\tilde{C}(x_0, \tilde{\rho}, \varphi)$ in the form

$$\tilde{C} = \tilde{C}_1 + \tilde{C}_2, \tag{5.10}$$

where

$$\tilde{C}_j\Big|_{x_0=0} = s_j(\varphi)\theta(\tilde{\rho}) e^{-i(\tilde{\alpha}_j \ln \tilde{\rho} - \int_{\varphi_{j0}}^{\varphi} \frac{\tilde{\alpha}_j + \tilde{B}_1^{(j)}(\varphi')\eta_0}{B_2^{(j)}(\varphi')} d\varphi' + \tilde{d}_j)} \frac{e^{-a\tilde{\rho}}\tilde{\rho}^{\varepsilon}}{\sqrt{\rho_i(\varphi) + \tilde{\rho}}}, \tag{5.11}$$

$$\frac{\partial \tilde{C}_j}{\partial x_0}\Big|_{x_0=0} = i\tilde{\beta}_j \tilde{C}_j\Big|_{x_0=0}, \tag{5.12}$$

where

$$\tilde{\beta}_j^i = \Big(\frac{A}{\rho} - \frac{B}{\rho^2}\rho_0'(\varphi)\Big)\frac{\tilde{\alpha}_j}{\tilde{\rho}} - \frac{B}{\rho^2}\frac{\partial S_j}{\partial \varphi} - \frac{\tilde{\alpha}_j}{\rho}, \ \varphi \in (\varphi_j, \varphi_{j1}), \tag{5.13}$$

$$\text{supp } s_j(\varphi) \subset (\varphi_j, \varphi_{j1}), \ j = 1, 2. \tag{5.14}$$

Denote

$$\tilde{\gamma}_m^{(j)} = \frac{1}{2\pi}\int_0^{2\pi} s_j(\varphi) e^{-i\big(-\int_{\varphi_{j0}}^{\varphi} \frac{\tilde{\alpha}_j + B_1^{(j)}(\varphi')\eta_0}{B_2^{(j)}} d\varphi' + \tilde{d}_j + m\varphi\big)} d\varphi.$$

By the Parseval's equality

$$\sum_{m=-\infty}^{\infty} |\tilde{\gamma}_m^{(j)}|^2 = \frac{1}{2\pi}\int_0^{2\pi} |s_j(\varphi)|^2 d\varphi. \tag{5.15}$$

Therefore as in §3 and §4 computing the average number of created particles we get

$$(5.16)\quad \langle 0|N(\tilde{C})|0\rangle = \sum_{j=1}^{2}\frac{1}{2\pi}\int_0^{2\pi}|s_j(\varphi)|^2 d\varphi$$
$$\cdot\frac{1}{2}\int_{-\infty}^{\infty}|\tilde{\alpha}_j - i\varepsilon|^2\Big|\frac{\eta_\rho}{(\eta_\rho^2+a^2)^{\frac{1}{4}}}+(\eta_\rho^2+a^2)^{\frac{1}{4}}\Big|^2,$$
$$\cdot\, e^{-2\pi\tilde{\alpha}_j}|\Gamma_1(i\tilde{\alpha}_j+\varepsilon)|^2 e^{2\tilde{\alpha}_j\arg(\eta_\rho+ia)}(\eta_\rho^2+a^2)^{-\varepsilon-1}d\eta_\rho$$
$$+\int_{-\infty}^{\infty}O(|\eta_\rho+ia|^{-2\varepsilon-2})d\eta_\rho.$$

Thus we proved the following theorem:

THEOREM 5.1. *The average number of particles created by the wave packet* (5.11)*,* (5.12) *is given by* (5.16)*.*

As in §4, replacing $\tilde{C}$ by the normalized wave packet $\tilde{C}_n$ and taking the limit as $a\to\infty$ we get that $\lim_{a\to\infty}\langle 0|N(\tilde{C}_n|0\rangle$ has an expression similar to (4.27) with $\eta_0|A|+\alpha_r$ replaced by $\tilde{\alpha}_r$.

References

[1] R. Banerjee and B. R. Majhi, *Hawking black body spectrum from tunneling mechanism*, Phys. Lett. B **675** (2009), no. 2, 243–245, DOI 10.1016/j.physletb.2009.04.005. MR2523572

[2] R. Brout, S. Massar, R. Parentani, and Ph. Spindel, *A primer for black hole quantum physics*, Phys. Rep. **260** (1995), no. 6, 329–446, DOI 10.1016/0370-1573(95)00008-5. MR1353182

[3] T.Damour and R.Ruffini, Black hole evaporation in the Klein-Sauter-Heisenberg-Euler formalism, Phys. Rev. D 14 (1976) 332.

[4] G. Eskin, *Inverse hyperbolic problems and optical black holes*, Comm. Math. Phys. **297** (2010), no. 3, 817–839, DOI 10.1007/s00220-010-1068-x. MR2653903

[5] G. Eskin and M. Hall, *Stationary black hole metrics and inverse problems in two space dimensions*, Inverse Problems **32** (2016), no. 9, 095006, 22, DOI 10.1088/0266-5611/32/9/095006. MR3543338

[6] G. Eskin, *Hawking radiation from acoustic black holes in two space dimensions*, J. Math. Phys. **59** (2018), no. 7, 072502, 17, DOI 10.1063/1.4996765. MR3826435

[7] G. Eskin, *Superradiance initiated inside the ergoregion*, Rev. Math. Phys. **28** (2016), no. 10, 1650025, 34, DOI 10.1142/S0129055X16500252. MR3572631

[8] K. Fredenhagen and R. Haag, *On the derivation of Hawking radiation associated with the formation of a black hole*, Comm. Math. Phys. **127** (1990), no. 2, 273–284. MR1037104

[9] G. W. Gibbons and S. W. Hawking, *Cosmological event horizons, thermodynamics, and particle creation*, Phys. Rev. D (3) **15** (1977), no. 10, 2738–2751, DOI 10.1103/PhysRevD.15.2738. MR0459479

[10] S. W. Hawking, *Particle creation by black holes*, Comm. Math. Phys. **43** (1975), no. 3, 199–220. MR0381625

[11] S. Hollands and R. M. Wald, *Quantum fields in curved spacetime*, Phys. Rep. **574** (2015), 1–35, DOI 10.1016/j.physrep.2015.02.001. MR3324615

[12] T. Jacobson, *Introduction to quantum fields in curved spacetime and the Hawking effect*, Lectures on quantum gravity, Ser. Cent. Estud. Cient., Springer, New York, 2005, pp. 39–89, DOI 10.1007/0-387-24992-3_2. MR2404955

[13] T. Jacobson, *Black holes and Hawking radiation in spacetime and its analogues*, Analogue gravity phenomenology, Lecture Notes in Phys., vol. 870, Springer, Cham, 2013, pp. 1–29, DOI 10.1007/978-3-319-00266-8_1. MR3183907

[14] M. K. Parikh and F. Wilczek, *Hawking radiation as tunneling*, Phys. Rev. Lett. **85** (2000), no. 24, 5042–5045, DOI 10.1103/PhysRevLett.85.5042. MR1812528

[15] M. Richartz, A. Prain, S. Liberati and S. Weinfurtner, Rotating black holes in a draining bathtub: superradiant scattering of gravity waves, Phys. Rev. D 91 (2015) no. 12, 124018

[16] T. Torres, S. Patrick, A, Coutant, M. Richartz, E. W.Tedford and S. Weinfurtner, Rotational superradiant scattering in a vortex flow, Nature Phys. 13, 833 (2017)

[17] W. Unruh, Notes on black hole evaporation, Phys. Rev. D 14 (1976) 871

[18] W. Unruh, Experimental Black Hole evaporation, Phys. Rev. Lett 46 (1981), 1351

[19] H. S. Vieira and V. B. Bezerra, *Acoustic black holes: massless scalar field analytic solutions and analogue Hawking radiation*, Gen. Relativity Gravitation **48** (2016), no. 7, Art. 88, 20, DOI 10.1007/s10714-016-2082-x. MR3510567

[20] M. Visser, *Essential and inessential features of Hawking radiation*, Internat. J. Modern Phys. D **12** (2003), no. 4, 649–661, DOI 10.1142/S0218271803003190. MR1975376

[21] M. Visser, *Acoustic black holes: horizons, ergospheres and Hawking radiation*, Classical Quantum Gravity **15** (1998), no. 6, 1767–1791, DOI 10.1088/0264-9381/15/6/024. MR1628019

[22] L.-C. Zhang, H.-F. Li, and R. Zhao, *Hawking radiation from a rotating acoustic black hole*, Phys. Lett. B **698** (2011), no. 5, 438–442, DOI 10.1016/j.physletb.2011.03.034. MR2787530

Department of Mathematics, University of California Los Angeles, Los Angeles, CA 90095-1555

Email address: eskin@math.ucla.edu

Contemporary Mathematics
Volume **734**, 2019
https://doi.org/10.1090/conm/734/14767

On multiplicative properties of determinants

Leonid Friedlander

To the centennial of Selim Grigorievich Krein.

ABSTRACT. Let A be an elliptic pseudodifferential operator of positive order on a compact closed manifold, and let T be a pseudodifferential operator of negative order such that T^m is of trace class. We compute $\log\det(A(I+T)) - \log\det A - \log\det_m(I+T)$ where first two determinants are zeta function regularized, and the last one is a regularized Fredholm determinant.

1. Introduction

Let A be an elliptic classical pseudodifferential operator of positive order k on a closed compact manifold M of dimension d. Suppose that A admits an Agmon angle, which is a solid angle $\{\lambda \in \mathbb{C} : \alpha \leq \arg\lambda \leq \beta\}$ that is free from values of the principal symbol of A. We assume that the null space of A is trivial. Then one can define complex powers A^z of A and the logarithm $\log A$. The operator $\log A$ is not a classical PDO but an operator with log-polyhomogeneous symbol ([L]); actually, the only logarithmic term in the expansion of its symbol is $\log|\xi|$. It is well known that the function

$$\zeta(z) = \mathrm{trace} A^z, \tag{1.1}$$

wich is a priori defined and analytic in the half-plane $\mathrm{Re} z < -d/k$, admits an analytic continuation to a meromorphic function in the whole complex plane; the point $z = 0$ is a regular point. Usually, the zeta-function is taken as function of $s = -z$. Then the determinant of A is defined according to the formula

$$\log\det A = \zeta'(0). \tag{1.2}$$

In the half-plane $\mathrm{Re} z < -d/k$ one can differentiate (1): $\zeta'(z) = \mathrm{trace}(\log A)A^z$, so

$$\log\det A = \mathrm{trace}(\log A)A^z\big|_{z=0}. \tag{1.3}$$

The expression on the right in (1.3) can be interpreted as a regularized trace of $\log A$; the operator A is used as the regularizer. It is known that the determinant of elliptic operators is not multiplicative; however the multiplicative anomaly $\log\det(AB) - \log\det A - \log\det B$ can be computed in terms of symbols of the operators A and B ([F], [KV]). In this paper, I discuss a different situation: what happens if one takes an operator $I + T$ for B where T is an operator of negative order. This problem may arise when one wants to compare determinants of two elliptic operators with

2010 *Mathematics Subject Classification.* Primary 58J52.

the same principal symbol. In the case when the order of T is smaller than $-d$, the operator T is of trace class, and it is not difficult to show that $\det(A(I+T)) = \det A \det(I+T)$ (e.g., see [F]). In general, the operator T belongs to the Schatten class $\mathfrak{S}_p$ for $p > d/s$ if T is a PDO of order $-s$.

Let us recall the notion of a regularized Fredholm determinant (e.g., see [GK], [S]). Let T be a compact operator in a Hilbert space, and $T \in \mathfrak{S}_p$. Let $m \geq p$ be an integer. Then

$$\det{}_m(I+T) = \prod_{j=1}^{\infty}\left((1+\lambda_j)\exp\sum_{p=1}^{m-1}\frac{(-1)^p}{p}\lambda_j^p\right) \tag{1.4}$$

where λ_j are eigenvalues of T; each eigenvalue is counted as many times as its multiplicity is. The product on the right in (1.4) converges. One can re-write (1.4) in the form

$$\log\det{}_m(I+T) = \operatorname{trace}\left(\log(I+T) + \sum_{p=1}^{m-1}\frac{(-1)^p}{p}T^p\right). \tag{1.5}$$

From this point, we assume that T is a classical pseudodifferential operator and the operator $I+T$ is invertible. If the operator T is not of trace class, there is no reason to expect the determinant of $A(I+T)$ to be equal to $\det A \det_m(I+T)$. My goal is to find an expression for

$$w_m(A,T) = \log\det(A(I+T)) - \log\det A - \log\det{}_m(I+T). \tag{1.6}$$

2. Variational formulas

Let $w(t) = w_m(A,tT)$. Clearly, $w(0) = 0$ and $w(1) = w_m(A,T)$. The variational formula for the determinant of an elliptic operator is well known:

$$\frac{d}{dt}\log\det(A(t)) = (\text{f.p.})_{z=0}\operatorname{trace}(A'(t)A(t)^{z-1}) \tag{2.1}$$

where $A'(t)$ is the derivative of the operator-valued function $A(t)$. The meromorphic function $\operatorname{trace}(A'(t)A(t)^{z-1})$ is either regular at $z=0$ or $z=0$ is its simple pole; the expression on the right in (2.1) is the constant term of its Laurent expansion at $z=0$. Therefore,

$$\begin{aligned}\frac{d}{dt}\log\det(A(I+tT)) &= (\text{f.p.})_{z=0}\operatorname{trace}[AT(A(I+tT))^{z-1}]\\ &= (\text{f.p.})_{z=0}\operatorname{trace}[AT(A(I+tT))^{z}(I+tT)^{-1}A^{-1}]\\ &= (\text{f.p.})_{z=0}\operatorname{trace}[(I+tT)^{-1}T(A(I+tT))^{z}]\\ &= \sum_{p=1}^{m-1}(-1)^{p-1}t^{p-1}(\text{f.p.})_{z=0}\operatorname{trace}[T^p(A(I+tT))^{z}]\\ &\quad+(-1)^{m-1}t^{m-1}\operatorname{trace}[T^m(I+tT)^{-1}].\end{aligned} \tag{2.2}$$

Here we assume the operator T^m to be of trace class. Formula (1.5) implies

$$\begin{aligned}\frac{d}{dt}\log\det{}_m(I+tT) &= \operatorname{trace}[T(I+tT)^{-1} - \sum_{p=1}^{m-1}(-1)^{p-1}t^{p-1}T^p]\\ &= (-1)^{m-1}t^{m-1}\operatorname{trace}[T^m(I+tT)^{-1}].\end{aligned} \tag{2.3}$$

From (2.2) and (2.3) we conclude

$$w'(t) = \sum_{p=1}^{m-1} (-1)^{p-1} t^{p-1} (\text{f.p.})_{z=0} \text{trace}[T^p (A(I+tT))^z]. \tag{2.4}$$

The operator valued function

$$\Phi_p(t,z) = T^p \frac{(A(I+tT))^z - A^z}{z}$$

is a holomorphic operator valued function in the sense of Guillemin [G]; operators $\Phi_p(t,z)$ are classical PDO of order $kz - ps$ (recall that T is an operator of order $-s$.) The function trace$\Phi_p(t,z)$ that is initially defined in the half-plane $\text{Re}(z) < (-d+ps)/k$ admits an analytic continuation to a meromorphic function in the whole complex plane, and it has simple poles. In particular, $z = 0$ may be a pole. Then,

$$\text{res}_{z=0} \text{trace}\Phi_p(t,z) = \frac{1}{k} \text{res} Q_p(t) \tag{2.5}$$

(see [G], [W]). Here res$Q_p(t)$ is the Guillemin-Wodzicki non-commutative residue of the operator $Q_p(t) = \Phi_p(t,0)$. The non-commutative residue of an operator Q

$$\text{res} Q = \int_{S^*M} q_{-d}(x,\xi)\mu \tag{2.6}$$

where $\mu = \alpha \wedge (d\alpha)^{d-1}$ with $\alpha = \xi dx$ being the Liouville form on T^*M, and $q_{-d}(x,\xi)$ is the homogeneous of degree $-d$ term in the complete symbol expansion of the operator Q. This term is not invariantly defined as a function on T^*M, so one has to use local coordinates and a partition of unity to get an expression in the right in (2.5). A remarkable fact that is due to Guillemin and Wodzicki is that the result is independent of the choice made. It follows from (2.5) that

$$(\text{f.p.})_{z=0} \text{trace}[T^p (A(I+tT))^z] = (\text{f.p.})_{z=0} \text{trace}(T^p A^z) + \frac{1}{k} \text{res} Q_p(t),$$

and formula (2.4) can be re-written as

$$w'(t) = \sum_{p=1}^{m-1} (-1)^{p-1} t^{p-1} (\text{f.p.})_{z=0} \text{trace}(T^p A^z) + \frac{1}{k} \sum_{p=1}^{m-1} (-1)^{p-1} t^{p-1} \text{res} Q_p(t). \tag{2.7}$$

The first sum on the right in (2.7) is a polynomial in t and hence can be integrated explicitly; the second sum contains local expressions only.

Notice that

$$Q_p(t) = T^p [\log(A(I+tT)) - \log A]. \tag{2.8}$$

Operators $\log(A(I+tT))$ and $\log A$ are not classical PDO; they are PDO with poly-logarithmic symbols; however their difference is a classical PDO. The order of Q_p is $-(p+1)s$. Keeping in mind that $w_m(A,T) = w(1)$ and $w(0) = 0$ one gets

THEOREM 2.1. *Let A be an elliptic pseudo-differential operator of positive order k on a d-dimensional closed manifold that posesses an Agmon angle. Let T ba a classical pseudo-differential operator on M of negative order $-s$. Assume that the* *null space of A is trivial and the operator $I+T$ is invertible. Let m be an integer number that is greater than d/s. Then*

$$\log\det(A(I+T)) - \log\det A - \log\det{}_m(I+T)$$
$$= \sum_{p=1}^{m-1} \frac{(-1)^{p-1}}{p}(f.p.)_{z=0} trace(T^p A^z) + \frac{1}{k}\sum_{p=1}^{m-1}(-1)^{p-1}\int_0^1 t^{p-1} resQ_p(t)dt \tag{2.9}$$

where operators Q_p are given in (2.8) and res is the Guillemin–Wodzicki non-commutative residue.

3. The structure of res$Q_p(t)$ and some special cases

Let

$$a(x,\xi) \sim \sum_{j=0}^{\infty} a_{k-j}(x,\xi) \quad \text{and} \quad \tau(x,\xi) \sim \sum_{j=0}^{\infty} \tau_{-s-j}(x,\xi)$$

be complete symbol expansions for operators A and T in local coordinates. An index shows the degree of homogeneity of the corresponding function. By $(\lambda - a)^{-1}$ I denote the complete symbol of the parametrix of the operator $\lambda I - A$ as a pseudo-differential operator with parameter (e.g., see [Sh]). It is the inverse of $\lambda - a$ in the non-commutative symbolic algebra. In the symbolic algebra of pseudo-differential operators with parameter, λ is treated as an additional dual variable, and one assigns weight k to it. Then

$$(\lambda - a(1+t\tau))^{-1} - (\lambda - a)^{-1} = \sum_{q=1}^{\infty} t^q (\lambda - a)^{-1}[a\tau(\lambda - a)^{-1}]^q \tag{3.1}$$

is the symbol of the difference between resolvents of the operators $A(I+tT)$ and A. All multiplications on the right in (3.1) are symbolic algebra multiplications. The q-th term in the sum in (3.1) is of order $-k-qs$, so the sum makes sense as an asymptotic sum. The principal symbol of the difference between resolvents equals $a_k(x,\xi)\tau_{-s}(x,\xi)/(\lambda - a_k(x,\xi))^2$, and all homogeneous terms in the symbolic expansion of the symbol (3.1) are rational functions of λ with the only pole at $\lambda = a_k(x,\xi)$; the residue at this pole vanishes.

Let ℓ be a ray in the complex plane that goes from 0 to ∞ and lies in the Agmon angle for A. We make a cut in the complex plane along ℓ to define $\log\lambda$. Let Γ be the contour that goes from $-\infty$ to the point ϵ-close to 0 along one side of the cut, then goes around 0 to the opposite side of the cut, and then goes back to ∞; here $\epsilon < |a_k(x,\xi)|$. Then

$$\log(a(1+t\tau)) - \log a = \sum_{q=1}^{\infty} t^q \frac{1}{2\pi i}\int_\Gamma \log\lambda(\lambda - a)^{-1}[a\tau(\lambda - a)^{-1}]^q d\lambda. \tag{3.2}$$

Once more, all operations in (3.2) are being done in the symbolic algebra sense. Despite of the presence of $\log\lambda$ in the integrands in (3.2) the result is a classical symbol because in each homogeneous term the residue at the only pole $\lambda = a_k(x,\xi)$ vanishes. The order of the q-th term in (3.2) equals $-qs$.Finally, for the symbop $q^{(p)}(t,x,\xi)$ of the operator $Q_p(t)$ one gets

$$q^{(p)}(t,x,\xi) = \sum_{j=1}^{\infty} t^j \tau^j \frac{1}{2\pi i}\int_\Gamma \log\lambda(\lambda - a)^{-1}[a\tau(\lambda - a)^{-1}]^j d\lambda. \tag{3.3}$$

The order of the q-th term on the right in (3.3) equals $-(q+p)s$; its residue vanishes if $q+p > d/s$ because the residue of a symbol of order smaller than $-d$ equals 0. Therefore,

$$\text{res}Q_p(t) = \sum_{j=1}^{[(d/s)-p]} t^j \text{res}\left(\tau^p \frac{1}{2\pi i}\int_\Gamma \log\lambda(\lambda-a)^{-1}[a\tau(\lambda-a)^{-1}]^j d\lambda.\right) \tag{3.4}$$

Here, by $[\alpha]$ I denote the floor of the number α.

Let us discuss two special cases. The first one is when $s > d/2$. An example would be the difference $\log\det(\Delta + u(x)) - \log\det\Delta$ on manifolds of dimension 2 or 3. Though the operator Δ is not invertible, its (modified) determinant equals the determinant of $A = \Delta + P$ where P is the orthogonal projection onto the null space of Δ. One can set $T = (u-P)(\Delta+P)^{-1}$. Notice that P is a pseudo-differential operator of order $-\infty$, and, for the purpose of computing symbols, can be disregarded. In this case $m = 2$, there is just one operator, Q_1 on the right in (2.9), its order equals $-2s < -d$, so its residue vanishes. One gets

$$\det(A(I+T)) - \det A - \det{}_2(I+T) = (\text{f.p.})_{z=0}\text{trace}(TA^z).$$

The second case is when $s = d/2$. Then $m = 3$. The operator Q_2 is of order $-3s < -d$, so its residue vanishes. However, the operator Q_1 is of order $-2s = -d$, and

$$\text{res}Q_1(t) = \int_{S^*(M)} q^{(1)}_{-d}(t,x,\xi)\mu.$$

Only the $q = 1$ term in (3.4) contributes to the principle symbol of Q_1. By evaluating the corresponding integral one gets

$$q^{(1)}_{-d}(t,x,\xi) = t\tau_{-d/2}(x,\xi)^2.$$

Then formuls (2.9) takes the form

$$\begin{aligned}\det(A(I+T)) - \det A - \det{}_3(I+T) &= (\text{f.p.})_{z=0}\text{trace}(TA^z)\\ &- \frac{1}{2}(\text{f.p.})_{z=0}\text{trace}(T^2A^z) + \frac{1}{2k}\int_{S^*M}\tau_{-d/2}(x,\xi)^2\mu.\end{aligned}$$

4. Acknowledgment

I am grateful to the anonymous referees for the careful reading of the manuscript and useful comments.

References

[F] L. Friedlander, *Determinants of elliptic operators*, ProQuest LLC, Ann Arbor, MI, 1989. Thesis (Ph.D.)–Massachusetts Institute of Technology. MR2941178

[FG] L. Friedlander and V. Guillemin, *Determinants of zeroth order operators*, J. Differential Geom. **78** (2008), no. 1, 1–12. MR2406263

[G] V. Guillemin, *A new proof of Weyl's formula on the asymptotic distribution of eigenvalues*, Adv. in Math. **55** (1985), no. 2, 131–160, DOI 10.1016/0001-8708(85)90018-0. MR772612

[GK] I. Gohberg, M. G. Krein, *Introduction into the theory of non-selfadjoint operators*, Springer, Berlin-Heidelberg-New York, 1971

[KV] M. Kontsevich, S. Vishik, *Determinants of elliptic pseudo=differential operators*, preprint, Max-Plank Institut für Mathematik, Bonn, 1994

[L] M. Lesch, *On the noncommutative residue for pseudodifferential operators with log-polyhomogeneous symbols*, Ann. Global Anal. Geom. **17** (1999), no. 2, 151–187, DOI 10.1023/A:1006504318696. MR1675408

[S] B. Simon, *Trace ideals and their applications*, 2nd ed., Mathematical Surveys and Monographs, vol. 120, American Mathematical Society, Providence, RI, 2005. MR2154153

[W] M. Wodzicki, *Noncommutative residue. I. Fundamentals*, K-theory, arithmetic and geometry (Moscow, 1984), Lecture Notes in Math., vol. 1289, Springer, Berlin, 1987, pp. 320–399, DOI 10.1007/BFb0078372. MR923140

Department of Mathematics, University of Arizona, Tucson, Arizona 85721
Email address: friedlan@math.arizona.edu

Contemporary Mathematics
Volume **734**, 2019
https://doi.org/10.1090/conm/734/14768

Spectral properties of the Neumann-Laplace operator in quasiconformal regular domains

V. Gol'dshtein, V. Pchelintsev, and A. Ukhlov

In memory of Selim Grigorievich Krein

ABSTRACT. In this paper we study spectral properties of the Neumann-Laplace operator in planar quasiconformal regular domains $\Omega \subset \mathbb{R}^2$. This study is based on the quasiconformal theory of composition operators on Sobolev spaces. We obtain estimates of constants in Poincaré-Sobolev inequalities and as a consequence lower estimates of the first non-trivial eigenvalue of the Neumann-Laplace operator in planar quasiconformal regular domains.

1. Introduction

We study the spectral problem for the Laplace operator with the Neumann boundary condition in planar quasiconformal regular domains $\Omega \subset \mathbb{R}^2$. The weak statement of this spectral problem is as follows: a function u solves this problem iff $u \in W_2^1(\Omega)$ and

$$\int\limits_{\Omega} \nabla u(x) \cdot \nabla v(x)\, dx = \mu \int\limits_{\Omega} u(x) v(x)\, dx$$

for all $v \in W_2^1(\Omega)$.

In this text, we prove discreteness of the spectrum of the Neumann–Laplace operator in quasiconformal β-regular domains and obtain the lower estimates of the first non-trivial eigenvalue in terms of the quasiconformal geometry of domains:

THEOREM A. *Let $\Omega \subset \mathbb{R}^2$ be a K-quasiconformal β-regular domain. Then the spectrum of the Neumann–Laplace operator in Ω is discrete, and can be written in the form of a non-decreasing sequence:*

$$0 = \mu_0(\Omega) < \mu_1(\Omega) \leq \mu_2(\Omega) \leq \ldots \leq \mu_n(\Omega) \leq \ldots,$$

and

$$\frac{1}{\mu_1(\Omega)} \leq \frac{4K}{\sqrt[\beta]{\pi}} \left(\frac{2\beta - 1}{\beta - 1}\right)^{\frac{2\beta-1}{\beta}} \big\|J_\varphi \mid L_\beta(\mathbb{D})\big\|,$$

where $\varphi : \mathbb{D} \to \Omega$ is the K-quasiconformal mapping.

2010 *Mathematics Subject Classification.* Primary 35P15, 46E35, 30C65.
Key words and phrases. Elliptic equations, Sobolev spaces, quasiconformal mappings.
The second author was supported by RFBR Grant No. 18-31-00011.

DEFINITION 1.1. Let $\Omega \subset \mathbb{R}^2$ be a simply connected planar domain. Then Ω is called a K-quasiconformal β-regular domain if there exists a K-quasiconformal mapping $\varphi : \mathbb{D} \to \Omega$ such that

$$\int_{\mathbb{D}} |J(x,\varphi)|^{\beta}\, dx < \infty \quad \text{for some} \quad \beta > 1,$$

where $J(x,\varphi)$ is the Jacobian of the mapping φ (a determinant of a formal Jacobi matrix) at a point $x \in \mathbb{D}$. The domain $\Omega \subset \mathbb{R}^2$ is called a K-quasiconformal regular domain if it is a K-quasiconformal β-regular domain for some $\beta > 1$.

The notion of quasiconformal regular domains is a generalization of the notion of conformal regular domains introduced in [**BGU1**], where it was used for studying conformal spectral stability of the Laplace operator (see also [**BGU2**]).

Recall that a homeomorphism $\varphi : \Omega \to \widetilde{\Omega}$ between planar domains $\Omega, \widetilde{\Omega} \subset \mathbb{R}^2$ is called a K-quasiconformal mapping if it preserves orientation, belongs to the Sobolev class $W^1_{2,\mathrm{loc}}(\Omega)$ and its directional derivatives D_v satisfy the distortion inequality

$$\max_{v:|v|=1} |D_v\varphi| \leq K \min_{v:|v|=1} |D_v\varphi| \quad \text{a.e. in } \Omega\,.$$

Notice that this class of quasiconformal regular domains includes the class of Gehring domains [**AK**] and can be described in terms of quasihyperbolic geometry [**KOT**].

REMARK 1.2. The notion of quasiconformal β-regular domains is more general then the notion of conformal α-regular domains. Consider, for example, the unit square $\mathbb{Q} \subset \mathbb{R}^2$. Then $\mathbb{Q}$ is a conformal α-regular domain for $2 < \alpha \leq 4$ ($\alpha = 2\beta$) [**GU5**] and it is a quasiconformal β-regular domain for all $1 < \beta \leq \infty$ because the unit square $\mathbb{Q}$ is quasiisometrically equivalent to the unit disc $\mathbb{D}$.

REMARK 1.3. Because $\varphi : \mathbb{D} \to \Omega$ is a quasiconformal mapping, then integrability of the derivative is equivalent to integrability of the Jacobian:

$$\int_{\mathbb{D}} |J(x,\varphi)|^{\beta}\, dx \leq \int_{\mathbb{D}} |D\varphi(x)|^{2\beta}\, dx \leq K^{\beta} \int_{\mathbb{D}} |J(x,\varphi)|^{\beta}\, dx.$$

In 1961 G. Polya [**P**] obtained upper estimates for eigenvalues of Neumann-Laplace operator in so-called plane-covering domains. Namely, for the first eigenvalue:

$$\mu_1(\Omega) \leq 4\pi|\Omega|^{-1}.$$

The lower estimates for $\mu_1(\Omega)$ were known before only for convex domains. In the classical work [**PW**] it was proved that if Ω is convex with diameter $d(\Omega)$ (see also [**ENT**, **FNT**, **V**]), then

$$\mu_1(\Omega) \geq \frac{\pi^2}{d(\Omega)^2}.$$

In [**GU5**] it was proved that if $\Omega \subset \mathbb{R}^2$ is a conformal regular domain, then the spectrum of Neumann-Laplace operator in Ω is discrete and the estimates of the first non-trivial Neumann eigenvalue depend on the hyperbolic geometry of the domain. Because quasiconformal mappings represent a more flexible class of mapping in the present paper we suggest an approach to the Poincaré-Sobolev inequalities which is based on the quasiconformal mappings theory in connection with the geometric theory of in Sobolev spaces.

The proof of Theorem A is based on the Poincaré–Sobolev inequalities in quasiconformal regular domains:

THEOREM B. *Let $\Omega \subset \mathbb{R}^2$ be a K-quasiconformal β-regular domain. Then:*

(1) *the embedding operator*

$$i_\Omega : W_2^1(\Omega) \hookrightarrow L_s(\Omega)$$

is compact for any $s \geq 1$;

(2) *for any function $f \in W_2^1(\Omega)$ and for any $s \geq 1$, the Poincaré–Sobolev inequality*

$$\inf_{c \in \mathbb{R}} \|f - c \mid L_s(\Omega)\| \leq B_{s,2}(\Omega) \|\nabla f \mid L_2(\Omega)\|$$

holds with the constant

$$B_{s,2}(\Omega) \leq K^{\frac{1}{2}} B_{\frac{\beta s}{\beta - 1},2}(\mathbb{D}) \|J_\varphi \mid L_\beta(\mathbb{D})\|^{\frac{1}{s}}.$$

Here $B_{r,2}(\mathbb{D}) \leq \left(2^{-1}\pi\right)^{\frac{2-r}{2r}} (r+2)^{\frac{r+2}{2r}}$, $r = \beta s/(\beta - 1)$ is the exact constant in the Poincaré-Sobolev inequality for the unit disc $\mathbb{D}$

$$\inf_{c \in \mathbb{R}} \|g - c \mid L_r(\mathbb{D})\| \leq B_{r,2}(\mathbb{D}) \|\nabla f \mid L_2(\mathbb{D})\|.$$

The description of compactness of Sobolev embedding operators in terms of capacity integrals was obtained in [**M**]. In the present work we give sufficient conditions of compactness of Sobolev embedding operators in terms of the quasiconformal geometry of domains.

The suggested method is based on the geometrical theory of composition operators in Sobolev spaces [**U**, **VG**, **VU2**] and its applications to the Sobolev type embedding theorems [**GG**, **GU1**].

The following diagram illustrates this idea:

$$\begin{array}{ccc} W_2^1(\Omega) & \xrightarrow{\varphi^*} & W_2^1(\mathbb{D}) \\ \downarrow & & \downarrow \\ L_s(\Omega) & \xleftarrow{(\varphi^{-1})^*} & L_r(\mathbb{D}). \end{array}$$

Here the operator φ^* defined by the composition rule $\varphi^*(f) = f \circ \varphi$ is a bounded composition operator on Sobolev spaces induced by a homeomorphism φ of $\mathbb{D}$ and Ω and the operator $(\varphi^{-1})^*$ defined by the composition rule $(\varphi^{-1})^*(f) = f \circ \varphi^{-1}$ is a bounded composition operator on Lebesgue spaces. This method allows to transfer Poincaré-Sobolev inequalities from regular domains (for example, from the unit disc $\mathbb{D}$) to Ω.

In recent works we study composition operators on Sobolev spaces defined on planar domains in connection with the conformal mappings theory [**GU2**]. This connection leads to weighted Sobolev embeddings [**GU3**, **GU4**] with the universal conformal weights. Another application of conformal composition operators was given in [**BGU1**] where the spectral stability problem for conformal regular domains was considered.

2. Composition operators and quasiconformal mappings

In this section we recall basic facts about composition operators on Lebesgue and Sobolev spaces and also the quasiconformal mappings theory.

The following theorem about composition operators on Lebesgue spaces is well known (see, for example [**VU2**]):

THEOREM 2.1. *Let $\varphi : \Omega \to \widetilde{\Omega}$ be a weakly differentiable homeomorphism between two domains Ω and $\widetilde{\Omega}$. Then the composition operator*

$$\varphi^* : L_r(\widetilde{\Omega}) \to L_s(\Omega), \ \ 1 \leq s \leq r < \infty,$$

is bounded, if and only if φ^{-1} possesses the Luzin N-property and

$$\left(\int\limits_{\widetilde{\Omega}} |J(y, \varphi^{-1})|^{\frac{r}{r-s}}\, dy \right)^{\frac{r-s}{rs}} = K < \infty, \ \ 1 \leq s < r < \infty,$$

$$\operatorname*{ess\,sup}_{y \in \widetilde{\Omega}} |J(y, \varphi^{-1})|^{\frac{1}{s}} = K < \infty, \ \ 1 \leq s = r < \infty.$$

The norm of the composition operator $\|\varphi^*\| = K$.

If Ω is an open subset of $\mathbb{R}^n$, the Sobolev space $W^1_p(\Omega)$, $1 \leq p < \infty$, is defined as a Banach space of locally integrable weakly differentiable functions $f : \Omega \to \mathbb{R}$ equipped with the following norm:

$$\|f \mid W^1_p(\Omega)\| = \|f \mid L_p(\Omega)\| + \|\nabla f \mid L_p(\Omega)\|.$$

The homogeneous seminormed Sobolev space $L^1_p(\Omega)$, $1 \leq p \leq \infty$, is defined as a space of locally integrable weakly differentiable functions $f : \Omega \to \mathbb{R}$ equipped with the following seminorm:

$$\|f \mid L^1_p(\Omega)\| = \|\nabla f \mid L_p(\Omega)\|.$$

By the standard definition, functions of $L^1_p(\Omega)$ are defined only up to a set of measure zero, but they can be redefined quasieverywhere i.e. up to a set of p-capacity zero. Indeed, every function $f \in L^1_p(\Omega)$ has a unique quasicontinuous representation $\tilde{f} \in L^1_p(\Omega)$. A function $\tilde{f}$ is termed quasicontinuous if for any $\varepsilon > 0$ there is an open set U_ε such that the p-capacity of U_ε is less than ε and on the set $\Omega \setminus U_\varepsilon$ the function $\tilde{f}$ is continuous (see, for example [**HKM, M**]).

Let Ω and $\widetilde{\Omega}$ be domains in $\mathbb{R}^n$. We say that a homeomorphism $\varphi : \Omega \to \widetilde{\Omega}$ induces by the composition rule $\varphi^*(f) = f \circ \varphi$ a bounded composition operator

$$\varphi^* : L^1_p(\widetilde{\Omega}) \to L^1_q(\Omega), \ \ 1 \leq q \leq p \leq \infty,$$

if the composition $\varphi^*(f) \in L^1_q(\Omega)$ is defined quasi-everywhere in Ω and there exists a constant $K_{p,q}(\Omega) < \infty$ such that

$$\|\varphi^*(f) \mid L^1_q(\Omega)\| \leq K_{p,q}(\Omega) \|f \mid L^1_p(\widetilde{\Omega})\|$$

for any function $f \in L^1_p(\widetilde{\Omega})$ [**VU3**].

Recall that a mapping $\varphi : \Omega \to \mathbb{R}^n$ belongs to $L^1_{p,\mathrm{loc}}(\Omega)$, $1 \leq p \leq \infty$, if its coordinate functions φ_j belong to $L^1_{p,\mathrm{loc}}(\Omega)$, $j = 1, \dots, n$. In this case the formal Jacobi matrix $D\varphi(x) = \left(\frac{\partial \varphi_i}{\partial x_j}(x)\right)$, $i, j = 1, \dots, n$, and its determinant (Jacobian) $J(x, \varphi) = \det D\varphi(x)$ are well defined at almost all points $x \in \Omega$. The norm $|D\varphi(x)|$

of the matrix $D\varphi(x)$ is the norm of the corresponding linear operator $D\varphi(x):\mathbb{R}^n\to\mathbb{R}^n$ defined by the matrix $D\varphi(x)$.

Let $\varphi:\Omega\to\widetilde{\Omega}$ be weakly differentiable in Ω. The mapping φ is the mapping of finite distortion if $|D\varphi(z)|=0$ for almost all $x\in Z=\{z\in\Omega : J(x,\varphi)=0\}$.

A mapping $\varphi:\Omega\to\mathbb{R}^n$ possesses the Luzin N-property if the image of any set of measure zero has measure zero. Note that any Lipschitz mapping possesses the Luzin N-property.

The following theorem gives the analytic description of composition operators on Sobolev spaces:

THEOREM 2.2. [**U**, **VU2**] *A homeomorphism $\varphi:\Omega\to\widetilde{\Omega}$ between two domains Ω and $\widetilde{\Omega}$ induces a bounded composition operator*

$$\varphi^*: L^1_p(\widetilde{\Omega})\to L^1_q(\Omega),\ \ 1\leq q<p<\infty,$$

if and only if $\varphi\in W^1_{1,\mathrm{loc}}(\Omega)$, has finite distortion, and

$$K_{p,q}(\Omega)=\left(\int\limits_{\Omega}\left(\frac{|D\varphi(x)|^p}{|J(x,\varphi)|}\right)^{\frac{q}{p-q}}\,dx\right)^{\frac{p-q}{pq}}<\infty.$$

In the Euclidean space $\mathbb{R}^n$, $n\geq 3$, a homeomorphism $\varphi:\Omega\to\widetilde{\Omega}$ is called a K-quasiconformal mapping if $\varphi\in W^1_{n,\mathrm{loc}}(\Omega)$ and there exists a constant $1\leq K<\infty$ such that

$$|D\varphi(x)|^n\leq K|J(x,\varphi)| \text{ for almost all } x\in\Omega.$$

Quasiconformal mappings have a finite distortion, i.e. $D\varphi(x)=0$ for almost all points x that belongs to set $Z=\{x\in\Omega : J(x,\varphi)=0\}$ because any quasiconformal mapping possesses the Luzin N-property and an inverse mapping is also quasiconformal.

If $\varphi:\Omega\to\widetilde{\Omega}$ is a K-quasiconformal mapping then φ is differentiable almost everywhere in Ω and

$$|J(x,\varphi)|=J_{\varphi}(x):=\lim_{r\to 0}\frac{|\varphi(B(x,r))|}{|B(x,r)|} \text{ for almost all } x\in\Omega.$$

For any planar K-quasiconformal homeomorphism $\varphi:\Omega\to\widetilde{\Omega}$, the following sharp results is known: $J(x,\varphi)\in L_{p,\mathrm{loc}}(\widetilde{\Omega})$ for any $p<K/(K-1)$ [**As**, **G1**].

If $K\equiv 1$ then 1-quasiconformal homeomorphisms are conformal mappings and in the space $\mathbb{R}^n$, $n\geq 3$, are exhausted by Möbius transformations.

DEFINITION 2.3. We call a bounded domain $\Omega\subset\mathbb{R}^2$ an (r,q)-Poincaré domain, $1\leq q,r\leq\infty$, if the Poincaré–Sobolev inequality

$$\inf_{c\in\mathbb{R}}||g-c\mid L_r(\Omega)||\leq B_{r,q}(\Omega)||\nabla g\mid L_q(\Omega)||$$

holds for any $g\in L^1_q(\Omega)$ with the Poincaré constant $B_{r,q}(\Omega)<\infty$. The unit disc $\mathbb{D}\subset\mathbb{R}^2$ is an example of the $(r,2)$-embedding domain for all $r\geq 1$.

The following theorem gives a characterization of composition operators in the classical Sobolev spaces W^1_p (see, for example [**GG, GU1, GU5**]):

THEOREM 2.4. *Let $\Omega\subset\mathbb{R}^n$ be an (r,q)-Poincaré domain for some $1\leq q\leq r\leq\infty$ and a domain $\widetilde{\Omega}$ has finite measure. Suppose that a homeomorphism $\varphi:\Omega\to\widetilde{\Omega}$ induces a bounded composition operator*

$$\varphi^*: L^1_p(\widetilde{\Omega})\to L^1_q(\Omega),\;\; 1\leq q\leq p<\infty,$$

and the inverse homeomorphism $\varphi^{-1}:\widetilde{\Omega}\to\Omega$ induces a bounded composition operator

$$(\varphi^{-1})^*: L_r(\Omega)\to L_s(\widetilde{\Omega}),\;\; 1\leq s\leq r<\infty,$$

for some $p\leq s\leq r$.

Then $\varphi:\Omega\to\widetilde{\Omega}$ induces a bounded composition operator

$$\varphi^*: W^1_p(\widetilde{\Omega})\to W^1_q(\Omega),\;\; 1\leq q\leq p<\infty.$$

This theorem allows us to obtain compactness of the Sobolev embedding operator in quasiconformal regular domains using weighted Poincaré-Sobolev inequalities.

3. Poincaré-Sobolev inequalities

Weighted Poincaré-Sobolev inequalities. Let $\Omega\subset\mathbb{R}^2$ be a planar domain and let $v:\Omega\to\mathbb{R}$ be a real valued function, $v>0$ a.e. in Ω. We consider the weighted Lebesgue space $L_p(\Omega,v)$, $1\leq p<\infty$, of measurable functions $f:\Omega\to\mathbb{R}$ with the finite norm

$$\|f\mid L_p(\Omega,v)\|:=\left(\int\limits_\Omega |f(x)|^p v(x)dx\right)^{\frac{1}{p}}<\infty.$$

It is a Banach space for the norm $\|f\mid L_p(\Omega,v)\|$.

The following lemma gives a connection between composition operators on Sobolev spaces and the quasiconformal mappings theory [**VG**].

LEMMA 3.1. *A homeomorphism $\varphi:\Omega\to\widetilde{\Omega}$ is a K-quasiconformal mapping if and only if φ generates, by the composition rule $\varphi^*(f)=f\circ\varphi$, an isomorphism of Sobolev spaces $L^1_n(\Omega)$ and $L^1_n(\widetilde{\Omega})$, i.e. the following inequality holds:*

$$\|\varphi^*(f)\mid L^1_n(\Omega)\|\leq K^{\frac{1}{n}}\|f\mid L^1_n(\widetilde{\Omega})\|$$

for any $f\in L^1_n(\widetilde{\Omega})$.

On the basis of this lemma we prove the universal weighted Poincaré-Sobolev inequality which holds for any simply connected planar domain with non-empty boundary.

THEOREM 3.2. *Suppose that $\Omega\subset\mathbb{R}^2$ is a simply connected domain with non-empty boundary and $h(y)=|J(y,\varphi^{-1})|$ is the quasiconformal weight defined by a K-quasiconformal mapping $\varphi:\mathbb{D}\to\Omega$. Then for every function $f\in W^1_2(\Omega)$, the inequality*

$$\inf_{c\in\mathbb{R}}\left(\int\limits_\Omega |f(y)-c|^r h(y)dy\right)^{\frac{1}{r}}\leq B_{r,2}(\Omega,h)\left(\int\limits_\Omega |\nabla f(y)|^2dy\right)^{\frac{1}{2}}$$

holds for any $r \geq 1$ with the constant

$$B_{r,2}(\Omega, h) \leq K^{\frac{1}{2}} \cdot B_{r,2}(\mathbb{D}) \leq \left(2^{-1}\pi\right)^{\frac{2-r}{2r}} (r+2)^{\frac{r+2}{2r}} K^{\frac{1}{2}}.$$

Here $B_{r,2}(\mathbb{D})$ is the best constant in the (non-weight) Poincaré-Sobolev inequality in the unit disc $\mathbb{D} \subset \mathbb{R}^2$ with the upper estimate (see, for example, [**GT**, **GU5**]):

$$B_{r,2}(\mathbb{D}) \leq \left(2^{-1}\pi\right)^{\frac{2-r}{2r}} (r+2)^{\frac{r+2}{2r}}.$$

PROOF. By [**A**] there exists a K-quasiconformal homeomorphism $\varphi : \mathbb{D} \to \Omega$. Then by Lemma 3.1 the inequality

$$||\nabla(f \circ \varphi) \mid L_2(\mathbb{D})|| \leq K^{\frac{1}{2}} ||\nabla f \mid L_2(\Omega)|| \tag{3.1}$$

holds for every function $f \in L^1_2(\Omega)$.

Let $f \in L^1_2(\Omega) \cap C^1(\Omega)$. Then the function $g = f \circ \varphi$ is defined almost everywhere in $\mathbb{D}$ and belongs to the Sobolev space $L^1_2(\mathbb{D})$ [**VGR**]. Hence, by the Sobolev embedding theorem $g = f \circ \varphi \in W^1_2(\mathbb{D})$ [**M**] and the classical Poincaré-Sobolev inequality,

$$\inf_{c \in \mathbb{R}} ||f \circ \varphi - c \mid L_r(\mathbb{D})|| \leq B_{r,2}(\mathbb{D}) ||\nabla(f \circ \varphi) \mid L_2(\mathbb{D})|| \tag{3.2}$$

holds for any $r \geq 1$.

Denote by $h(y) := |J(y, \varphi^{-1})|$ the quasiconformal weight in Ω. Using the change of variable formula for the quasiconformal mappings [**VGR**], the classical Poincaré-Sobolev inequality for the unit disc

$$\inf_{c \in \mathbb{R}} \left(\int_{\mathbb{D}} |g(x) - c|^r dx \right)^{\frac{1}{r}} \leq B_{r,2}(\mathbb{D}) \left(\int_{\mathbb{D}} |\nabla g(x)|^2 dx \right)^{\frac{1}{2}},$$

and inequality (3.1), we get

$$\begin{aligned}
\inf_{c \in \mathbb{R}} \left(\int_{\Omega} |f(y) - c|^r h(y) dy \right)^{\frac{1}{r}} &= \inf_{c \in \mathbb{R}} \left(\int_{\Omega} |f(y) - c|^r |J(y, \varphi^{-1})| dy \right)^{\frac{1}{r}} \\
&= \inf_{c \in \mathbb{R}} \left(\int_{\mathbb{D}} |g(x) - c|^r dx \right)^{\frac{1}{r}} \leq B_{r,2}(\mathbb{D}) \left(\int_{\mathbb{D}} |\nabla g(x)|^2 dx \right)^{\frac{1}{2}} \\
&\leq K^{\frac{1}{2}} B_{r,2}(\mathbb{D}) \left(\int_{\Omega} |\nabla f(y)|^2 dy \right)^{\frac{1}{2}}.
\end{aligned}$$

Approximating an arbitrary function $f \in W^1_2(\Omega)$ by smooth functions we have

$$\inf_{c \in \mathbb{R}} \left(\int_{\Omega} |f(y) - c|^r h(y) dy \right)^{\frac{1}{r}} \leq B_{r,2}(\Omega, h) \left(\int_{\Omega} |\nabla f(y)|^2 dy \right)^{\frac{1}{2}},$$

with the constant

$$B_{r,2}(\Omega, h) \leq K^{\frac{1}{2}} \cdot B_{r,2}(\mathbb{D}) \leq \left(2^{-1}\pi\right)^{\frac{2-r}{2r}} (r+2)^{\frac{r+2}{2r}} K^{\frac{1}{2}}.$$

□

The property of the quasiconformal β-regularity implies the integrability of a Jacobian of quasiconformal mappings and therefore for any quasiconformal β-regular domain we have the embedding of weighted Lebesgue spaces $L_r(\Omega, h)$ into non-weighted Lebesgue spaces $L_s(\Omega)$ for $s = \frac{\beta-1}{\beta}r$.

LEMMA 3.3. *Let Ω be a K-quasiconformal β-regular domain. Then for any function $f \in L_r(\Omega, h)$, $\beta/(\beta-1) \leq r < \infty$, the inequality*

$$||f \mid L_s(\Omega)|| \leq \left(\int\limits_{\mathbb{D}} \left|J(x,\varphi)\right|^{\beta} \, dx \right)^{\frac{1}{\beta}\cdot\frac{1}{s}} ||f \mid L_r(\Omega, h)||$$

holds for $s = \frac{\beta-1}{\beta}r$.

PROOF. By the assumptions of the lemma there exists a K-quasiconformal mapping $\varphi : \mathbb{D} \to \Omega$ such that

$$\int\limits_{\mathbb{D}} \left|J(x,\varphi)\right|^{\beta} \, dx < +\infty.$$

Let $s = \frac{\beta-1}{\beta}r$. Then using the change of variable formula for quasiconformal mappings [**VGR**], Hölder's inequality with exponents $(r/s, r/(r-s))$ and the equality $|J(y,\varphi^{-1})| = h(y)$, we obtain

$$\begin{aligned}
&||f \mid L_s(\Omega)|| \\
&= \left(\int\limits_{\Omega} |f(y)|^s dy \right)^{\frac{1}{s}} = \left(\int\limits_{\Omega} |f(y)|^s \left|J(y,\varphi^{-1})\right|^{\frac{s}{r}} \left|J(y,\varphi^{-1})\right|^{-\frac{s}{r}} \, dy \right)^{\frac{1}{s}} \\
&\leq \left(\int\limits_{\Omega} |f(y)|^r |J(y,\varphi^{-1})| dy \right)^{\frac{1}{r}} \left(\int\limits_{\Omega} \left|J(y,\varphi^{-1})\right|^{-\frac{s}{r-s}} \, dy \right)^{\frac{r-s}{rs}} \\
&\leq \left(\int\limits_{\Omega} |f(y)|^r h(y) \, dy \right)^{\frac{1}{r}} \left(\int\limits_{\mathbb{D}} \left|J(x,\varphi)\right|^{\frac{r}{r-s}} \, dx \right)^{\frac{r-s}{rs}} \\
&= \left(\int\limits_{\Omega} |f(y)|^r h(y) \, dy \right)^{\frac{1}{r}} \left(\int\limits_{\mathbb{D}} \left|J(x,\varphi)\right|^{\beta} \, dx \right)^{\frac{1}{\beta}\cdot\frac{1}{s}}.
\end{aligned}$$

□

The following theorem is the main technical tool of this paper:

THEOREM B. *Let $\Omega \subset \mathbb{R}^2$ be a K-quasiconformal β-regular domain. Then:*

(1) *the embedding operator*

$$i_{\Omega} : W_2^1(\Omega) \hookrightarrow L_s(\Omega)$$

is compact for any $s \geq 1$;

(2) *for any function $f \in W_2^1(\Omega)$ and for any $s \geq 1$, the Poincaré–Sobolev inequality*

$$\inf_{c \in \mathbb{R}} \|f - c \mid L_s(\Omega)\| \leq B_{s,2}(\Omega) \|\nabla f \mid L_2(\Omega)\|$$

holds with the constant

$$B_{s,2}(\Omega) \le K^{\frac{1}{2}} B_{\frac{\beta s}{\beta-1},2}(\mathbb{D}) \|J_\varphi \mid L_\beta(\mathbb{D})\|^{\frac{1}{s}}.$$

PROOF. Let $s \ge 1$. Since Ω is a K-quasiconformal β-regular domain then there exists a K-quasiconformal mapping $\varphi : \mathbb{D} \to \Omega$ such that

$$\int_{\mathbb{D}} |J(x,\varphi)|^\beta \, dx < \infty \quad \text{for some} \quad \beta > 1.$$

By Theorem 2.1 the composition operator $(\varphi^{-1})^* : L_r(\mathbb{D}) \to L_s(\Omega)$ is bounded if

$$\left(\int_{\mathbb{D}} |J(x,\varphi)|^{\frac{r}{r-s}} \, dx \right)^{\frac{r-s}{rs}} < \infty.$$

Because Ω is a K-quasiconformal β-regular domain this integrability condition holds for $r/(r-s) = \beta$ i.e. for $r = \beta s/(\beta - 1)$.

Since the mapping $\varphi : \mathbb{D} \to \Omega$ induces a bounded composition operator $\varphi^* : L^1_2(\Omega) \to L^1_2(\mathbb{D})$, then by Theorem 2.4 the composition operator $\varphi^* : W^1_2(\Omega) \to W^1_2(\mathbb{D})$, is bounded also. In the unit disc $\mathbb{D}$ the embedding operator $i_{\mathbb{D}} : W^1_2(\mathbb{D}) \hookrightarrow L_r(\mathbb{D})$, is compact (see, for example [**M**]) for any $r \ge 1$.

Therefore the embedding operator

$$i_\Omega : W^1_2(\Omega) \to L_s(\Omega)$$

is compact as a composition of bounded composition operators φ^*, $(\varphi^{-1})^*$ and the compact embedding operator $i_{\mathbb{D}}$.

Let $f \in W^1_2(\Omega)$. Then by Theorem 3.2 and Lemma 3.3 we obtain

$$\begin{aligned}
&\inf_{c\in\mathbb{R}} \left(\int_\Omega |f(y) - c|^s dy \right)^{\frac{1}{s}} \\
&\qquad \le \left(\int_{\mathbb{D}} |J(x,\varphi)|^\beta \, dx \right)^{\frac{1}{\beta}\cdot\frac{1}{s}} \inf_{c\in\mathbb{R}} \left(\int_\Omega |f(y) - c|^r h(y) dy \right)^{\frac{1}{r}} \\
&\qquad \le K^{\frac{1}{2}} B_{r,2}(\mathbb{D}) \left(\int_{\mathbb{D}} |J(x,\varphi)|^\beta \, dx \right)^{\frac{1}{\beta}\cdot\frac{1}{s}} \left(\int_\Omega |\nabla f(y)|^2 dy \right)^{\frac{1}{2}}
\end{aligned}$$

for $s \ge 1$. □

The following theorem gives compactness of the embedding operator in the limit case $\beta = \infty$:

THEOREM 3.4. *Let Ω is a K-quasiconformal ∞-regular domain. Then:*

(1) *The embedding operator*

$$i_\Omega : W^1_2(\Omega) \hookrightarrow L_2(\Omega),$$

is compact.

(2) *For any function $f \in W^1_2(\Omega)$, the Poincaré–Sobolev inequality*

$$\inf_{c\in\mathbb{R}} \|f - c \mid L_2(\Omega)\| \le B_{2,2}(\Omega) \|\nabla f \mid L_2(\Omega)\|$$

holds.

(3) *The following estimate is correct:* $B_{2,2}(\Omega) \leq K^{\frac{1}{2}} B_{2,2}(\mathbb{D}) \| J_\varphi \mid L_\infty(\mathbb{D}) \|^{\frac{1}{2}}$, *where* $B^2_{2,2}(\mathbb{D}) = 1/\mu_1(\mathbb{D})$ *is the exact constant in the Poincaré inequality in the unit disc.*

PROOF. Since Ω is a K-quasiconformal ∞-regular domain then there exists a K-quasiconformal mapping $\varphi : \mathbb{D} \to \Omega$ such that

$$\| J_\varphi \mid L_\infty(\mathbb{D}) \| = \operatorname*{ess\,sup}_{x \in \mathbb{D}} |J(x,\varphi)| < \infty.$$

Hence by Theorem 2.1 the composition operator $(\varphi^{-1})^* : L_2(\mathbb{D}) \to L_2(\Omega)$ is bounded.

Since the mapping $\varphi : \mathbb{D} \to \Omega$ induced a bounded composition operator $\varphi^* : L^1_2(\Omega) \to L^1_2(\mathbb{D})$, then by Theorem 2.4 the composition operator $\varphi^* : W^1_2(\Omega) \to W^1_2(\mathbb{D})$, will be also bounded.

For the unit disc $\mathbb{D}$, the embedding operator $i_{\mathbb{D}} : W^1_2(\mathbb{D}) \hookrightarrow L_2(\mathbb{D})$, is compact (see, for example [**M**]). Therefore the embedding operator $i_\Omega : W^1_2(\Omega) \to L_2(\Omega)$, is compact as a composition of bounded composition operators φ^*, $(\varphi^{-1})^*$ and the compact embedding operator $i_{\mathbb{D}}$. The first part of this theorem is proved.

For every function $f \in W^1_2(\Omega) \cap C^1(\Omega)$ and $g = f \circ \varphi \in W^1_2(\mathbb{D})$, the following inequality are correct:

$$\inf_{c \in \mathbb{R}} \left(\int_\Omega |f(y) - c|^2 dy \right)^{\frac{1}{2}} = \inf_{c \in \mathbb{R}} \left(\int_\Omega |f(y) - c|^2 |J(y,\varphi^{-1})|^{-1} |J(y,\varphi^{-1})| \, dy \right)^{\frac{1}{2}}$$
$$\leq \| J_{\varphi^{-1}} \mid L_\infty(\Omega) \|^{-\frac{1}{2}} \inf_{c \in \mathbb{R}} \left(\int_\Omega |f(y) - c|^2 |J(y,\varphi^{-1})| \, dy \right)^{\frac{1}{2}}.$$

Because quasiconformal mappings possess the Luzin N-property, then $|J(y,\varphi^{-1})|^{-1} = |J(x,\varphi)|$ for almost all $x \in \mathbb{D}$ and for almost all $y = \varphi(x) \in \Omega$. Hence

$$\inf_{c \in \mathbb{R}} \left(\int_\Omega |f(y) - c|^2 dy \right)^{\frac{1}{2}}$$
$$\leq \| J_\varphi \mid L_\infty(\mathbb{D}) \|^{\frac{1}{2}} \inf_{c \in \mathbb{R}} \left(\int_\Omega |f(y) - c|^2 |J(y,\varphi^{-1})| \, dy \right)^{\frac{1}{2}}.$$

By the change of variable formula for quasiconformal mappings [**VGR**]

$$\inf_{c \in \mathbb{R}} \left(\int_\Omega |f(y) - c|^2 |J(y,\varphi^{-1})| \, dy \right)^{\frac{1}{2}} = \inf_{c \in \mathbb{R}} \left(\int_{\mathbb{D}} |g(x) - c|^2 \, dx \right)^{\frac{1}{2}}.$$

Using the Poincaré inequality in the unit disc and the inequality (3.1) finally we obtain

$$\inf_{c\in\mathbb{R}}\left(\int_{\Omega}|f(y)-c|^2dy\right)^{\frac{1}{2}} \leq \|J_\varphi \mid L_\infty(\mathbb{D})\|^{\frac{1}{2}} \inf_{c\in\mathbb{R}}\left(\int_{\mathbb{D}}|g(x)-c|^2\,dx\right)^{\frac{1}{2}}$$
$$\leq \|J_\varphi \mid L_\infty(\mathbb{D})\|^{\frac{1}{2}} B_{2,2}(\mathbb{D})\left(\int_{\mathbb{D}}|\nabla g(x)|^2\,dx\right)^{\frac{1}{2}}$$
$$\leq K^{\frac{1}{2}} B_{2,2}(\mathbb{D})\|J_\varphi \mid L_\infty(\mathbb{D})\|^{\frac{1}{2}}\left(\int_{\Omega}|\nabla f(y)|^2\,dy\right)^{\frac{1}{2}}.$$

□

4. Eigenvalue Problem for Neumann-Laplacian

The eigenvalue problem for the free vibrating membrane is equivalent to the corresponding spectral problem for the Neumann–Laplace operator. By the Min–Max Principle (see, for example, [**D**]) the first non-trivial Neumann eigenvalue $\mu_1(\Omega)$ for the Laplacian can be characterized as

$$\mu_1(\Omega)=\min\left\{\frac{\int_{\Omega}|\nabla u(x)|^2\,dx}{\int_{\Omega}|u(x)|^2\,dx} : u\in W_2^1(\Omega)\setminus\{0\},\ \int_{\Omega}u\,dx=0\right\}.$$

Hence $\mu_1(\Omega)^{-\frac{1}{2}}$ is the best constant $B_{2,2}(\Omega)$ in the following Poincaré inequality

$$\inf_{c\in\mathbb{R}}\|u-c \mid L_2(\Omega)\| \leq B_{2,2}(\Omega)\|\nabla u \mid L_2(\Omega)\|, \quad u\in W_2^1(\Omega).$$

THEOREM A. *Let $\Omega\subset\mathbb{R}^2$ be a K-quasiconformal β-regular domain. Then the spectrum of the Neumann–Laplace operator in Ω is discrete, and can be written in the form of a non-decreasing sequence:*

$$0=\mu_0(\Omega)<\mu_1(\Omega)\leq\mu_2(\Omega)\leq\ldots\leq\mu_n(\Omega)\leq\ldots,$$

and

$$\frac{1}{\mu_1(\Omega)}\leq KB_{\frac{2\beta}{\beta-1},2}(\mathbb{D})\|J_\varphi \mid L_\beta(\mathbb{D})\|\leq\frac{4K}{\sqrt[\beta]{\pi}}\left(\frac{2\beta-1}{\beta-1}\right)^{\frac{2\beta-1}{\beta}}\|J_\varphi \mid L_\beta(\mathbb{D})\|,$$

where $\varphi:\mathbb{D}\to\Omega$ is the K-quasiconformal mapping.

PROOF. By Theorem B in the case $s=2$, the embedding operator

$$i_\Omega : W_2^1(\Omega)\to L_2(\Omega)$$

is compact.

Therefore the spectrum of the Neumann–Laplace operator is discrete and can be written in the form of a non-decreasing sequence.

By the same theorem and the Min-Max principle we have

$$\inf_{c\in\mathbb{R}}\left(\int_{\Omega}|f(y)-c|^2dy\right)\leq B_{2,2}^2(\Omega)\int_{\Omega}|\nabla f(y)|^2dy,$$

where

$$B_{2,2}(\Omega) \leq K^{\frac{1}{2}} B_{r,2}(\mathbb{D}) \left(\int\limits_{\mathbb{D}} \left|J(x,\varphi)\right|^{\beta} \, dx \right)^{\frac{1}{2\beta}}.$$

Hence

$$\frac{1}{\mu_1(\Omega)} \leq K B^2_{r,2}(\mathbb{D}) \left(\int\limits_{\mathbb{D}} \left|J(x,\varphi)\right|^{\beta} \, dy \right)^{\frac{1}{\beta}}.$$

By the upper estimate of the Poincaré constant in the unit disc (see, for example, [**GT**, **GU5**])

$$B_{r,2}(\mathbb{D}) \leq \left(2^{-1}\pi\right)^{\frac{2-r}{2r}} (r+2)^{\frac{r+2}{2r}}.$$

Recall that by Theorem B, $r = 2\beta/(\beta-1)$. In this case

$$B_{\frac{2\beta}{\beta-1},2}(\mathbb{D}) \leq 2\pi^{-\frac{1}{2\beta}} \left(\frac{2\beta-1}{\beta-1}\right)^{\frac{2\beta-1}{2\beta}}.$$

Thus

$$\frac{1}{\mu_1(\Omega)} \leq \frac{4K}{\sqrt[\beta]{\pi}} \left(\frac{2\beta-1}{\beta-1}\right)^{\frac{2\beta-1}{\beta}} \left\|J_{\varphi} \mid L_{\beta}(\mathbb{D})\right\|.$$

□

In the case of K-quasiconformal ∞-regular domains we have:

THEOREM 4.1. *Let $\Omega \subset \mathbb{R}^2$ be a K-quasiconformal β-regular domain for $\beta = \infty$. Then the spectrum of the Neumann–Laplace operator in Ω is discrete, and can be written in the form of a non-decreasing sequence:*

$$0 = \mu_0(\Omega) < \mu_1(\Omega) \leq \mu_2(\Omega) \leq \ldots \leq \mu_n(\Omega) \leq \ldots,$$

and

$$\frac{1}{\mu_1(\Omega)} \leq K B^2_{2,2}(\mathbb{D}) \left\|J_{\varphi} \mid L_{\infty}(\mathbb{D})\right\| = \frac{K}{(j'_{1,1})^2} \left\|J_{\varphi} \mid L_{\infty}(\mathbb{D})\right\|, \tag{4.1}$$

where $j'_{1,1} \approx 1.84118$ denotes the first positive zero of the derivative of the Bessel function J_1, and $\varphi : \mathbb{D} \to \Omega$ is the K-quasiconformal mapping.

As an application of Theorem 4.1, we obtain the lower estimates of the first non-trivial eigenvalue on the Neumann eigenvalue problem for the Laplace operator in non-convex domains with a non-smooth boundaries.

EXAMPLE 4.2. The homeomorphism

$$w = \left(|z|^{k-1}z + 1\right)^2, \quad z = x + iy, \quad k \geq 1,$$

is k-quasiconformal and maps the unit disc $\mathbb{D}$ onto the interior of the cardioid

$$\Omega_c = \left\{(x,y) \in \mathbb{R}^2 : (x^2+y^2-2x)^2 - 4(x^2+y^2) = 0\right\}.$$

We calculate the Jacobian of mapping w by the formula

$$J(z,w) = |w_z|^2 - |w_{\overline{z}}|^2.$$

Here

$$w_z = \frac{1}{2}\left(\frac{\partial w}{\partial x} - i\frac{\partial w}{\partial y}\right) \quad \text{and} \quad w_{\overline{z}} = \frac{1}{2}\left(\frac{\partial w}{\partial x} + i\frac{\partial w}{\partial y}\right).$$

A straightforward calculation yields

$$w_z = (k+1)|z|^{k-1}\left(|z|^{k-1}z+1\right) \quad \text{and} \quad w_{\overline{z}} = (k-1)|z|^{k-3}z^2\left(|z|^{k-1}z+1\right).$$

Hence

$$J(z,w) = 4k|z|^{2k-2}\left(|z|^{2k}+|z|^{k-1}(z+\overline{z})+1\right).$$

Then by Theorem 4.1 we have

$$\frac{1}{\mu_1(\Omega_c)} \leq \frac{K}{(j'_{1,1})^2} \operatorname*{ess\,sup}_{|z|\leq 1} \left|J(z,w)\right| \leq \frac{16k^2}{(j'_{1,1})^2} \approx 4.7k^2.$$

EXAMPLE 4.3. The homeomorphism

$$w = |z|^k z, \quad z = x+iy, \quad k \geq 0,$$

is $(k+1)$-quasiconformal and maps the square

$$Q := \left\{(x,y) \in \mathbb{R}^2 : -\frac{\sqrt{2}}{2} < x < \frac{\sqrt{2}}{2},\ -\frac{\sqrt{2}}{2} < y < \frac{\sqrt{2}}{2}\right\}$$

onto star-shaped domains Ω^*_ε with vertices $(\pm\sqrt{2}/2, \pm\sqrt{2}/2)$, $(\pm\varepsilon, 0)$ and $(0, \pm\varepsilon)$, where $\varepsilon = (\sqrt{2}/2)^{k+1}$.

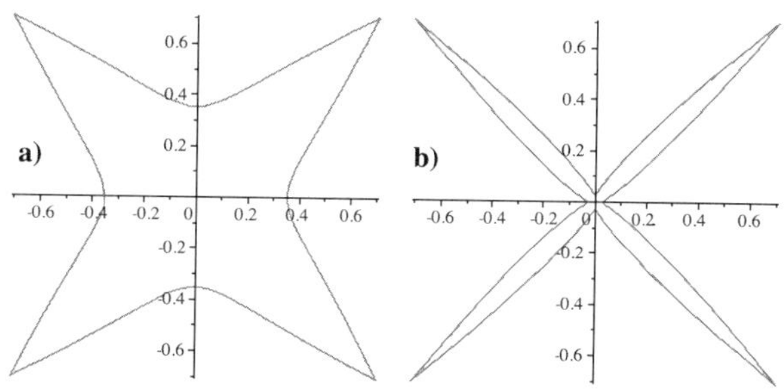

FIGURE 1. Domains Ω^*_ε under $\varepsilon = \frac{1}{2\sqrt{2}}$ and $\varepsilon = \frac{1}{32}$.

We calculate the partial derivatives of mapping w

$$w_z = (\frac{k}{2}+1)|z|^k \quad \text{and} \quad w_{\overline{z}} = \frac{k}{2}|z|^{k-2}z^2.$$

Thus

$$J(z,w) = (k+1)|z|^{2k}.$$

Because the square Q is the quasiconformal ∞-regular domain, by Theorem 4.1 we have

$$\frac{1}{\mu_2(\Omega^*_\varepsilon)} \leq B^2_{2,2}(Q) \cdot K \cdot \operatorname*{ess\,sup}_{|z|\leq 1} |J(z,w)| \leq \frac{2(k+1)^2}{\pi^2}. \tag{4.2}$$

Here $B_{2,2}(Q) = \sqrt{2}/\pi$ (see, for example, [**KN**]) is the exact constant for the Poincaré inequality in the square Q.

In [**VV**] (see, example 4.2) obtained estimates of constants in weighted Poincaré inequality for stars Ω^T_ϵ:

$$C_{p,\lambda}(\Omega^T_\epsilon) \leq 14\bar{C}_p\left[\frac{8}{7}\left(\frac{1}{2}+\frac{p}{2\bar{C}_p}\right)\right]^{\frac{1}{p}}, \tag{4.3}$$

where $p \geq 1$ and

$$\bar{C}_1 = \frac{1}{2}, \quad \bar{C}_2 = \frac{1}{\pi}, \quad \bar{C}_p \leq 2\left(\frac{p}{2}\right)^{\frac{1}{p}}.$$

Note that estimate (4.2) under $0 \leq k < 6$ is better than estimate (4.3) for $p = 2$ ($C_{2,\lambda}(\Omega_\epsilon^T) \leq 4\sqrt{7(1+2\pi)}/\pi \approx 9,09117$).

Acknowledgment

The first author was supported by the United States-Israel Binational Science Foundation (BSF Grant No. 2014055).

References

[A] L. V. Ahlfors, *Lectures on quasiconformal mappings*, Manuscript prepared with the assistance of Clifford J. Earle, Jr. Van Nostrand Mathematical Studies, No. 10, D. Van Nostrand Co., Inc., Toronto, Ont.-New York-London, 1966. MR0200442

[As] K. Astala, *Area distortion of quasiconformal mappings*, Acta Math. **173** (1994), no. 1, 37–60, DOI 10.1007/BF02392568. MR1294669

[AK] K. Astala and P. Koskela, *Quasiconformal mappings and global integrability of the derivative*, J. Anal. Math. **57** (1991), 203–220, DOI 10.1007/BF03041070. MR1191747

[BM] A. F. Beardon and D. Minda, *The hyperbolic metric and geometric function theory*, Quasiconformal mappings and their applications, Narosa, New Delhi, 2007, pp. 9–56. MR2492498

[B] D. Bertilsson, *On Brennan's conjecture in conformal mapping*, Royal Institute of Technology, Stockholm, 1999. Dissertation, Royal Institute of Technology, Stockholm, 1999. MR1728890

[BCT] B. Brandolini, F. Chiacchio, and C. Trombetti, *Optimal lower bounds for eigenvalues of linear and nonlinear Neumann problems*, Proc. Roy. Soc. Edinburgh Sect. A **145** (2015), no. 1, 31–45, DOI 10.1017/S0308210513000371. MR3304574

[BI] B. Bojarski and T. Iwaniec, *Analytical foundations of the theory of quasiconformal mappings in* $\mathbf{R}^n$, Ann. Acad. Sci. Fenn. Ser. A I Math. **8** (1983), no. 2, 257–324, DOI 10.5186/aasfm.1983.0806. MR731786

[BGU1] V. I. Burenkov, V. Gol'dshtein, and A. Ukhlov, *Conformal spectral stability estimates for the Dirichlet Laplacian*, Math. Nachr. **288** (2015), no. 16, 1822–1833, DOI 10.1002/mana.201400253. MR3417872

[BGU2] V. I. Burenkov, V. Gol'dshtein, and A. Ukhlov, *Conformal spectral stability estimates for the Neumann Laplacian*, Math. Nachr. **289** (2016), no. 17-18, 2133–2146, DOI 10.1002/mana.201500439. MR3583260

[D] E. B. Davies, *Spectral theory and differential operators*, Cambridge Studies in Advanced Mathematics, vol. 42, Cambridge University Press, Cambridge, 1995. MR1349825

[ENT] L. Esposito, C. Nitsch, and C. Trombetti, *Best constants in Poincaré inequalities for convex domains*, J. Convex Anal. **20** (2013), no. 1, 253–264. MR3086452

[FNT] V. Ferone, C. Nitsch, and C. Trombetti, *A remark on optimal weighted Poincaré inequalities for convex domains*, Atti Accad. Naz. Lincei Rend. Lincei Mat. Appl. **23** (2012), no. 4, 467–475, DOI 10.4171/RLM/640. MR2999557

[G] F. W. Gehring, *The definitions and exceptional sets for quasiconformal mappings*, Ann. Acad. Sci. Fenn. Ser. A I No. **281** (1960), 28. MR0124488

[GT] D. Gilbarg and N. S. Trudinger, *Elliptic partial differential equations of second order*, Springer-Verlag, Berlin-New York, 1977. Grundlehren der Mathematischen Wissenschaften, Vol. 224. MR0473443

[G1] V. M. Gol'dshtein, *The degree of summability of generalized derivatives of quasiconformal homeomorphisms*, Siberian Math. J. 22 (1981)(6), 821–836.

[GG] V. Gol′dshteĭn and L. Gurov, *Applications of change of variables operators for exact embedding theorems*, Integral Equations Operator Theory **19** (1994), no. 1, 1–24, DOI 10.1007/BF01202289. MR1271237

[GU1] V. Gol′dshtein and A. Ukhlov, *Weighted Sobolev spaces and embedding theorems*, Trans. Amer. Math. Soc. **361** (2009), no. 7, 3829–3850, DOI 10.1090/S0002-9947-09-04615-7. MR2491902

[GU2] V. Gol′dshtein and A. Ukhlov, *Brennan's conjecture for composition operators on Sobolev spaces*, Eurasian Math. J. **3** (2012), no. 4, 35–43. MR3040685

[GU3] V. Gol'dshtein and A. Ukhlov, *Conformal weights and Sobolev embeddings*, J. Math. Sci. (N.Y.) **193** (2013), no. 2, 202–210, DOI 10.1007/s10958-013-1447-x. Problems in mathematical analysis. No. 71. MR3077182

[GU4] V. Gol'dshtein and A. Ukhlov, *Brennan's conjecture and universal Sobolev inequalities*, Bull. Sci. Math. **138** (2014), no. 2, 253–269, DOI 10.1016/j.bulsci.2013.06.005. MR3175022

[GU5] V. Gol'dshtein and A. Ukhlov, *On the first eigenvalues of free vibrating membranes in conformal regular domains*, Arch. Ration. Mech. Anal. **221** (2016), no. 2, 893–915, DOI 10.1007/s00205-016-0988-9. MR3488539

[GU6] V. Gol'dshtein and A. Ukhlov, *Spectral estimates of the p-Laplace Neumann operator in conformal regular domains*, Trans. A. Razmadze Math. Inst. **170** (2016), no. 1, 137–148, DOI 10.1016/j.trmi.2016.03.002. MR3538994

[HS] H. Hedenmalm and S. Shimorin, *Weighted Bergman spaces and the integral means spectrum of conformal mappings*, Duke Math. J. **127** (2005), no. 2, 341–393, DOI 10.1215/S0012-7094-04-12725-3. MR2130416

[HK] J. Heinonen and P. Koskela, *Weighted Sobolev and Poincaré inequalities and quasiregular mappings of polynomial type*, Math. Scand. **77** (1995), no. 2, 251–271, DOI 10.7146/math.scand.a-12564. MR1379269

[HKM] J. Heinonen, T. Kilpeläinen, and O. Martio, *Nonlinear potential theory of degenerate elliptic equations*, Oxford Mathematical Monographs, The Clarendon Press, Oxford University Press, New York, 1993. Oxford Science Publications. MR1207810

[KOT] P. Koskela, J. Onninen, and J. T. Tyson, *Quasihyperbolic boundary conditions and Poincaré domains*, Math. Ann. **323** (2002), no. 4, 811–830, DOI 10.1007/s00208-002-0331-7. MR1924280

[KN] N. Kuznetsov and A. Nazarov, *Sharp constants in the Poincaré, Steklov and related inequalities (a survey)*, Mathematika **61** (2015), no. 2, 328–344, DOI 10.1112/S0025579314000229. MR3343056

[M] V. Maz'ya, *Sobolev spaces with applications to elliptic partial differential equations*, Second, revised and augmented edition, Grundlehren der Mathematischen Wissenschaften [Fundamental Principles of Mathematical Sciences], vol. 342, Springer, Heidelberg, 2011. MR2777530

[P] G. Pólya, *On the eigenvalues of vibrating membranes*, Proc. London Math. Soc. (3) **11** (1961), 419–433, DOI 10.1112/plms/s3-11.1.419. MR0129219

[PW] L. E. Payne and H. F. Weinberger, *An optimal Poincaré inequality for convex domains*, Arch. Rational Mech. Anal. **5** (1960), 286–292 (1960), DOI 10.1007/BF00252910. MR0117419

[U] A. D. Ukhlov, *Mappings that generate embeddings of Sobolev spaces* (Russian, with English and Russian summaries), Sibirsk. Mat. Zh. **34** (1993), no. 1, 185–192, 227, 233, DOI 10.1007/BF00971252; English transl., Siberian Math. J. **34** (1993), no. 1, 165–171. MR1216847

[V] D. Valtorta, *Sharp estimate on the first eigenvalue of the p-Laplacian*, Nonlinear Anal. **75** (2012), no. 13, 4974–4994, DOI 10.1016/j.na.2012.04.012. MR2927560

[VV] A. Veeser and R. Verfürth, *Poincaré constants for finite element stars*, IMA J. Numer. Anal. **32** (2012), no. 1, 30–47, DOI 10.1093/imanum/drr011. MR2875242

[VG] S. K. Vodop′janov and V. M. Gol′dšteĭn, *Lattice isomorphisms of the spaces* W_n^1 *and quasiconformal mappings* (Russian), Sibirsk. Mat. Ž. **16** (1975), 224–246, 419. MR0379841

[VGR] S. K. Vodop′yanov, V. M. Gol′dshtein, and Yu. G. Reshetnyak, *The geometric properties of functions with generalized first derivatives* (Russian), Uspekhi Mat. Nauk **34** (1979), no. 1(205), 17–65. MR525650

[VU1] S. K. Vodop′yanov and A. D. Ukhlov, *Sobolev spaces and* (P,Q)*-quasiconformal mappings of Carnot groups* (Russian, with Russian summary), Sibirsk. Mat. Zh. **39** (1998), no. 4, 776–795, i, DOI 10.1007/BF02673052; English transl., Siberian Math. J. **39** (1998), no. 4, 665–682. MR1654140

[VU2] S. K. Vodop′yanov and A. D. Ukhlov, *Superposition operators in Sobolev spaces* (Russian), Izv. Vyssh. Uchebn. Zaved. Mat. **10** (2002), 11–33; English transl., Russian Math. (Iz. VUZ) **46** (2002), no. 10, 9–31 (2003). MR1956487

[VU3] S. K. Vodop′yanov and A. D. Ukhlov, *Set functions and their applications in the theory of Lebesgue and Sobolev spaces. I [Translation of Mat. Tr.* **6** *(2003), no. 2, 14–65; MR2033646]*, Siberian Adv. Math. **14** (2004), no. 4, 78–125 (2005). MR2125997

Department of Mathematics, Ben-Gurion University of the Negev, P.O.Box 653, Beer Sheva, 8410501, Israel

Email address: vladimir@math.bgu.ac.il

Department of Higher Mathematics and Mathematical Physics, Tomsk Polytechnic University, 634050 Tomsk, Lenin Ave. 30, Russia; Department of General Mathematics, Tomsk State University, 634050 Tomsk, Lenin Ave. 36, Russia

Current address: Department of Mathematics, Ben-Gurion University of the Negev, P.O.Box 653, Beer Sheva, 8410501, Israel

Email address: vpchelintsev@vtomske.ru

Department of Mathematics, Ben-Gurion University of the Negev, P.O.Box 653, Beer Sheva, 8410501, Israel

Email address: ukhlov@math.bgu.ac.il

Contemporary Mathematics
Volume **734**, 2019
https://doi.org/10.1090/conm/734/14769

First steps of the global bifurcation theory in the plane

Yu. Ilyashenko

Dedicated to the memory of Selim Grigorievich Krein

ABSTRACT. We present here a survey of the first results and problems in the global bifurcation theory in the plane. For sake of compactness we consider the Sphere S^2 rather than the plane. The results of this paper are obtained in collaboration with my students N.Goncharuk, Yu.Kudriashov, I.Schurov, N.Solodovnikov, L.Starichkova.

In [**I16**], the author outlined the future development of the global bifurcation theory in the plane. This paper summarizes the progress achieved in this direction. The results are not yet published, but most part of them is presented in papers either submitted, or in the final study of preparation. This text is based on the lectures given by the author in the Summer School "Dynamical Systems " at Stola, Slovakia, June 25 – July 8, 2017. I am grateful to A. Fishkin, the organizer of the school, to O. Romaskevich, who coordinated writing of the lecture notes, and to all the participants of the school.

1. Class $Vect^*(S^2)$ vs generic families

Bifurcation theory considers robust changes of the topology of a phase portrait in families of vector fields that occur when the parameter of the family changes. We consider C^∞ families of the form

$$\dot{x} = v_\varepsilon(x), \ x \in S^2, \ \varepsilon \in (\mathbb{R}^k, 0),$$

with an arbitrary number of parameters k, sometimes called the *dimension* of the family. First of all, we want the vector fields of the families not to be too wild.

DEFINITION 1. *A vector field on S^2 belongs to the class Vect$^*(S^2)$ provided that it has but a finite number of limit cycles, and all its singular points have finite multiplicity.*

CONJECTURE 1. *For any number of parameters k there exists $N = N(k)$ such that a residual set of C^N families of vector fields on S^2 contains vector fields of class Vect$^*(S^2)$ only.*

2010 *Mathematics Subject Classification.* 34C23, 37G99, 37E35.
Key words and phrases. Bifurcations, polycycles, sparkling saddle connections.
The author was supported in part by the grant RFBR 16-01-00748.

To the best of my knowledge, this conjecture is not yet proved. We suggest here a plan of the future proof. Very roughly speaking, we expect that the complement to $\mathrm{Vect}^*(S^2)$ in the space of all vector fields on S^2 has codimension infinity. Hence, generic finite dimensional manifold does not intersect it at all.

In more detail, for any natural μ, let $M(\mu)$ be the set of vector fields having a singular point or a limit cycle of multiplicity greater than μ. It is expected to be a stratified Banach submanifold of the space $\mathrm{Vect}^{N(\mu)}(S^2)$ ($C^{N(\mu)}$ - smooth vector fields on S^2) of codimension $d(\mu)$. For $N(\mu)$ large enough, $d(\mu)$ does not depend on $N(\mu)$ and tends to infinity as $\mu \to \infty$. By Thom's transversality theorem, a generic family of dimension $k < d(\mu)$ does not intersect this manifold. Hence, generic families of this dimension have singular points and limit cycles of multiplicity less that μ.

Consider now the set I of vector fields on S^2 that has an infinite number of limit cycles. As understood even by Poincaré, they must accumulate to a monodromic polycycle of the corresponding vector field. Roughly speaking, a monodromic polycycle is a separatrix polygone with a finite number of vertexes and a Poincaré map defined on the side of it, see, for instance, [**I90**] for details. By the famous Dulac theorem, [**Du**], [**I85**], the monodromy map of this polycycle either is strongly attracting, or repelling, or admits an asymptotic series with an infinite number of the numeric coefficients. If the vector field has an infinite number of limit cycles accumulating to this polycycle, than the first two cases are impossible, and the coefficients of the asymptotic series in the third case are all zeros except for the first one. This is a countable number of independent relations that may not hold simultaneously for vector fields found in generic finite-parameter families.

Making these agreements precise is a subject of a future investigation. In what follows, we consider vector fields of class $\mathrm{Vect}^*(S^2)$ only.

2. Basic definitions.

We present here necessary definitions from [**AAIS**] and earlier sources, in the form they are presented in [**IKS**].

Denote by $Vect(M)$ the set of C^∞-smooth vector fields on M.

DEFINITION 2. *A* family of vector fields *on M with the base B is a smooth map $V : B \to Vect(M)$. The dimension of the base is called the number of the parameters or the dimension of the family.*

Denote by $Vect_k(M)$ the space of k-parameter families of vector fields on M which are C^∞ smooth both in the phase variable and the parameter.

DEFINITION 3. *A family of vector fields is* transversal *to a Banach submanifold* $\mathbf{T}$ *of $Vect(M)$ provided that the corresponding map $V : B \to Vect(M)$ is transversal to* $\mathbf{T}$.

DEFINITION 4. *Two vector fields v and $\tilde{v}$ on a manifold M are called* orbitally topologically equivalent, *if there exists a homeomorphism $M \to M$ that links the phase portraits of v and $\tilde{v}$, that is, sends orbits of v to orbits of $\tilde{v}$ and preserves their time orientation.*

DEFINITION 5. *Two families of vector fields $\{v_\alpha | \alpha \in B\}$, $\{\tilde{v}_{\tilde{\alpha}} | \tilde{\alpha} \in \tilde{B}\}$ on M are called* weakly topologically equivalent *if there exists a map*

$$H\colon B \times M \to \tilde{B} \times M, \quad H(\alpha, x) = (h(\alpha), H_\alpha(x)) \tag{1}$$

such that $h : B \to \tilde{B}$ *is a homeomorphism, and for each* $\alpha \in B$ *the map* $H_\alpha \colon M \to M$ *is a homeomorphism that links the phase portraits of* v_α *and* $\tilde{v}_{h(\alpha)}$. *The map* H *is called* weak topological equivalence *of the corresponding families.*

Two families are *topologically equivalent* or *strongly equivalent* if there exists a *homeomorphism* H with above properties. Topological classification of families with a very simple dynamics may have functional invariants that occur due to the requirement of the continuity of H_α in α [**R**]. Thus, the topological equivalence is too rigid. On the other hand, weak topological equivalence introduced in [**AAIS**] is too loose: some families with apparently different bifurcations occur to be weakly topologically equivalent. For this reason a definition of moderate topological equivalence for generic families of vector fields on S^2 and in domains of S^2 is given in [**GI***], work in progress. It is presented in Section 6 below.

3. Moduli of strong equivalence

Consider a C^∞ vector field with a non-hyperbolic cycle and a cross-section Γ to this cycle. By Takens theorem [**T**], the Poincaré map P of this cycle defined on Γ has a smooth generator, that is, a C^∞ vector field u on Γ for which P is the time-one shift:

$$P = g_u^1.$$

This generator determines a time distance on the complement of the cross section Γ to its intersection point O with the cycle, on each side of this point. Namely, the distance between the two points $a, b \in \Gamma$ equals the integral

$$d(a, b) = \int_a^b \omega,$$

where ω is a dual form to u on Γ:

$$(\omega, u) \equiv 1.$$

Now, let two one-parameter local families of vector fields $V = \{v_\varepsilon\}$ and $N = \{w_\delta\}$ have a parabolic cycle for the critical parameter values $0 : \gamma$ for v_0 and $\tilde{\gamma}$ for w_0. Let these two families be strongly equivalent, and the conjugacy H_0 brings γ to $\tilde{\gamma}$. Then the family of maps H_ε generates family of the homeomorphisms g_ε of the cross section Γ that conjugate the parameter depending Poincaré map defined for γ and $\tilde{\gamma}$ respectively. These families of the Poincaré map have a parabolic fixed point for the parameter value zero. For such families the Newhouse-Palis-Takens theorem holds [**NPT**]. This theorem claims that the map g_0 is not only a homeomorphism but rather a diffeomorphism that conjugates the generators of the Poincaré map for v_0, γ and $w_0, \tilde{\gamma}$. Thus, this map preserves the time distance on the cross-section.

Now, there are distinguished points on the cross sections Γ and $\tilde{\Gamma}$, that have to be mapped to each other by g_0. Namely, these are intersection points of separatrices of the hyperbolic saddles of v_0 and w_0 with Γ and $\tilde{\Gamma}$. If there are two points of this kind on each cross-section, then the time distance between them is the same, because g_0 preserves this distance. Hence, this distance is a numeric modulus of the strong classification of one-parameter families under considerations [**MP**].

In a similar way, functional moduli in a strong classification of two-parameter families may be obtained [**GI1***], in preparation.

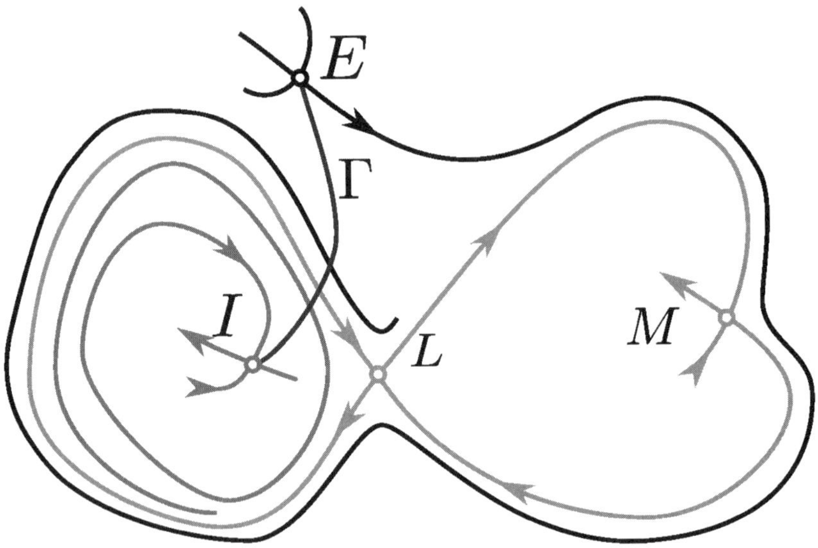

FIGURE 1. The phase portrait of the degenerate vector field v

4. Structurally unstable families

In [**AAIS**], Arnold posed several questions about generic families of vector fields on S^2. These questions determined the development of the global bifurcation theory on S^2, though the answers appeared to be different from those expected by Arnold. In particular, *generic structurally unstable families on vector fields on S^2 exist.* This effect was discovered in [**IKS**].

Let us describe structurally unstable local families of dimension three. Consider a vector field v_0 with three saddle connection, see Fig 1.

These connections form a polycycle γ with two hyperbolic saddles L and M as vertexes; two edges between L and M and a separatrix loop l of L. There are two extra saddles, one outside γ, named E, and another inside γ, named I (E and I stand for *exterior* and *interior*). The eigenvalues of L and M may be so chosen that the polycycle γ would attract the nearby orbits from outside, and the loop will repel the inside orbits. So we may assume that one separatrix of E winds towards γ, and one separatrix of I winds towards l in the negative time.

Consider now a one-parameter subfamily of a generic unfolding of v_0 defined in the following way. Vector fields of this subfamily still have connections between L and M unbroken, but the saddle loop is broken. Let ε be the parameter of this subfamily. Suppose that for $\varepsilon > 0$ the unstable separatrix of L enters inside the destroyed loop. Eventually it may form saddle connections between the saddles L and I; let $\varepsilon_k \to 0$ be the corresponding sequence of values of ε. In the same way, an unstable separatrix of E after making several turns around γ, may eventually form a saddle connection between E and L; let δ_m be the sequence of the corresponding parameter values.

Standard tools of the bifurcation theory allow us to prove that $\log(-\log \varepsilon_k)$ and $\log(-\log \delta_m)$ form two sequences equal to arithmetic progressions modulo a reminder term that tends to zero as k, m tend to infinity.

The relative density of these progressions, equal to the ratio of their differences, is a topological invariant of the unfolding. These ratios are functions of eigenvalues of the saddles L and M. Changing these eigenvalues we may change these ratios, and thus obtain two nearby families that are topologically non-equivalent. Hence, neither of them is structurally stable.

Two problems arise.

(1) Distinguish structurally unstable three-parameter families from structurally stable ones.

(2) Prove that generic one and two-parameter families are structurally stable.

For one parameter families this problem is solved by joint efforts of N. Goncharuk, N.Solodovnikov, V.Starichkova and the author [**IS**, **GIS***, **S***].

5. Global bifurcations in one-parameter families

5.1. Basic list. Local and non-local one-parameter bifurcations in the plane seem to be well known. There is a prejudice that the bifurcations in the typical one-parameter families are thus completely described. Yet the discovery of sparkling saddle connections by Malta-Palis [**MP**] shows that this is not the fact.

The following degeneracies may be met in an unavoidable way in typical one-parameter families:

AH:: a non-hyperbolic singular point with a pair of non-zero pure imaginary eigenvalues;

SN:: a saddle-node singular point;

SC:: a saddle connection between two different hyperbolic saddles;

HC:: a homoclinic curve of a saddle-node;

SL:: a separatrix loop of a hyperbolic saddle;

PC:: a parabolic cycle, that is, a non-hyperbolic limit cycle.

For future references, this list of degeneracies will be called *a basic list.* AH stays for Andronov – Hopf. Other abbreviations correspond to the first two words in the description of a degeneracy.

The bifurcations in these classes are described in many sources [**AAIS**] and references therein. Local bifurcations are described near non-hyperbolic singular points (classes AH and SN). The semi-local ones, that is, happening in a neighborhood of a homoclinic curve or a limit cycle are described for the last four classes. No interesting bifurcation happens in these neighborhoods for the classes SN, SC and PC. In the classes SN and PC the non-hyperbolic singular point or cycle either splits in two on one side of the critical parameter value, or vanishes on the other side. The saddle connection simply breaks. In the classes AH, HC, SL a limit cycle is generated under the bifurcation.

5.2. Sotomayor's theorems. Vector fields from the basic list that satisfy some extra genericity assumptions (that we do not list here) are called *quasigeneric.*

The first steps in the global bifurcation theory were done by Sotomayor in [**S**].

THEOREM 1. *Degenerate vector fields that occur in generic one-parameter families are quasigeneric.*

THEOREM 2. *Quasigeneric vector fields form an open and dense subset of the set Σ of structurally unstable vector fields. Quasigeneric vector fields are structurally stable in the corresponding classes from the basic list.*

The latter statement means that for any quasigeneric vector field there exists a neighborhood in the corresponding class from the basic list such that any vector field from this neighborhood is orbitally topologically equivalent to v.

5.3. Description of global bifurcations in generic one-parameter families. The following questions should be answered.

A. Is generic one-parameter family structurally stable?

B. Is the bifurcation in a local one-parameter family determined by the phase portrait of the degenerated vector field corresponding to the critical parameter value?

C. Classify the bifurcations in the local one-parameter families.

The answer to A is YES.

The answer to B is YES for all the classes except for PC, and NO in the class PC. Note, that for local and semi local one-parameter bifurcations investigated in the classical works, the answer to B is always YES.

In what follows, we motivate these answers, and answer question C.

THEOREM 3. *Generic one-parameter families of vector fields on the sphere S^2 are structurally stable.*

THEOREM 4. *If two vector fields not of class PC met in generic one-parameter families are orbitally topologically equivalent then their unfoldings are weakly equivalent. This is wrong for vector fields of class PC.*

These theorems summarize the results of [**IS**], [**GIS***], [**S***].

Global bifurcations in the first four classes from the basic list coincide with the local and semi-local ones.

Bifurcations in the last two classes are more sophisticated. Let us call generic one-parameter families that unfold vector fields of class SL and PC SL- and PC-families respectively.

5.4. Global bifurcations in the SL families. Consider a vector field v_0 with a hyperbolic saddle L having a characteristic value $\lambda < 1$, and a separatrix loop γ. Let Γ be a cross section to γ oriented inside $\gamma, O = \Gamma \cap \gamma$, and x be a chart on Γ such that $x(O) = 0$. Let $U(0)$, $S(0)$ be stable and unstable separatices of L that cross Γ at O. Consider a local family $V = \{v_\varepsilon\}$, $\varepsilon \in (\mathbb{R}^1, 0)$, that unfolds v_0. Let $L(\varepsilon)$ be a saddle of v_ε continuous in ε and such that $L(0) = L$. Let $U(\varepsilon)$ and $S(\varepsilon)$ be the unstable and stable separatrices of the saddle $L(\varepsilon)$ such that the germs $(U(\varepsilon), L(\varepsilon))$, $(S(\varepsilon), L(\varepsilon))$ depend continuously on ε, and coincide with $(U(0), L)$, $(S(0), L)$ for $\varepsilon = 0$. Let $u(\varepsilon) = U(\varepsilon) \cap \Gamma$, $u(\varepsilon)$ depends continuously on ε, and $u(O) = 0$, $u(\varepsilon)$ is the first intersection point $u(\varepsilon) \cap \Gamma$ on the time oriented orbit $U(\varepsilon)$. Let $s(\varepsilon)$ be the last intersection point of $S(\varepsilon)$ with Γ on the time oriented orbit $S(\varepsilon)$; $s(\varepsilon)$ depends continuously on ε, and $s(O) = 0$.

For generic unfoldings the function $\delta(\varepsilon) = x(s(\varepsilon)) - x(u(\varepsilon))$ has a non-zero derivative at 0. Then the difference $\delta(\varepsilon)$ may and will be taken as a new parameter of the family.

It is well known that for $\varepsilon < 0$ the loop γ generates a limit cycle; for $\varepsilon > 0$ the separatix $U(\varepsilon)$ in a sense enters the domain formerly bounded by γ, and nothing intersecting happens in a neighborhood of γ.

But in the whole sphere, sparkling saddle connections may occur. Indeed, suppose that there are several saddles I_k, $k = 1, \dots, K$ of the vector field v_0 inside

γ whose separatrices wind towards γ in the negative time. These separatrices may have a complicated mutual location.

When $\varepsilon > 0$ decreases, these separatrices may eventually coincide with $U(\varepsilon)$.

The sequence of such bifurcations called the *bifurcation scenario*, is characterized by a so called *marked finite set* on a circle. Namely, take an arbitrary transversal loop C inside γ such that the vector field v_0 has no singular points between C and γ. Suppose that C is oriented as the boundary of this domain. Let A be the set of all the intersection points of C with the separatices of the saddles I_k. Say that two points are equivalent if the corresponding separatices belong to the same saddle.

Let us enumerate the points of A in the order as they follow on the oriented curve C. The corresponding saddles are numerated in the same way. If the points $a_k, a_m \in A$ belong to the separatices of the same saddle, then we set: $I_k = I_m$.

The marked set determines the bifurcation scenario in the family V. It is described as follows. Denote by $a(\varepsilon)$ the (unique) intersection point of $U(\varepsilon)$ and C. As $\varepsilon > 0$ tends to 0, the point $a(\varepsilon)$ rotates along C with the growing speed, as to make an infinite number of winds along C. In this motion, it hits the points of the set A. Sparkling saddle connections in the family V_ε occur in the same order: if $a(\varepsilon)$ hits a_k then a_{k+1}, then the connection $L(\varepsilon)I_k(\varepsilon)$ occurs first, and the connection $L(\varepsilon)I_{k+1}(\varepsilon)$ occurs next after that. This is an approximate description: the set A depends on ε. But this dependence is smooth near zero; hence, the order of occurrence of saddle connections is the same as if A would not depend on ε. This describes the bifurcations scenario in the unfoldings of vector fields of the class SL.

5.5. Global bifurcations in the PC families. Consider a vector field v_0 with a parabolic cycle γ of multiplicity 2. This means that for any smooth cross section Γ to γ, with the intersection point $O = \Gamma \cap \gamma$ and a special chart x on Γ, $x(O) = 0$, the Poincaré map of v on Γ is

$$P_0(x) = x + x^2 + \dots$$

Consider a generic one-parameter unfolding $V = \{v_\varepsilon\}$ of v_0. Under a suitable choice of the coordinate and the parameter in the family, the Poincaré map of v_ε on Γ has the form

$$P_\varepsilon(x) = x + (x^2 + \varepsilon) f(\varepsilon, x), \ f(0,0) = 1. \tag{2}$$

For $\varepsilon < 0$ this map has two fixed points $\pm\sqrt{-\varepsilon}$ that correspond to two hyperbolic limit cycles; for $\varepsilon > 0$ this map is monotonic, that is $P_\varepsilon(x) > x$, and no limit cycles occur. But the most interesting global bifurcations occur exactly when $\varepsilon > 0$.

Suppose that the vector field v_0 has several hyperbolic saddles $E_1, \dots, E_M$ outside γ, whose separatrices l_k^+ wind towards γ in the positive time, and several hyperbolic saddles $I_1, \dots, I_K$ inside γ, whose separatrices l_k^- wind towards γ in the negative time. Let $K > 0$, $M > 0$. When $\varepsilon > 0$ tends to 0, the saddles E_m, I_k slightly change and form families of saddles $E_m(\varepsilon), I_k(\varepsilon)$ with families of germs of separatrices $(l_m^+(\varepsilon), E_m(\varepsilon))$, $(l_k^-(\varepsilon), I_k(\varepsilon))$ continuous in ε. For isolated values of ε these separatrices eventually coincide and form *sparkling saddle connections* between $E_m(\varepsilon), I_k(\varepsilon)$. The order in which these connections occur form a *bifurcation scenario* in the family V. Again, as in case of SL families, marked finite sets on a circle occur. But for the PC families they come in couples, and the circles on which these sets lie have a certain coordinate.

5.6. Marked sets on coordinate circles. The following heuristic principle is important in the local dynamics: *a dynamical system near an equilibrium point gives rise to a special chart near this point.* For the family (2) called a *saddle-node local family*, this principle is formalized by the following theorem

THEOREM 5. **[IYa]** *Let P_ε be a generic one-parameter C^∞ unfolding of a parabolic germ*

$$P_\varepsilon(x) = x + x^2 + \varepsilon + \dots .$$

Then in the domain $\{\varepsilon \geq 0\} \setminus \{0,0\}$, the family P_ε is C^∞ equivalent to the time one phase flow transformation of the field

$$u_\varepsilon(x) = \frac{x^2+\varepsilon}{1+\alpha(\varepsilon)x}, \tag{3}$$

where $\alpha(\varepsilon)$ is a C^∞ function; the equivalence is infinitely smooth both in x and ε.

Consider now the family of vector fields u_ε on Γ for $\varepsilon \geq 0$, and let ω_ε be the dual one-form on Γ:

$$(\omega_\varepsilon, u_\varepsilon) \equiv 1.$$

Fix two points b^-, b^+ on Γ, b^- outside, b^+ inside γ. Let $T_\varepsilon^\pm$ be the time functions on Γ:

$$T_\varepsilon^\pm(x) = \int_{b^\pm}^x \omega_\varepsilon.$$

The functions T_0^+ and T_0^- are defined only on half neighborhoods $x > 0$ and $x < 0$ respectively because $\omega_0(0) = \infty$. For $\varepsilon > 0$ they are well defined in a whole neighborhood of 0 on Γ. Let us choose two transversal loops for the vector field v_0 near γ, C^- outside, C^+ inside γ. Define the coordinates φ_ε^+ on C^+ and φ_ε^- on C^- induced by the time functions $T_\varepsilon^\pm$ in the following way. Consider first C^-. Take a point $a \in C^-$ and emerge a forward orbit of v_ε from it, see Figure 2. Let $b \in \Gamma^-$ be its first intersection point with Γ. Take

$$\varphi_\varepsilon^-(a) = T_\varepsilon^-(b).$$

Note that $\varphi_\varepsilon^-(b^-) = 0$; as a tends to b^-, one of the one-sided limits of φ_ε^- at b^- is 0 and the other is 1. Thus φ_ε^- maps C^- onto the coordinate circle denoted by S_-^1. The same construction provides a function

$$\varphi_\varepsilon^+ : C^+ \to S_+^1,\ a' \mapsto T_\varepsilon^+(b'),$$

see Figure 2. These ε-dependent coordinates $\varphi_\varepsilon^\pm$ on the transversal loops $C^\pm$ are called *canonical.*

Let us now define two marked sets on the circles $C^\pm$ with the coordinates $\varphi_0^\pm$ on them. Consider all the separatrices of the saddles I_k of v_0 that wind towards γ in the negative time. Each one of them intersects C^+ at exactly one point. Denote the set of all these intersection points by A^+. Let us choose one point of A^+ to be the first, and order the set of points of A^+ on C^+ as they are met when C^+ is run from a_1 in the counterclockwise direction. We get the set

$$A^+ = \{a_1, \dots, a_K\}.$$

The corresponding saddles are numbered in the same way $I_1, \dots, I_K$. Two points of A^+ are equivalent provided that they belong to the separatrices of the same

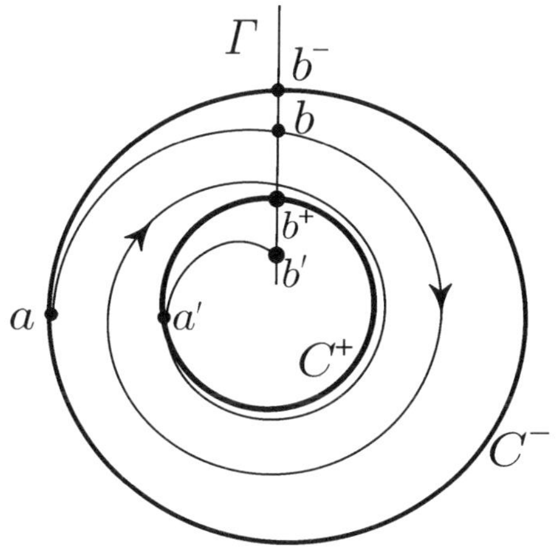

FIGURE 2. Canonical coordinates and the Poincaré map on the transversal loops

saddle. The set A^+ with this equivalence relation is called *marked*. It is a set on the coordinate circle C^+. In the same way the set $A^- \subset C^-$ may be defined. If a_k and a_m are equivalent we set $I_k = I_m$.

Let us identify C^- and C^+ by identifying the points with the same coordinate: $a \equiv a' \Leftrightarrow \varphi^-(a) = \varphi^+(a')$.

DEFINITION 6. *Two sets* $A^+, A^- \subset \mathbb{R}/\mathbb{Z}$ *satisfy* the non-synchronization condition *iff for any* a

$$\#(A^- + a) \cap A^+ \leq 1.$$

The pair $A^+, A^- \subset \mathbb{R}/\mathbb{Z}$ is called a *characteristic pair* for the vector field $v_0 \in PC$.

THEOREM 6. *Generic vector field $v_0 \in PC$ has non-synchronized characteristic pair. The bifurcation scenario in the unfolding of such a generic v_0 is completely determined by corresponding characteristic pair. Any non-synchronized characteristic pair of marked sets $A^+, A^- \subset \mathbb{R}/\mathbb{Z}$ may be realized as a characteristic pair for some vector field v_0 of class PC.*

Let us now describe the bifurcation scenario for a generic unfolding of a vector field of class PC.

LEMMA 1. *For $\varepsilon > 0$, let*

$$\tau(\varepsilon) = \int_{b^-}^{b^+} \omega_\varepsilon.$$

Let Δ_ε be the monodromy map along the orbits of v_ε of the circle C^- to C^+ written in the coordinates φ^-, φ^+. Then

$$\varphi^+(\Delta_\varepsilon(a)) = \varphi^-(a) - \tau(\varepsilon)(mod\ \ \mathbb{Z}).$$

The intersection points of separatrices of $I_k(\varepsilon), E_m(\varepsilon)$ with $C^\pm$ depend smoothly on ε. This generates two sets $A^\pm(\varepsilon) \subset \mathbb{R}/\mathbb{Z}$. The transformation of the circle: $a \mapsto a - \tau(\varepsilon)(\mathrm{mod}\ \ \mathbb{Z})$ is a rotation by an angle that tends to infinity as $\varepsilon \searrow 0$. Sparkling saddle connections occur when the sets $A^+(\varepsilon)$ and $A^-(\varepsilon) - \tau(\varepsilon)$ get a non-empty intersection. The non-synchronization condition implies that for $\varepsilon > 0$ small this intersection is by a unique point. It corresponds to a unique saddle connection. The bifurcation scenario is as follows. The set A^- rotates with a growing speed that tends to infinity as $\varepsilon \searrow 0$. It hits the set A^+; any time when the intersection of rotated A^- with A^+ is non-empty a saddle connection occurs. This is an approximate description: the sets A^- and A^+ depend on ε. But this dependence is smooth near zero; hence, the order of occurrence of saddle connections is the same as if A^- and A^+ would not depend on ε. This completes the description of the bifurcation scenario in a PC family.

Note that the topology of the phase portrait of v_0 does not determine the bifurcation scenario. The latter is determined by the geometry of the sets A^+ and A^-. Namely, this geometry determines the order in which saddle connections occur as ε decreases. It may be $E_1I_1, E_1I_2, E_2I_1 \dots$ or $E_1I_1, E_2I_1, E_1I_2 \dots$. These different bifurcation scenarios may correspond to orbitally topologically equivalent vector fields v_0 and w_0.

6. Large bifurcation supports

In [AAIS], Arnold wrote:

Although even local bifurcations in high codimensions (at least three) on a disc are not fully investigated, it is natural to discuss nonlocal bifurcations in multiparameter families of vector fields on a two-dimensional sphere. For their description, it is necessary to single out the set of trajectories defining perestroikas in these families.

This goal is achieved by the notion of so called large bifurcation supports. There are two definitions of these supports: an axiomatic and a constructive one. We need first a notion of *moderate equivalence* of local families. Definitions of this section are borrowed from [**GI***].

For a family $V = \{v_\varepsilon \mid \varepsilon \in B\}$ of vector fields, let Sing V, Per V, and Sep V be subsets of $B \times S^2$ formed by all singular points, all limit cycles, and all separatrices of v_ε respectively. We will also use notations Sing $v \subset S^2$, Per $v \subset S^2$, and Sep $v \subset S^2$ for the union of all singular points, limit cycles, and separatrices of an individual vector field v.

DEFINITION 7. *Two families of vector fields on S^2, $V = \{v_\varepsilon, \varepsilon \in B\}$ and $W = \{w_\varepsilon, \varepsilon \in B'\}$ are* equivalent *at $\varepsilon = 0$ if there exists a map* (1) *such that h is a homeomorphism, $h(0) = 0$, and for each $\varepsilon \in B$ the map $H_\varepsilon : S^2 \to S^2$ is a homeomorphism that links the phase portraits of v_ε and $w_{h(\varepsilon)}$. They are* strongly equivalent *provided that H is a homeomorphism on $B \times S^2$. They are* weakly equivalent *if we do not pose any additional requirements on H. They are* moderately equivalent *provided that H is continuous with respect to (ε, x) on the set*

$$\text{(4)} \qquad Sing\ v_0 \cup Per\ v_0 \cup Sep\ v_0 \cup \partial((\overline{Per\ V} \cup \overline{Sep\ V}) \cap \{\varepsilon = 0\})$$

and H^{-1} is continuous with respect to (ε, x) on the set

$$\text{(5)} \qquad Sing\ w_0 \cup Per\ w_0 \cup Sep\ w_0 \cup \partial((\overline{Per\ W} \cup \overline{Sep\ W}) \cap \{\varepsilon = 0\})$$

DEFINITION 8 (axiomatic definition of large bifurcation supports). *Suppose that for any local smooth family of vector fields $V = \{v_\varepsilon\}_{\varepsilon \in (B,0)} \subset Vect^*(S^2)$, a closed v_0-invariant subset $L(V) \subset S^2$ is defined. This set is called the* large bifurcation support *of V if it has the following property.*

Let two vector fields v_0 and w_0 be orbitally topologically equivalent on S^2. Let V and W be unfoldings of v_0 and w_0 that are moderately equivalent in some neighborhoods of $L(V)$, $L(W)$; let this moderate equivalence agree with the topological equivalence for $\varepsilon = 0$. Then the families V and W are weakly topologically equivalent on the whole sphere.

No doubt, the whole sphere satisfies this definition. But we are interested in smaller sets. A constructive definition of the large bifurcation support follows.

DEFINITION 9. *A* non-interesting limit cycle *is a limit cycle whose nest*

- *Either contains a limit cycle of odd multiplicity;*
- *Or contains limit cycles of even multiplicity only, but inside the inner one or outside the outer one there exists only one singular point, and this point is either a hyperbolic repeller or a hyperbolic attractor.*

DEFINITION 10. *An α or ω-limit set is* non-interesting *provided that it is a hyperbolic repeller or a hyperbolic attractor, or a non-interesting limit cycle. In the opposite case, the α or ω-limit set is called* interesting.

DEFINITION 11. *Extra large bifurcation support of a vector field is the union of all its non-hyperbolic singular points and limit cycles, together with the closure of the set of all the non-singular points whose α and ω-limit sets are both interesting.*

DEFINITION 12. Large bifurcation support *of a local family V of vector fields is* $LBS(V) = ELBS(v_0) \cap \big(Sing\ v_0 \cup (\overline{Per\ V} \cup \overline{Sep\ V}) \cap \{\varepsilon = 0\}\big)$.

THEOREM 7. [**GI***] *For any k there is an open and dense set in the space of k-parameter local families of vector fields in the two sphere such that for any fixed family from this set the following holds. There exists a neighborhood of the fixed family such that for any two local families from this neighborhood the moderate topological equivalence of these two families in some neighborhoods of their large supports implies the weak equivalence of the families on the whole sphere, provided that the vector fields corresponding to the critical parameter values are orbitally topologically equivalent.*

This is a powerful theorem that allows us to investigate bifurcations in neighborhoods of large bifurcation supports only. This provides a strategy of the proof of structural stability of generic two parameter families, and to the complete investigation of global bifurcations in such families.

The first step is to classify the degeneracies in these families, thus to obtain an analogue of the basic list for dimension two.

The second step is to describe all possible bifurcation supports that may occur for this list. This is a widely extended analogue of the characteristic sets defined in dimension one.

The third step is to classify, up to moderate equivalence, the bifurcations for each of the large bifurcation supports obtained, in a neighborhood of this support only.

7. Two-parameter families with separatrix polygons of codimension two

We consider here only families named in the title. There is about 30 generic two parameter families with two independent degeneracies of codimension one corresponding to the critical parameter value, or families from the basic list with an additional degeneracy. They are subject to another investigation.

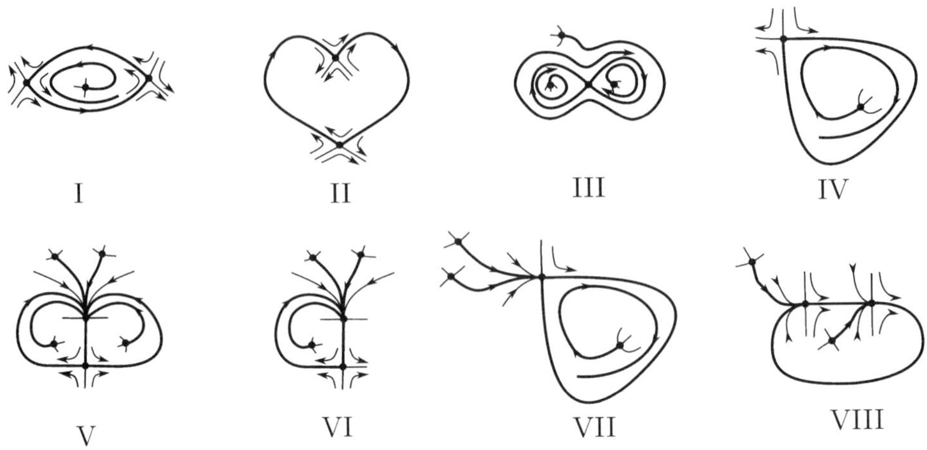

FIGURE 3. Conjectured large bifurcation supports in eight classes of degeneracies of codimension two

7.1. Kotova zoo in codimension two. At the beginning of 90's my student Anna Kotova collected a list of all polycycles that may occur in generic two and three-parameter families on the sphere [**KS**], the so called Kotovs Zoo. The zoo consists of more than 30 species, only 8 of them met in codimension two. The list of these polycycles is: I. a lune, II a heart, III an eight shaped figure, IV a separatrix loop of a hyperbolic saddle with the characteristic value 1, V an apple, VI a half-apple, VII a separatrix loop of a saddle-node, VIII two subsequent saddlenodes, see Figure 3. Our goal here is to outline the study of the global bifurcations in generic two parameter families where these polycycles occur.

7.2. Large bifurcation supports in the Kotova zoo. In the classes I – IV listed above, all the singular points are hyperbolic, and the second part of the large bifurcation support does not occur. Possible large bifurcation supports for these classes are shown at Figure 3; the comments follow.

All the singular points and limit cycles of the vector fields of the classes I – IV without extra degeneracies are hyperbolic. Hence, they do not belong to the large bifurcation support except for those singular points that are shown on Figure 3. The unperturbed vector field , as well as the nearby ones, have no saddle connections disjoint from the polycycles. Hence, the separatrices shown at the figure, and only those, belong to the large bifurcation supports of the unfoldings of the vector fields of classes I – IV.

What about the last four classes, the large bifurcation supports shown on Figure 3 are mere conjectures. We stress that the separatrices that enter the parabolic sectors of the saddle-nodes of the classes V, VI, VII and VIII do belong to the large bifurcation supports.

7.3. Some results and conjectures. Semilocal bifurcations in the classes I, III, IV are investigated long ago, see [**AAIS**] and references therein. In the classes V, VI, they are described in [**G**]. We have no reference for the study of the semilocal bifurcations in the classes VII and VIII; we have a heuristic description of those. Bifurcations in the polycycle "heart", class II, were recently investigated by my student A. Dukov [**D**]. He discovered a new effect: sparkling saddle connections occurred for unfoldings of a vector field that has no saddles outside the polycycle. The large bifurcation support for a vector field of this class consists of the polycycle only.

Conjectures.

1. Sparkling saddle connections occur in generic families of classes I, III, IV, V, VI, VII. The corresponding curves on the bifurcation diagram accumulate to each of SL and PC curves in the diagram. Here SL and PC stand for separatrix loop and parabolic cycle respectively. No sparkling saddle connections occur in the Class VIII.

Occurrence of sparkling saddle connections for the bifurcations of class II is proved by A. Dukov.

2. All the local families that unfold the vector fields with a polycycle of codimension two are structurally stable.

The author is grateful to N. Goncharuk who read the manuscript, and made many valuable comments, and to N. Solodovnikov for the help with the figures.

References

[AAIS] V. I. Arnold, V. S. Afrajmovich, Yu. S. Il′yashenko, and L. P. Shil′nikov, *Bifurcation theory and catastrophe theory*, Springer-Verlag, Berlin, 1999. Translated from the 1986 Russian original by N. D. Kazarinoff; Reprint of the 1994 English edition from the series Encyclopaedia of Mathematical Sciences [*Dynamical systems. V*, Encyclopaedia Math. Sci., 5, Springer, Berlin, 1994; MR1287421 (95c:58058)]. MR1733750

[D] A. V. Dukov, *Bifurcations of the 'heart' polycycle in generic 2-parameter families*, Trans. Moscow Math. Soc. **79** (2018), 209–229, DOI 10.1090/mosc/284. MR3881466

[Du] H. Dulac, *Sur les cycles limites* (French), Bull. Soc. Math. France **51** (1923), 45–188. MR1504823

[GI*] N. Goncharuk, Yu. Ilyashenko, *Large bifurcation supports*, in preparation

[GI1*] N. Goncharuk, Yu. Ilyashenko, *Equivalence relations in the bifurcation theory*, in preparation

[GIS*] N. Goncharuk, Yu. Ilyashenko, N. Solodovnikov, *Global bifurcations in generic one-parameter families with a parabolic cycle on* S^2, submittted.

[G] T. M. Grozovskiĭ, *Bifurcations of "apple" and "half-apple" polycycles in generic two-parameter families* (Russian, with Russian summary), Differ. Uravn. **32** (1996), no. 4, 458–469, 572; English transl., Differential Equations **32** (1996), no. 4, 459–470. MR1436982

[I85] Yu. S. Il′yashenko, *Dulac's memoir "On limit cycles" and related questions of the local theory of differential equations* (Russian), Uspekhi Mat. Nauk **40** (1985), no. 6(246), 41–78, 199. MR815489

[I90] Yu. S. Il′yashenko, *Finiteness theorems for limit cycles* (Russian), Uspekhi Mat. Nauk **45** (1990), no. 2(272), 143–200, 240, DOI 10.1070/RM1990v045n02ABEH002335; English transl., Russian Math. Surveys **45** (1990), no. 2, 129–203. MR1069351

[I16] Yu. Ilyashenko, *Towards the general theory of global planar bifurcations*, in the book "Mathematical Sciences with Multidisciplinary Applications, In Honor of Professor Christiane Rousseau. And In Recognition of the Mathematics for Planet Earth Initiative", Springer 2016, pp 269 – 299.

[IKS] Yu. Ilyashenko, Yu. Kudryashov, and I. Schurov, *Global bifurcations in the two-sphere: a new perspective*, Invent. Math. **213** (2018), no. 2, 461–506, DOI 10.1007/s00222-018-0793-1. MR3827206

[IS] Yu. Ilyashenko and N. Solodovnikov, *Global bifurcations in generic one-parameter families with a separatrix loop on* S^2, Mosc. Math. J. **18** (2018), no. 1, 93–115, DOI 10.17323/1609-4514-2018-18-1-93-115. MR3778561

[IYa] Yu. S. Il′yashenko and S. Yu. Yakovenko, *Nonlinear Stokes phenomena in smooth classification problems*, Nonlinear Stokes phenomena, Adv. Soviet Math., vol. 14, Amer. Math. Soc., Providence, RI, 1993, pp. 235–287. MR1206045

[KS] A. Kotova and V. Stanzo, *On few-parameter generic families of vector fields on the two-dimensional sphere*, Concerning the Hilbert 16th problem, Amer. Math. Soc. Transl. Ser. 2, vol. 165, Amer. Math. Soc., Providence, RI, 1995, pp. 155–201, DOI 10.1090/trans2/165/05. MR1334343

[MP] I. P. Malta and J. Palis, *Families of vector fields with finite modulus of stability*, Dynamical systems and turbulence, Warwick 1980 (Coventry, 1979/1980), Lecture Notes in Math., vol. 898, Springer, Berlin-New York, 1981, pp. 212–229. MR654891

[NPT] S. Newhouse, J. Palis, and F. Takens, *Bifurcations and stability of families of diffeomorphisms*, Inst. Hautes Études Sci. Publ. Math. **57** (1983), 5–71. MR699057

[R] R. Roussarie, *Weak and continuous equivalences for families on line diffeomorphisms*, Dynamical systems and bifurcation theory (Rio de Janeiro, 1985), Pitman Res. Notes Math. Ser., vol. 160, Longman Sci. Tech., Harlow, 1987, pp. 377–385. MR907899

[S] J. Sotomayor, *Generic one-parameter families of vector fields on two-dimensional manifolds*, Inst. Hautes Études Sci. Publ. Math. **43** (1974), 5–46. MR0339279

[S*] V. Starichkova, *Global bifurcations in generic one-parameter families on* $\mathbb{S}^2$, Regul. Chaotic Dyn. **23** (2018), no. 6, 767–784, DOI 10.1134/S1560354718060102. MR3890813

[T] F. Takens, *Normal forms for certain singularities of vectorfields* (English, with French summary), Ann. Inst. Fourier (Grenoble) **23** (1973), no. 2, 163–195. Colloque International sur l'Analyse et la Topologie Différentielle (Colloques Internationaux du Centre National de la Recherche Scientifique, Strasbourg, 1972). MR0365620

National Research University Higher School of Economics, Russia & Independent University of Moscow

Email address: yulijs@gmail.com

Contemporary Mathematics
Volume **734**, 2019
https://doi.org/10.1090/conm/734/14770

Spectral asymptotics for fractional Laplacians

Victor Ivrii

ABSTRACT. In this article we consider fractional Laplacians which seem to be of interest to probability theory. This is a rather new class of operators for us but our methods works (with a twist, as usual). Our main goal is to derive a two-term asymptotics since one-term asymptotics is easily obtained by R. Seeley's method.

In this section we consider fractional Laplacians.

This is a rather new class of operators for us but our methods works (with a twist, as usual). Our main goal is to derive a two-term asymptotics since one-term asymptotics is rather easily obtained by R. Seeley's method.

1. Problem set-up

Let us consider a bounded domain $X \subset \mathbb{R}^d$ with the smooth boundary $\partial X \in \mathscr{C}^\infty$ [1]. In this domain we consider a fractional Laplacian $\Lambda_m = (\Delta^{m/2})_{\mathsf{D}}$ with $m > 0$ originally defined on functions $u \in \mathscr{C}^\infty(\mathbb{R}^d) : \theta_X u \in \mathscr{H}^{m/2}(\mathbb{R}^d)$ by

$$\Lambda_{m,X} := (\Delta^{m/2})_{\mathsf{D}} u = R_X \Delta^{m/2}(\theta_X u) \tag{1.1}$$

where θ_X is a characteristic function of X, R_X is an operator of restriction to X and $\Delta^{m/2}$ is a standard pseudodifferential operator in $\mathbb{R}^d$ with the Weyl symbol $g(x,\xi)^{m/2}$ where as usual $g(x,\xi)$ is non-degenerate Riemannian metrics.

REMARK 1.1. (i) We consider $\Lambda_{m,X}$ as an unbounded operator in $\mathscr{L}^2(X)$ with domain $\mathfrak{D}(\Lambda_{m,X}) = \{u \in \mathscr{L}^2(\mathbb{R}^d) : \operatorname{supp}(u) \subset \bar{X}, R_X\Lambda_m \in \mathscr{L}^2(X)\} \subset \mathscr{H}_0^{m/2}(X)$.
(ii) This operator can also be introduced through positive quadratic form with domain $\{u \in \mathscr{H}^m(\mathbb{R}^d), \operatorname{supp}(u) \subset \bar{X}\}$ and is a positive self2-adjoint operator which is Friedrichs extension of operator originally defined on $\mathscr{H}_0^m(X)$.
(iii) We can consider this operator as a bounded operator from $\mathscr{H}_0^{m/2}(X)$ to $\mathscr{H}^{-m/2}(X) := \mathscr{H}_0^{m/2\,*}(X)$.
(iv) Let $0 < m \notin 2\mathbb{Z}$. Then $\mathfrak{D}(\Lambda_{m,X}) \subset \mathscr{H}^m(X)$ if and only if $m \in (0,1)$; otherwise even eigenfunctions of $\Lambda_{m,X}$ may not belong to $\mathscr{H}^m(X)$.

2010 *Mathematics Subject Classification.* Primary 35P20.
This research was supported in part by National Science and Engineering Research Council (Canada) Discovery Grant RGPIN 13827.

[1] Alternatively consider a bounded domain X on the Riemannian manifold $\mathcal{X}$.

(v) Since Λ_m does not possess transmission property for $m \notin 2\mathbb{Z}$, we are not in the framework of the Boutet-de-Monvel algebra, but pretty close: Λ_m possess μ-transmission property introduced by L. Hörmander and systematically studied by G. Grubb in [**5**, **6**]. We provide definition in Subsection 6.2.

We are interested in the asymptotics of the eigenvalue counting function $\mathsf{N}(\lambda)$ for $\Lambda_{m,X}$ as $\lambda \to +\infty$.

2. Preliminary analysis

As usual we reduce problem to a semiclassical one. Let $A = A_h := h^m \Lambda_{m,X} - 1$ with $h = \lambda^{-1/m}$, $e_h(x, y, \tau)$ the Schwartz kernel of $\theta(\tau - A_h)$ a spectral projector of A_h, and N_h^- the number of negative eigenvalues of A_h.

PROPOSITION 2.1. *Let $\bar{x} \in X$, $B(\bar{x}, 2\gamma) \subset X$, $\gamma \geq h$. Then*

$$|e_h(x, x, 0) - \mathsf{Weyl}(x)| \leq Ch^{1-d}\gamma^{-1} \qquad \forall x \in B(\bar{x}, \gamma) \tag{2.1}$$

and

$$|\int \psi((x - \bar{x})/\gamma)\Bigl(e_h(x, x, 0) - \mathsf{Weyl}(x)\Bigr)\, dx| \leq Ch^{1-d}\gamma^{d-1}\bigl(\gamma^{\delta} + h^{\delta}\gamma^{-\delta}\bigr) \tag{2.2}$$

as $\psi \in \mathscr{C}_0^\infty(B(0,1))$ and $\delta > 0$, where $\mathsf{Weyl}(x) = (2\pi h)^{-d} \operatorname{mes}(\{\xi \colon g(x, \xi) \leq 1\})$ is the standard pointwise Weyl expression.

PROOF. Estimate (2.1) is easily proven by just rescaling as modulo $O(h^s\gamma^{-s})$ we get a $\hbar$-pseudodifferential operator with $\hbar = h\gamma^{-1}$.

Estimate (2.2) is easily proven by rescaling plus R. Seeley's method as described in Subsection 7.5.1 of [**9**]. We leave easy details to the reader. □

Then we immediately arrive to

COROLLARY 2.2. (i) *Contribution of the inner zone $\{x : \operatorname{dist}(x, \partial X) \geq h\}$ to the Weyl remainder does not exceed Ch^{1-d}.*
(ii) *Contribution of the intermediate strip $\{x : \varepsilon \geq \operatorname{dist}(x, \partial X) \geq \varepsilon^{-1}h\}$ to the Weyl remainder does not exceed $\eta(\varepsilon)h^{1-d}$ with $\eta(\varepsilon) \to 0$ as $\varepsilon \to 0$.*

Here and in what follows $\varepsilon > 0$ is an arbitrarily small constant.

PROPOSITION 2.3. *The following estimate holds:*

$$|e_h(x, x, 0)| \leq Ch^{-d}. \tag{2.3}$$

PROOF. The standard proof we leave to the reader. □

THEOREM 2.4. (i) *For operator A the Weyl remainder in the asymptotics for N_h^- does not exceed Ch^{1-d}.*
(ii) *For operator $\Lambda_{m,X}$ the following asymptotics holds*

$$\mathsf{N}(\lambda) = \kappa_0 \lambda^{\frac{d}{m}} + O(\lambda^{\frac{d-1}{m}}) \qquad \text{as } \lambda \to +\infty, \tag{2.4}$$

where $\kappa_0 = (2\pi)^{-d}\varpi_d \operatorname{Vol}(X)$, ϖ_d is a volume of the unit ball in $\mathbb{R}^d$ and $\operatorname{Vol}(X)$ means the Riemannian volume of X.

PROOF. Statement (i) follows immediately from Corollary 2.2(i) and Proposition 2.3. Statement (ii) follows immediately from (i) as $d \geq 2$. □

REMARK 2.5. (i) Therefore, we extended the result, well-known for $m \in 2\mathbb{Z}^+$ to $m \in \mathbb{R}^+$.

(ii) This was easy but to recover the second term (which is also known for $m \in 2\mathbb{Z}^+$ under non-periodicity condition) is a much more daunting task requiring first to improve the contribution of the *near boundary strip* $\{x : \operatorname{dist}(x, \partial X) \le \varepsilon^{-1}h\}$ and also of the *inner zone* $\{x : \operatorname{dist}(x, \partial X) \ge \varepsilon\}$.

3. Propagation of singularities near boundary

Without any loss of the generality one can assume that

$$X = \{x : x_1 > 0\}, \qquad g^{jk} = \delta_{1j} \quad \forall j = 1, \dots, d. \tag{3.1}$$

First let us study the propagation of singularities along the boundary:

THEOREM 3.1. *On the energy level* $\tau : |\tau| \le \epsilon_0$
(i) *Singularities (with respect to* x'*) propagate with the speed not exceeding* c *with respect to* (x, ξ').
(ii) *For* $|\xi'| \asymp \rho \ge Ch^{\frac{1}{2}-\delta}$, *and* $|t| \le T = \epsilon\rho$, *singularities (with respect to* x'*) move from* $x' = y'$ *with the speed* $\asymp \rho$ *with respect to* x.

PROOF. In the terminology of [**9**] both statements mean that $u = u_h(x, y, t)$, the Schwartz kernel of $e^{-ih^{-1}tA}$, satisfies

$$F_{t\to h^{-1}\tau}\chi_T(t)Q_{1,x}u\,{}^t\!Q_{2,y} = O(h^s), \tag{3.2}$$

where $F_{t\to h^{-1}\tau}$ is h-Fourier transform, $Q_1 = q_1(x, hD')$ and $Q_2 = q_2(x, hD')$ are h-pseudodifferential operators, ${}^t\!Q_{2,y}$ is a dual operator, acting with respect to y (and we write it to the right of the function, it is applied to), $\chi_T(t) = \chi(t/T)$, where

(i) in the Statement (i) $h^{\frac{1}{2}-\delta} \le T_0$, T_0 is the small constant, the distance between $\operatorname{supp}(q_1)$ and $\operatorname{supp}(q_2)$ is at least cT, and $\chi \in \mathscr{C}_0^\infty([-1, 1])$.
(ii) in the Statement (ii) the diameter of $\operatorname{supp}(q_1) \cup \operatorname{supp}(q_2)$ is does not exceed $\epsilon\rho$, ϵ is the small constant, and $\chi \in \mathscr{C}_0^\infty([-1, -\frac{1}{2}] \cup [\frac{1}{2}, 1])$;

s is an arbitrarily large exponent, $\delta > 0$ is arbitrarily small,

The proof is standard, by means of the positive commutator method, like those proofs in the Chapters 2 and 3 of [**9**], since it involves only pseudodifferential operators $q_j(x, hD')$ and their commutators with A, but one can see easily that those commutators do not bring any troubles as the energy level is $\asymp 1$. We leave all easy details to the reader. ☐

COROLLARY 3.2. (i) *Let* $\rho \ge Ch^{\frac{1}{2}-\delta}$, $h^{1-\delta} \le \gamma \le \epsilon$. *Then the contribution of the zone* $\{(x, \xi') : x_1 \le \gamma, |\xi'| \asymp \rho\}$ *to the Tauberian remainder with* $T \in (T_*(\rho), T^*(\rho))$, *where* $T_*(\rho) = h^{1-\delta}\rho^{-2}$, $T^*(\rho) = \epsilon\rho$ *does not exceed*

$$C\rho^{d-1}h^{-d} \times \gamma \times h^{1-\delta}\rho^{-2} \times \rho^{-1}. \tag{3.3}$$

(ii) *The total contribution of the zone* $\{(x, \xi') : x_1 \le \gamma = h^{1-\delta}\}$ *to the Tauberian error with* $T = h^{1-3\delta}$ *does not exceed* $Ch^{-d+1+\delta}$.

PROOF. In the terminology of [**9**] the *Tauberian error* is the difference between N_h^- and the *Tauberian expression*

$$\mathsf{N}_h^{\mathsf{T}} := h^{-1} \int_{-\infty}^{0} \Bigl(\int F_{t\to h^{-1}\tau} \bar{chi}_T(t) u(x, x, t)\, dx \Bigr)\, d\tau, \tag{3.4}$$

where $\bar{\chi} \in \mathscr{C}_0^\infty([-1, 1])$, $\bar{\chi} = 1$ on $[-\frac{1}{2}, \frac{1}{2}]$.

The easy and standard proof of Statement (i) is left to the reader; it is like those proofs in Chapter 7 of [**9**].

Then the contribution of the zone $x_1 \le h^{1-\delta}$, $|\xi'| \ge Ch^\delta$ to the Tauberian error with $T = h^{1-3\delta}$ does not exceed $Ch^{-d+1+\delta}$. Since the contribution of the zone $\{(x,\xi') : x_1 \le \gamma = h^{1-\delta}, |\xi'| \le Ch^{2\delta}\}$ to the asymptotics does not exceed $C\rho^{d-1}h^{-d} \times \gamma \le Ch^{-d+1+\delta}$ we arrive to Statement (ii). □

Therefore in this zone $\{(x,\xi') : x_1 \le \gamma = h^{1-\delta}\}$ all we need is to pass from the Tauberian expression to the Weyl expression.

However the inner zone should be reexamined and we need to describe what happens with the propagation along Hamiltonian trajectory in the zone $\{(x,\xi) : x_1 \le h^{1-\delta}\}$. We can assume that $|\xi_1| \ge \varepsilon$ since the measure of the remaining trajectories is small, here $\varepsilon > 0$ is an arbitrarily small constant.

4. Reflection of singularities from the boundary

4.1. Toy model. We start from the pilot-model which will be used to prove the main case. Namely, let us consider 1-dimensional operator on half-line $\mathbb{R}^+$ with Euclidean metrics

$$B := B_{m,a,h} = ((h^2D_x^2 + a^2)^{m/2})_{\mathsf{D}} \tag{4.1}$$

with $a \ge 0$. We denote $\mathbf{e}_{m,a,h}(x_1, y_1, \tau)$ the Schwartz kernel of its spectral projector.

Observe that scaling $x \mapsto x\gamma^{-1}$, $\tau \mapsto \tau\rho^{-m}$ transforms operator to one with $h \mapsto h/(\rho\gamma)$, $\tau \mapsto \tau\rho^{-m}$; because of this we can assume that $h = 1$ and the second scaling implies that we can assume that either $a = 1$ or $\tau = 1$.

PROPOSITION 4.1. (i) *The spectrum of operator $\Lambda_{m,a}$ is absolutely continuos and it coincides with $[a^m, \infty)$.*
(ii) *The following equalities hold:*

$$\begin{aligned} \mathbf{e}_{m,a,h}(x,y,\lambda) &= \mathbf{e}_{m,1,ha^{-1}}(ax,ay,\lambda a^{-m}) = \\ &\lambda^{1/m}\mathbf{e}_{m,a\lambda^{-1/m},h}(\lambda^{1/m}x, \lambda^{1/m}y, 1) = a\mathbf{e}_{m,1,h}(ax,ay,\lambda a^{-m}). \end{aligned} \tag{4.2}$$

PROPOSITION 4.2. *Let $\psi \in \mathscr{C}_0^\infty([-1,1])$, $\psi_\gamma(x) = \psi(x/\gamma)$ and $\phi \in \mathscr{C}_0^\infty([-1,1])$, $0 \le a \le 1-\epsilon_0$. Then as $\gamma \ge h^{1-\delta}$, $T \ge C_0\gamma$, $h^\delta \ge \eta \ge h^{1-\delta}T^{-1}$*

$$\|\phi(\eta^{-1}(hD_t - 1))\psi_\gamma e^{i(mh)^{-1}tB}\psi_\gamma|_{t=T}\| \le CT^{-1}\gamma + Ch^{\delta'}. \tag{4.3}$$

PROOF. Observe first that if for u supported in $\mathbb{R}^+$ and $L = x_1hD_1 - ih/2 = L^*$

$$\operatorname{Re} i(BLu,u) = \frac{1}{2}(i[B,L]u,u), \tag{4.4}$$

and then

$$\operatorname{Re}(Lu, u_t - ih^{-1}Bu) = \frac{1}{2}(ih^{-1}[B,L]u,u) + \frac{1}{2}\partial_t \operatorname{Re}(Lu,u) \tag{4.5}$$

and

$$\begin{aligned} &\operatorname{Re}(ktu + Lu, u_t - ih^{-1}Bu) = \\ &\qquad \frac{1}{2}\partial_t\bigl(kt\|u\|^2 + (Lu,u)\bigr) + \frac{1}{2}\bigl((ih^{-1}[B,L]u,u) - k\|u\|^2\bigr). \end{aligned} \tag{4.6}$$

Let us plug

$$u = \phi(\eta^{-1}hD_t - 1)e^{ih^{-1}TB}\psi_\gamma v \tag{4.7}$$

with $\|v\| = 1$. Then the left hand expression in (4.6) is 0 and

$$\frac{1}{2}\partial_t\bigl(kt\|u\|^2 + (Lu,u)\bigr) \le \frac{1}{2}\bigl(-(ih^{-1}[B,L]u,u) + k\|u\|^2\bigr). \tag{4.8}$$

Let us estimate from above the right-hand expression; obviously

$$ih^{-1}[B_m, L] = m(B_m - a^2 B_{m-2}). \tag{4.9}$$

(a) Assume first that $m > 2$. Then since

$$(1 - C\eta)\|u\| \le \|B_m u\| \le (1 + C\eta)\|u\| \tag{4.10}$$

due to cutoff by ϕ and

$$B_{m-2} \le B_m^{(m-2)/m} \tag{4.11}$$

in virtue of Corollary 6.2 we conclude that as $k = m(1-a^2)$ the right-hand expression does not exceed $C\eta$ and therefore

$$m(1 - a^2)t\|u\|^2 + (Lu,u) \le C\gamma + C\eta T \tag{4.12}$$

since the value of the left-hand expression as $t = 0$ does not exceed $C\gamma$.

Further, observe that on the energy levels from $(1 - C_0\eta, 1 + C_0\eta)$ the singularities propagate with a speed (with respect to x_1) not exceeding $m(1-a^2)^{\frac{1}{2}}(1+C_0\eta)$. Therefore we conclude that u is negligible as $|x_1| \ge m(1 - a^2)^{\frac{1}{2}}(1 + C_0\eta)T + C\gamma$ and therefore since

$$\|D_1 u\| \le (B_2 u, u) \le ((1 - a^2)^{\frac{1}{2}} + C_0\eta), \tag{4.13}$$

we conclude using (4.10) and (4.11) that

$$|(Lu,u)| \le m(1 - a^2 + C_0\eta)T + C\gamma - \epsilon_0 T\|\psi_\gamma(x_1)u\|^2 \tag{4.14}$$

and the left-hand expression of (4.12) is greater than $\epsilon_0 T\|\psi_\gamma u\|^2 - C(\eta T + \gamma)$ and we arrive to (4.3).

(b) Assume now that $0 < m < 2$. Then our above proof fails short in both estimating $\operatorname{Re} ih^{-1}([B,L]u,u)$ from below and $|(Lu,u)|$ from above and we need to remedy it.

Note first that away from $x_1 = 0$ only symbols are important and therefore the right-hand expression of (4.8) does not exceed $\frac{m}{2}(a^m - a^2)\|\psi_\sigma u\|^2 + Ch^2\sigma^{-2}$ since $\|B_{m-2}\| \le a^{m-2}$. Indeed, we need just to decompose $1 = \psi_\sigma^2 + \psi_\sigma'^2$ and use our standard arguments to rewrite the right-hand expression of (4.8) as the sum of the same expressions for $\psi_\sigma u$ and $\psi'_\sigma u$ plus $Ch^2\sigma^{-2}\|u\|^2\|$.

Similarly we deal $\operatorname{Re}(Lu,u) = \operatorname{Re}(L\psi_\sigma u, \psi_\sigma u) + \operatorname{Re}(L\psi'_\sigma u, \psi'_\sigma u)$ and the absolute value of the second term does not exceed $m(1 - a^2)^{\frac{1}{2}}T\|B_m^{1/m}\psi'_\sigma\|^2$.

We claim that

$$\operatorname{Re}(L\psi_\sigma u, \psi_\sigma u) \le C\sigma\|\psi_\sigma u\|^2 + C\sigma h^\delta; \tag{4.15}$$

we prove it later but now instead of (4.12) we arrive for $\sigma = \epsilon_0 t$ to

$$((1 - a^2) - \epsilon)P(T) \le Ch^{\delta'} + (a^m - a^2)T^{-1}\int_\gamma^T P(t)\,dt + C\gamma T^{-1}$$

with $P(t) = \|\psi_{\epsilon_0 t}u(.,t)\|^2$. Then since $\nu = (a^m - a^2)/((1 - a^2) - \epsilon) < 0$ this inequality implies (4.3) again. ☐

Proof of (4.15). Indeed, as $h=1$, $\|B_m^{1/m}u\|\le 1$ we from G. Grubb [**5**, **6**] conclude that $|u(x_1)|\le Cx_1^{(m-1)/2}\|uB_m^{1/m}u\|$ and $|Lu(x_1)|\le Cx_1^{(m-1)/2}$ and therefore $|(Lu,u)|\le C\sigma^m\|B_m^{1/m}u\|^2$. Take $\sigma=1$.

Scaling returns (4.15) as $\sigma=h$. □

Proposition 4.3. *Let $\Lambda=\Lambda_{m,X}$ be a d-dimensional operator* (1.1) *on the half-space $X=\{x\in\mathbb{R}^d, x_1>0\}$ with Euclidean metrics ($d\ge 2$) and $A=h\Lambda^{1/m}-1$.*

Let $\psi\in\mathscr{C}_0^\infty([-1,1])$, $\psi_\gamma(x)=\psi(x_1/\gamma)$, $\phi\in\mathscr{C}_0^\infty([-1,1])$, $\varphi\in\mathscr{C}_0^\infty(\mathbb{R}^{d-1})$ supported in $\{|\xi'|\le 1-\epsilon\}$ with $\epsilon>0$. Finally, let $\gamma\ge h^{1-\delta}$, $T\ge Ch^{-\delta}\gamma$, $h^\delta\ge\eta\ge h^{1-\delta}T^{-1}$. Then

$$\|\phi(\eta^{-1}hD_t-1)\varphi(hD')\psi_\gamma(x_1)e^{ih^{-1}tA}\psi_\gamma(x_1)|_{t=T}\|=O(h^s)\tag{4.16}$$

with arbitrarily large s.

Proof. By making Fourier transform $F_{x'\to h^{-1}\xi'}$ we reduce the general case to $d=1$ and operator B.

According to Proposition 4.2 $\|\phi(\eta^{-1}hD_t)\psi_\gamma e^{ih^{-1}TA}\psi\|\le h^\delta$. Thus for $s=\delta'$ (4.16) has been proven (we reduce δ' if necessary).

Without any loss of the generality we assume that $\psi(x_1)=1$ as $x_1\le 1$, $\psi(x_1)=0$ as $x_1\ge 2$. Observe that due to propagation as $t\le T$ and $x_1\ge\gamma$ we see that $\phi(\eta^{-1}hD_t)Q^+(hD_1)(1-\psi_\gamma)e^{ih^{-1}TA}\psi_\gamma$ is negligible where $Q^\pm\in\mathscr{C}^\infty(\mathbb{R})$ is supported in $\{\pm\xi_1>\epsilon\}$. Furthermore, from the standard ellipticity arguments we conclude that $\phi(\eta^{-1}hD_t)Q^0(hD_1)(1-\psi_\gamma)e^{ih^{-1}TA}\psi_\gamma$ is also negligible for $Q^0\in\mathscr{C}_0^\infty([-2\epsilon,2\epsilon])$.

Finally, due to propagation as $t\ge T$ and $x_1\ge\gamma$ we conclude that $\phi(\eta^{-1}hD_t)\psi_\gamma e^{ih^{-1}(t-T)A}Q^-(hD_1)(1-\psi_\gamma)e^{ih^{-1}TA}\psi_\gamma$ is negligible for $t\ge T$.

What is left is $\phi(\eta^{-1}hD_t)\psi_\gamma e^{ih^{-1}(t-T)A}Q^-(hD_1)\psi_\gamma e^{ih^{-1}TA}\psi_\gamma$ and since (4.16) holds for $s=\delta'$ we conclude that it holds for $s=2\delta'$ and T replaced by $2T$.

Continuing this process we see that (4.16) holds for $s=n\delta'$ and T replaced by nT. Therefore, as we redenote nT by T (and T by T/n respectively), we acquire factor $(\gamma n/T)^n$ in our estimate and it is $O(h^s)$ for any s as h is sufficiently small, $\gamma/T\le h^\delta$ and $n=s/\delta'$. □

4.2. General case.

Theorem 4.4. *Let $(\bar{x},\bar{\xi})$ be a point on the energy level 1. Consider a Hamiltonian trajectory $\Psi_t(\bar{x},\bar{\xi})$ with $\pm t\in[0,mT]$ (one sign only) with $T\ge\epsilon_0$ and assume that for each t indicated it meets ∂X transversally i.e.*

$$\begin{aligned}&\operatorname{dist}(\pi_x\Psi_t(x,\xi),\partial X)\le\epsilon\implies\\ &\qquad|\frac{d}{dt}\operatorname{dist}(\pi_x\Psi_t(x,\xi),\partial X)|\ge\epsilon\qquad\forall t:\pm t\in[0,mT].\end{aligned}\tag{4.17}$$

Also assume that

$$\operatorname{dist}(\pi_x\Psi_t(x,\xi),\partial X)\ge\epsilon_0\qquad \text{as}\ \ t=0,\ \pm t=mT.\tag{4.18}$$

Let $\epsilon>0$ be a small enough constant, Q be supported in ϵ-vicinity of (x,ξ) and $Q_1\equiv 1$ in $C_0\epsilon$-vicinity of $\Psi_t(x,\xi)$ as $t=\pm mT$. Then operator $(I-Q_1)e^{-ih^{-1}tH}Q$ is negligible as $t=\pm mT$.

PROOF. (a) Obviously without any loss of the generality one can assume that there is just one reflection from ∂X (and this reflection is transversal) and that (3.1) is fulfilled in its vicinity.

Further, without any loss of the generality one can assume that Q is supported in ε-vicinity of $(\bar{x},\bar{\xi})$, $Q_1\equiv 1$ in ε-vicinity of $\Psi_{mT}(\bar{x},\bar{\xi})$ and $T\asymp\varepsilon$ with $\varepsilon=h^{\frac{1}{2}-\delta'}$. Then both $\bar{x}$ and $\pi_x\Psi_{mT}(\bar{x},\bar{\xi})$ belong to $C_0\varepsilon$-vicinity of ∂X.

Indeed, it follows from the propagation inside of domain.

(b) Then instead of isotropic vicinities we can consider anisotropic ones: ε with respect to (x',ξ'), $h^{1-3\delta'}$ with respect to x_1 and $h^{\delta'}$ with respect to ξ_1. Let now Q and Q_1 be corresponding operators.

In this framework from the propagation inside of domain it follows that without any loss of the generality one can assume that $T\asymp\gamma=h^{1-\delta''}$ and both $\bar{x}$ and $\pi_x\Psi_{mT}(\bar{x},\bar{\xi})$ belong to $C_0\gamma$-vicinity of ∂X.

(c) Then one can employ the method of the successive approximations freezing coefficients at point $\bar{x}$ and in this case the statement of the theorem follows from the construction of Section 7.2 of **[9]**[2] and Proposition 4.3. We leave easy details to the reader. □

Then we arrive immediately to

COROLLARY 4.5. *Under standard non-periodicity condition*[3] N_h^- *is given with* $o(h^{1-d})$*-error by the Tauberian expression with* $T=h^{1-\delta}$.

PROOF. Easy details are left to the reader. □

5. Main results

5.1. From Tauberian to Weyl asymptotics. Now we can apply the method of successive approximations as described in Section 7.2 [2] and prove that for operator A the Tauberian expression with $T=h^{1-\delta}$ (with sufficiently small $\delta>0$) equals to Weyl expression $\mathsf{N}_h^{\mathsf{W}}$ with $O(h^{2-d-\delta''})$ error,

$$\mathsf{N}_h^{\mathsf{W}}=\kappa_0h^{-d}+\kappa_{1,m}h^{1-d}+o(h^{1-m}) \tag{5.1}$$

with the standard coefficient $\kappa_0=(2\pi)^{-d}\varpi_d\operatorname{Vol}_d(X)$ and with

$$\kappa_{1,m}=(2\pi)^{1-d}\varpi_{d-1}\varkappa_m\operatorname{Vol}_{d-1}(\partial X), \tag{5.2}$$

where

$$\varkappa_m=\frac{d-1}{m}\iint_1^\infty\lambda^{-(d-1)/m-1}\Bigl(\mathbf{e}_m(x_1,x_1,\lambda)-\pi^{-1}(\lambda-1)^{1/m}\Bigr)\,dx_1d\lambda \tag{5.3}$$

with $\mathbf{e}_m(x_1,y_1,\tau)=\mathbf{e}_{m,1,1}(x_1,y_1,\tau)$ the Schwartz kernel of the spectral projector of operator $\mathbf{a}_m:=B_{m,1,1}$ introduced by (4.1):

$$\mathbf{a}_m=((D_x^2+1)^{m/2})_{\mathsf{D}} \tag{5.4}$$

Recall that Vol_d and Vol_{d-1} are Riemannian volumes corresponding to metrics g and its restriction to ∂X respectively and $\pi^{-1}(\lambda-1)^{1/m}$ is a Weyl approximation to $\mathbf{e}_{m,1}(x_1,x_1,\lambda)$. Thus we arrive to

---

[2] Insignificant and rather obvious modifications are required.

[3] The set of all periodic billiards has measure zero.

THEOREM 5.1. *Under standard non-periodicity condition the following asymptotics holds:*

$$\mathsf{N}(\tau) = \kappa_0 \tau^{\frac{d}{m}} + \kappa_{1,m}\tau^{\frac{d-1}{m}} + o(\tau^{\frac{d-1}{m}}) \qquad as\ \ \tau \to +\infty. \tag{5.5}$$

PROOF. First we establish as described above asymptotics

$$\mathsf{N}^-_h = \kappa_0 h^{-d} + \kappa_{1,m} h^{1-d} + o(h^{1-d}) \qquad \text{as}\ \ h \to +0, \tag{5.6}$$

which immediately implies (5.5). □

REMARK 5.2. Based on our analysis one can prove easily also the asymptotics for the Riesz means:

$$\Bigl(\mathsf{N}(\tau) - \sum_{k<r+1} \kappa_{k,m}\tau^{\frac{d-k}{m}}\Bigr) * \tau_+^{r-1} = O(\tau^{\frac{d-1}{m}}) \qquad \text{as}\ \ \tau \to +\infty, \tag{5.7}$$

and under standard non-periodicity condition

$$\Bigl(\mathsf{N}(\tau) - \sum_{k\le r+1} \kappa_{k,m}\tau^{\frac{d-k}{m}}\Bigr) * \tau_+^{r-1} = o(\tau^{\frac{d-1}{m}}) \qquad \text{as}\ \ \tau \to +\infty, \tag{5.8}$$

with some coefficients $\kappa_{k,m}$.

5.2. Discussion. The following problems seem to be challenging

PROBLEM 5.3. As $m_1 > 0$, $m_2 > 0$ consider

$$K = K_{m_1,m_2,X} := \Lambda_{m,X} - \Lambda_{m_1,X}\Lambda_{m_2,X} \tag{5.9}$$

on $\mathfrak{D}(\Lambda_m)$ with $m = m_1 + m_2$. From Corollary 6.2 we conclude that this is a non-negative operator. Furthermore, due to [**12, 13**] it is a positive operator. Obviously singularities of its Schwartz kernel $K(x,y)$ belong to $\partial X \times \partial X$.

(i) Provide an effective estimate for this operator from below.
(ii) Prove that as $X = \{x \in \mathbb{R}^d : x_1 > 0\}$ with Euclidean metrics its Schwartz kernel $K(x,y) = k(x_1, y_1, x'-y')$ which is positive homogeneous of degree $-m-d$ satisfies

$$|D_x^\alpha D_y^\beta K(x,y)| \le C_{\alpha\beta} x_1^{-\frac{m}{2}-\alpha_1} y_1^{-\frac{m}{2}-\beta_1} (x_1 + y_1 + |z|)^{-d-|\alpha'|+|\beta'|}. \tag{5.10}$$

(iii) In the general case in the local coordinates in which $X = \{x : x_1 > 0\}$ and $x_1 = \operatorname{dist}(x, \partial X)$ not only (5.10) holds but also

$$\begin{aligned} &|D_x^\alpha D_y^\beta \bigl(K(x,y) - K^0(x,y)\bigr)| \\ &\qquad\qquad \le C_{\alpha\beta}(x_1+y_1)^{-m-\alpha_1-\beta_1}(x_1+y_1+|z|)^{-d-|\alpha'|+|\beta'|+1} \end{aligned} \tag{5.11}$$

where $K^0(x,y) = k(x_1, y_1, x'-y')$ and $g^{jk} = \delta_{jk}$ at point $(0, \frac{1}{2}(x'+y'))$.

PROBLEM 5.4. For $m > 0$, $n > 0$ consider operator

$$K = K_{m,n,X} := \Lambda_{m,X} - \Lambda_{n,X}^{m/n} \tag{5.12}$$

on $\mathfrak{D}(\Lambda_k)$ with $k = \max(m,n)$. Then this is a non-negative (non-positive) operator as $m > n$ ($m < n$ respectively. Furthermore, due to [**12, 13**] it is a positive (negative) operator respectively.

(i) Provide an effective estimate for this operator from below.
(ii) Prove that if $X = \{x \in \mathbb{R}^d : x_1 > 0\}$ with Euclidean metrics its Schwartz kernel $K(x,y) = k(x_1, y_1, x'-y')$ which is positive homogeneous of degree $-(m+d)$ and satisfies (5.10).

(iii) Prove that in the framework of Problem 5.3(iii) both (5.10) and (5.11) hold.

PROBLEM 5.5. (i) Consider operators $(\Delta^{m/2})_{\mathsf{D}}$ with $m < 0$ and the asymptotics of eigenvalues tending to $+0$.
(ii) Consider operators with degenerations like $A_{m,X} = h^m \Lambda_{m,X} + V(x)$.
(iii) Consider more general operators where instead of Δ general elliptic (matrix) operator is used.

PROBLEM 5.6. (i) Consider Neumann boundary conditions: having smooth metrics g in the vicinity of $\bar{X}$ for each point $x \notin X$ in the vicinity of ∂X we can assign a mirror point $j(x) \in \bar{X}$ such that x and $j(x)$ are connected by a (short) geodesics orthogonal to Y at the point of intersection. Each u defined in X we can continue to the vicinity of $\bar{X}$ as $Ju(x) = \psi(x)u(j(x))$ with ψ supported in the vicinity of $\bar{X}$ and $\psi = 1$ in the smaller vicinity of $\bar{X}$. Then $\Lambda_m u = R_X \Delta^{m/2} Ju$.
- Establish eigenvalue asymptotics for this operator.
- Surely we need to prove that the choice neither of metrics outside of X nor ψ is important.

(ii) One can also try $Ju(x) = -\psi(x)u(j(x))$ and prove that eigenvalue asymptotics for this operator do not differ from what we got just for continuation by 0.

PROBLEM 5.7. Consider manifolds with all geodesic billiards closed as in Section 8.3 of [**9**]. To do this we need to calculate the "phase shift" at the transversal reflection point itself seems to be an extremely challenging problem.

6. Appendices

6.1. Variational estimates for fractional Laplacian. We follow here R. Frank and L. Geisinger [**4**]. This is Lemma 19 and the next paragraph of their paper:

LEMMA 6.1. (i) *Let B be a non-negative operator with* $\operatorname{Ker} B = \{0\}$ *and let P be an orthogonal projection. Then for any operator monotone function $\phi : (0,\infty) \to \mathbb{R}$,*

$$P\phi(PBP)P \ge P\phi(B)P. \tag{6.1}$$

(ii) *If, in addition, B is positive definite and ϕ is not affine linear, then $\phi(PBP) = P\phi(B)P$ implies that the range of P is a reducing subspace of B.*

We recall that, by definition, the range of P is a reducing subspace of a non-negative (possibly unbounded) operator if $(B+\tau)^{-1} \operatorname{Ran} P \subset \operatorname{Ran} P$ for some $\tau > 0$. We note that this is equivalent to $(B+\tau)^{-1}$ commuting with P, and we see that the definition is independent of τ since

$$\begin{aligned}(B+\tau')^{-1}P - P(B+\tau')^{-1}\\ = (B+\tau)(B+\tau')^{-1}\left((B+\tau)^{-1}P - P(B+\tau)^{-1}\right)(B+\tau)(B+\tau')^{-1}.\end{aligned}$$

We refer to the proof given there.

COROLLARY 6.2. *The following inequality holds*

$$\Lambda_{m,X} \le \Lambda_{n,X}^{m/n} \qquad \text{as } 0 < m < n. \tag{6.2}$$

PROOF. Plugging into (6.1) $B = \Delta^{n/2}$ in $\mathbb{R}^d$, $P = \theta_X(x)$ and $\phi(\lambda) = \lambda^{m/n}$ we get (6.2). ∎

Repeating arguments of Proposition 20 and following it Subsection 6.4 of R. Frank and L. Geisinger [**4**] (powers of operators will be different but also negative) we conclude that

PROPOSITION 6.3. *Let $d\ge 2$. Then $-\varkappa_m$ is positive strictly monotone increasing function of $m>0$.*

We leave details to the reader.

6.2. μ-transmission property. Proposition 1 of G. Grubb [**6**] claims that

PROPOSITION 6.4. *A necessary and sufficient condition in order that $R_X Pu\in \mathscr{C}^\infty(\bar{X})$ for all $u\in \mathscr{E}_\mu(\bar{X})$ is that P satisfies the* μ-transmission condition *(in short: is of type μ), namely that*

$$\partial_x^\beta\partial_\xi^\alpha p_j(x,-N)=e^{\pi i(m-2\mu-j-|\alpha|)}\partial_x^\beta\partial_\xi^\alpha p_j(x,N)\qquad \forall x\in\partial\Omega, \tag{6.3}$$

for all j,α,β, where N denotes the interior normal to ∂X at x, m is an order of classical pseudo-differential operator P and for $\mu\in\mathbb{C}$ with $\operatorname{Re}\mu>-1$.

Here $\mathscr{E}_\mu(\bar{X})$ denotes the space of functions u such that $u=E_Xd(x)^\mu v$ with $v\in\mathscr{C}^\infty(\bar{X})$ where E_X is an operator of extension by 0 to $\mathbb{R}^d\setminus X$ and $d(x)=\operatorname{dist}(x,\partial X)$.

Observe that for $\mu=0$ we have an ordinary transmission property (see Definition 1.4.3) of [**9**].

7. Global theory

Let us discuss fractional Laplacians defined by (1.1) in domain $X\subset\mathbb{R}^d$. Then under additional condition

$$\operatorname{dist}(x,y)\le C_0|x-y|\qquad \forall x,y\in X \tag{7.1}$$

(where $\operatorname{dist}(x,y)$ is a "connected" distance between x and y) everything seems to work. We leave to the reader:

PROBLEM 7.1. Under assumption (7.1)
(i) prove Lieb-Cwikel-Rozeblioum estimate (9.A.11) of [**9**].
(ii) Restore results of Chapter 9 of [**9**].
(iii) Reconsider examples of Sections 11.2 and 11.3 of [**9**].

REMARK 7.2. Obviously domains with cuts and inner spikes (inner angles of 2π) do not fit (7.1). On the other hand, in the case of the domain with the cut due to non-locality of Δ^r with $r\in\mathbb{R}^+\setminus\mathbb{Z}$ both sides of the cut "interact" and at least coefficient in the second term of two-term asymptotics may be wrong; in the case of the inner spike some milder effects are expected.

The following problem seems to be very challenging:

PROBLEM 7.3. (i) Investigate fractional Laplacians in domains with cuts and inner spikes and save whatever is possible.
(ii) Generalize these results to higher dimensions.

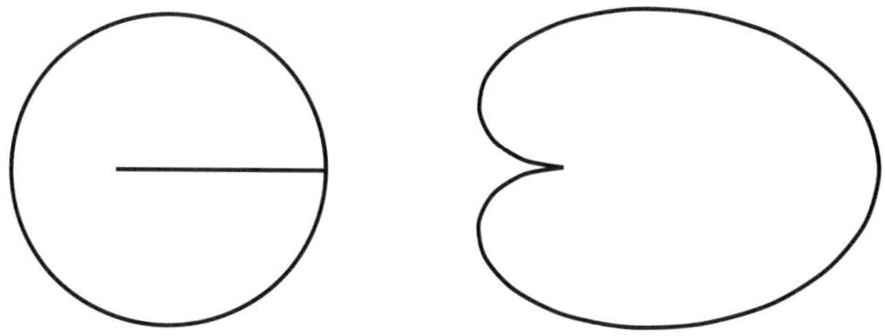

(A) (B)

FIGURE 1. Domain with a cut (A) and an inner spike (B).

Comments

This project started when I learned to my surprise that fractional Laplacians are of the interest to probability theory: which seem to be of interest to probability theory starting from R. M. Blumenthal, R. M. and R. K. Getoor [**3**] and then by R. Bañuelos and T. Kulczycki [**1**], R. Bañuelos, T. Kulczycki and B. Siudeja [**2**], M. Kwaśnicki [**11**]; some of these authors were interested in "Ivrii-type results" (i.e. generalizations from $m = 2$ to $m \in (0, 2)$

Those operators were formulated in the framework of stochastic processes and thus were not accessible for me until I found paper R. Frank and L. Geisinger [**4**] provided definition we follow here. They showed that the trace has a two-term expansion regardless of dynamical assumptions[4], and the second term in their expansion paper [**4**] defined by (3.2)–(3.3) is closely related to $\kappa_{1,m}$. It corresponds to $r = 1$ in Remark 5.2.

Furthermore, I learned that one-term asymptotics for more general operators (albeit without remainder estimate) was obtained by G. Grubb [**7**].

Very recently I used the ideas of Section 4 to study sharp spectral asymptotics for Dirichlet-to-Neumann operator in [**10**].

I express my gratitudes to G. Grubb and R. Frank for pointing to rather nasty errors in the previous version of this article and very useful comments, and to R. Bañuelos for very useful comments. I also express my gratitudes to the referee of this paper for several useful remarks.

References

[1] R. Bañuelos and T. Kulczycki, *Trace estimates for stable processes*, Probab. Theory Related Fields **142** (2008), no. 3-4, 313–338, DOI 10.1007/s00440-007-0106-x. MR2438694

[2] R. Bañuelos, T. Kulczycki, and B. Siudeja, *On the trace of symmetric stable processes on Lipschitz domains*, J. Funct. Anal. **257** (2009), no. 10, 3329–3352, DOI 10.1016/j.jfa.2009.06.037. MR2568694

[3] R. M. Blumenthal and R. K. Getoor, *Sample functions of stochastic processes with stationary independent increments*, J. Math. Mech. **10** (1961), 493–516. MR0123362

[4] R. L. Frank and L. Geisinger, *Refined semiclassical asymptotics for fractional powers of the Laplace operator*, J. Reine Angew. Math. **712** (2016), 1–37, DOI 10.1515/crelle-2013-0120. MR3466545

[4] The fact that R. Frank and L. Geisinger obtain a second term regardless of dynamical assumptions is simply due to the fact that they study $\mathrm{Tr}(f(\Lambda_{m,X}))$ with $f(\lambda) = -\lambda\theta(-\lambda)$, which is one order smoother than $f(\lambda) = \theta(-\lambda)$.

[5] G. Grubb, *Local and nonlocal boundary conditions for μ-transmission and fractional elliptic pseudodifferential operators*, Anal. PDE **7** (2014), no. 7, 1649–1682, DOI 10.2140/apde.2014.7.1649. MR3293447

[6] G. Grubb, *Fractional Laplacians on domains, a development of Hörmander's theory of μ-transmission pseudodifferential operators*, Adv. Math. **268** (2015), 478–528, DOI 10.1016/j.aim.2014.09.018. MR3276603

[7] G. Grubb, *Spectral results for mixed problems and fractional elliptic operators*, J. Math. Anal. Appl. **421** (2015), no. 2, 1616–1634, DOI 10.1016/j.jmaa.2014.07.081. MR3258341

[8] G. Grubb and L. Hörmander, *The transmission property*, Math. Scand. **67** (1990), no. 2, 273–289, DOI 10.7146/math.scand.a-12338. MR1096462

[9] V. Ivrii, Microlocal Analysis and Sharp Spectral Asymptotics, available online at `http://www.math.toronto.edu/ivrii/monsterbook.pdf`

[10] V. Ivrii, Spectral asymptotics for Dirichlet to Neumann operator, arXiv:1802.07524, 1–14, (2018).

[11] M. Kwaśnicki, *Eigenvalues of the fractional Laplace operator in the interval*, J. Funct. Anal. **262** (2012), no. 5, 2379–2402, DOI 10.1016/j.jfa.2011.12.004. MR2876409

[12] R. Musina and A. I. Nazarov, *On fractional Laplacians*, Comm. Partial Differential Equations **39** (2014), no. 9, 1780–1790, DOI 10.1080/03605302.2013.864304. MR3246044

[13] R. Musina and A. I. Nazarov, *On fractional Laplacians—2*, Ann. Inst. H. Poincaré Anal. Non Linéaire **33** (2016), no. 6, 1667–1673, DOI 10.1016/j.anihpc.2015.08.001. MR3569246

Department of Mathematics, University of Toronto, 40 St. George St., Toronto, ON, M5S 2E4, Canada

Email address: `ivrii@math.toronto.edu`

Contemporary Mathematics
Volume **734**, 2019
https://doi.org/10.1090/conm/734/14771

Spectral analysis and decomposition of normal operators related with the multi-interval finite Hilbert transform

Alexander Katsevich, Marco Bertola, and Alexander Tovbis

Dedicated to the centennial of Selim Grigorievich Krein, a great scientist and person and the admired teacher of one of the authors

ABSTRACT. Let $I_{\rm in}$ be a finite collection of bounded intervals, and $I_{\rm ex} = \mathbb{R}\backslash I_{\rm in}$. In this paper we consider the Finite Hilbert transform as a map $\mathcal{H} : L^2(I_{\rm in}) \to L^2(I_{\rm ex})$. More precisely, we study the nature of the spectrum of the normal operators $\mathcal{H}^*\mathcal{H}$ and $\mathcal{H}\mathcal{H}^*$ and find their spectral resolution. We also compute alternative diagonalizations of $\mathcal{H}^*\mathcal{H}$ and $\mathcal{H}\mathcal{H}^*$, which are more convenient from the numerical point of view.

1. Introduction

Let $I_{\rm in}, I_{\rm ex} \subset \mathbb{R}$ be two Lebesgue measurable sets. Consider the Finite Hilbert transform (FHT) and its adjoint:

$$\begin{aligned} &\mathcal{H} : L^2(I_{\rm in}) \to L^2(I_{\rm ex}) \ \ (\mathcal{H}f)(y) = \frac{1}{\pi}\int_{I_{\rm in}} \frac{f(x)}{x-y}dx; \\ &\mathcal{H}^* : L^2(I_{\rm ex}) \to L^2(I_{\rm in}),\ (\mathcal{H}^*g)(x) = \frac{1}{\pi}\int_{I_{\rm ex}} \frac{g(y)}{x-y}dy. \end{aligned} \tag{1.1}$$

We will call the operator $\mathcal{H}$ (as well as $\mathcal{H}^*$) a multi-interval FHT.

An important problem is to obtain diagonalization of $\mathcal{H}$ and the corresponding normal operators $\mathcal{H}^*\mathcal{H}$ and $\mathcal{H}\mathcal{H}^*$. In the case when $I_{\rm in} = I_{\rm ex}$, this and related problems were thoroughly studied starting in the 50-s and 60-s, see, for example, [**11**–**13**, **15**, **17**, **20**, **21**]. More recently, the problem of diagonalization of $\mathcal{H}^*\mathcal{H}$ and $\mathcal{H}\mathcal{H}^*$ occured when solving the problem of image reconstruction from incomplete tomographic data, e.g. when solving the interior problem of tomography [**2**, **5**, **14**, **23**–**25**]. Inspired by applications in tomography, we call the set $I_{\rm in}$ - interior, and the set $I_{\rm ex}$ - exterior. In these applications, $I_{\rm in}$ and $I_{\rm ex}$ are the unions of finitely many intervals. Different arrangements of $I_{\rm in}$ and $I_{\rm ex}$ are possible, and they lead to different spectral properties of the associated multi-interval FHT.

2010 *Mathematics Subject Classification.* Primary 44A15, 47A10.

The first and third authors were supported in part by NSF grants DMS-1211164 and DMS-1615124.

The second author was supported in part by the Natural Sciences and Engineering Research Council of Canada grant RGPIN-2016-06660.

In several settings when the endpoints of the intervals that make up I_{in} and the endpoints of the intervals that make up I_{ex} are two disdjoint sets, the spectrum of the normal operators is discrete. Spectral asymptotics for various arrangements of I_{in}, I_{ex}, where each consists of a single interval (the intervals can be disjoint or have a partial overlap), was obtained in [**1**, **2**, **9**]. The analysis of a general multi-interval case, obtained in [**4**], required a new approach, based on the nonlinear steepest descent method for asymptotic solution of a matrix Riemann-Hilbert Problem (RHP). The approach of the current work follows an outline similar to that of [**4**].

The case when I_{in} and I_{ex} consist of a single bounded interval each and they share a common endpoint was considered in [**10**]. Endpoints that are shared by both I_{in} and I_{ex} are called *double* points. We used the technique based on a commuting differential operator to study the spectral properties of the normal operators $\mathcal{H}^*\mathcal{H}$ and $\mathcal{H}\mathcal{H}^*$. In particular, we showed that the spectrum is purely continuous, and we obtained unitary transformations diagonalizing the operators and calculated their asymptotics in the limit when the spectral parameter $\lambda \to 0$. This result is consistent with the well-known case when I_{in} and I_{ex} are half-lines that share a common endpoint, where the spectrum is also continuous (see e.g. [**6**]). In the current work, we give a complete spectral description in the case of multi-intervals I_{in}, I_{ex} that have $2n$, $n \in \mathbb{N}$, double points subject to one essential constraint: the intervals fill the whole real axis, i.e. $I_{\text{in}} \cup I_{\text{ex}} = \mathbb{R}$. Just as in [**10**], the spectrum turns out to be continuous.

Suppose one is given a collection of $2n$ points $b_j \in \mathbb{R}$, $1 \le j \le 2n$ (i.e., all b_j are double points). We assume that they are arranged in ascending order: $b_j < b_{j+1}$, $1 \le j < 2n$. Define

(1.2)
$$I_{\text{in}} = [b_1, b_2] \cup [b_3, b_4] \cup \cdots \cup [b_{2n-1}, b_{2n}], \qquad I_{\text{ex}} = (-\infty, b_1] \cup [b_2, b_3] \cup \cdots \cup [b_{2n}, \infty),$$

i.e. $I_{\text{in}} \cup I_{\text{ex}} = \mathbb{R}$. In this case, the solution of the underlying RHP can be expressed explicitly in terms of elementary functions. Using this solution, we first calculate the resolution of identity $\widehat{E}_\lambda$, again in elementary functions, and then obtain the main results of this paper, stated in Theorems 5.1 and 5.2:

- *the spectral set of operators $\mathcal{H}^*\mathcal{H}$ and $\mathcal{H}\mathcal{H}^*$ consists of the segment $[0, 1]$;*
- *the spectrum consists of only the absolutely continuous component that has multiplicity $2n$;*
- *the unitary operators, diagonalizing $\mathcal{H}^*\mathcal{H}$ and $\mathcal{H}\mathcal{H}^*$, are given explicitly by* (5.6), (5.23).

Additionally, (Theorem 5.3*), we found another unitary transformation that conjugates $\mathcal{H}^*\mathcal{H}$ with multiplication by $1/\cosh^2(\xi\pi/2)$ in the corresponding Fourier domain. A similar result is also valid for $\mathcal{H}\mathcal{H}^*$.*

The paper is organized as follows. In Section 2 we review two simple cases with $n = 1$: $I_{\text{in}} = [0, \infty)$, $I_{\text{ex}} = (-\infty, 0]$ and $I_{\text{in}} = [-b, b]$, $I_{\text{ex}} = (-\infty, -b] \cup [b, \infty)$. In these cases, the resolution of identity $\widehat{E}_\lambda$ for $\mathcal{H}^*\mathcal{H}$ is obtained via an appropriate change of variables. Even though the derivations in this section are quite straightforward, they help to illustrate the main concepts used in the rest of the paper. In the case $n > 1$, there does not seem to be any easy change of variables, so instead we use the method of [**4**]. Namely, in Section 3 we consider an integral operator $\widehat{K} : L^2(\mathbb{R}) \to L^2(\mathbb{R})$, which, roughly speaking, coincides with $\mathcal{H}$ and $\mathcal{H}^*$ when $\widehat{K}$ is restricted to $L^2(I_{\text{in}})$ and $L^2(I_{\text{ex}})$, respectively. This self-adjoint operator turns out to have the so-called *integrable* kernel so that, according to the technique of Its et. al. of [**7**], the resolvent $\widehat{S}$ of $\widehat{K}$ can be expressed in terms of the solution Γ of the corresponding RHP 3.1, see Lemma 3.5. In Section 4, we first construct the RHP solution Γ and then obtain the resolvent $\widehat{S}$.

The main difference between this work and [**4**] is that in the latter the spectrum is purely discrete, so the spectral projectors for $\mathcal{H}^*\mathcal{H}$ and $\mathcal{H}\mathcal{H}^*$ are obtained by computing the residues of $\pi_{in}\widehat{S}\pi_{in}$ and $\pi_{ex}\widehat{S}\pi_{ex}$, respectively. Here $\pi_{\text{in}} : L^2(\mathbb{R}) \to L^2(I_{\text{in}})$ and $\pi_{\text{ex}} : L^2(\mathbb{R}) \to L^2(I_{\text{ex}})$ are orthogonal projections (restrictions). In contrast, in this work we deal with absolutely continuous spectra. Therefore, instead of projectors onto the eigenspaces, we need to calculate more general resolution of the identity operators for $\mathcal{H}^*\mathcal{H}$ and for $\mathcal{H}\mathcal{H}^*$. The kernels of these operators are expressed in terms of the jumps of the kernels of $\pi_{in}\widehat{S}\pi_{in}$ and $\pi_{ex}\widehat{S}\pi_{ex}$, respectively, across the continuous spectrum in the spectral plane. The jump of $\pi_{in}\widehat{S}\pi_{in}$ is calculated explicitly in Theorem 4.7. It is shown, in particular, that this jump is an integral operator with a degenerate kernel of rank $2n$. The spectral properties of $\mathcal{H}^*\mathcal{H}$, including its spectral resolution, are obtained in Section 5. We show that the operator is absolutely continuous, its spectrum is the interval $[0,1]$, and the multiplicity of the spectrum is $2n$ (i.e., twice the number of intervals or the number of double points). Note that the absolute continuity of $\mathcal{H}^*\mathcal{H}$ follows from Theorem 2 in [**17**], which is proven using a completely different approach. We also establish an alternative diagonalization of $\mathcal{H}^*\mathcal{H}$, which is simpler and more convenient from numerical point of view than its spectral resolution. Completely analogous results can be obtained for $\mathcal{H}\mathcal{H}^*$. The derivation is similar, so we indicate only the main differences between the two cases. We demonstrate also that the more general results obtained in Section 5 coincide with those obtained using different means in Section 2 for $n=1$. Finally, some technical results are proven in the Appendix.

2. Two simple cases

First we consider the case $I_{\text{in}} = [0,\infty)$ and $I_{\text{ex}} = (-\infty, 0]$. Consider the operators

$$(\mathcal{H}f)(y) = \frac{1}{\pi}\int_0^\infty \frac{f(x)}{x-y}dx,\ (\mathcal{H}^*g)(x) = \frac{1}{\pi}\int_{-\infty}^0 \frac{g(y)}{x-y}dy. \tag{2.1}$$

Let T_{in} and T_{ex} denote the operators

$$T_{\text{in}} : f(x) \to \phi(t) := f(e^{2t})e^t,\ T_{\text{ex}} : g(y) \to \psi(s) := g(-e^{2s})e^s. \tag{2.2}$$

Clearly,

$$T_{\text{in}} : L^2([0,\infty)) \to L^2(\mathbb{R}),\ T_{\text{ex}} : L^2((-\infty,0]) \to L^2(\mathbb{R}), \tag{2.3}$$

are unitary. Then

(2.4)
$$(T_{\text{ex}}\mathcal{H}T_{\text{in}}^{-1}\phi)(s) = \frac{1}{\pi}\int_{-\infty}^\infty \frac{\phi(t)}{\cosh(t-s)}dt, \quad (T_{\text{in}}\mathcal{H}^*T_{\text{ex}}^{-1}\psi)(t) = \frac{1}{\pi}\int_{-\infty}^\infty \frac{\psi(s)}{\cosh(t-s)}ds.$$

Let $\mathcal{F}$ denote the Fourier transform:

$$\tilde{\phi}(\xi) := (\mathcal{F}\phi)(\xi) = \frac{1}{\sqrt{2\pi}}\int_{\mathbb{R}} \phi(t)e^{i\xi t}dt. \tag{2.5}$$

Then

$$\phi(t) = (\mathcal{F}^{-1}\tilde{\phi})(t) = \frac{1}{\sqrt{2\pi}}\int_{\mathbb{R}} \tilde{\phi}(\xi)e^{-i\xi t}d\xi, \tag{2.6}$$

and $\|\mathcal{F}\phi\| = \|\phi\|$. From (2.4) (using, for example, integral 2.5.46.5 in [**16**]):

(2.7)

$$(\mathcal{F}T_{\text{ex}})\mathcal{H}(\mathcal{F}T_{\text{in}})^{-1}\tilde{\phi}(\xi) = \frac{\tilde{\phi}(\xi)}{\cosh(\xi\pi/2)},\ (\mathcal{F}T_{\text{in}})\mathcal{H}^*(\mathcal{F}T_{\text{ex}})^{-1}\tilde{\psi}(\xi) = \frac{\tilde{\psi}(\xi)}{\cosh(\xi\pi/2)}.$$

Consequently,

$$\begin{aligned} (\mathcal{F}T_{\text{in}})(\mathcal{H}^*\mathcal{H})(\mathcal{F}T_{\text{in}})^{-1}\tilde{\phi}(\xi) &= \frac{\tilde{\phi}(\xi)}{\cosh^2(\xi\pi/2)},\\ (\mathcal{F}T_{\text{ex}})(\mathcal{H}\mathcal{H}^*)(\mathcal{F}T_{\text{ex}})^{-1}\tilde{\psi}(\xi) &= \frac{\tilde{\psi}(\xi)}{\cosh^2(\xi\pi/2)}. \end{aligned} \tag{2.8}$$

From this it is obvious that the spectrum of the operators $\mathcal{H}^*\mathcal{H}$ and $\mathcal{H}\mathcal{H}^*$ is the interval $[0,1]$, the spectrum is absolutely continuous and of multiplicity 2. A similar derivation in a slightly different context is in [**6**]. A more general class of integral transforms on the half-line, which include the Hilbert transform, is diagonalized in [**19**].

Next we consider the case

$$I_{\text{in}} = [-b,b], \quad I_{\text{ex}} = (-\infty,-b]\cup[b,\infty) \tag{2.9}$$

and the operators

$$(\mathcal{H}f)(y) = \frac{1}{\pi}\int_{-b}^{b}\frac{f(x)}{x-y}dx,\ (\mathcal{H}^*g)(x) = \frac{1}{\pi}\int_{\mathbb{R}\setminus[-b,b]}\frac{g(y)}{x-y}dy, \tag{2.10}$$

for some $b>0$. Similarly to (2.2) and (2.3), consider the following unitary operators:

(2.11)

$$\begin{aligned} &T_{\text{in}}:\ f(x)\to\phi(t) := \sqrt{b}f(-b\tanh t)/\cosh t,\ T_{\text{in}}:\ L^2([-b,b])\to L^2(\mathbb{R}),\\ &T_{\text{ex}}:\ g(y)\to\psi(s) := \sqrt{b}g(-b\coth s)/\sinh s,\ T_{\text{ex}}:\ L^2((-\infty,-b]\cup[b,\infty))\to L^2(\mathbb{R}). \end{aligned}$$

Similarly to (2.4)–(2.8), we obtain

(2.12)

$$(T_{\text{ex}}\mathcal{H}T_{\text{in}}^{-1}\phi)(s) = \frac{1}{\pi}\int_{-\infty}^{\infty}\frac{\phi(t)}{\cosh(t-s)}dt,\quad (T_{\text{in}}\mathcal{H}^*T_{\text{ex}}^{-1}\psi)(t) = \frac{1}{\pi}\int_{-\infty}^{\infty}\frac{\psi(s)}{\cosh(t-s)}ds$$

and

$$\begin{aligned} (\mathcal{F}T_{\text{in}})(\mathcal{H}^*\mathcal{H})(\mathcal{F}T_{\text{in}})^{-1}\tilde{\phi}(\xi) &= \frac{\tilde{\phi}(\xi)}{\cosh^2(\xi\pi/2)},\\ (\mathcal{F}\mathcal{M}_{\text{ex}}T_{\text{ex}})(\mathcal{H}\mathcal{H}^*)(\mathcal{F}\mathcal{M}_{\text{ex}}T_{\text{ex}})^{-1}\tilde{\psi}(\xi) &= \frac{\tilde{\psi}(\xi)}{\cosh^2(\xi\pi/2)}, \end{aligned} \tag{2.13}$$

where T_{in} and T_{ex} are given by (2.11), and $\mathcal{M}_{\text{ex}}$ is the operator of multiplication by $-\text{sgn}(s)$. As in the first example, (2.13) implies that the spectrum of $\mathcal{H}^*\mathcal{H}$ and $\mathcal{H}\mathcal{H}^*$ is the interval $[0,1]$, the spectrum is absolutely continuous and of multiplicity 2. The fact that we obtained the same equations (and, thus, the same spectral properties) is not surprising, since one case can be transformed to the other by the transformation $1/(z-q)$ for some $q<0$, and then shifting. This transformation maps the interior interval $I_{\text{in}} = [0,\infty)$ in the first case to the interior interval $I_{\text{in}} = [-b,b]$ in the second case.

Let A denote $\mathcal{H}^*\mathcal{H}$ or $\mathcal{H}\mathcal{H}^*$ in any of the two cases. By (2.8) and (2.13), A is unitarily equivalent to a multiplication operator in $L^2(\mathbb{R})$. Denoting the latter by $\tilde{A}$, we find its resolvent

$$\widehat{R}_\lambda = (\tilde{A} - \lambda \mathrm{Id})^{-1} = (a(\xi)-\lambda)^{-1}, \ a(\xi) := \frac{1}{\cosh^2(\xi\pi/2)}. \tag{2.14}$$

Here and below Id denotes the identity operator, and the space on which it acts is clear from the context. Let $\widehat{V}_\lambda$ be the resolution of the identity for $\tilde{A}$. The kernel of $\widehat{V}_\lambda$, which we denote V_λ, is obtained by finding the jump of R_λ (see e.g. Sections 7.2 and 7.3 in [**22**] for this well-known derivation):

$$\begin{aligned} \Delta R_\lambda &= \lim_{\epsilon\to 0^+}\left(\frac{1}{a(\xi)-(\lambda+i\epsilon)} - \frac{1}{a(\xi)-(\lambda-i\epsilon)}\right) = 2\pi i \delta(\lambda - a(\xi)), \\ \frac{dV_\lambda}{d\lambda} &= \frac{1}{2\pi i}\Delta R_\lambda = \delta(\lambda - a(\xi)). \end{aligned} \tag{2.15}$$

Here the limit is understood in the sense of distributions (that is, as a kernel of the integral operator with independent variable λ), and the operators are understood as acting by multiplication on functions of ξ.

In the paper we use the following convention. For an integral operator and its kernel we use the same letter (e.g., V, E, R). To distinguish between the two, the operator is denoted with a hat (e.g., $\widehat{V}$), and its kernel has no hat (e.g., V).

To prove that the spectrum of A consists only of the absolutely continuous part, we need to show that the function $\sigma_\phi(\lambda) := (\widehat{V}_\lambda\phi,\phi)$ is absolutely continuous for all $\phi \in L^2(\mathbb{R})$. The function σ_ϕ is called the spectral measure associated with the vector ϕ (see e.g. [**18**], Section VII.2). From (2.15), we have

$$\widehat{V}_\lambda \phi = \theta(\lambda - a(\xi))\phi(\xi), \tag{2.16}$$

where θ is the Heaviside step function. Thus,

$$\sigma_\phi(\lambda) = (\widehat{V}_\lambda\phi,\phi) = \int \theta(\lambda - a(\xi))|\phi(\xi)|^2 d\xi = \int_{a(\xi)<\lambda} |\phi(\xi)|^2 d\xi. \tag{2.17}$$

Note that

$$\sigma_\phi(\lambda) \equiv 0, \ \lambda<0; \quad \sigma_\phi(\lambda) \equiv \|\phi\|^2, \ \lambda > 1. \tag{2.18}$$

Since $\cosh(\xi\pi/2)$ does not stay constant on a set of positive measure, from (2.8) we see that $\tilde{A}$ (and, of course, A) does not have eigenvalues. Hence $\sigma_\phi(\lambda)$ is continuous. To establish absolute continuity of $\sigma_\phi(\lambda)$, it remains to show that the following integral defines an absolutely continuous function:

$$\int_{a^{-1}(\lambda)}^{\infty} |\phi(\xi)|^2 d\xi, \ \lambda \in [\epsilon, 1-\epsilon], \tag{2.19}$$

for any $\epsilon > 0$. Here $a^{-1}(\lambda)$ is the inverse of $a(\xi)$ with the range $[0,\infty)$. The other half (on the negative half-axis) is considered analogously. Clearly, $|\phi(\xi)|^2 \in L^1(\mathbb{R})$ and a^{-1} is smooth on $[\epsilon, 1-\epsilon]$. The desired assertion is now obvious.

Consider the case (2.9). Let $\widehat{E}_\lambda$ be the resolution of the identity of the operator $\mathcal{H}^*\mathcal{H} : L^2([-b,b]) \to L^2([-b,b])$. By (2.13),

$$\frac{d\widehat{E}_\lambda}{d\lambda} = (\mathcal{F}T_{\text{in}})^{-1}\frac{d\widehat{V}_\lambda}{d\lambda}(\mathcal{F}T_{\text{in}}). \tag{2.20}$$

Substituting (2.5), (2.11), and (2.15) into (2.20) gives

(2.21)

$$\begin{aligned}\left(\frac{d\widehat{E}_\lambda}{d\lambda}f\right)(y) &= \left((\mathcal{F}T_{\text{in}})^{-1}\frac{d\widehat{V}_\lambda}{d\lambda}(\mathcal{F}T_{\text{in}})f\right)(y)\\ &= \frac{\cosh s(y)}{\sqrt{b}}\int\frac{\delta(\lambda-a(\xi))}{2\pi}\int\sqrt{b}\frac{f(-b\tanh t)}{\cosh t}e^{i\xi t}dt\, e^{-i\xi s(y)}d\xi,\end{aligned}$$

where $s(y) = \tanh^{-1}(-y/b)$, $|y| < b$, is determined by solving

$$y = -b\tanh s,\ |y| < b. \tag{2.22}$$

Since $a(\xi)$ is even (cf. (2.13)), we find

(2.23)

$$\begin{aligned}\int\delta(\lambda-a(\xi))e^{i\xi(t-s)}d\xi &= 2\int_0^\infty\delta(\lambda-a(\xi))\cos(\xi(t-s))d\xi = \frac{2\cos(\xi(t-s))}{|a'(\xi)|}\\ &= \frac{2}{\pi}\frac{\cos(\xi(t-s))\cosh^3(\xi\pi/2)}{\sinh(\xi\pi/2)},\quad \xi = a^{-1}(\lambda) = \frac{2}{\pi}\ln\frac{1+\sqrt{1-\lambda}}{\sqrt{\lambda}}.\end{aligned}$$

In (2.23) and below, $a^{-1}(\lambda)$ computes the positive root. As is easily checked,

$$\sinh(\xi\pi/2) = \frac{\sqrt{1-\lambda}}{\sqrt{\lambda}},\ \cosh(\xi\pi/2) = \frac{1}{\sqrt{\lambda}}. \tag{2.24}$$

Thus, the integral in (2.23) becomes

$$\frac{2}{\pi}\frac{\cos(\xi(t-s))}{\lambda\sqrt{1-\lambda}},\ \xi = a^{-1}(\lambda). \tag{2.25}$$

Similarly to (2.22), change variables $t = t(x) = \tanh^{-1}(-x/b)$, $|x| < b$, in (2.21). Then, from (2.21) and (2.25)

(2.26)

$$\frac{dE_\lambda}{d\lambda} = \frac{\cosh s\cosh t}{\pi^2 b\lambda\sqrt{1-\lambda}}\cos(\xi(t-s)),\ s = \tanh^{-1}(-y/b), t = \tanh^{-1}(-x/b), \xi = a^{-1}(\lambda).$$

Solving for s and t and simplifying gives

(2.27)

$$\frac{dE_\lambda}{d\lambda} = \frac{b}{\pi^2\lambda\sqrt{1-\lambda}\sqrt{(b^2-x^2)(b^2-y^2)}}\cos\left(\frac{1}{\pi}\ln\frac{1+\sqrt{1-\lambda}}{\sqrt{\lambda}}\ln\left(\frac{b-x}{b+x}\cdot\frac{b+y}{b-y}\right)\right),$$

which matches what we will get after dividing (4.33) by $2\pi i$.

3. Resolvent operator in the case when I_{in} consists of two or more subintervals

Recall that the points b_j (cf. (1.2)) are called double points. Similarly to Section 2, consider the operators

$$(\mathcal{H}f)(y) = \frac{1}{\pi}\int_{I_{\text{in}}}\frac{f(x)}{x-y}dx,\ (\mathcal{H}^*g)(x) = \frac{1}{\pi}\int_{I_{\text{ex}}}\frac{g(y)}{x-y}dy. \tag{3.1}$$

Define the integral operator $(\widehat{K}\phi)(z) = \int_I K(z,x)\phi(x)dx$ from $L^2(I)$ to $L^2(I)$, where $I = I_{\text{in}} \cup I_{\text{ex}} = \mathbb{R}$ and

$$K(z,x) = \frac{\chi_e(z)\chi_i(x) + \chi_i(z)\chi_e(x)}{2\pi i(x-z)}. \tag{3.2}$$

Here $\chi_i(z)$ and $\chi_e(z)$ denote the characteristic functions of I_{in} and I_{ex}, respectively. It is straightforward to check that $\widehat{K}$ is a self-adjoint, bounded operator and that in the direct sum decomposition $L^2(I) = L^2(I_{\text{in}}) \oplus L^2(I_{\text{ex}})$ the operator $\widehat{K}$ has the block structure

$$\widehat{K} = \begin{bmatrix} 0 & \frac{i}{2}\mathcal{H}^* \\ -\frac{i}{2}\mathcal{H} & 0 \end{bmatrix}. \tag{3.3}$$

Note that we now consider the case when exterior and interior intervals, I_{ex} and I_{in}, touch each other at every double point b_j, $j = 1, \ldots, 2n$. A case when the endpoints of the interior and exterior sub-intervals (called then single points) are separated by some positive gaps was studied in [**4**], where it was shown that $\widehat{K}$ is an operator with "integrable" kernel (see [**7**]). Following [**4**], we present here some necessary information about operators with "integrable" kernel. It is known that spectral properties of such operators are intimately related to a suitable Riemann–Hilbert Problem (RHP). In particular, the kernel of the resolvent integral operator $\widehat{S} = \widehat{S}(\mu) : L^2(I) \to L^2(I)$, defined by

$$(\mathrm{Id} + \widehat{S})\left(\mathrm{Id} - \frac{1}{\mu}\widehat{K}\right) = \mathrm{Id}, \tag{3.4}$$

can be expressed through the solution Γ of the following RHP (as explained in Lemma 3.5 below).

Riemann-Hilbert Problem 3.1. Find a 2×2 matrix-function $\Gamma = \Gamma(z;\mu)$, $\mu \in \mathbb{C} \setminus \{0\}$, which (i) is analytic in $\mathbb{C} \setminus I$, where $I = I_{\text{in}} \cup I_{\text{ex}}$, (ii) is bounded near $z = \infty$, (iii) admits non-tangential boundary values from the upper/lower half-planes that belong to L^2_{loc} on I, and (iv) satisfies

$$\begin{aligned} &\Gamma_+(z;\mu) = \Gamma_-(z;\mu)\begin{bmatrix} 1 & 0 \\ \frac{i}{\mu} & 1 \end{bmatrix}, \quad z \in I_{\text{in}}; \\ &\Gamma_+(z;\mu) = \Gamma_-(z;\mu)\begin{bmatrix} 1 & -\frac{i}{\mu} \\ 0 & 1 \end{bmatrix}, \quad z \in I_{\text{ex}}, \end{aligned} \tag{3.5}$$

$$\Gamma(z;\mu) = \mathbf{1} + O(z-i) \text{ as } z \to i. \tag{3.6}$$

We will frequently omit the dependence on μ from notation for convenience.

We will refer informally to the conditions (3.5), as well as similar conditions to be introduced later, as jump conditions or simply "jumps".

Remark 3.2. Since the jump matrices in RHP 3.1 are analytic in z at all points in the interior of I (in fact, they are constant), the solution of the RHP can be easily shown to admit *analytic* boundary values. A similar observation applies to all subsequent RHPs.

For convenience of matrix calculations, here and henceforth we use the Pauli matrices

$$\sigma_1 = \begin{bmatrix} 0 & 1 \\ 1 & 0 \end{bmatrix}, \ \sigma_2 = \begin{bmatrix} 0 & -i \\ i & 0 \end{bmatrix}, \ \sigma_3 = \begin{bmatrix} 1 & 0 \\ 0 & -1 \end{bmatrix}.$$

PROPOSITION 3.3. *If a solution to the RHP* 3.1 *exists, then it is unique.*

This proposition is a well known fact about solutions of RHPs. Remark 3.4 below follows from Proposition 3.3.

REMARK 3.4. The function $\Gamma(z;\mu)$ has the symmetry

$$\Gamma(z;-\mu) = \sigma_3\Gamma(z;\mu)\sigma_3, \tag{3.7}$$

which follows by noticing that the jumps have the same symmetry. In particular the RHP for $\Gamma(z;\mu)$ is solvable if and only if the one for $\Gamma(z;-\mu)$ is. This is a reflection of the symmetry of the spectrum of the problem.

Lemma 3.5 below is a variation of the resolvent formula derived in [**7**]. In the interest of self-containedness, we present it with a proof.

LEMMA 3.5. *If μ is such that the solution $\Gamma(z;\mu)$ of the RHP* 3.1 *exists, then the kernel S of the resolvent $\widehat{S}$ defined by* (3.4) *is given by*

$$S(z,x;\mu) = \frac{\vec{g}^t(x)\Gamma^{-1}(x;\mu)\Gamma(z;\mu)\vec{f}(z)}{2\pi i\mu(z-x)}, \tag{3.8}$$

$$\vec{f}(z) := \begin{bmatrix} i\chi_e(z) \\ \chi_i(z) \end{bmatrix},\ \vec{g}(x) := \begin{bmatrix} -i\chi_i(x) \\ \chi_e(x) \end{bmatrix}, \tag{3.9}$$

where $\vec{g}^t$ denotes the transposition of $\vec{g}$.

PROOF. Let $\mathbf{1}$ denote the 2×2 identity matrix. The jumps of the RHP 3.1 can be written in the form

$$\Gamma_+(z;\mu) = \Gamma_-(z;\mu)\left(\mathbf{1} - \frac{1}{\mu}\vec{f}(z)\vec{g}^t(z)\right),\ z\in I, \tag{3.10}$$

so, by the Sokhotski-Plemelj formula,

$$\Gamma(z;\mu) = \mathbf{1} - \int_{\mathbb{R}}\left[\frac{1}{\zeta-z} - \frac{1}{\zeta-i}\right]\cdot\frac{\Gamma_-(\zeta;\mu)\vec{f}(\zeta)\vec{g}^t(\zeta)d\zeta}{2\pi i\mu}. \tag{3.11}$$

It is clear that the integral in (3.11) is absolutely convergent. Also note that

$$K(z,x) = \frac{\vec{f}^t(z)\vec{g}(x)}{2\pi i(z-x)} \text{ and } \vec{f}^t(z)\vec{g}(z)\equiv 0,\ z\in I. \tag{3.12}$$

The latter equation implies that the boundary value in the integrand of (3.11) is irrelevant because

$$\Gamma_+(\zeta;\mu)\vec{f}(\zeta)\vec{g}^t(\zeta) = \Gamma_-(\zeta;\mu)\left(\mathbf{1} - \frac{\vec{f}(\zeta)\vec{g}^t(\zeta)}{\mu}\right)\vec{f}(\zeta)\vec{g}^t(\zeta) = \Gamma_-(\zeta;\mu)\vec{f}(\zeta)\vec{g}^t(\zeta). \tag{3.13}$$

The same is true for the boundary values in (3.8).

Equation (3.4) can be written as

$$\widehat{K}\widehat{S} = \mu\widehat{S} - \widehat{K}. \tag{3.14}$$

To complete the proof, it is sufficient to show that the kernel of $\widehat{K}\widehat{S}$ is equal to $\mu S(x,y;\mu) - K(x,y)$, where K, S are given by (3.8), (3.2) respectively. Indeed, taking into account (3.12) and (3.11), we calculate the kernel of $\widehat{K}\widehat{S}$ as

(3.15)

$$\int_{\mathbb{R}} d\zeta \frac{\vec{g}^t(y)\Gamma^{-1}(y;\mu)\Gamma(\zeta;\mu)\vec{f}(\zeta)}{2\pi i\mu(\zeta-y)} \frac{\vec{g}^t(\zeta)\vec{f}(x)}{2\pi i(x-\zeta)} =$$
$$= \int_{\mathbb{R}} d\zeta \frac{\vec{g}^t(y)\Gamma^{-1}(y;\mu)\Gamma(\zeta;\mu)\vec{f}(\zeta)\vec{g}^t(\zeta)\vec{f}(x)}{(2\pi i)^2\mu(x-y)} \left[\left(\frac{1}{\zeta-y}-\frac{1}{\zeta-i}\right)-\left(\frac{1}{\zeta-x}-\frac{1}{\zeta-i}\right)\right]$$
$$= \frac{\vec{g}^t(y)\Gamma^{-1}(y;\mu)\left(\Gamma(x;\mu)-\Gamma(y;\mu)\right)\vec{f}(x)}{2\pi i(x-y)} = \mu S(x,y;\mu) - K(x,y).$$

The condition L^2_{loc} from the RHP 3.1 guarantees that the integrals above are well defined. □

REMARK 3.6. Note that the particular normalization (3.6) does not affect the resolvent in (3.8) as changing the normalization will result in a constant (in z) invertible factor on the left of $\Gamma(z;\mu)$.

LEMMA 3.7. *The resolvent of* $\frac{1}{4}\mathcal{H}^*\mathcal{H} : L^2(I_{in}) \to L^2(I_{in})$ *is given by the formula*

$$\mathcal{R}(\mu^2) := \left(\mathrm{Id} - \frac{1}{4\mu^2}\mathcal{H}^*\mathcal{H}\right)^{-1} = \mathrm{Id} + \pi_{in}\widehat{S}(\mu)\pi_{in} := \mathrm{Id} + \widehat{R}(\mu) \tag{3.16}$$

where $\pi_{in} : L^2(\mathbb{R}) \to L^2(I_{in})$ *is the orthogonal projection (restriction).*

PROOF. According to (3.4),

$$\widehat{S}(\mu) = \sum_{j=1}^{\infty} \left(\frac{\widehat{K}}{\mu}\right)^j, \tag{3.17}$$

where, according to (3.3), all the even powers in (3.17) are block diagonal, whereas all the odd powers in (3.17) are block off-diagonal. Therefore, the expansion of $\pi_{\rm in}\widehat{S}(\mu)\pi_{\rm in}$ contains only first diagonal blocks of the even powers in (3.17) and, thus, we obtain the left hand side of (3.16). □

As follows from Lemma 3.5, the kernel $\mathbf{R}(z,x;\mu)$ of $\widehat{R}(\mu)$, is given by

$$\mathbf{R}(z,x;\mu) = \frac{-i(\Gamma^{-1}(x;\mu)\Gamma(z;\mu))_{12}}{2\pi i\mu(z-x)} = \frac{\det\begin{bmatrix}\Gamma_{12}(z) & \Gamma_{12}(x)\\ \Gamma_{22}(z) & \Gamma_{22}(x)\end{bmatrix}}{2\pi\mu(x-z)}, \quad x,z\in I_{\rm in}. \tag{3.18}$$

REMARK 3.8. Note that $\Gamma_{2+} = \Gamma_{2-}$ on $I_{\rm in}$, where Γ_2 is the second column of the matrix Γ. So, Γ_2 is analytic on the interior of $I_{\rm in}$, whereas Γ_1 is analytic on the interior of $I_{\rm ex}$.

REMARK 3.9. The standard definition of the resolvent of the operator $\mathcal{H}^*\mathcal{H}$ is $\widehat{\mathcal{S}}(\lambda) = (\mathcal{H}^*\mathcal{H} - \lambda Id)^{-1}$. Comparison with (3.16) shows that $\lambda = 4\mu^2$ and

$$\widehat{\mathcal{S}}(\lambda) = -\frac{1}{\lambda}\left(\mathbf{1} - \frac{\mathcal{H}^*\mathcal{H}}{\lambda}\right)^{-1} = -\frac{1}{4\mu^2}\mathcal{R}(\mu^2). \tag{3.19}$$

4. Solution of the RHP 3.1 and the jump of the resolvent kernel

Consider first the case of the RHP 3.1 where $I_{\text{in}} = \varnothing$ and $I_{\text{ex}} = \mathbb{R}$. Solution $\Gamma_\infty(z;\mu)$ of such RHP can be trivially constructed; we choose it to be

$$\Gamma_\infty(z;\mu) = \begin{cases} \begin{bmatrix} 1 & -\frac{i}{2\mu} \\ 0 & 1 \end{bmatrix}, & \Im z > 0, \\ \begin{bmatrix} 1 & \frac{i}{2\mu} \\ 0 & 1 \end{bmatrix}, & \Im z < 0. \end{cases} \tag{4.1}$$

Note that the solution (4.1) does not satisfy the normalization (3.6) but has the symmetry $\overline{\Gamma(\overline{z};\overline{\mu})} = \Gamma(z;\mu)$.

Our next step is to study $\Gamma_0 = \Gamma\Gamma_\infty^{-1}$, where Γ solves the RHP 3.1 with the exception of the normalization (3.6), which is replaced by $\Gamma_0(\infty) = \mathbf{1}$. We have

$$\begin{aligned} \Gamma_{0+} = \Gamma_+\Gamma_{\infty+}^{-1} = \Gamma_-V(z)V_e^{-1}\Gamma_{\infty-}^{-1} &= \begin{cases} \Gamma_-\Gamma_{\infty-}^{-1}, & z \in I_{\text{ex}} \\ \Gamma_-\Gamma_{\infty-}^{-1}\Gamma_{\infty-}V_{\text{in}}V_e^{-1}\Gamma_{\infty-}^{-1}, & z \in I_{\text{in}} \end{cases} \\ &= \begin{cases} \Gamma_{0-}, & z \in I_{\text{ex}} \\ \Gamma_{0-}W, & z \in I_{\text{in}}. \end{cases} \end{aligned} \tag{4.2}$$

Here $V(z)$ is a piecewise constant jump matrix from the RHP 3.1: V_{in}, V_{ex} are the constant values of $V(z)$ on $I_{\text{in}}, I_{\text{ex}}$, respectively, and

$$W := V_{\text{ex}}^{-\frac{1}{2}}V_{\text{in}}V_e^{-1}V_{\text{ex}}^{\frac{1}{2}} = V_{\text{ex}}^{-\frac{1}{2}}V_{\text{in}}V_{\text{ex}}^{-\frac{1}{2}} = \begin{bmatrix} 1-\frac{1}{2\mu^2} & \frac{i}{\mu}\left(1-\frac{1}{4\mu^2}\right) \\ \frac{i}{\mu} & 1-\frac{1}{2\mu^2} \end{bmatrix}. \tag{4.3}$$

Here we have used the fact that $\Gamma_{\infty\pm}(z;\mu) = V_{\text{ex}}^{\pm\frac{1}{2}}$.

We now solve the RHP for Γ_0, which consists of the jump condition (4.2), the usual normalization condition $\Gamma_0(\infty) = \mathbf{1}$ and the usual requirement that $\Gamma_{0\pm}(z) \in L^2_{loc}$. The characteristic equation for the eigenvalues $\nu_\pm$ of W is

$$\nu^2 + (\mu^{-2} - 2)\nu + 1 = 0, \tag{4.4}$$

so that $\nu_\pm < 0$ when $\mu \in \mathbb{R}$ and $|\mu| < \frac{1}{2}$. More precisely,

$$\nu_\pm = \frac{2\mu^2 - 1 \pm \sqrt{1-4\mu^2}}{2\mu^2}. \tag{4.5}$$

Here and henceforth the branch cut in $\sqrt{1-4\mu^2}$ is taken to be $[-\frac{1}{2}, \frac{1}{2}]$. Notice that $\nu_\pm$ are complex conjugate numbers on the unit circle if $|\mu| > 1/2$. Then $W = FDF^{-1}$, where $D = \operatorname{diag}(\nu_+, \nu_-)$ and

$$\begin{aligned} F &= \begin{bmatrix} \frac{\nu_+-\nu_-}{2} & \frac{\nu_--\nu_+}{2} \\ i/\mu & i/\mu \end{bmatrix} = \operatorname{diag}\left(\frac{\nu_+-\nu_-}{2}, \frac{i}{\mu}\right)(\mathbf{1} - i\sigma_2), \\ F^{-1} &= (\mathbf{1} + i\sigma_2)\operatorname{diag}\left(\frac{1}{\nu_+-\nu_-}, \frac{\mu}{2i}\right). \end{aligned} \tag{4.6}$$

If $\tilde{\Gamma} = F^{-1}\Gamma_0F$, then $\tilde{\Gamma}$ is the solution of the RHP

$$\tilde{\Gamma}_+ = \tilde{\Gamma}_-D \text{ on } I_{\text{in}}, \ \tilde{\Gamma}(\infty) = \mathbf{1}. \tag{4.7}$$

Thus,

$$\tilde{\Gamma}(z,\mu) = e^{\frac{\ln \nu_+}{2\pi i}\mathring{\phi}(z)\sigma_3}, \tag{4.8}$$

where

$$\mathring{\phi}(z) = \int_{I_{\text{in}}} \frac{d\zeta}{\zeta - z} = \ln \prod_{j=1}^{n} \frac{z - b_{2j}}{z - b_{2j-1}} \tag{4.9}$$

solves the scalar RHP

$$\mathring{\phi}_+(z) - \mathring{\phi}_-(z) = 2\pi i \text{ on } I_{\text{in}}, \; \mathring{\phi}(\infty) = 0. \tag{4.10}$$

So,

$$\Gamma = \Gamma_0\Gamma_\infty = Fe^{\frac{\ln \nu_+}{2\pi i}\mathring{\phi}(z)\sigma_3}F^{-1}V_{\text{ex}}^{\pm\frac{1}{2}} \text{ when } \pm\Im z > 0 \tag{4.11}$$

solves the RHP 3.1 with the exception of the normalization (3.6), which was traded for the symmetry, stated in Corollary 4.2 below.

REMARK 4.1. According to (4.11), the normalization of the obtained solution $\Gamma(z;\mu)$ can be stated as $\Gamma(z;\mu) \to V_{\text{ex}}^{\pm\frac{1}{2}}$ as $z \to \infty$ when $\pm\Im z > 0$, respectively.

COROLLARY 4.2. *For any $\mu \in \mathbb{C}$, the solution $\Gamma(z;\mu)$ given by* (4.11) *has the Schwarz symmetry*

$$\overline{\Gamma(\overline{z};\overline{\mu})} = \Gamma(z;\mu). \tag{4.12}$$

This statement is true because (as the reader may verify) the matrix $\overline{\Gamma(\overline{z};\overline{\mu})}$ solves the same RHP as $\Gamma(z;\mu)$.

REMARK 4.3. The requirement that $\tilde{\Gamma}_\pm(z,\mu)$ is in L^2_{loc} near the double points requires the choice of

$$\arg\nu_+ = -\pi \text{ when } \arg\mu = 0^+ \text{ and } \arg\nu_+ = \pi \text{ when } \arg\mu = 0^-. \tag{4.13}$$

LEMMA 4.4. *The spectrum of the operator $\widehat{K}$ is a subset of the segment $[-\frac{1}{2},\frac{1}{2}]$.*

PROOF. First note that $\Gamma = \Gamma(z;\mu)$ is analytic in μ when $\mu \in \mathbb{C} \setminus [-\frac{1}{2},\frac{1}{2}]$. Then, according to Remark 4.3, the L^2_{loc} requirement in the RHP 3.1 is satisfied. Now the assertion follows from Lemma 3.5 and Remark 3.6. □

We now want to study the jump of the resolvent kernel $\mathbf{R}(z,x;\mu)$ over the segment $[-\frac{1}{2},\frac{1}{2}]$. In terms of Γ from (4.11), the solution to the RHP 3.1 is given by $\Gamma^{-1}(i;\mu)\Gamma(z;\mu)$. However, in view of Remark 3.6, the left constant factor does not affect the resolvent kernel $S(z,x;\mu)$, whereas the symmetry (4.12) will be beneficial. Therefore, in the rest of the paper we will use Γ given by (4.11) in the calculations of the resolvent kernels $S(z,x;\mu)$ and $\mathbf{R}(z,x;\mu)$.

The kernel $\mathbf{R}(z,x;\mu)$ of the resolvent $\widehat{R}$, given by (3.18), is a regular kernel (bounded at $x = z$). Since the second column $\Gamma_2(z,\mu)$ is analytic on I_{in}, see Remark 3.8, the symmetry condition (4.12) yields

$$\Gamma_2(z,\mu) = \overline{\Gamma_2(z,\bar{\mu})}, \; z \in I_{\text{in}}, \tag{4.14}$$

so the jump $\Delta_\mu\Gamma_2(z,\mu)$ of $\Gamma_2(z,\mu)$ over $\mathbb{R}$ in the complex μ plane becomes

$$\Delta_\mu\Gamma_2(z,\mu) = 2i\Im\Gamma_2(z,\mu), \; z \in I_{\text{in}}. \tag{4.15}$$

Applying the same considerations to the kernel $\mathbf{R}(z,x;\mu)$, see (3.18), we obtain

$$\Delta_\mu \mathbf{R}(z,x;\mu) = \frac{2i\Im \det \begin{bmatrix} \Gamma_{12}(z) & \Gamma_{12}(x) \\ \Gamma_{22}(z) & \Gamma_{22}(x) \end{bmatrix}}{2\pi\mu(x-z)}, \tag{4.16}$$

where the numerator can be represented as

$$2i\left(\det\left[\Re\Gamma_2(z), \Im\Gamma_2(x)\right] + \det\left[\Re\Gamma_2(x), -\Im\Gamma_2(z)\right]\right). \tag{4.17}$$

Since $\Gamma_{2+} = \Gamma_{2-}$ on I_{in}, a particular choice of boundary values of Γ_2 in (4.16) does not matter. Let us choose Γ_{2+} in (4.16). Then, according to (4.11),

$$\Gamma_2 = Fe^{\frac{\ln\nu_+}{2\pi i}\mathring{\phi}(z)\sigma_3}U_2,$$

where U_2 denotes the second column of the matrix

$$U := F^{-1}V_{\text{ex}}^{\frac{1}{2}} = \begin{bmatrix} 1 & 1+\frac{1}{\sqrt{1-4\mu^2}} \\ -1 & 1-\frac{1}{\sqrt{1-4\mu^2}} \end{bmatrix} \cdot \operatorname{diag}\left(\frac{1}{\nu_+-\nu_-}, -\frac{i\mu}{2}\right). \tag{4.18}$$

Now we have

$$\begin{aligned} d := |\Gamma_2(z), \Gamma_2(x)| &= \left|F\cdot\left(\tilde{\Gamma}(z)U_2, \tilde{\Gamma}(x)U_2\right)\right| \\ &= \frac{-i\sqrt{1-4\mu^2}}{4\mu}\left(1-\frac{1}{1-4\mu^2}\right)\left|\tilde{\Gamma}(z)\begin{pmatrix}1\\1\end{pmatrix}, \tilde{\Gamma}(x)\begin{pmatrix}1\\1\end{pmatrix}\right| \\ &= \frac{2i\mu}{\sqrt{1-4\mu^2}}\sinh\left(\frac{\ln\nu_+}{2\pi i}[\mathring{\phi}(z)-\mathring{\phi}(x)]\right). \end{aligned} \tag{4.19}$$

Let us consider μ on the upper shore of $(0,\frac{1}{2})$. Then, according to Remark 4.3,

$$\Im d = \frac{2\mu\Re\sinh\left(\frac{\ln\nu_+}{2\pi i}[\mathring{\phi}(z)-\mathring{\phi}(x)]\right)}{\sqrt{1-4\mu^2}} = \frac{-2\mu\sinh\left(\frac{\mathring{\phi}(z)-\mathring{\phi}(x)}{2}\right)\cos\left(\frac{\ln|\nu_+|}{2\pi}[\mathring{\phi}(z)-\mathring{\phi}(x)]\right)}{\sqrt{1-4\mu^2}}, \tag{4.20}$$

where, as it was mentioned above, x and z here and henceforth are taken on the positive side of I_{in}.

REMARK 4.5. It is easy to see that for $\mu\in(\frac{1}{2},+\infty)$ we have $\frac{\ln\nu_+}{2\pi i}\in\mathbb{R}$; therefore d, defined by (4.19), is real and $\Delta_\mu\mathbf{R}(z,x;\mu)$, according to (4.16), is zero. Similar considerations work for $\mu<0$.

Let us define the polynomials and their ratio

$$\beta_{\text{od}}(z) = \prod_{j=1}^n (z-b_{2j-1}), \quad \beta_{\text{ev}}(z) = \prod_{j=1}^n (z-b_{2j}), \quad \beta(z) = \beta_{\text{ev}}(z)/\beta_{\text{od}}(z). \tag{4.21}$$

According to (4.9), $\mathring{\phi}(z) = \ln\beta(z)$. Note that $\arg\beta(z) = \pi$ when $z\in I_{\text{in}}$ and $\arg\beta(z) = 0$ when $z\in I_{\text{ex}}$. Therefore, $\mathring{\phi}(z)-\mathring{\phi}(x)\in\mathbb{R}$ for any $z,x\in I_{\text{in}}$. Moreover

$$\mathring{\phi}(z)-\mathring{\phi}(x) = \phi(z)-\phi(x), \tag{4.22}$$

where $\phi(z) := \Re\mathring{\phi}(z)$, provided that both z,x are either in I_{in} or in I_{ex}. According to (4.22), in what follows we will replace $\mathring{\phi}(z)-\mathring{\phi}(x)$ with $\phi(z)-\phi(x)$. Now we calculate

$$2\sinh\left(\frac{\phi(z)-\phi(x)}{2}\right)=\sqrt{\frac{\beta(z)}{\beta(x)}}-\sqrt{\frac{\beta(x)}{\beta(z)}}=\frac{\beta_{\rm ev}(z)\beta_{\rm od}(x)-\beta_{\rm ev}(x)\beta_{\rm od}(z)}{\sqrt{\prod_{j=1}^{2n}(x-b_j)(z-b_j)}}$$
(4.23)
$$=\frac{(z-x)\sum_{i,j=1}^{n}\beta_{ij}z^{i-1}x^{j-1}}{\sqrt{\prod_{j=1}^{2n}(x-b_j)(z-b_j)}},$$

(4.24)
$$\cos\left(\frac{\ln|\nu_+|}{2\pi}[\phi(z)-\phi(x)]\right)=\cos\left(\frac{\phi(z)\ln|\nu_+|}{2\pi}\right)\cos\left(\frac{\phi(x)\ln|\nu_+|}{2\pi}\right)$$
$$+\sin\left(\frac{\phi(z)\ln|\nu_+|}{2\pi}\right)\sin\left(\frac{\phi(x)\ln|\nu_+|}{2\pi}\right),\quad x,z\in I_{\rm in},$$

where the matrix $B_n(\beta_{\rm ev},\beta_{\rm od})=(\beta_{ij})$ is called the Bézout matrix of the polynomials $\beta_{\rm ev}(z),\beta_{\rm od}(z)$. Note that $\prod_{j=1}^{2n}(x-b_j)(z-b_j)>0$ for any z,x in the interior of $I_{\rm in}$, and the square root in (4.23) is computed according to (4.30) below.

LEMMA 4.6. *For polynomials β_{ev},β_{od} of order $n\in\mathbb{N}$, defined above, the Bézout matrix $B_n(\beta_{ev},\beta_{od})=(\beta_{ij})$ is symmetric and positive definite and, therefore,*

$$\sum_{i,j=1}^{n}\beta_{ij}z^{i-1}x^{j-1}=\sum_{j=1}^{n}\rho_jP_j(z)P_j(x), \tag{4.25}$$

where $\rho_j>0$, $j=1,\dots,n$, are the eigenvalues of the matrix $B_n(\beta_{ev},\beta_{od})$ and $P_j(z)$, $j=1,\dots,n$ are polynomials of degree not exceeding $n-1$.

PROOF. The symmetry of the matrix $B_n(\beta_{\rm ev},\beta_{\rm od})$ follows directly from the definition. Invertibility of $B_n(\beta_{\rm ev},\beta_{\rm od})$ follows from the fact that the polynomials $\beta_{\rm od}(z),\beta_{\rm ev}(z)$ have no common roots. The fact that B_n is positive-definite is proven in Lemma A.2 below. As a symmetric matrix, $B_n=B_n(\beta_{\rm ev},\beta_{\rm od})$ can be represented as follows

$$B_n(\beta_{\rm ev},\beta_{\rm od})=\Omega^t{\rm diag}(\rho_1,\dots,\rho_n)\Omega, \tag{4.26}$$

where Ω is an orthogonal matrix. Then
(4.27)
$$\sum_{i,j=1}^{n}\beta_{ij}z^{i-1}x^{j-1}=\vec{z}_n^{\,t}B_n\vec{x}_n=\vec{z}_n^{\,t}\Omega^t{\rm diag}(\rho_1,\dots,\rho_n)\Omega\vec{x}_n=\sum_{j=1}^{n}\rho_jP_j(z)P_j(x),$$

where $\vec{z}_n^{\,t}=(1,z,\dots z^{n-1}),\vec{x}_n^{\,t}=(1,x,\dots x^{n-1})$ and $(P_1(z),\dots,P_n(z))=\vec{z}_n^{\,t}\Omega^t$. □

Let us return to the jump of $\mathbf{R}(z,x;\mu)$ given by (4.16). Applying Lemma 4.6 to $\Im d$ from (4.20), we obtain
(4.28)
$$\Delta_\mu\mathbf{R}(z,x;\mu)=\frac{2i\Im d}{2\pi\mu(x-z)}=\frac{i\cos\left(\frac{\ln|\nu_+|}{2\pi}[\phi(z)-\phi(x)]\right)}{\pi\sqrt{1-4\mu^2}\sqrt{\prod_{j=1}^{2n}(x-b_j)(z-b_j)}}\sum_{j=1}^{n}\rho_jP_j(z)P_j(x).$$

We now observe that for any $z\in I_{\rm in}$

$$\prod_{j=1}^{2n}(z-b_j)=e^{i(\pi+2\arg\beta_{\rm od}(z))}\prod_{j=1}^{2n}|z-b_j|. \tag{4.29}$$

As $e^{i\arg\beta_{\rm od}(z)} = \mathrm{sgn}\beta_{\rm od}(z)$, we obtain

$$\sqrt{\prod_{j=1}^{2n}(z-b_j)(x-b_j)} = -\mathrm{sgn}\beta_{\rm od}(z)\mathrm{sgn}\beta_{\rm od}(x)\prod_{j=1}^{2n}|(z-b_j)(x-b_j)|^{\frac{1}{2}}. \tag{4.30}$$

Now, (4.28) becomes

$$\Delta_\mu\mathbf{R}(z,x;\mu) = \frac{-i}{\pi\sqrt{1-4\mu^2}}\sum_{j=1}^{n}\left[f_j(z,\mu)f_j(x,\mu)+g_j(z,\mu)g_j(x,\mu)\right], \tag{4.31}$$

where

$$\begin{aligned} f_j(z,\mu) &= \frac{P_j(z)}{\mathrm{sgn}(\beta_{\rm od}(z))}\sqrt{\frac{\rho_j}{\prod_{m=1}^{2n}|z-b_m|}}\cos\left(\frac{\phi(z)\ln|\nu_+|}{2\pi}\right), \\ g_j(z,\mu) &= \frac{P_j(z)}{\mathrm{sgn}(\beta_{\rm od}(z))}\sqrt{\frac{\rho_j}{\prod_{m=1}^{2n}|z-b_m|}}\sin\left(\frac{\phi(z)\ln|\nu_+|}{2\pi}\right). \end{aligned} \tag{4.32}$$

Thus, we see that the jump of the resolvent kernel $\Delta_\mu\mathbf{R}(z,x;\lambda)$ over the continuous spectrum is a degenerate kernel of rank $2n$. Remark 3.9 implies the following theorem.

THEOREM 4.7. *The jump of the kernel* $\mathbf{S}$ *of the resolvent* $\widehat{\mathcal{S}}(\lambda)$ *from Remark 3.9 over the oriented interval* $(0,1)$ *is a degenerate kernel of rank* $2n$ *of the form*

(4.33)

$$\Delta_\lambda\mathbf{S}(z,x;\lambda) = \frac{i}{\pi\lambda\sqrt{1-\lambda}}\sum_{j=1}^{n}\left[f_j\left(z,\sqrt{\frac{\lambda}{2}}\right)f_j\left(x,\sqrt{\frac{\lambda}{2}}\right)+g_j\left(z,\sqrt{\frac{\lambda}{2}}\right)g_j\left(x,\sqrt{\frac{\lambda}{2}}\right)\right],$$

where f_j, g_j *are given in* (4.32) *and* $\lambda = 4\mu^2$.

REMARK 4.8. As an example, suppose $n=1$ and $b_1=-b$, $b_2=b$ for some $b>0$. In this case, $\rho_1=2b$, $\phi(z)=\ln[(b-z)/(z+b)]$, $P_1\equiv 1$, and $\Delta_\lambda\mathbf{S}(z,x;\lambda)/(2\pi i)$ matches exactly (2.27) calculated in Section 2 (see also (A.1)–(A.3) in the Appendix, which shows the equality of the cosine terms).

REMARK 4.9. Our calculations of the jump of the resolvent kernel of the operator $\mathcal{H}^*\mathcal{H}$, summarized in Theorem 4.7, can be repeated verbatim for the operator $\mathcal{H}\mathcal{H}^*$. The main difference in the calculations is that the numerator $-i(\Gamma^{-1}(x;\mu)\Gamma(z;\mu))_{12}$ in (3.18) should be replaced with $i(\Gamma^{-1}(x;\mu)\Gamma(z;\mu))_{21}$. Correspondingly, $\det[\Gamma_2(z),\Gamma_2(x)]$ in (4.16) should be replaced with $\det[\Gamma_1(z),\Gamma_1(x)]$, but the rest of the formula stays the same. This difference causes the replacement of the column U_2 with the column U_1 in (4.19), but direct calculations show the net effect is just a change of sign in (4.19). However, when $z\in I_{\rm ex}$, the right hand side of (4.30) also changes sign. Thus, at the end, the jump $\Delta_\lambda\mathbf{S}(z,x;\lambda)$ given by (4.33) remains the same for both $\mathcal{H}\mathcal{H}^*$ and $\mathcal{H}^*\mathcal{H}$.

5. Spectral properties of $\mathcal{H}^*\mathcal{H}$ and $\mathcal{H}\mathcal{H}^*$.

We show first that $A := \mathcal{H}^*\mathcal{H}$ has only the absolutely continuous spectrum. Since the norm of $\mathcal{H}$ is bounded by 1, the spectrum of $\mathcal{H}^*\mathcal{H}$ is a subset of $[0,1]$. First, we establish that $\lambda = 0, 1$ are not the eigenvalues of A. The standard argument based on analytic continuation shows that $\lambda = 0$ is not an eigenvalue. If $\lambda = 1$ is an eigenvalue, and $f \in L^2(I_{\text{in}})$ is an associated eigenfunction, then, by definition $\|\mathcal{H}^*\mathcal{H}f\|_{I_{\text{in}}} = \|f\|_{I_{\text{in}}}$. Let $\mathcal{H}_G$ denote the "global" Hilbert transform, which computes values for all $x \in \mathbb{R}$. Since $\mathcal{H}^*$ is just (-1) times the FHT acting from $L^2(I_{\text{ex}}) \to L^2(I_{\text{in}})$, and the norm of the "global" Hilbert transform $\mathcal{H}_G$ equals 1, we have $\|\mathcal{H}^*\mathcal{H}f\|_{I_{\text{in}}} \le \|\mathcal{H}f\|_{I_{\text{ex}}}$. Using again that the norm of $\mathcal{H}_G$ equals 1, we have

$$\|f\|_{I_{\text{in}}} = \|\mathcal{H}^*\mathcal{H}f\|_{I_{\text{in}}} \le \|\mathcal{H}f\|_{I_{\text{ex}}} \le \|\mathcal{H}_G f\|_{\mathbb{R}} = \|f\|_{I_{\text{in}}}. \tag{5.1}$$

Hence, $\mathcal{H}f \equiv 0$ on I_{in}. Using the uniqueness property of the solution to a RHP in the L^2 class, we conclude that $f \equiv 0$.

Our next goal is to show that $\sigma_\phi(\lambda)$ is absolutely continuous on $[\epsilon, 1-\epsilon]$ for any $\epsilon > 0$. Let ϕ be any smooth function with compact support in the interior of I_{in}. It is obvious from (4.32) and (4.33) that $\sigma_\phi(\lambda)$ has the desired property. Let X_{ac} be the set of absolute continuity of H^*H. As is known (see e.g. Theorem 1.5 in Section X.1.2 of [**8**]), X_{ac} is a closed linear manifold. The functions ϕ we just considered are in X_{ac}, and they are dense in $L^2(I_{\text{in}})$ Hence $X_{ac} = L^2(I_{\text{in}})$. We just established the following result.

THEOREM 5.1. *The operator $\mathcal{H}^*\mathcal{H} : L^2(I_{in}) \to L^2(I_{in})$ has absolutely continuous spectrum, which coincides with the interval $[0,1]$.*

For convenience, let us rewrite the expressions for $|\nu_+|$ and f_j, g_j from Section 4 in terms of the variable λ (instead of μ) and, simultaneously, slightly modify the functions f_j and g_j. We obtain:

$$|\nu_+| = \frac{2-\lambda-2\sqrt{1-\lambda}}{\lambda}, \tag{5.2}$$

$$\begin{aligned} f_j(z,\lambda) &= \frac{P_j(z)\text{sgn}(\beta_{\text{od}}(z))}{\pi\sqrt{2\lambda}(1-\lambda)^{1/4}}\sqrt{\frac{\rho_j}{\prod_{m=1}^{2n}|z-b_m|}}\cos\left(\frac{\phi(z)\ln|\nu_+|}{2\pi}\right), \\ g_j(z,\lambda) &= \frac{P_j(z)\text{sgn}(\beta_{\text{od}}(z))}{\pi\sqrt{2\lambda}(1-\lambda)^{1/4}}\sqrt{\frac{\rho_j}{\prod_{m=1}^{2n}|z-b_m|}}\sin\left(\frac{\phi(z)\ln|\nu_+|}{2\pi}\right), \end{aligned} \tag{5.3}$$

and, using (4.33),

(5.4)
$$\frac{dE_\lambda}{d\lambda} = \frac{\Delta_\lambda \mathbf{S}(z,x;\lambda)}{2\pi i} = \sum_{j=1}^{n}[f_j(z,\lambda)f_j(x,\lambda)+g_j(z,\lambda)g_j(x,\lambda)] = \sum_{j=1}^{2n}\varphi_j(z,\lambda)\varphi_j(x,\lambda),$$
$$\varphi_j := f_j,\ \varphi_{j+n} := g_j,\ 1 \le j \le n.$$

As before, we denote by $\widehat{E}$ the integral operator with a kernel E. Since $\widehat{E}_\lambda$ is the resolution of the identity for A, we have

(5.5)
$$f(z) = \int_0^1 \left(\frac{d\widehat{E}_\lambda}{d\lambda} f\right) d\lambda = \sum_{j=1}^{2n} \int_0^1 \varphi_j(z,\lambda) \left(\int_{I_{\text{in}}} \varphi_j(x,\lambda) f(x) dx\right) d\lambda,$$
$$Af(z) = \int_0^1 \lambda \left(\frac{d\widehat{E}_\lambda}{d\lambda} f\right) d\lambda = \sum_{j=1}^{2n} \int_0^1 \lambda\varphi_j(z,\lambda) \left(\int_{I_{\text{in}}} \varphi_j(x,\lambda) f(x) dx\right) d\lambda, \ f \in L^2(I_{\text{in}}).$$

Consider the map $\mathcal{F}_{\text{in}} : L^2(I_{\text{in}}) \to L^2_{2n}([0,1])$:

$$(\mathcal{F}_{\text{in}} f)(\lambda) := \tilde{\mathbf{f}}(\lambda) := (\tilde{\mathbf{f}}_1(\lambda), \dots, \tilde{\mathbf{f}}_{2n}(\lambda)) \\ := \left(\int_{I_{\text{in}}} \varphi_1(x,\lambda) f(x) dx, \dots, \int_{I_{\text{in}}} \varphi_{2n}(x,\lambda) f(x) dx\right). \tag{5.6}$$

Here $L^2_{2n}([0,1])$ is the direct sum of $2n$ copies of $L^2([0,1])$, $L^2_{2n}([0,1]) = \oplus_{j=1}^{2n} L^2([0,1])$. Similar notation is used in what follows. The map $\mathcal{F}_{\text{in}}$ is an isometry, because

(5.7)
$$\|\mathcal{F}_{\text{in}} f\|^2 = \int_0^1 \sum_j \tilde{\mathbf{f}}_j^2(\lambda) d\lambda = \sum_j \int_0^1 \left(\int_{I_{\text{in}}} \varphi_j(x,\lambda) f(x) dx\right) \left(\int_{I_{\text{in}}} \varphi_j(y,\lambda) f(y) dy\right) d\lambda$$
$$= \int_{I_{\text{in}}} f(x) \left[\sum_j \int_0^1 \varphi_j(x,\lambda) \left(\int_{I_{\text{in}}} \varphi_j(y,\lambda) f(y) dy\right) d\lambda\right] dx$$
$$= \int_{I_{\text{in}}} f^2(x) dx = \|f\|^2.$$

The second to last equality follows from the fact that $\widehat{E}_\lambda$ is the resolution of the identity (cf. (5.5)).

To better understand the transform $\mathcal{F}_{\text{in}}$ in (5.6) and, in particular, establish that it is onto, we introduce three additional transforms. The first one is given by

(5.8)
$$\tilde{\mathbf{f}}(\lambda) \in \mathbf{L^2_{2n}([0,1])} \to \mathring{\mathbf{f}}(\xi) := T_1 \tilde{\mathbf{f}} := \sqrt{\pi\lambda}(1-\lambda)^{1/4} \tilde{\mathbf{f}}(\lambda)\Big|_{\lambda=\lambda(\xi)} \in L^2_{2n}((-\infty,0]),$$
$$\xi = \ln|\nu_+|/\pi.$$

Note that $0 < |\nu_+| < 1$ for $\lambda \in (0,1)$. From (5.2), $d\xi/d\lambda = (\pi\lambda\sqrt{1-\lambda})^{-1}$, so T_1 is an isometry. By construction, T_1 is just a change of variables divided by the square root of the derivative $d\xi/d\lambda$ to preserve the L^2-norm. Hence T_1 is onto. The second isometry is given by

$$\mathring{\mathbf{f}}(\xi) \in L^2_{2n}((-\infty,0]) \to \hat{\mathbf{f}}(\xi) := (T_2 \mathring{\mathbf{f}})(\xi) \in L^2_n(\mathbb{R}),$$
$$\hat{\mathbf{f}}_j(\xi) = \frac{1}{\sqrt{2}} \begin{cases} \mathring{\mathbf{f}}_j(\xi) + i\mathring{\mathbf{f}}_{j+n}(\xi), & \xi < 0, \\ \mathring{\mathbf{f}}_j(-\xi) - i\mathring{\mathbf{f}}_{j+n}(-\xi), & \xi > 0, \end{cases} \quad 1 \le j \le n, \tag{5.9}$$

and it is also onto. Note that T_2 commutes with T_1.

Using (5.3), (5.11), (5.6), and the definitions of the above transformations, we calculate

$$\mathcal{T}\mathcal{F}_{\text{in}}f(\xi) = \frac{1}{2\sqrt{\pi}}\left(\int_{I_{\text{in}}} f(x)p_1(x)e^{i\xi\frac{\phi(x)}{2}}dx, \dots, \int_{I_{\text{in}}} f(x)p_n(x)e^{i\xi\frac{\phi(x)}{2}}dx\right), \tag{5.10}$$

where

$$\mathcal{T} := T_2T_1, \quad p_j(z) = \frac{P_j(z)}{\text{sgn}(\beta_{\text{od}}(z))}\sqrt{\frac{\rho_j}{\prod_{m=1}^{2n}|z-b_m|}}. \tag{5.11}$$

Finally, the third map is given by

$$\begin{aligned} &f(x) \in L^2(I_{\text{in}}) \to \check{\mathbf{f}}(t) := (T_{\text{in}}f)(t) \\ &:= \sqrt{2}\left(\frac{\text{sgn}(\beta_{\text{od}}(x))f(x)}{\sqrt{|\phi'(x)|}}\bigg|_{x=\phi_1^{-1}(2t)}, \dots, \frac{\text{sgn}(\beta_{\text{od}}(x))f(x)}{\sqrt{|\phi'(x)|}}\bigg|_{x=\phi_n^{-1}(2t)}\right) \in L_n^2(\mathbb{R}). \end{aligned} \tag{5.12}$$

Here ϕ_k^{-1} is the inverse of $\phi(x)$ on the k-th interval $[b_{2k-1}, b_{2k}]$. By Lemma A.1, $\phi'(x) < 0$ on I_{in}. Clearly, T_{in} is also an isometry and onto. From (4.9),

$$\phi'(x) = \frac{Q(x)}{\beta_{\text{od}}(x)\beta_{\text{ev}}(x)}, \quad Q(x) := \beta_{\text{ev}}'(x)\beta_{\text{od}}(x) - \beta_{\text{ev}}(x)\beta_{\text{od}}'(x). \tag{5.13}$$

It follows from (5.13) and Lemma A.1 that $Q(x)$ is positive and bounded away from zero on I_{in}. Using these three maps, we obtain from (5.3), (5.4), and (5.6):

$$\begin{aligned} (\mathcal{T}\mathcal{F}_{\text{in}}T_{\text{in}}^{-1}\check{\mathbf{f}})_j(\xi) =& \sum_{k=1}^{n}\frac{1}{\sqrt{2\pi}}\int_{\mathbb{R}}\check{\mathbf{f}}_k(t)M_{jk}^{(\text{in})}(t)e^{i\xi t}dt, \ 1 \le j \le n, \\ M_{jk}^{(\text{in})}(t) :=& P_j(x_k)\sqrt{\frac{\rho_j}{Q(x_k)}}, \ x_k = \phi_k^{-1}(2t). \end{aligned} \tag{5.14}$$

The formula (5.14) can be written in compact form as follows:

$$\mathcal{T}\mathcal{F}_{\text{in}}T_{\text{in}}^{-1} = \mathcal{F}\mathcal{M}_{\text{in}} : L_n^2(\mathbb{R}) \to L_n^2(\mathbb{R}), \tag{5.15}$$

where $\mathcal{F}$ is the conventional Fourier transform of (2.5) applied component-wise, and $\mathcal{M}_{\text{in}}$ is the operator of multiplication by the matrix $M^{(\text{in})}$. Since $Q(x)$ is bounded away from zero on I_{in}, the factor in (5.14) is continuous and bounded on $\mathbb{R}$. Moreover, it is shown in Lemma A.3 that the matrix $M^{(\text{in})}$ is orthogonal for all t. Hence, by the well-known properties of the Fourier transform, it is easy to see that the range of $\mathcal{F}_{\text{in}}$ is all of $L_{2n}^2([0,1])$. Combining with (5.7) and (5.5) we obtain the spectral decomposition of A.

THEOREM 5.2. *Let $A = \mathcal{H}^*\mathcal{H}$. The operator $\mathcal{F}_{in} : L^2(I_{in}) \to L_{2n}^2([0,1])$ defined in (5.6) is unitary. Its inverse is given by*

$$(\mathcal{F}_{in}^{-1}\tilde{\mathbf{f}})(x) = (\mathcal{F}_{in}^*\tilde{\mathbf{f}})(x) = \sum_{j=1}^{2n}\int_0^1 \varphi_j(x,\lambda)\tilde{\mathbf{f}}_j(\lambda)d\lambda. \tag{5.16}$$

The multiplicity of the spectrum of $\mathcal{H}^*\mathcal{H}$ is $2n$, and one has

$$\mathcal{H}^*\mathcal{H} = \mathcal{F}_{in}^{-1}\lambda\mathcal{F}_{in}. \tag{5.17}$$

PROOF. From (5.7), $\mathcal{F}_{\rm in}$ is an isometry. From (5.15) and the properties of $\mathcal{T}$, $T_{\rm in}$, and $\mathcal{M}_{\rm in}$, $\mathcal{F}_{\rm in}$ is onto. Hence it is unitary, which proves the first equality in (5.16). The second equality in (5.16) is obtained by computing the adjoint operator $\mathcal{F}_{\rm in}^*$. The equation (5.17) follows from the second equation in (5.5), (5.6), and (5.16). From (5.17), $\mathcal{H}^*\mathcal{H}$ is unitarily equivalent to the operator of multiplication by λ from $L^2_{2n}([0,1])$ into itself. Hence the multiplicity of the spectrum equals $2n$. $\square$

Inverting (5.15), one obtains

$$T_{\rm in}\mathcal{F}_{\rm in}^{-1}\mathcal{T}^{-1} = \mathcal{M}_{\rm in}^T\mathcal{F}^{-1}. \tag{5.18}$$

Since $\widehat{E}_\lambda$ is the resolution of the identity for $A = \mathcal{H}^*\mathcal{H}$ (cf. (5.5)), equations (5.15) and (5.18) imply

$$A = \mathcal{F}_{\rm in}^{-1}\lambda\mathcal{F}_{\rm in} = [T_{\rm in}^{-1}\mathcal{M}_{\rm in}^T\mathcal{F}^{-1}\mathcal{T}]\lambda[\mathcal{T}^{-1}\mathcal{F}\mathcal{M}_{\rm in}T_{\rm in}]. \tag{5.19}$$

From (5.8) and (5.9), $\mathcal{T}\lambda\mathcal{T}^{-1}$ is the same as the multiplication with $\lambda(\xi)$. From (5.2) and (5.8),

$$\lambda(\xi) = 1/\cosh^2(\xi\pi/2). \tag{5.20}$$

Combining with (5.19) gives the analogue of the first formula in (2.13)

$$\mathcal{H}^*\mathcal{H} = (\mathcal{F}\mathcal{M}_{\rm in}T_{\rm in})^{-1}\frac{1}{\cosh^2(\xi\pi/2)}\mathcal{F}\mathcal{M}_{\rm in}T_{\rm in}. \tag{5.21}$$

Thus we obtain another convenient diagonalization of $A = \mathcal{H}^*\mathcal{H}$.

THEOREM 5.3. *One has*

$$\mathcal{H}^*\mathcal{H} = (\mathcal{F}\mathcal{M}_{in}T_{in})^{-1}\frac{1}{\cosh^2(\xi\pi/2)}\mathcal{F}\mathcal{M}_{in}T_{in}. \tag{5.22}$$

Here T_{in} is defined in (5.12), $\mathcal{M}_{in}$ is the operator of multiplication by the orthogonal matrix $M^{(in)}$ defined in (5.14), and $\mathcal{F}$ is the conventional component-wise Fourier transform on $L^2_n(\mathbb{R})$.

We believe that the diagonalization in (5.22) is more convenient for analyzing the ill-posedness of inversion and from the computational perspective than the one in (5.17).

A similar derivation works for the second normal operator $\mathcal{H}\mathcal{H}^*$. Since $\beta(x) > 0$ on $I_{\rm ex}$, we have $\phi(x) = \mathring{\phi}(x)$. Similarly to the proof of Lemma A.1, we can show that $\phi' > 0$ on $I_{\rm ex}$. Since $\beta > 0$ on $I_{\rm ex}$, (5.13) implies that $Q > 0$ on $I_{\rm ex}$. In view of the Remark 4.9, we define $\mathcal{F}_{\rm ex} : L^2(I_{\rm ex}) \to L^2_{2n}([0,1])$ similarly to (5.6):

$$\begin{aligned}(\mathcal{F}_{\rm ex}f)(\lambda) &:= \tilde{\mathbf{f}}(\lambda) := (\tilde{\mathbf{f}}_1(\lambda), \dots, \tilde{\mathbf{f}}_{2n}(\lambda)) \\ &:= \left(\int_{I_{\rm ex}} \varphi_1(x,\lambda)f(x)dx, \dots, \int_{I_{\rm ex}} \varphi_{2n}(x,\lambda)f(x)dx\right).\end{aligned} \tag{5.23}$$

The analogue of (5.12) becomes

$$\begin{aligned}&f(x) \in L^2(I_{\rm in}) \to \check{\mathbf{f}}(s) := (T_{\rm ex}f)(s) \\ &:= \sqrt{2}\left(\left.\frac{\operatorname{sgn}(\beta_{\rm od}(x))f(x)}{\sqrt{\phi'(x)}}\right|_{x=\phi_1^{-1}(2s)}, \dots, \left.\frac{\operatorname{sgn}(\beta_{\rm od}(x))f(x)}{\sqrt{\phi'(x)}}\right|_{x=\phi_n^{-1}(2s)}\right) \in L^2_n(\mathbb{R}),\end{aligned} \tag{5.24}$$

where $\phi_k^{-1}(2s)$ is the inverse of ϕ that gives values on the k-th sub-interval of $I_{\rm ex}$. The way the sub-intervals of $I_{\rm ex}$ are enumerated is irrelevant. Note that one of these sub-intervals contains the point at infinity.

Let $M_{jk}^{(\rm ex)}$ be defined as in (5.14), but with $\phi_k^{-1}(2s)$ defined as in (5.24). Then all the results in this section hold for $A = \mathcal{H}\mathcal{H}^*$. The only change is that $\mathcal{F}_{\rm in}$, $T_{\rm in}$, and $\mathcal{M}_{\rm in}$ need to be replaced by $\mathcal{F}_{\rm ex}$, $T_{\rm ex}$, and $\mathcal{M}_{\rm ex}$, respectively.

As an example, consider the case $n = 1$ (cf. Remark 4.8). It is easy to see that the matrices $M^{(\rm in)}$ and $M^{(\rm ex)}$ are the identity, and the maps $T_{\rm in}$ and $T_{\rm ex}$ defined in (5.12), (5.24), coincide with $T_{\rm in}$ and $T_{\rm ex}$ defined in (2.11), respectively. Hence, (5.22) and the same equation for $\mathcal{H}\mathcal{H}^*$ coincide with (2.13) in this case.

Appendix A. Some auxiliary calculations

Transforming (2.27) Since cos is even we use a minus in front of the square root (i.e., use $-a^{-1}(\lambda)$ instead of $a^{-1}(\lambda)$ in (2.23)):

$$\text{(A.1)}\qquad \ln\frac{1-\sqrt{1-\lambda}}{\sqrt{\lambda}} = \frac{1}{2}\ln\frac{2-\lambda-2\sqrt{1-\lambda}}{\lambda} = \frac{1}{2}\ln(-\nu_+) = \frac{1}{2}\ln|\nu_+|.$$

In our case (cf. (4.9)),

$$\text{(A.2)}\qquad \phi(z) = \Re\mathring{\phi}(z) = \ln\frac{b-z}{z+b},$$

so the cos term in (2.27), see (4.22), becomes

$$\text{(A.3)}\qquad \cos\left(\frac{1}{2\pi}\ln|\nu_+|[\phi(x)-\phi(y)]\right),$$

which matches the left-hand side of (4.24).

LEMMA A.1. *With $\phi(x) = \Re\mathring{\phi}(x)$, where $\mathring{\phi}(x)$ is given by (4.9), we have $\phi'(x) < 0$ for all $x \in I_{in}$. Also, $\phi(x) \to +\infty$ as $x \to b_{2j-1}^+$ and $\phi(x) \to -\infty$ as $x \to b_{2j}^-$.*

PROOF. As follows from (4.9)

$$\text{(A.4)}\qquad -\phi'(x) = \left[\frac{1}{x-b_1} - \frac{1}{x-b_{2n}}\right] - \frac{1}{x-b_2} + \cdots + \frac{1}{x-b_{2k-1}} - \frac{1}{x-b_{2k}} + \cdots - \frac{1}{x-b_{2n-2}} + \frac{1}{x-b_{2n-1}}.$$

Suppose $b_{2k-1} < x < b_{2k}$. If $k < n$, the right-hand side of (A.4) gets smaller when we shift b_{2n-1} as far to the left as possible, i.e. to b_{2n-2}. In this case the last term two terms on the right cancel each other and we get:

(A.5)

$$-\phi'(x) \geq \left[\frac{1}{x-b_1} - \frac{1}{x-b_{2n}}\right] - \frac{1}{x-b_2} + \cdots + \frac{1}{x-b_{2k-1}} - \frac{1}{x-b_{2k}} + \cdots + \frac{1}{x-b_{2n-3}}.$$

Similarly, if $k > 1$, to make the right-hand side of (A.5) smaller, we should shift b_2 as far to the right as possible, i.e. to b_3 to get

(A.6)

$$-\phi'(x) > \left[\frac{1}{x-b_1} - \frac{1}{x-b_{2n}}\right] - \frac{1}{x-b_4} + \cdots + \frac{1}{x-b_{2k-1}} - \frac{1}{x-b_{2k}} + \cdots + \frac{1}{x-b_{2n-3}}.$$

We continue shifting on the left and on the right as needed. When we reach the kth subinterval all terms except the two in brackets cancel out and we get $-\phi'(x) > 0$ because $b_1 < x < b_{2n}$.

The other assertions of the lemma are obvious. □

LEMMA A.2. *The Bézout matrix $B_n(\beta_{ev}, \beta_{od}) = (\beta_{ij})$ of Lemma 4.6 is positive-definite.*

PROOF. The result follows from the fact that the roots of $\beta_{\rm od}$ and $\beta_{\rm ev}$ are simple and separated, see e.g. [**3**]. Here, for completeness, we sketch the proof. Denote

$$B(x,z) := \sum_{i,j=1}^{n} \beta_{ij} x^{i-1} z^{j-1}. \tag{A.7}$$

Dividing both sides of (4.23) by $z-x$ and letting $z \to x$ for any fixed $x \in I_{\rm in}$, we get (cf. (4.30)):

$$-\phi'(x) \prod_{j=1}^{2n} |x-b_j| = B(x,x). \tag{A.8}$$

Lemma A.1 implies that $B(x,x) > 0$ for any $x \in I_{\rm in}$. It follows from (4.9) and (A.8) that $B(b_j,b_j) > 0$ for all $j, 1 \le j \le 2n$. Since $B(x,z) = (\beta_{\rm ev}(z)\beta_{\rm od}(x) - \beta_{\rm ev}(x)\beta_{\rm od}(z))/(z-x)$, we have $B(b_{2j}, b_{2k}) = 0$ for any $1 \le j,k \le n$, $j \ne k$. The vectors $\vec{b}_j := (1, b_{2j}, \dots, b_{2j}^{n-1})^t$, $1 \le j \le n$, form a basis in $\mathbb{R}^n$. Also,

$$\left(B_n \sum_{j=1}^{n} c_j \vec{b}_j, \sum_{k=1}^{n} c_k \vec{b}_k \right) = \sum_{j=1}^{n} c_j^2 B(b_{2j}, b_{2j}) > 0, \tag{A.9}$$

which finishes the proof. □

LEMMA A.3. *The matrix $M_{in} = (M_{jk}^{(in)}(t))_{1 \le j,k \le n}$ is orthogonal for all $t \in \mathbb{R}$.*

PROOF. We need to check the formula

$$\sum_{j=1}^{n} M_{jk}^{(\rm in)}(t) M_{jm}^{(\rm in)}(t) = \delta_{km}, \ 1 \le k,m \le n. \tag{A.10}$$

By (4.23), (4.25) and (5.13), we have

(A.11)

$$\begin{aligned}
\sum_{j=1}^{n} M_{jk}^{(\rm in)}(t) M_{jm}^{(\rm in)}(t) &= \sum_{j=1}^{n} P_j(z) \sqrt{\frac{\rho_j}{Q(z)}} P_j(x) \sqrt{\frac{\rho_j}{Q(x)}} \\
&= \frac{\beta_{\rm ev}(z)\beta_{\rm od}(x) - \beta_{\rm ev}(x)\beta_{\rm od}(z)}{z-x} \frac{1}{\sqrt{Q(z)Q(x)}} \\
&= \frac{2\sinh\left(\frac{\phi(z)-\phi(x)}{2}\right)}{z-x} \sqrt{\prod_{j=1}^{2n} (x-b_j)(z-b_j)} \frac{1}{\sqrt{Q(z)Q(x)}}, \\
z = \phi_k^{-1}(2t),\ x &= \phi_m^{-1}(2t).
\end{aligned}$$

Recall that ρ_j, $j = 1, \dots, n$, are the eigenvalues of the Bezout matrix B_n, see Lemma 4.6. If $k \ne m$, then $z \ne x$, but $\phi(z) = \phi(x) = 2t$, and the last expression on the right in (A.11) equals zero. If $k = m$, then $z = x$. Since $M_{jk}^{(\mathrm{in})}(t)$ is continuous, we have

$$\sum_{j=1}^{n} M_{jk}^{(\mathrm{in})}(t) M_{jk}^{(\mathrm{in})}(t) = \lim_{\tilde{t}\to t} \sum_{j=1}^{n} M_{jk}^{(\mathrm{in})}(\tilde{t}) M_{jk}^{(\mathrm{in})}(t). \tag{A.12}$$

Therefore, to evaluate (A.10) we replace z with $\tilde{z}$ and take the limit as $\tilde{z} \to x$ in the second expression on the right in (A.11). Elementary calculations give that the limit of the right-hand side of (A.11) becomes (cf. (4.30) and (5.13)):

$$\phi'(x)\left(-\prod_{j=1}^{2n} |x - b_j|\right) / Q(x) = 1. \tag{A.13}$$

Here we have used that $Q(x) > 0, x \in I_{\mathrm{in}}$. □

Acknowledgment

The authors express their gratitude to the reviewer for useful comments.

References

[1] R. Alaifari, M. Defrise, and A. Katsevich, *Asymptotic analysis of the SVD for the truncated Hilbert transform with overlap*, SIAM J. Math. Anal. **47** (2015), no. 1, 797–824, DOI 10.1137/140952296. MR3313824

[2] R. Al-Aifari and A. Katsevich, *Spectral analysis of the truncated Hilbert transform with overlap*, SIAM J. Math. Anal. **46** (2014), no. 1, 192–213, DOI 10.1137/130910798. MR3148644

[3] M. Álvarez and G. Sansigre, *On polynomials with interlacing zeros*, Orthogonal polynomials and applications (Bar-le-Duc, 1984), Lecture Notes in Math., vol. 1171, Springer, Berlin, 1985, pp. 255–258, DOI 10.1007/BFb0076551. MR838991

[4] M. Bertola, A. Katsevich, and A. Tovbis, *Singular value decomposition of a finite Hilbert transform defined on several intervals and the interior problem of tomography: the Riemann-Hilbert problem approach*, Comm. Pure Appl. Math. **69** (2016), no. 3, 407–477, DOI 10.1002/cpa.21547. MR3455591

[5] M. Courdurier, F. Noo, M. Defrise, and H. Kudo, *Solving the interior problem of computed tomography using a priori knowledge*, Inverse Problems **24** (2008), no. 6, 065001, 27, DOI 10.1088/0266-5611/24/6/065001. MR2456948

[6] C. L. Epstein and J. Schotland, *The bad truth about Laplace's transform*, SIAM Rev. **50** (2008), no. 3, 504–520, DOI 10.1137/060657273. MR2429447

[7] A. R. It· s, A. G. Izergin, V. E. Korepin, and N. A. Slavnov, *Differential equations for quantum correlation functions*, Proceedings of the Conference on Yang-Baxter Equations, Conformal Invariance and Integrability in Statistical Mechanics and Field Theory, Internat. J. Modern Phys. B **4** (1990), no. 5, 1003–1037, DOI 10.1142/S0217979290000504. MR1064758

[8] T. Kato, *Perturbation theory for linear operators*, Classics in Mathematics, Springer-Verlag, Berlin, 1995. Reprint of the 1980 edition. MR1335452

[9] A. Katsevich and A. Tovbis, *Finite Hilbert transform with incomplete data: null-space and singular values*, Inverse Problems **28** (2012), no. 10, 105006, 28, DOI 10.1088/0266-5611/28/10/105006. MR2987909

[10] A. Katsevich and A. Tovbis, *Diagonalization of the finite Hilbert transform on two adjacent intervals*, J. Fourier Anal. Appl. **22** (2016), no. 6, 1356–1380, DOI 10.1007/s00041-016-9458-x. MR3572905

[11] W. Koppelman, *On the spectral theory of singular integral operators*, Trans. Amer. Math. Soc. **97** (1960), 35–63, DOI 10.2307/1993363. MR0119099

[12] W. Koppelman, *Spectral multiplicity theory for a class of singular integral operators*, Trans. Amer. Math. Soc. **113** (1964), 87–100, DOI 10.2307/1994092. MR0164256

[13] W. Koppelman and J. D. Pincus, *Spectral representations for finite Hilbert transformations*, Math. Z. **71** (1959), 399–407, DOI 10.1007/BF01181411. MR0107144

[14] H. Kudo, M. Courdurier, F. Noo, and M. Defrise. *Tiny a priori knowledge solves the interior problem in computed tomography. Phys. Med. Biol.*, 53:2207–2231, 2008.
[15] J. D. Pincus, *On the spectral theory of singular integral operators*, Trans. Amer. Math. Soc. **113** (1964), 101–128, DOI 10.2307/1994093. MR0164257
[16] A. P. Prudnikov, Yu. A. Brychkov, and O. I. Marichev, *Integrals and series. Vol. 1*, Gordon & Breach Science Publishers, New York, 1986. Elementary functions; Translated from the Russian and with a preface by N. M. Queen. MR874986
[17] C. R. Putnam, *The spectra of generalized Hilbert transforms*, J. Math. Mech. **14** (1965), 857–872. MR0181912
[18] M. Reed and B. Simon. *Functional Analysis, Volume I.* Academic Press, San Diego, 1981.
[19] M. Rosenblum, *On the Hilbert matrix. II*, Proc. Amer. Math. Soc. **9** (1958), 581–585, DOI 10.2307/2033212. MR0099599
[20] M. Rosenblum, *A spectral theory for self-adjoint singular integral operators*, Amer. J. Math. **88** (1966), 314–328, DOI 10.2307/2373195. MR0198294
[21] H. Widom, *Singular integral equations in L_p*, Trans. Amer. Math. Soc. **97** (1960), 131–160, DOI 10.2307/1993367. MR0119064
[22] J. Weidmann, *Linear operators in Hilbert spaces*, Graduate Texts in Mathematics, vol. 68, Springer-Verlag, New York-Berlin, 1980. Translated from the German by Joseph Szücs. MR566954
[23] Y. Ye, H. Yu, Y. Wei, and G. Wang. A general local reconstruction approach based on a truncated Hilbert transform. *International Journal of Biomedical Imaging*, 2007. Article ID 63634.
[24] Y. B. Ye, H. Y. Yu, and G. Wang. Exact interior reconstruction with cone-beam CT. *International Journal of Biomedical Imaging*, 2007. Article ID 10693.
[25] Y. B. Ye, H. Y. Yu, and G. Wang. Local reconstruction using the truncated Hilbert transform via singular value decomposition. *Journal of X-Ray Science and Technology*, 16:243–251, 2008.

Department of Mathematics, University of Central Florida, P.O. Box 161364, 4000 Central Florida Blvd, Orlando, FL 32816-1364
Email address: `Alexander.Katsevich@ucf.edu`

Department of Mathematics and Statistics, Concordia University, 1455 de Maisonneuve W., Montréal, Québec, Canada H3G 1M8 –and– SISSA, International School for Advanced Studies, via Bonomea 265, Trieste, Italy
Email address: `marco.bertola@concordia.ca`

Department of Mathematics, University of Central Florida, P.O. Box 161364, 4000 Central Florida Blvd, Orlando, FL 32816-1364
Email address: `alexander.tovbis@ucf.edu`

Contemporary Mathematics
Volume **734**, 2019
https://doi.org/10.1090/conm/734/14772

Polynomial-like elements in vector spaces with group actions

Minh Kha and Vladimir Lin

To the memory of Selim Grigorievich Krein, a great man and mathematician

ABSTRACT. In this paper, we study polynomial-like elements in vector spaces equipped with group actions. We first define these elements via iterated difference operators. In the case of a full rank lattice acting on an Euclidean space, these polynomial-like elements are exactly polynomials with periodic coefficients, which are closely related to solutions of periodic differential equations. Our main theorem confirms that if the space of polynomial-like elements of degree zero, is of finite dimension then for any $n \in \mathbb{Z}_+$, the space consisting of all polynomial-like elements of degree at most n is also finite dimensional.

Introduction

In 1984, T. Lyons and D. Sullivan [**13**] used the connections between the theory of harmonic functions and the theory of stochastic processes to prove that on a nilpotent covering of a compact Riemannian manifold, there are no nonconstant positive (and a fortiori no nonconstant bounded) harmonic functions. In [**10**], a new approach was proposed, applicable both to bounded holomorphic functions on nilpotent coverings of complex spaces and to bounded harmonic functions on such coverings of Riemannian manifolds.

In the case of a compact base with a fixed Riemannian metric, the question naturally arises of the structure of spaces of holomorphic or harmonic functions on coverings. In particular, it is natural to expect that on nilpotent coverings the spaces of the corresponding functions of bounded polynomial growth are finite-dimensional. In complex-analytic case, some results for *abelian* coverings were obtained by A. Brudnyi [**3**] and then, in both complex-analytic and harmonic cases in the paper of P. Kuchment and Y. Pinchover [**9**].

On the other hand, in the series of three papers [**4**]-[**6**], T. Colding and W. Minicozzi studied harmonic functions of restricted growth on Riemannian manifolds. In particular, it follows from their results that the spaces of harmonic or holomorphic functions of restricted polynomial growths on nilpotent coverings of *Kähler* manifolds are of finite dimension.

Another part of our motivation in studying polynomial-like elements via difference operators approach comes from the related studies of periodic equations,

---

2010 *Mathematics Subject Classification.* Primary 54C40, 14E20; Secondary 46E25, 20C20.
Key words and phrases. Algebraic geometry, Group actions, Periodic differential operators.
The first author acknowledges support of the NSF Grant DMS-1517938.

e.g., Liouville type results for elliptic equations of second-order with periodic coefficients (in divergence form) on Euclidean spaces, which appeared in the work [**2**]. It is also worthwhile to note that analogous Liouville type results have been established in [**11**, **14**]. Such Liouville type results show that every solution with polynomial growth of the equation admits a representation as a linear combination of polynomials whose coefficients are periodic functions and moreover, each of these polynomials is also a solution. Hence, the spaces of such solutions (with a fixed polynomial growth) are finite dimensional. As a first attempt to generalize some of these results, our first step is to give a definition of polynomial-like elements in vector spaces equipped with group actions. Rather than using explicit formulas, we choose to define in a more invariant way by using difference operators.

Let us give a brief outline of the paper. Let G be a group that acts linearly on a vector space A. Section 1 is devoted to defining polynomial-like elements (or G-polynomials) in A through the iterated difference operators D^n ($n \in \mathbb{Z}_+$). In Subsection 1.1, these iterated difference operators, which can be considered as analogs of the usual derivatives or the difference operators for G-moduli spaces, are introduced inductively via the action of G on A. In Subsection 1.2, G-polynomials of degree at most n in A are defined as elements in the kernel of the $(n+1)^{th}$-iterated difference operator D^{n+1}.

Motivated by Liouville type results, we would like to understand the finite dimensionality of the spaces of polynomial-like elements under certain conditions. Our main aim is to prove the following theorem:

MAIN THEOREM. *Let F be a field of characteristic 0, G be a group, and A be a F-vector space endowed with a linear right G-action. Let A^G be the space consisting of all G-invariant elements in A and $\mathcal{P}_n(G, A)$ be the space consisting of all G-polynomials of degree at most n in A. If the group $\widetilde{G} = G/[G,G]$ is finitely generated and $\dim_F A^G < \infty$ then*

$$\dim_F \mathcal{P}_n(G, A) < \infty \quad \textit{for every } n \in \mathbb{Z}_+ .$$

We will develop the necessary tools, and then use them to prove this theorem in the rest of Section 1.

In Section 2, we study further properties of the iterated difference operators if A has an additional ring structure that is compatible with the group action. These results are needed for the next section.

In Section 3, we apply the above Main Theorem to the case when G is a lattice acting on A. We consider here an important example when A is a G-invariant subspace of the algebra of continuous functions on the Euclidean space $\mathbb{R}^r$, where r is the rank of the lattice G. In this example, we characterize G-polynomials (Proposition 3.2): these are exactly polynomials with G-invariant coefficients [1], which we introduce at the beginning of Subsection 3.1. Then in Subsection 3.2, we give a useful interpretation of the Main Theorem when A is the space of all classical global solutions of a G-periodic linear differential operator $\mathcal{D}$. Finally, Subsection 3.3 provides some remarks related to periodic operators acting on co-compact regular Riemannian coverings whose deck transformation groups G are not necessarily abelian.

[1] These polynomials are called Floquet functions, which play an important role in studying the spectral theory of periodic differential operators (see e.g., [**8**]).

1. Polynomial-like elements

1.1. Iterated difference operators in G-moduli. Let G be a group with the unity $\mathbf{e}$ and A be an additive abelian group.

DEFINITION 1.1. For $n \in \mathbb{Z}_+$, let us denote by $\mathcal{C}^n(G, A)$ the additive group of all *normalized* n-cochains of G with the values in A; that is, $\mathcal{C}^0(G, A) = A$ and for $n \in \mathbb{N}$ the group $\mathcal{C}^n(G, A)$ consists of all functions

$$c\colon\ G^n = \underbrace{G \times \cdots \times G}_{n} \ni (g_1, ..., g_n) \mapsto c(g_1, ..., g_n) \in A$$

such that $c(g_1, ..., g_n) = 0$ whenever at least one of the elements $g_1, ..., g_n$ equals $\mathbf{e}$. Here the group operation is the addition between two functions in $\mathcal{C}^n(G, A)$.

Suppose that A is endowed with a right G-module structure

$$A \ni a \mapsto a^g \in A, \quad g \in G. \tag{1.1}$$

Let A^G denote the subgroup of A consisting of all G-invariant elements, i.e.,

$$A^G = \{a \in A \mid a^g = a, \quad \forall g \in G\}.$$

If A is a vector space and the given G-action in A is linear then A^G is a vector subspace of A.

Such a structure induces the following right G-actions on cochain groups $\mathcal{C}^n(G, A)$:

$$\mathcal{C}^n(G, A) \ni c \mapsto c^g \in \mathcal{C}^n(G, A)\,, \quad c^g(g_1, ..., g_n) = [c(g_1, ..., g_n)]^g\,,$$
$$(g, g_1, ..., g_n \in G\,, \ n \in \mathbb{Z}_+)\,.$$

These actions give rise to group homomorphisms

$$D^n\colon A \to \mathcal{C}^n(G, A) \quad (n \in \mathbb{Z}_+)$$

defined as follows. First, we define homomorphisms

$$d_n\colon\ \mathcal{C}^{n-1}(G, A) \to \mathcal{C}^n(G, A) \quad (n \geq 1)$$

by the formulas

$$\begin{aligned}(d_n c)(g_1, \ldots, g_{n-1}, g_n) &= c^{g_n}(g_1, \ldots, g_{n-1}) - c(g_1, \ldots, g_{n-1})\\ &= [c(g_1, \ldots, g_{n-1})]^{g_n} - c(g_1, \ldots, g_{n-1})\\ &\qquad (c \in \mathcal{C}^{n-1}(G, A)\,, \quad g_1, \ldots, g_n \in G\,, \quad n > 1)\,.\end{aligned} \tag{1.2}$$

Using these homomorphisms, we define homomorphisms $D^n\colon A \to \mathcal{C}^n(G, A)$ by the recursion relations

$$\begin{aligned}&D^0 = \mathbf{id}_A \text{ (the identity operator in } A) \text{ and } D^n = d_n D^{n-1} \text{ for } n \in \mathbb{N}\,,\\ &\text{or, equivalently, } D^n = d_n \cdots d_1 D^0 \text{ for all } n \in \mathbb{Z}_+.\end{aligned} \tag{1.3}$$

These homomorphisms D^n are called the *iterated difference operators.*

NOTATION 1.2. Let $n, s, i_1, ..., i_s \in \mathbb{N}$, where $1 \leq s \leq n$ and $1 \leq i_1 < ... < i_s \leq n$. For any $g_1, ..., g_n \in G$, set

$$\pi_{i_1,...,i_s}(g_1, ..., g_n) := g_1 \cdot \ldots \cdot \widehat{g_{i_1}} \cdot \ldots \cdot \widehat{g_{i_2}} \cdot \ldots \cdot \widehat{g_{i_s}} \cdot \ldots \cdot g_n$$

(terms with hats in the right hand side must be omitted, and the empty product is defined to be equal to $\mathbf{e}$, the unity of G).

LEMMA 1.3. *For any $a \in A$, $n \in \mathbb{N}$, and $g_1, ..., g_n \in G$, we have*

$$[D^n a](g_1, ..., g_n) = a^{g_1 \cdots g_n} + \\ + \sum_{s=1}^{n-1} (-1)^s \sum_{1 \le i_1 < ... < i_s \le n} a^{\pi_{i_1,...,i_s}(g_1,...,g_n)} + (-1)^n a\,. \tag{1.4}$$

PROOF. Induction in n. □

REMARK 1.4. Let us define

$$[\Delta^n a](g_1, ..., g_n) := \sum_{s=1}^{n} (-1)^s S_s(g_1, ..., g_n), \text{ where} \\ S_s(g_1, ..., g_n) := \sum_{1 \le i_1 < ... < i_s \le n} a^{\pi_{i_1,...,i_s}(g_1,...,g_n)}\,. \tag{1.5}$$

Then formula (1.4) can be written in the form

$$[D^n a](g_1, ..., g_n) = a^{g_1 \cdots g_n} + [\Delta^n a](g_1, ..., g_n)\,, \tag{1.6}$$

Each sum $S_s(g_1, ..., g_n)$ in (1.5) is a symmetric function of $g_1, ..., g_n$, that is,

$$S_s(g_{j_1}, ..., g_{j_n}) = S_s(g_1, ..., g_n)$$

for any $g_1, ..., g_n \in G$ and any permutation $(j_1, ..., j_n)$ of the indices $1, ..., n$. Therefore, *the function* $[\Delta^n a](g_1, ..., g_n) = [D^n a](g_1, ..., g_n) - a^{g_1 \cdots g_n}$ *is symmetric in* $g_1, ..., g_n$. ○

For any *abelian* G, the function $(g_1, ..., g_n) \mapsto a^{g_1 \cdots g_n}$ is symmetric, and the above remark implies the symmetry of $[D^n a](g_1, ..., g_n)$. For non-abelian G this may be wrong; however, the following property holds in general case.

LEMMA 1.5. *Suppose that $D^{n+1} a = 0$. Then, for any natural $k \le n+1$, the functions $a^{g_1 \cdots g_k}$ and $[D^k a](g_1, ..., g_k)$ are symmetric in $g_1, ..., g_k \in G$.*

PROOF. The assumption $D^{n+1} a = 0$ implies that

$$\begin{aligned} a^{g_1 \cdots g_{n+1}} &= [D^{n+1} a](g_1, ..., g_{n+1}) - [\Delta^{n+1} a](g_1, ..., g_{n+1}) \\ &= -[\Delta^{n+1} a](g_1, ..., g_{n+1})\,. \end{aligned}$$

By Remark 1.4, $[\Delta^{n+1} a](g_1, ..., g_{n+1})$ is symmetric in $g_1, ..., g_{n+1}$; therefore, $a^{g_1 \cdots g_{n+1}}$ is symmetric as well. For any $k \le n+1$, we have

$$a^{g_1 \cdots g_k} = a^{g_1 \cdots g_{n+1}}\big|_{g_{k+1} = ... = g_{n+1} = \mathbf{e}}\;;$$

hence $a^{g_1 \cdots g_k}$ is symmetric in $g_1, \ldots, g_k$ and, by (1.6), $[D^k a](g_1, ..., g_k)$ is symmetric. □

LEMMA 1.6. *Let $n \in \mathbb{Z}_+$ and $a \in A$. Then*

$$[D^{n+1} a](g_1, g_2, ..., g_{n+1}) = [D^n (a^{g_1})](g_2, ..., g_{n+1}) - [D^n a](g_2, ..., g_{n+1}) \\ \text{for all } \; g_1, g_2, ..., g_{n+1} \in G\,. \tag{1.7}$$

PROOF. The proof is by induction in n. For $n = 0, 1$, formula (1.7) looks as follows:

$$n = 0: \quad [D^1 a](g_1) = a^{g_1} - a\,; \tag{1.8}$$

$$n = 1: \quad [D^2 a](g_1, g_2) = [D^1 (a^{g_1})](g_2) - [D^1 a](g_2)\,. \tag{1.9}$$

Formula (1.8) is just the definition of D^1. Furthermore, we have:

$$\begin{aligned}[D^2a](g_1, g_2) &= [a^{g_1} - a]^{g_2} - [a^{g_1} - a] = a^{g_1g_2} - a^{g_2} - a^{g_1} + a \\ &= [(a^{g_1})^{g_2} - a^{g_1}] - (a^{g_2} - a) = [D^1(a^{g_1})](g_2) - [D^1a](g_2),\end{aligned}$$

which coincides with (1.9). This provides us with the base of induction.

Suppose now that for some $m \geq 2$, the formula (1.7) holds for all $n \leq m - 1$. Let us show that it holds true for $n = m$. According to (1.3),

$$\begin{aligned}[D^{m+1}a](g_1, g_2, ..., g_{m+1}) &= \{[D^m a](g_1, g_2, ..., g_m)\}^{g_{m+1}} - [D^m a](g_1, g_2, ..., g_m), \\ [D^m a](g_2, ..., g_{m+1}) &= \{[D^{m-1}a](g_2, ..., g_m)\}^{g_{m+1}} - [D^{m-1}a](g_2, ..., g_m),\end{aligned}$$

which implies

$$\begin{aligned}&[D^{m+1}a](g_1, g_2, ..., g_{m+1}) + [D^m a](g_2, ..., g_{m+1}) \\ &\quad = \{[D^m a](g_1, g_2, ..., g_m)\}^{g_{m+1}} - [D^m a](g_1, g_2, ..., g_m) \\ &\qquad + \{[D^{m-1}a](g_2, ..., g_m)\}^{g_{m+1}} - [D^{m-1}a](g_2, ..., g_m) \\ &\quad = \{[D^m a](g_1, g_2, ..., g_m) + [D^{m-1}a](g_2, ..., g_m)\}^{g_{m+1}} \\ &\qquad - \{[D^m a](g_1, g_2, ..., g_m) + [D^{m-1}a](g_2, ..., g_m)\}.\end{aligned} \tag{1.10}$$

By the induction hypothesis,

$$[D^m a](g_1, g_2, ..., g_m) = [D^{m-1}(a^{g_1})](g_2, ..., g_m) - [D^{m-1}a](g_2, ..., g_m). \tag{1.11}$$

It follows from (1.10), (1.11), and (1.4) that

$$\begin{aligned}&[D^{m+1}a](g_1, g_2, ..., g_{m+1}) + [D^m a](g_2, ..., g_{m+1}) \\ &\quad = \{[D^m a](g_1, ..., g_m) + [D^{m-1}a](g_2, ..., g_m)\}^{g_{m+1}} \\ &\qquad - \{[D^m a](g_1, ..., g_m) + [D^{m-1}a](g_2, ..., g_m)\} \\ &= \{[D^{m-1}(a^{g_1})](g_2, ..., g_m) - [D^{m-1}a](g_2, ..., g_m) + [D^{m-1}a](g_2, ..., g_m)\}^{g_{m+1}} \\ &\quad - \{[D^{m-1}(a^{g_1})](g_2, ..., g_m) - [D^{m-1}a](g_2, ..., g_m) + [D^{m-1}a](g_2, ..., g_m)]\} \\ &= \{[D^{m-1}(a^{g_1})](g_2, ..., g_m)\}^{g_{m+1}} - [D^{m-1}(a^{g_1})](g_2, ..., g_m) \\ &= [D^m(a^{g_1})](g_2, ..., g_{m+1}),\end{aligned}$$

which yields

$$[D^{m+1}a](g_1, g_2, ..., g_{m+1}) = [D^m(a^{g_1})](g_2, ..., g_{m+1}) - [D^m a](g_2, ..., g_{m+1}).$$

This completes the proof. □

1.2. Polynomial-like elements. From now on, whenever A is a vector space we assume that the given right G-action in A is *linear*, i.e., all the mappings (1.1) are linear operators. Then the homomorphisms d_m and D^n defined in (1.2) and (1.3) respectively, are linear operators.

DEFINITION 1.7. Let us denote by $\mathcal{P}_n = \mathcal{P}_n(G, A)$ the kernel of the homomorphism $D^{n+1}\colon A \to \mathcal{C}^{n+1}(G, A)$, i.e.,

$$\begin{aligned}\mathcal{P}_n &= \mathcal{P}_n(G, A) = \{a \in A \mid D^{n+1}a = 0\} \\ &= \{a \in A \mid [D^{n+1}a](g_1, ..., g_{n+1}) = 0 \ \ \forall\, g_1, ..., g_{n+1} \in G\} \quad (n \in \mathbb{Z}_+).\end{aligned}$$

Clearly, $\mathcal{P}_n(G, A)$ is a subgroup of A (a vector subspace whenever A is a vector space with a linear G-action).

An element $p \in \mathcal{P}_n$ is said to be a *G-polynomial in A of order at most n*, or a *polynomial-like element* in A (of order at most n). It follows from (1.3) that

$$\mathcal{P}_0 \subseteq \mathcal{P}_1 \subseteq \ldots \subseteq \mathcal{P}_n \subseteq \mathcal{P}_{n+1} \subseteq \ldots . \tag{1.12}$$

EXAMPLE 1.8. Let G be an additive subgroup of $\mathbb{R}^r$ acting by translations in the space $C(\mathbb{R}^r)$ of all (real or complex) continuous functions on $\mathbb{R}^r$:

$$f^g(x) = f(x+g) \quad \text{for all } f \in C(\mathbb{R}^r),\ x \in \mathbb{R}^r,\ g \in G \subseteq \mathbb{R}^r$$

(all the mappings $f \mapsto f^g$ are, in fact, automorphisms of the algebra $C(\mathbb{R}^r)$). Let $A \subseteq C(\mathbb{R}^r)$ be a G-invariant subspace of $C(\mathbb{R}^r)$. For instance, the following natural cases seem interesting:

(a) $G = \mathbb{R}^r$ and $A = C(\mathbb{R}^r)$;
(b) $G \cong \mathbb{Z}^r$ is a lattice of rank r in $\mathbb{R}^r$ and $A = C(\mathbb{R}^r)$;
(c) G is as in (b) and $A = \mathcal{H}(\mathbb{R}^r)$ is the space of all harmonic functions on $\mathbb{R}^r$;
(d) $r = 2m$ is even, $\mathbb{R}^r = \mathbb{C}^m$, $G \cong \mathbb{Z}^{2m}$ is a lattice of rank $2m$ in $\mathbb{C}^m$, and $A = \mathcal{O}(\mathbb{C}^m)$ is the space of all holomorphic functions on $\mathbb{C}^m$.

In case (a) the space A^G of all G-invariant elements of A is just the field of constants (that is, either $A^G = \mathbb{R}$ or $A^G = \mathbb{C}$). In cases (b) – (d), A^G is the space of all G-invariant functions that are, respectively, (b) continuous, (c) harmonic, (d) holomorphic.

In fact, with help of Proposition 1.24 proven in Subsection 1.6 below, one can see that in cases (a) – (c) the space $\mathcal{P}_n(G, A)$ coincides with the space $A^G_n[x_1, ..., x_r]$ of all polynomials in the standard coordinates $x_1, ..., x_r$ of degree at most n with coefficients in A^G; in case (d) the space $A^G_n[z_1, ..., z_m]$ of polynomials in *complex* coordinates $z_1, ..., z_m$ should be taken instead.

EXAMPLE 1.9. Let $G \subset \mathbb{C}$ be a lattice of rank 2 and $\mathcal{M} = \mathcal{M}(\mathbb{C})$ be the field of meromorphic functions of one complex variable $z \in \mathbb{C}$. Then $\mathcal{P}_0(G, \mathcal{M}) = \mathcal{M}^G$ is just the field of elliptic functions associated with the lattice G. It can be proven that $\mathcal{P}_n(G, \mathcal{M}) = \mathcal{M}^G_n[z]$, where $\mathcal{M}^G_n[z]$ stands for the space of all polynomial in z of degree at most n with coefficients in the field of elliptic functions $\mathcal{M}^G$.

EXAMPLE 1.10. It is interesting to take $A = \mathcal{M}^* = \mathcal{M} \setminus \{0\}$, the multiplicative group of the field $\mathcal{M}$, with a lattice $G \subset \mathbb{C}$ of rank 2 acting by translations. Clearly, $\mathcal{P}_0(G, \mathcal{M}^*) = (\mathcal{M}^*)^G$, the multiplicative group of the field $\mathcal{M}^G$ of G-elliptic functions.

PROPOSITION 1.11. *Each $\mathcal{P}_n(G, A) \subseteq A$ is a G-invariant subgroup* (*respectively, a G-invariant vector subspace whenever A is a vector space*).

PROOF. Since $\mathcal{P}_n(G, A)$ is the kernel of a homomorphism (respectively, linear operator), we only need to check that $a^g \in \mathcal{P}_n(G, A)$ whenever $a \in \mathcal{P}_n(G, A)$ and $g \in G$. In view of Lemma 1.6 (with $n+2$ instead of n), the assumption $D^{n+1}a = 0$ implies

$$[D^{n+1}(a^g)](g_1, ..., g_{n+1}) = [D^{n+1}a](g_1, ..., g_{n+1}) + [D^{n+2}a](g, g_1, ..., g_{n+1}) = 0+0 = 0,$$

which concludes the proof. □

NOTATION 1.12. Let $[G, G]$ denote the *commutator subgroup* of a group G and $\widetilde{G} = G/[G, G]$ be the **abelianization** of G. For every $g \in G$, let $\widetilde{g}$ denote the image of g in $\widetilde{G}$ under the natural epimorphism $G \to \widetilde{G}$.

Our main aim in the remainder of this Section is to establish the following result:

THEOREM 1.13. *Let F be a field of characteristic 0, G be a group, and A be a F-vector space endowed with a linear right G-action. If $\widetilde{G}$ is finitely generated and $\dim_F A^G < \infty$ then $\dim_F \mathcal{P}_n(G,A) < \infty$ for every $n \in \mathbb{Z}_+$.*

To prove this theorem, we show that the quotient spaces $\mathcal{P}_n(G,A)/\mathcal{P}_{n-1}(G,A)$ ($n \in \mathbb{N}$) can be embedded into an appropriate vector space of A^G-valued *symmetric polylinear forms* on the group G. Under certain conditions, the latter space is of finite dimension, which makes it possible to use the induction in n. To develop this idea, we need to define some spaces of polylinear forms on G and establish relations between the iterated difference operators D^n and the coboundary operators δ related to the *trivial* left G action in A. We will do this in the next Subsections 1.3, 1.4, 1.5 and 1.6.

1.3. Group polymorphisms. We introduce here the concept of polylinear forms on the group G and its related properties.

DEFINITION 1.14. Let G be a group and B be an abelian group. A B-valued *polymorphism*, or, more precisely, *n-morphism* of G is a function

$$L\colon \underbrace{G \times \cdots \times G}_{n} \to B$$

such that for each $i \in \{1, ..., n\}$, the relation

$$\begin{aligned} L(g_1, ..., g_{i-1}, g'g'', g_{i+1}, ..., g_n) \\ = L(g_1, ..., g_{i-1}, g', g_{i+1}, ..., g_n) + L(g_1, ..., g_{i-1}, g'', g_{i+1}, ..., g_n) \end{aligned} \tag{1.13}$$

holds for any $g_1, ..., g_{i-1}, g', g'', g_{i+1}, ..., g_n \in G$.

The set $\mathcal{L}_n(G,B)$ of all B-valued n-morphisms of G is an abelian group. It is convenient to set $\mathcal{L}_0(G,B) = B$. If B is a vector space over a field F then $\mathcal{L}_n(G,B)$ is a vector space over F as well; in this case polymorphisms $L \in \mathcal{L}_n(G,B)$ are also called B-valued *polylinear* (or *n-linear*) *forms* on G.

We skip the proof of the following elementary lemma, leaving this as an exercise for the reader.

LEMMA 1.15. *(a) Let G be a free abelian group of rank r and*

$$L\colon \underbrace{G \times \cdots \times G}_{n} \to B \tag{1.14}$$

be a n-morphism of G to an abelian group B. Let $\mathbf{h}_1, ..., \mathbf{h}_r$ be a free basis of G and set

$$b_{i_1,...,i_n} = L(\mathbf{h}_{i_1}, ..., \mathbf{h}_{i_n}), \quad 1 \le i_1, ..., i_n \le r. \tag{1.15}$$

Then for any $g_1, ..., g_n \in G$ we have

$$\begin{aligned} L(g_1, ..., g_n) &= \sum_{1 \le i_1,...,i_n \le r} g_{1,i_1} \cdots g_{n,i_n} \cdot L(\mathbf{h}_{i_1}, ..., \mathbf{h}_{i_n}) \\ &= \sum_{1 \le i_1,...,i_n \le r} g_{1,i_1} \cdots g_{n,i_n} \cdot b_{i_1,...,i_n}, \end{aligned}$$

where $g_{k,1}, ..., g_{k,r} \in \mathbb{Z}$ are the coordinates of the element $g_k \in G$ represented in the basis $\mathbf{h}_1, ..., \mathbf{h}_r$ $(k = 1, ..., n)$.

(b) For any choice of r^n elements $b_{i_1,\dots,i_n} \in B$ $(1 \le i_1, \dots, i_n \le r)$, there exists a unique n-morphism (1.14) *satisfying* (1.15).

(c) Let $G \subset \mathbb{R}^r$ be a lattice of rank r, $\mathbf{h}_1, \dots, \mathbf{h}_r$ be a basis of G, and L be a n-morphism of G to a vector space B. Then there exists a unique B-valued n-linear form

$$\mathfrak{L}\colon \underbrace{\mathbb{R}^r \times \cdots \times \mathbb{R}^r}_{n} \to B$$

that extends L from G to $\mathbb{R}^r$ (i. e., $\mathfrak{L}$ satisfies $\mathfrak{L}|_G = L$). This form $\mathfrak{L}$ is defined by

$$\begin{aligned}\mathfrak{L}(v_1, \dots, v_n) &= \sum_{1 \le i_1, \dots, i_n \le r} v_{1,i_1} \cdots v_{n,i_n} L(\mathbf{h}_{i_1}, \dots, \mathbf{h}_{i_n}) \\ &= \sum_{1 \le i_1, \dots, i_n \le r} v_{1,i_1} \cdots v_{n,i_n} b_{i_1,\dots,i_n},\end{aligned}$$

where $v_{k,1}, \dots, v_{k,r} \in \mathbb{R}$ are the coordinates of the vector $v_k \in \mathbb{R}^r$ represented in the basis $\mathbf{h}_1, \dots, \mathbf{h}_r$ $(k = 1, \dots, n)$.

The next lemma collects some simple and useful properties of B-valued n-morphisms on a (possibly non-abelian) group G.

LEMMA 1.16. *(a) $\mathcal{L}_n(G, B) \subseteq \mathcal{C}^n(G, B)$ for every $n \in \mathbb{N}$; in other words, every n-morphism L is also a n-cochain, i.e., $L(g_1, \dots, g_n) = 0$ whenever one of the elements $g_1, \dots, g_n$ equals* $\mathbf{e}$.

(b) For any $L \in \mathcal{L}_n(G, B)$, $L(g_1, \dots, g_n) = 0$ whenever one of the elements $g_1, \dots, g_n$ belongs to the commutator subgroup $[G, G]$.

(c) Consider any $L \in \mathcal{L}_n(G, B)$ and $\widetilde{g_1}, \dots, \widetilde{g_n} \in \widetilde{G} = G/[G, G]$ and an arbitrary representatives $g_i \in \widetilde{g_i}$, $1 \le i \le n$. Let

$$\widetilde{L}(\widetilde{g_1}, \dots, \widetilde{g_n}) := L(g_1, \dots, g_n).$$

Then $\widetilde{L}$ is a well-defined B-valued n-morphism of the abelianization $\widetilde{G}$ of G. The correspondence $L \mapsto \widetilde{L}$ defines an isomorphism of groups

$$\mu\colon \mathcal{L}_n(G, B) \ni L \mapsto \mu(L) = \widetilde{L} \in \mathcal{L}_n(\widetilde{G}, B).$$

Moreover, μ is an isomorphism of vector spaces whenever B is a vector space.

(d) Suppose that B is a vector space over a field F of characteristic 0, $\dim_F B < \infty$, and the abelianization $\widetilde{G}$ of G is finitely generated. Then [2]

$$\dim_F \mathcal{L}_n(G, B) = \dim_F \mathcal{L}_n(\widetilde{G}, B) < \infty \quad \textit{for every } n \in \mathbb{N}.$$

PROOF. (a) It suffices to show that $L(\mathbf{e}, g_2, \dots, g_n) = 0$. The latter relation follows from the following identities:

$$L(\mathbf{e}, g_2, \dots, g_n) = L(\mathbf{e} \cdot \mathbf{e}, g_2, \dots, g_n) = L(\mathbf{e}, g_2, \dots, g_n) + L(\mathbf{e}, g_2, \dots, g_n).$$

(b) Using (1.13) and the fact that B is abelian, we get that for any $h_1, h_2, g_2, \dots, g_n \in G$, the following identities hold:

$$L(h_1 h_2 h_1^{-1} h_2^{-1}, g_2, \dots, g_n) = L(h_1 h_1^{-1} h_2 h_2^{-1}, g_2, \dots, g_n) = L(\mathbf{e}, g_2, \dots, g_n) = 0.$$

This implies (*b*).

(c) This is an immediate consequence of (*b*).

[2] Although this fact is simple and well-known, we treat it here just for completeness of the proof of our main result. Notice that the assumption char $F = 0$ could be omitted.

(d) Since $\widetilde{G}$ is a finitely generated abelian group, we have $\widetilde{G} \cong \mathbb{Z}^r \oplus K$, where $r \in \mathbb{Z}_+$ and K is a finite abelian group. For any $\widetilde{L} \in \mathcal{L}_n(\widetilde{G}, B)$ we have $\widetilde{L}(\widetilde{g_1}, ..., \widetilde{g_n}) = 0$ whenever at least one of the elements $\widetilde{g_1}, ..., \widetilde{g_n}$ belongs to the subgroup K. Indeed, assuming, for instance, that $\widetilde{g_1}$ is an element of finite order $m \in \mathbb{N}$, we see that

$$m \cdot \widetilde{L}(\widetilde{g_1}, ..., \widetilde{g_n}) = \widetilde{L}(\widetilde{g_1}^m, ..., \widetilde{g_n}) = \widetilde{L}(\widetilde{\mathbf{e}}, ..., \widetilde{g_n}) = 0\,,$$

which implies $\widetilde{L}(\widetilde{g_1}, ..., \widetilde{g_n}) = 0$ (note that $\operatorname{char} F = 0$). It follows from (c) and the above fact

$$\mathcal{L}_n(G, B) \cong \mathcal{L}_n(\widetilde{G}, B) \cong \mathcal{L}_n(\mathbb{Z}^r, B)\,.$$

Therefore, it suffices to show that $\dim_F \mathcal{L}_n(\mathbb{Z}^r, B) < \infty$.

Let $\mathbf{h}_1 = (1, 0, ..., 0), ..., \mathbf{h}_r = (0, ..., 0, 1)$ be the standard basis of $\mathbb{Z}^r$ and $\mathbf{b}^1, ..., \mathbf{b}^s$ be a basis in B. Let $\mathcal{I}$ be the family of all finite ordered sets $I = (i_1, ..., i_n)$ consisting of n natural numbers $i_1, ..., i_n$ such that $1 \le i_1, ..., i_n \le r$ (clearly, $\#\mathcal{I} = r^n$). For each $I = (i_1, ..., i_n) \in \mathcal{I}$ and each $i = 1, ..., s$, we define the n-linear form $\ell_I^i \in \mathcal{L}^n(\mathbb{Z}^r, B)$ as follows:

$$(1.16) \qquad \ell_I^i(\mathbf{h}_{j_1}, ..., \mathbf{h}_{j_n}) = \begin{cases} \mathbf{b}^i & \text{if } (j_1, ..., j_n) = I\,, \\ 0 & \text{otherwise}\,; \end{cases}$$

These sr^n forms ℓ_I^i constitute a basis [3] of $\mathcal{L}_n(\mathbb{Z}^r, B)$. Indeed, for any $L \in \mathcal{L}_n(\mathbb{Z}^r, B)$, it suffices to show that L is a linear combination of ℓ_I^i. Let us consider $I = (i_1, ..., i_n) \in \mathcal{I}$. Since $L(\mathbf{h}_{i_1}, ..., \mathbf{h}_{i_n}) \in B$ and $\{\mathbf{b}^1, ..., \mathbf{b}^s\}$ is a basis of B, we have

$$(1.17) \qquad L(\mathbf{h}_{i_1}, ..., \mathbf{h}_{i_n}) = \sum_{i=1}^{s} a_I^i\, \mathbf{b}^i \quad \text{for some } a_I^1, ..., a_I^s \in F\,.$$

It follows from (1.16) and (1.17) that

$$L(\mathbf{h}_{i_1}, ..., \mathbf{h}_{i_n}) = \sum_{i=1}^{s} a_I^i\, \ell_I^i(\mathbf{h}_{i_1}, ..., \mathbf{h}_{i_n}) = \sum_{i=1}^{s} \sum_{J \in \mathcal{I}} a_J^i\, \ell_J^i(\mathbf{h}_{i_1}, ..., \mathbf{h}_{i_n}).$$

This proves that

$$L = \sum_{I \in \mathcal{I}} \sum_{i=1}^{s} a_I^i\, \ell_I^i\,,$$

which concludes the proof.

□

1.4. Embeddings $\left[\mathcal{P}^n(G, A)/\mathcal{P}^{n-1}(G, A)\right] \hookrightarrow \mathcal{C}^n(G, A^G)$. By Definition 1.7, the iterated difference operator D^{n+1} annihilates the subgroup $\mathcal{P}_n = \mathcal{P}_n(G, A) = \ker D^{n+1} \subseteq A$ of all polynomial-like elements of order at most n. However, the preceding iterated difference operator D^n may be non-zero on $\mathcal{P}_n$. In this section, we study properties of the restriction

$$D^n_{\mathcal{P}_n} \overset{\text{def}}{=\!=} D^n\,|_{\mathcal{P}_n} : \mathcal{P}_n \to \mathcal{C}^n(G, A)$$

of the operator $D^n \colon A \to \mathcal{C}^n(G, A)$ to the subgroup $\mathcal{P}_n \subseteq A$.

According to our notation, $\mathcal{C}^n(G, A^G)$ is the group of all n-cochains of G with values in the subgroup $A^G \subseteq A$ consisting of all G-invariant elements in

[3] Obviously, these forms ℓ_I^i are linearly independent.

A. Clearly, $\mathcal{C}^n(G, A^G)$ is a subgroup of the group $\mathcal{C}^n(G, A)$ (if A is a vector space then $\mathcal{C}^n(G, A^G)$ is a vector subspace of the vector space $\mathcal{C}^n(G, A)$).

LEMMA 1.17. *The image of* $D^n_{\mathcal{P}_n}$ *is contained in the subgroup* $\mathcal{C}^n(G, A^G) \subseteq \mathcal{C}^n(G, A)$, *i.e.*,

$$D^n(\mathcal{P}_n) \subseteq \mathcal{C}^n(G, A^G).$$

That is, $D^n_{\mathcal{P}_n}$ *may be regarded as a group homomorphism (a linear operator in case of vector spaces)*

$$D^n_{\mathcal{P}_n} : \mathcal{P}_n(G, A) \to \mathcal{C}^n(G, A^G)$$

acting from the group $\mathcal{P}_n(G, A)$ *to the group* $\mathcal{C}^n(G, A^G)$.

By (1.12), (1.4), *and* Definition 1.7, *the kernel* $\ker D^n_{\mathcal{P}_n}$ *of* $D^n_{\mathcal{P}_n}$ *coincides with* $\mathcal{P}_{n-1}(G, A)$. *Thus, the homomorphism* $D^n_{\mathcal{P}_n}$ *gives rise to the exact sequence*

$$0 \longrightarrow \mathcal{P}_{n-1}(G, A) \longrightarrow \mathcal{P}_n(G, A) \xrightarrow{D^n_{\mathcal{P}_n}} \mathcal{C}^n(G, A^G), \tag{1.18}$$

and the embedding of the quotient group

$$i_n : [\mathcal{P}_n(G, A)/\mathcal{P}_{n-1}(G, A)] \hookrightarrow \mathcal{C}^n(G, A^G). \tag{1.19}$$

PROOF. Let $n \in \mathbb{Z}_+$. Take any $a \in \mathcal{P}_n(G, A)$. By Definition 1.7, $D^{n+1}a = 0$. According to (1.3), we have

$$0 = D^{n+1}a = d_{n+1}(D^n a).$$

In other words, we get

$$\begin{aligned} 0 &= [d_{n+1}(D^n a)](g_1, ..., g_{n+1}) = [(D^n a)^{g_{n+1}} - D^n a](g_1, ..., g_n) \\ &= \{[D^n a](g_1, ..., g_n)\}^{g_{n+1}} - [D^n a](g_1, ..., g_n) \quad \text{for all } g_1, ..., g_{n+1} \in G. \end{aligned}$$

The latter relation may be written as

$$\{[D^n a](g_1, ..., g_n)\}^g = [D^n a](g_1, ..., g_n) \quad \text{for all } g, g_1, ..., g_n \in G,$$

which actually means that all the values $[D^n a](g_1, ..., g_n)$ of the n-cochain $D^n a$ are G-invariant.

Hence, for every $a \in \mathcal{P}_n$ we may regard the n-cochain $D^n_{\mathcal{P}_n}(a) = D^n a$ (which is originally an A-valued one) as a n-cochain of G *with values in* A^G: $D^n_{\mathcal{P}_n}(a) = D^n a \in \mathcal{C}^n(G, A^G)$. So $D^n_{\mathcal{P}_n}(\mathcal{P}_n) \subseteq \mathcal{C}^n(G, A^G)$ and $D^n_{\mathcal{P}_n}$ may be regarded as a homomorphism (linear operator) acting from $\mathcal{P}_n(G, A)$ to $\mathcal{C}^n(G, A^G)$.

The rest statements of Lemma 1.17 follow immediately from what we have proved above. □

1.5. Coboundary operators and iterated difference operators. In this subsection, we want to study more carefully the image $D^n_{\mathcal{P}_n}(\mathcal{P}_n) \subseteq \mathcal{C}^n(G, A^G)$ of the homomorphism $D^n_{\mathcal{P}_n}$. We will see that this image is contained in the group $\mathcal{L}_n(G, A^G)$ of all A^G-valued n-morphism G (according to Lemma 1.16(a), $\mathcal{L}_n(G, A^G) \subseteq \mathcal{C}^n(G, A^G)$). To prove this, we will introduce the *coboundary operators* $\delta^n : \mathcal{C}^n(G, A) \to \mathcal{C}^{n+1}(G, A)$ (related to the *trivial* left G-action in A) and study their relation to the iterated difference operators D^n.

DEFINITION 1.18. Let us consider A as a trivial left G-module and define the *coboundary operator*

$$\delta^n : \mathcal{C}^n(G, A) \to \mathcal{C}^{n+1}(G, A)$$

by the standard formula

$$[\delta^n c](g_1, \ldots, g_{n+1}) = c(g_2, \ldots, g_{n+1}) - c(g_1 g_2, g_3, \ldots, g_{n+1}) + c(g_1, g_2 g_3, \ldots, g_{n+1}) + \ldots + (-1)^n c(g_1, \ldots, g_{n-1}, g_n g_{n+1}) + (-1)^{n+1} c(g_1, \ldots, g_n).$$

Notice that $\delta^0 = 0$, since we use the trivial left G-action in A. Indeed, $\mathcal{C}^0(G, A) = A$ and for any $a \in A$ and $g \in G$ we have $(\delta^0 a)(g) = {}^g a - a = a - a = 0$.

LEMMA 1.19. *Any n-morphism is a n-cocycle.*

PROOF. According to Lemma 1.16(a), any n-morphism c is a n-cochain. A straightforward calculation shows that $\delta^n c(g_1, \ldots, g_{n+1}) = 0$ for all $g_1, \ldots, g_{n+1}$. □

In the following two lemmas, we establish some simple algebraic properties of the operators δ, d, and D. We omit a simple computation that proves the first lemma.

LEMMA 1.20. *Let $n \in \mathbb{N}$ and $c \in \mathcal{C}^n(G, A)$. For each $h \in G$, we define the $(n-1)$-cochain c_h by*

$$c_h(h_1, \ldots, h_{n-1}) := c(h, h_1, \ldots, h_{n-1}).$$

Then for every $h, g_1, \ldots, g_n \in G$, we have

$$[\delta^n c](h, g_1, \ldots, g_n) = c(g_1, g_2, \ldots, g_n) - c(hg_1, g_2, \ldots, g_n) + c(h, g_2, \ldots, g_n) - [\delta^{n-1} c_h](g_1, \ldots, g_n).$$

LEMMA 1.21. *For any $n \geq 1$ we have:*

$$\delta^n d_{n-1} = d_n \delta^{n-1} + (-1)^n d_n d_{n-1}, \tag{1.20}$$

$$\delta^n D^n = d_n \delta^{n-1} D^{n-1} + (-1)^n D^{n+1}, \tag{1.21}$$

$$\delta^n D^n = \begin{cases} -D^{n+1} & \text{for } n \text{ odd}, \\ 0 & \text{for } n \text{ even}. \end{cases} \tag{1.22}$$

PROOF. Let $c \in \mathcal{C}^{n-1}(G, A)$ and $g_1, \ldots, g_{n+1} \in G$. Put $c' := d_{n-1} c$. Then we have:

$$\begin{aligned}
&(\delta^n d_{n-1} c)(g_1, \ldots, g_{n+1}) = (\delta^n c')(g_1, \ldots, g_{n+1}) \\
&= c'(g_2, \ldots, g_{n+1}) - c'(g_1 g_2, g_3, \ldots, g_{n+1}) \\
&\quad + c'(g_1, g_2 g_3, g_4, \ldots, g_{n+1}) + \ldots + (-1)^{n-1} c'(g_1, \ldots, g_{n-1} g_n, g_{n+1}) \\
&\quad + (-1)^n c'(g_1, \ldots, g_{n-1}, g_n g_{n+1}) + (-1)^{n+1} c'(g_1, \ldots, g_{n-1}, g_n) \\
&= (d_{n-1} c)(g_2, \ldots, g_{n+1}) - (d_{n-1} c)(g_1 g_2, g_3, \ldots, g_{n+1}) \\
&\quad + (d_{n-1} c)(g_1, g_2 g_3, g_4, \ldots, g_{n+1}) + \ldots + (-1)^{n-1} (d_{n-1} c)(g_1, \ldots, g_{n-1} g_n, g_{n+1}) \\
&\quad + (-1)^n (d_{n-1} c)(g_1, \ldots, g_{n-1}, g_n g_{n+1}) + (-1)^{n+1} (d_{n-1} c)(g_1, \ldots, g_{n-1}, g_n) \\
&= [c^{g_{n+1}}(g_2, \ldots, g_n) - c(g_2, \ldots, g_n)] - [c^{g_{n+1}}(g_1 g_2, g_3, \ldots, g_n) - c(g_1 g_2, g_3, \ldots, g_n)] \\
&\quad + [c^{g_{n+1}}(g_1, g_2 g_3, g_4, \ldots, g_n) - c(g_1, g_2 g_3, g_4, \ldots, g_n)] \\
&\quad + \ldots + (-1)^{n-1} [c^{g_{n+1}}(g_1, \ldots, g_{n-2}, g_{n-1} g_n) - c(g_1, \ldots, g_{n-2}, g_{n-1} g_n)] \\
&\quad + (-1)^n [c^{g_n g_{n+1}}(g_1, \ldots, g_{n-1}) - c(g_1, \ldots, g_{n-1})] \\
&\quad + (-1)^{n+1} [c^{g_n}(g_1, \ldots, g_{n-1}) - c(g_1, \ldots, g_{n-1})]
\end{aligned}$$

Adding the zero sum $(-1)^n c^{g_{n+1}}(g_1, ..., g_{n-1}) - (-1)^n c^{g_{n+1}}(g_1, ..., g_{n-1})$ to the right hand side of the above equality and rearranging the obtained expression, we see that

$$\begin{aligned}
&(\delta^n d_{n-1} c)(g_1, ..., g_{n+1}) \\
&= [c(g_2, ..., g_n) - c(g_1 g_2, ..., g_n) + c(g_1, g_2 g_3, ..., g_n) \\
&\qquad + ... + (-1)^{n-1} c(g_1, ..., g_{n-1} g_n) + (-1)^n c(g_1, ..., g_{n-1})]^{g_{n+1}} \\
&\qquad - [c(g_2, ..., g_n) - c(g_1 g_2, ..., g_n) + c(g_1, g_2 g_3, ..., g_n) \\
&\qquad + ... + (-1)^{n-1} c(g_1, ..., g_{n-1} g_n) + (-1)^n c(g_1, ..., g_{n-1})] \\
&\qquad + (-1)^n [c^{g_n g_{n+1}}(g_1, ..., g_{n-1}) - c^{g_{n+1}}(g_1, ..., g_{n-1}) \\
&\qquad - c^{g_n}(g_1, ..., g_{n-1}) + c(g_1, ..., g_{n-1})] \\
&= [(\delta^{n-1} c)(g_1, ..., g_n)]^{g_{n+1}} - (\delta^{n-1} c)(g_1, ..., g_n) \\
&\qquad + (-1)^n \{[(d_{n-1} c)(g_1, ..., g_n)]^{g_{n+1}} - (d_{n-1} c)(g_1, ..., g_n)\} \\
&= [d_n(\delta^{n-1} c)](g_1, ..., g_{n+1}) + (-1)^n [d_n(d_{n-1} c)](g_1, ..., g_{n+1}) \\
&= [(d_n \delta^{n-1} + (-1)^n d_n d_{n-1}) c](g_1, ..., g_{n+1}),
\end{aligned}$$

which proves (1.20). Furthermore, by (1.3) and (1.20), we have

$$\begin{aligned}
\delta^n D^n &= \delta^n d_{n-1} d_{n-2} \cdots d_0 = (\delta^n d_{n-1})\, d_{n-2} \cdots d_0 \\
&= (d_n \delta^{n-1} + (-1)^n d_n d_{n-1})\, d_{n-2} \cdots d_0 \\
&= d_n \delta^{n-1} d_{n-2} \cdots d_0 + (-1)^n d_n d_{n-1} d_{n-2} \cdots d_0 \\
&= d_n \delta^{n-1} D^{n-1} + (-1)^n D^{n+1}.
\end{aligned}$$

This proves (1.21). Finally, to prove (1.22), we use (1.21) and induction in n. Since $\delta^0 = 0$, we have $\delta^0 D^0 = 0$. Assume that for some $m \in \mathbb{N}$, relation (1.22) is already proven for all $n \leq m - 1$. If m is odd then $m - 1$ is even and

$$\delta^m D^m = d_m \delta^{m-1} D^{m-1} + (-1)^m D^{m+1} = -D^{m+1}.$$

If m is even then $m - 1$ is odd and

$$\begin{aligned}
\delta^m D^m &= d_m \delta^{m-1} D^{m-1} + (-1)^m D^{m+1} \\
&= d_m(-D^m) + D^{m+1} = -D^{m+1} + D^{m+1} = 0.
\end{aligned}$$

This competes the proof of the lemma. □

Corollary 1.22. *Let $n \in \mathbb{Z}_+$ and $a \in \mathcal{P}_n(G, A)$. Then $D^n a$ is a n-cocycle of G with values in A^G.*

Proof. By Lemma 1.17, $D^n a \in \mathcal{C}^n(G, A^G)$. Thus, we only need to check that $\delta^n D^n a = 0$. By (1.22),

$$\delta^n D^n = \begin{cases} -D^{n+1} & \text{for } n \text{ odd}, \\ 0 & \text{for } n \text{ even}; \end{cases}$$

this implies $\delta^n D^n a = 0$ since $D^{n+1} a = 0$. □

1.6. Iterated difference operators and polymorphisms. Here we establish that polynomial-like elements are closely related to polymorphisms. This connection provides us with an appropriate tool for the proving that the spaces of polynomial-like elements are of finite dimension, i.e., the Main Theorem.

NOTATION 1.23. Let $\mathcal{L}_n^S(G, A^G)$ denote the subgroup of the group $\mathcal{L}_n(G, A^G)$ consisting of all *symmetric* A^G-valued n-morphisms of G. In more details, a n-morphism

$$L\colon \underbrace{G \times \cdots \times G}_{n} \to A^G$$

belongs to $\mathcal{L}_n^S(G, A^G)$ whenewer $L(g_{j_1}, ..., g_{j_n}) = L(g_1, ..., g_n)$ for any $g_1, ..., g_n \in G$ and any permutation $(j_1, ..., j_n)$ of the indices $1, ..., n$. ○

Suppose $a \in \mathcal{P}_0(G, A)$, that is, $a \in A$ and $[D^1 a](g) = a^g - a = 0$ for all all $g \in G$. This shows that the element a is in A^G and hence its image $D^0 a = a$ belongs to $A^G = \mathcal{L}_0(G, A^G)$ (see Definition 1.14).

Suppose now that $a \in \mathcal{P}_1(G, A)$. Then, according to Corollary 1.22, $D^1 a$ is a 1-cocycle of G with values in A^G. Hence

$$D^1 a\colon G \to A^G \quad \text{and} \quad \delta^1(D^1 a) = 0\,, \quad \text{that is,}$$
$$[D^1 a](g_2) - [D^1 a](g_1 g_2) + [D^1 a](g_1) = 0 \text{ for all } g_1, g_2 \in G\,,$$

which shows that the mapping $D^1 a\colon G \to A^G$ *is a group homomorphism whenever* $a \in \mathcal{P}_1(G, A)$.

Combining this simple observation with Lemma 1.5, we come to the following property:

PROPOSITION 1.24. *Let $n \in \mathbb{Z}_+$ and $a \in \mathcal{P}_n(G, A)$. Then $D^n a \in \mathcal{L}_n^S(G, A^G)$, that is, $D^n a$ is a symmetric A^G-valued n-morphism of G.*

PROOF. The symmetry of $D^n a$ and the inclusion $[D^n a](g_1, ..., g_n) \in A^G$ for any $a \in \mathcal{P}_n(G, A)$ and all $g_1, ..., g_n \in G$ have been established in Lemmas 1.5 and 1.17. Thus, we only need to prove that $D^n a$ is a polymorphism.

The proof is by induction in n. The cases $n = 0, 1$ have been already handled above. Suppose that $m \geq 2$ and the statement is already proven for $n = m - 1$. To complete the proof, we need to show that $D^m a \in \mathcal{L}_m(G, A^G)$ whenever $D^{m+1} a = 0$.

Define cochains $c \in \mathcal{C}^m(G, A^G)$ and $c_h \in \mathcal{C}^{m-1}(G, A^G)$ $(h \in G)$ by

$$\begin{aligned} c(g_1, ..., g_m) &= [D^m a](g_1, g_2, ..., g_m) \quad \text{and} \\ c_h(g_2, ..., g_m) &= c(h, g_2, ..., g_m) = [D^m a](h, g_2, ..., g_m)\,. \end{aligned} \tag{1.23}$$

We start with the following claim:

CLAIM 1. For any $h \in G$, the $(m - 1)$-cochain c_h,

$$c_h\colon \underbrace{G \times \cdots \times G}_{m-1} \ni (g_2, ..., g_m) \mapsto c_h(g_2, ..., g_m) = [D^m a](h, g_2, ..., g_m) \in A^G\,,$$

is an A^G-valued $(m - 1)$-homomorphism.

To justify this claim, we apply Lemma 1.6 to the $(m + 1)$-cochain $D^{m+1} a$. Taking into account that $D^{m+1} a = 0$ for any $h_1, ..., h_m \in G$, we obtain

$$\begin{aligned} [D^m(a^h - a)](h_1, ..., h_m) &= [D^m(a^h)](h_1, ..., h_m) - [D^m a](h_1, ..., h_m) \\ &= [D^{m+1} a](h, h_1, ..., h_m) = 0\,. \end{aligned}$$

The latter relation means that the element $b_h \stackrel{\text{def}}{=\!=} a^h - a$ is a polynomial-like element in A of order at most $m-1$. Hence, by the induction hypothesis, $D^{m-1}b_h \in \mathcal{L}_{m-1}(G, A^G)$. Using Lemma 1.6 again, we obtain that for any $g_2, ..., g_m \in G$

$$\begin{aligned} c_h(g_2, ..., g_m) &= [D^m a](h, g_2, ..., g_m) = [D^{m-1}(a^h)](g_2, ..., g_m) - [D^{m-1}a](g_2, ..., g_m) \\ &= [D^{m-1}(a^h - a)](g_2, ..., g_m) = [D^{m-1}b_h](g_2, ..., g_m)\,. \end{aligned}$$

Thereby, $c_h = D^{m-1}b_h \in \mathcal{L}_{m-1}(G, A^G)$, which proves Claim 1.

In view of Claim 1, the following statement is an immediate consequence of Lemma 1.19:

CLAIM 2. For any $h \in G$, the $(m-1)$-cochain $c_h \in \mathcal{L}_{m-1}(G, A^G)$ defined in (1.23) is a $(m-1)$-cocycle, that is, $[\delta^{m-1}c_h](g_1, ..., g_m) = 0$ for every $g_1, ..., g_m \in G$.

CLAIM 3. For any $g_2, ..., g_m \in G$ the mapping

$$F_{g_2,...,g_m}\colon G \ni g \mapsto [D^m a](g, g_2, ..., g_m) \in A^G \tag{1.24}$$

is a group homomorphism.

To justify Claim 3, we apply Lemma 1.20 (with $n = m+1$) to the m-cochain $c = D^m a$ and $(m-1)$-cochain c_h defined in (1.23). This gives the relation

$$\begin{aligned} [\delta^m D^m a](h, g_1, ..., g_m) &= [\delta^m c](h, g_1, ..., g_m) \\ &= c(g_1, g_2, \dots, g_m) - c(hg_1, g_2, \dots, g_m) \\ &\quad + c(h, g_2, \dots, g_m) - [\delta^{m-1}c_h](g_1, \dots, g_m) \\ &= [D^m a](g_1, g_2, \dots, g_m) - [D^m a](hg_1, g_2, \dots, g_m) \\ &\quad + [D^m a](h, g_2, \dots, g_m) - [\delta^{m-1}c_h](g_1, \dots, g_m)\,. \end{aligned} \tag{1.25}$$

Since $D^m a$ is a m-cocycle (Corollary 1.22) and c_h is a $(m-1)$-cocycle (Claim 2), relation (1.25) takes the form

$$[D^m a](hg_1, g_2, \dots, g_m) = [D^m a](h, g_2, \dots, g_m) + [D^m a](g_1, g_2, \dots, g_m)\,,$$

which proves (1.24).

Combining Claim 1 with Claim 3, we complete the induction step, which proves Proposition 1.24. □

REMARK 1.25. The above proof of Proposition 1.24 can be simplified: whenever Claim 1 is proven, one may use the symmetry of $D^n a$ (Lemma 1.5) to conclude that for any $n \geq 2$, the mapping $D^n a\colon \underbrace{G \times \cdots \times G}_{n} \to A^G$ is a group polymorphism. ○

MAIN THEOREM. *(a) $D^n(\mathcal{P}_n(G, A)) \subseteq \mathcal{L}_n^{\mathcal{S}}(G, A^G)$ for every $n \in \mathbb{N}$. That is, the image $D^n(\mathcal{P}_n(G, A))$ of the subgroup $\mathcal{P}_n(G, A) \subseteq A$ of all polynomial-like elements of order at most n under the homomorphism $D^n\colon \mathcal{P}_n(G, A) \to \mathcal{C}^n(G, A)$ is contained in the subgroup $\mathcal{L}_n^{\mathcal{S}}(G, A^G) \subseteq \mathcal{C}^n(G, A)$ consisting of all symmetric A^G-valued n-morphisms of the group G.*

(b) Exact sequence (1.18) *and embedding* (1.19) *may be written in the form*

$$0 \longrightarrow \mathcal{P}_{n-1}(G, A) \longrightarrow \mathcal{P}_n(G, A) \xrightarrow{D^n_{\mathcal{P}_n}} \mathcal{L}_n^{\mathcal{S}}(G, A^G) \tag{1.26}$$

and

$$i_n\colon [\mathcal{P}_n(G, A)/\mathcal{P}_{n-1}(G, A)] \hookrightarrow \mathcal{L}_n^{\mathcal{S}}(G, A^G)\,.$$

(*c*) *Suppose that* F *is a field of characteristic* 0, G *is a group, and* A *is a* F-*vector space endowed with a linear right* G-*action. If the group* $\widetilde{G} = G/[G,G]$ *is finitely generated and* $\dim_F A^G < \infty$ *then*

$$\dim_F \mathcal{P}_n(G,A) < \infty \quad \textit{for every } n \in \mathbb{Z}_+ .$$

PROOF. (a) is proven in Proposition 1.24. (b) follows from (a) and Lemma 1.17. To proof (c), we use induction in n, statement (b), and Lemma 1.16(d). Indeed, by assumption, the space $\mathcal{P}_0(G,A) = A^G$ is of finite dimension. Suppose that we already have that $\dim_F \mathcal{P}_{n-1}(G,A) < \infty$. By Lemma 1.16($d$), we also have that $\dim_F \mathcal{L}_n^{\mathcal{S}}(G,A^G) \leq \dim_F \mathcal{L}_n(G,A^G) < \infty$. Therefore, the induction hypothesis and the exact sequence (1.26) would imply that $\dim_F \mathcal{P}_n(G,A) < \infty$. □

We immediately obtain the following estimate:

PROPOSITION 1.26. *Under the same assumption of part (c) of Main Theorem, the following estimate holds:*

$$\dim_F \mathcal{P}_n(G,A) \leq s \cdot \binom{n+r}{r},$$

where $s = \dim_F A^G$ *and* r *is the rank of the abelian group* $\widetilde{G}$.

PROOF. We prove it by induction in n. The estimate is obviously true for $n = 0$. So we assume that $n \geq 1$ and the estimate holds for $n-1$, i.e.,

$$\dim_F \mathcal{P}_{n-1}(G,A) \leq s \cdot \binom{n+r-1}{r}. \tag{1.27}$$

According to the proof of Main Theorem, we have

$$\begin{aligned}\dim_F \mathcal{P}_n(G,A) &= \dim_F \mathcal{P}_n(G,A)/\mathcal{P}_{n-1}(G,A) + \dim_F \mathcal{P}_{n-1}(G,A) \\ &\leq \dim_F \mathcal{L}_n^{\mathcal{S}}(G,A^G) + \dim_F \mathcal{P}_{n-1}(G,A).\end{aligned} \tag{1.28}$$

Using Lemma 1.16(d), we have

$$\dim_F \mathcal{L}_n^{\mathcal{S}}(G,A^G) = \dim_F \mathcal{L}_n^{\mathcal{S}}(\mathbb{Z}^r,A^G).$$

Moreover, we recall from the proof of Lemma 1.16(d) that the polymorphisms $\{\ell_I^i\}_{1\leq i\leq s, I\in\mathcal{I}}$ constitute a vector basis for $\mathcal{L}_n(\mathbb{Z}^r,A^G)$, where $\mathcal{I}$ is the family of all finite ordered sets I consisting of n natural numbers in $\{1,\ldots,r\}$. For any $I = (i_1,\ldots,i_n) \in \mathcal{I}$, we define $\gamma(I) := (\gamma_1(I),\ldots,\gamma_r(I))$, where for each $1 \leq k \leq r$, the notation $\gamma_k(I)$ is the number of indices $j \in \{1,\ldots,n\}$ such that $i_j = k$. If $I, J \in \mathcal{I}$ and $\gamma(I) = \gamma(J)$, we say that I and J belong to the same class. Consider any $L \in \mathcal{L}_n^{\mathcal{S}}(\mathbb{Z}^r,A^G)$ and write it as follows

$$L = \sum_{i=1}^{s}\sum_{I\in\mathcal{I}} a_I^i \ell_I^i.$$

Due to the symmetry of L, whenever I, J are in the same class, we must have $a_I^i = a_J^i$ for any $i \in \{1,\ldots,s\}$. Thus, to determine L completely, one just needs to know the coefficients a_I^i for only one representative I in each class. By definition, the cardinality of these classes is equal to the number of the r-tuples consisting of r nonnegative intergers $(\gamma_1,\ldots,\gamma_r)$ satisfying $\gamma_1 + \ldots \gamma_r = n$. Therefore, we have

$$\dim_F \mathcal{L}_n^{\mathcal{S}}(G,A^G) = (\dim_F A^G) \cdot \binom{n+r-1}{r-1}, \tag{1.29}$$

for any $n \in \mathbb{Z}_+$. Using (1.27), (1.28), (1.29), we finish the proof. □

2. Polynomial-like elements in a ring

Suppose that a group G acts on the right side on a ring A in such a way that for any $g \in G$ the mapping $A \ni a \mapsto a^g \in A$ is a ring automorphism. Under this assumption, we prove now some additional properties of the iterated difference operators and the set $\mathcal{P}(G, A)$ of all polynomial-like elements in A.

PROPOSITION 2.1. *(a)* ***Leibniz-type formula:*** *For any $a, b \in A$ and $g \in G$,*

$$[D^1(ab)](g) = a^g \cdot \{[D^1 b](g)\} + \{[D^1 a](g)\} \cdot b\,. \tag{2.1}$$

(b) Let $m, n \in \mathbb{Z}_+$ and $m + n \geq 1$. Then

$$D^{m+n-1}(ab) = 0 \quad \textit{whenever} \quad a, b \in A \quad \textit{and} \ D^m a = D^n b = 0\,, \tag{2.2}$$

Equivalently, $ab \in \mathcal{P}_{m+n-1}(G, A)$ whenever $a \in \mathcal{P}_m(G, A)$ and $b \in \mathcal{P}_n(G, A)$. In particular, the set $\mathcal{P}(G, A)$ of all polynomial-like elements in A is a subring of the ring A.

(c) Each operator $D^n \colon A \to \mathcal{C}^n(G, A)$, $n \in \mathbb{N}$, is A^G-linear, i.e., whenever $a \in A$ is a G-invariant element, we have that $D^n(ab) = aD^n b$ for all $b \in A$.

PROOF. *(a)* (2.1) is straightforward:

$$\begin{aligned}\left[D^1(ab)\right](g) &= (ab)^g - ab = a^g b^g - ab = a^g \cdot (b^g - b) + (a^g - a) \cdot b \\ &= a^g \cdot \{[D^1 b](g)\} + \{[D^1 a](g)\} \cdot b\,.\end{aligned}$$

(b) The proof is by induction in $N = m + n$. Since $D^0 = \mathrm{id}_A$, statement (2.2) is trivial whenever one of the numbers m, n is 0; hence (2.2) is true for $N = 1$ and, from now on, we may assume $m, n \in \mathbb{N}$. Suppose that for some $N \geq 1$, statement (2.2) is true for all $m, n \in \mathbb{N}$ with $m+n = N$. Take any $m, n \in \mathbb{N}$ with $m+n = N+1$ and assume $D^m a = D^n b = 0$. By Lemma 1.6,

$$\begin{aligned}[D^{m-1}(a^{g_1} - a)](g_2, ..., g_m) &= [D^{m-1}(a^{g_1})](g_2, ..., g_m) - [D^{m-1} a](g_2, ..., g_m) \\ &= [D^m a](g_1, ..., g_m) = 0\,,\end{aligned}$$

that is,

$$D^{m-1}(a^{g_1} - a) = 0 \ \text{ for any } \ g_1 \in G\,. \tag{2.3}$$

Hence, taking into account the assumption $D^n b = 0$ and the induction hypothesis, we obtain

$$D^{m+n-2}[(a^{g_1} - a)b] = 0\,. \tag{2.4}$$

Similarly, the assumption $D^n b = 0$ implies

$$D^{n-1}(b^{g_1} - b) = 0 \ \text{ for any } \ g_1 \in G\,. \tag{2.5}$$

combining this with the assumption $D^m a = 0$ and the induction hypothesis, we get

$$D^{m+n-2}[a(b^{g_1} - b)] = 0\,. \tag{2.6}$$

Summing up relations (2.4) and (2.6) we see that $D^{m+n-2}[(a^{g_1} - a)b + a(b^{g_1} - b)] = 0$, that is,

$$D^{m+n-2}(a^{g_1} b + ab^{g_1} - 2ab) = 0\,. \tag{2.7}$$

Relation (2.3) certainly implies that $D^m(a^{g_1} - a) = 0$; in view of (2.5) and the induction hypothesis, this shows that

$$D^{m+n-2}[(a^{g_1} - a)(b^{g_1} - b)] = 0 .$$

Thus,

$$D^{m+n-2}(a^{g_1} b^{g_1} - ab^{g_1} - a^{g_1} b + ab) = 0 . \tag{2.8}$$

Summing up (2.7) and (2.8), we get

$$D^{m+n-2}(a^{g_1} b^{g_1} - ab) = 0 . \tag{2.9}$$

Finally, we apply Lemma 1.6 to the element $D^{m+n-1}(ab)$ and take into account (2.9) to see that for every $g_1, g_2, ..., g_{m+n-1} \in G$, we have

$$\begin{aligned}
&[D^{m+n-1}(ab)](g_1, g_2, ..., g_{m+n-1}) \\
&\qquad = \{D^{m+n-2}[(ab)^{g_1}]\}(g_2, ..., g_{m+n-1}) - [D^{m+n-2}(ab)](g_2, ..., g_{m+n-1}) \\
&\qquad = [D^{m+n-2}(a^{g_1} b^{g_1} - ab)](g_2, ..., g_{m+n-1}) = 0 .
\end{aligned}$$

Therefore, $D^{m+n-1}(ab) = 0$, which completes the induction step and proves (b).

(c) This follows immediately from Lemma 1.3 and the fact that for any $g_1, \ldots, g_n \in G$ and $1 \le i_1 < \ldots < i_s \le n$, one has

$$(ab)^{\pi_{i_1,\ldots,i_s}(g_1,\ldots,g_n)} = ab^{\pi_{i_1,\ldots,i_s}(g_1,\ldots,g_n)} .$$

□

3. Polynomial-like elements related to lattices

In this section, we now present an application of the Main Theorem to an important special case, when G is a lattice [4] that acts naturally [5] on an Euclidean space whose dimension is the same as the rank of the abelian group G. In this case, A is a G-invariant subalgebra of the algebra of continuous functions defined on the Euclidean space (see Example 1.8). We will first prove that polynomial-like elements of order n are exactly polynomials of the same order with G-periodic coefficients. As a consequence, we obtain Corollary 3.5 which states roughly that given a periodic elliptic operator $\mathcal{D}$, all of the dimensions of the spaces of polynomial-like solutions of $\mathcal{D}$ of order n ($n \in \mathbb{Z}_+$) are finite provided that there are only finitely many linealy independent periodic solutions. Notice that the same results should hold for the case of periodic elliptic operators on co-compact abelian coverings; however, for simplicity, here we focus first on the Euclidean case and then consider briefly the covering case.

3.1. Polynomials with periodic coefficients. Let $G \cong \mathbb{Z}^r$ be a lattice of rank r in $\mathbb{R}^r$. This lattice acts naturally in the space $C(\mathbb{R}^r)$ of all (real or complex) continuous functions on $\mathbb{R}^r$; namely [6],

$$\begin{aligned}
&C(\mathbb{R}^r) \ni f \mapsto f^g \to C(\mathbb{R}^r) , \\
&\qquad \text{where } f^g(x) = f(T_g x) = f(x+g) , \quad x \in \mathbb{R}^r , \quad g \in G .
\end{aligned} \tag{3.1}$$

[4] About lattices in $\mathbb{R}^r$, one can find a definition and some basics in [**8**, Section 2]

[5] I.e., the action is isometric, free, properly discontinuous, and co-compact.

[6] If we wish, this action may be regarded as a right one; this is of no importance since G is commutative.

Actually, all mappings (3.1) are automorphisms of the algebra $C(\mathbb{R}^r)$. Denote by C^G the subalgebra of $C(\mathbb{R}^r)$ consisting of all G-invariant (or, which is the same, G-periodic) functions $f \in C(\mathbb{R}^r)$. For $n \in \mathbb{Z}_+$, set

$$P^G = C^G[x_1, ..., x_r]\,,$$
$$P_n^G = C_n^G[x_1, ..., x_r] = \{p \in C^G[x_1, ..., x_r] \mid \deg p \le n\}\,;$$

here $x_1, ..., x_r$ are the standard coordinates in $\mathbb{R}^r$. In other words, P^G is the algebra (over $\mathbb{C}$ or $\mathbb{R}$) of all polynomials in the coordinates $x_1, ..., x_r$ with continuous G-periodic coefficients; P_n^G is the vector subspace of P_n^G consisting of all polynomials in $x_1, ..., x_r$ of degree at most n with continuous G-periodic coefficients. By misuse of language, we call elements in P^G as *G-periodic polynomials* (or more specifically, *G-periodic polynomials of degree at most n* for elements in P_n^G). A G-periodic polynomial of the form $Q_n = f(x) \cdot x_{i_1} \cdots x_{i_n}$ $(1 \le i_1, ..., i_n \le r)$ with G-periodic coefficient $f \ne 0$ is said to be *G-periodic monomial of degree n*; such a monomial may also be written in the form $Q_n = f(x) \cdot x_1^{j_1} \cdots x_r^{j_r}$, where $j_1, ..., j_r \in \mathbb{Z}_+$ and $j_1 + ... + j_r = n$. Any G-periodic polynomial of degree at most n is a sum of G-periodic monomials of degree at most n; such a representation is unique (up to an order of summands). Clearly, all P_n^G are G-invariant subspaces of $C(\mathbb{R}^r)$, and P^G itself is a G-invariant subalgebra of $C(\mathbb{R}^r)$.

The G-action in $C(\mathbb{R}^r)$ gives rise to iterated difference operators

$$D^n \colon C(\mathbb{R}^r) \to \mathcal{C}^n(G, C(\mathbb{R}^r))$$

(see (1.2) and (1.3)). As in Section 1, we define the subspaces $\mathcal{P}_n(G, C(\mathbb{R}^r)) \subseteq C(\mathbb{R}^r)$ of *polynomial-like elements in $C(\mathbb{R}^r)$ of order at most n* by

$$\mathcal{P}_n(G, C(\mathbb{R}^r)) = \ker D^{n+1}$$

(see Definition 1.7).

LEMMA 3.1. *(a) $D^{n+1}Q_m = 0$ for any G-periodic monomial of degree $m \le n$.*

(b) Let $Q_n = x_{i_1} \cdots x_{i_n} = x_1^{j_1} \cdots x_r^{j_r}$, where $1 \le i_1, ..., i_n \le r$, $j_1, ..., j_r \in \mathbb{Z}_+$, $j_1 + ... + j_r = n$, and each coordinate x_k $(k = 1, ..., r)$ enters the product $x_{i_1} \cdots x_{i_n}$ precisely j_k times. Let $g_1, ..., g_n \in G$ and

$$g_i = (g_{i,1}, ..., g_{i,r}) = \sum_{k=1}^{r} g_{i,k}\mathbf{e}_k \quad (1 \le i \le n)\,,$$

where $\{\mathbf{e}_1, ..., \mathbf{e}_r\}$ is the standard basis in $\mathbb{R}^r$. Then

$$\begin{aligned}[D^n Q_n](g_1, ..., g_n) &= \sum_{\boldsymbol{\sigma} \in \mathbf{S}(n)} g_{1,i_{s_1}} g_{2,i_{s_2}} \cdots g_{n,i_{s_n}} \\ &= j_1! \cdots j_r! \cdot \sum_{\boldsymbol{\kappa}} g_{1,k_1} g_{2,k_2} \cdots g_{n,k_n}\,;\end{aligned} \tag{3.2}$$

here $\boldsymbol{\sigma} = (s_1, ..., s_n) \in \mathbf{S}(n)$ in the first sum runs over all the permutations of the indices $1, ..., n$, whereas $\boldsymbol{\kappa} = (k_1, ..., k_n)$ in the second sum runs over all finite sequences of length n containing each of the indices $k \in \{1, ..., r\}$ precisely j_k times.

PROOF. (a) The proof is by induction in n. If f is a G-periodic function then $f^g = f$ for all $g \in G$, that is,

$$D^1 f = 0 \quad \text{for any } G\text{-periodic } f\,. \tag{3.3}$$

For each coordinate function $\phi_i(x) = x_i$ ($1 \le i \le r$) and any $g = (g_1, ..., g_r) \in G$, we have

$$\phi_i^g(x) = \phi_i(x+g) = x_i + g_i \quad \text{and} \quad \{[D^1\phi_i](g)\}(x) = \phi_i(x+g) - \phi_i(x) = g_i.$$

Since the right hand side g_i of the latter relation does not depend on x, it is G-invariant and, therefore,

$$\{[D^2\phi_i](g', g'')\} = \{[D^1\phi_i](g')\}^{g''} - \{[D^1\phi_i](g')\} = 0,$$

which means that $D^2\phi_i = 0$. Due to Proposition 2.1, this implies that

$$D^2(a\phi_i) = 0 \quad \text{for all } i = 1, ..., r \text{ whenever } a \in C(\mathbb{R}^r)^G. \tag{3.4}$$

Relations (3.3), (3.4) provide us with a base of induction.

Suppose that statement (a) is already proven for some $n \ge 1$. Let $Q_m = f(x) \cdot x_{i_1} \cdots x_{i_m}$ be a G-periodic monomial of degree $m \le n+1$. If $m \le n$ then, by the induction hypothesis, $D^{n+1}Q_m = 0$, which certainly implies $D^{n+2}Q_m = 0$. On the other hand, if $m = n+1$, we may represent the monomial $Q_{n+1} = Q_m$ as a product $Q_{n+1} = Q_n \cdot x_{i_{n+1}}$, where Q_n is a G-periodic monomial of degree n. By the induction hypothesis and (3.4), we have

$$D^{n+1}Q_n = 0 \quad \text{and} \quad D^2 x_{i_{n+1}} = D^2\phi_{i_{n+1}} = 0.$$

Using this and Proposition 2.1(b) (with the pair $(n+1, 2)$ instead of (m, n)), we obtain

$$D^{n+2}(Q_{n+1}) = D^{(n+1)+2-1}(Q_n \cdot x_{i_{n+1}}) = 0,$$

which concludes the induction step and proves (a).

(b) The proof of the first equality in (3.2) is by induction on n (the coincidence of the two sums in (3.2) under the condition $x_{i_1} \cdots x_{i_n} = x_1^{j_1} \cdots x_r^{j_r}$ is an elementary combinatorial fact; we skip its proof).

First, if $n = 1$ we have $Q_1 = x_i$ for some $i \in \{1, ..., r\}$, and hence $[D^1Q_1](g_1) = (x_i + g_{1,i}) - x_i = g_{1,i}$. On the other hand, it is easily seen that the first sum in (3.2) is equal to $g_{1,i}$ whenever $n = 1$, $j_i = 1$, and $j_k = 0$ for $k \ne i$. This proves formula (3.2) for $n = 1$.

Suppose that for some $n \ge 2$, the formula (3.2) is already proven for all $m \le n-1$. Let us show that then (3.2) holds true for $m = n$. Take a monomial $Q_n = x_{i_1} \cdots x_{i_n}$ $(1 \le i_1, ..., i_n \le r)$ and write it in the form $Q_n = x_{i_1} \cdot Q_{n-1}$, where $Q_{n-1} = x_{i_2} \cdots x_{i_n}$. By Lemma 1.6, we have

$$\begin{aligned} [D^nQ_n](g_1, g_2, ..., g_n) &= [D^{n-1}(Q_n^{g_1})](g_2, ..., g_n) - [D^{n-1}Q_n](g_2, ..., g_n) \\ &= \{D^{n-1}\left([D^1Q_n](g_1)\right)\}(g_2, ..., g_n) \\ &= \{D^{n-1}\left([D^1(x_{i_1} \cdot Q_{n-1})](g_1)\right)\}(g_2, ..., g_n). \end{aligned} \tag{3.5}$$

By the Leibniz-type formula (Proposition 2.1(a)),

$$\begin{aligned} &[D^1(x_{i_1} \cdot Q_{n-1})](g_1) = (x_{i_1}^{g_1}) \cdot [D^1Q_{n-1}](g_1) + \{[D^1x_{i_1}](g_1)\} \cdot Q_{n-1} \\ &= (x_{i_1} + g_{1,i_1}) \cdot [D^1Q_{n-1}](g_1) + g_{1,i_1} \cdot Q_{n-1} \\ &= (x_{i_1} + g_{1,i_1}) \cdot [(x_{i_2} + g_{1,i_2}) \cdots (x_{i_n} + g_{1,i_n}) - x_{i_2} \cdots x_{i_n}] + g_{1,i_1} \cdot x_{i_2} \cdots x_{i_n}, \end{aligned}$$

which shows that

$$
\begin{aligned}
[D^1(x_{i_1}\cdot Q_{n-1})](g_1) &= x_{i_1}\cdot\sum_{s=2}^{n} g_{1,i_s}\cdot x_{i_2}\cdots\widehat{x_{i_s}}\cdots x_{i_n} + g_{1,i_1}\cdot x_{i_2}\cdots x_{i_n}\\
&\qquad + P_{n-2}(x_1,...,x_r;\,g_1)\\
&= \sum_{s=1}^{n} g_{1,i_s}\cdot x_{i_1}\cdots\widehat{x_{i_s}}\cdots x_{i_n} + P_{n-2}(x_1,...,x_r;\,g_1)\,,
\end{aligned}
\tag{3.6}
$$

where $P_{n-2}(x_1,...,x_r)$ is a polynomial in $x_1,...,x_r$ of degree at most $n-2$ (with constant coefficiens depending on the element $g_1\in G$). It follows from statement (a) that $D^{n-1}P_{n-2}=0$; therefore, formulae (3.5) and (3.6) imply

$$
\begin{aligned}
[D^nQ_n](g_1,g_2,...,g_n) &= \Big[D^{n-1}\sum_{s=1}^{n} g_{1,i_s}\cdot x_{i_1}\cdots\widehat{x_{i_s}}\cdots x_{i_n}\Big](g_2,...,g_n)\\
&= \sum_{s=1}^{n} g_{1,i_s}\cdot\big[D^{n-1}\big(x_{i_1}\cdots\widehat{x_{i_s}}\cdots x_{i_n}\big)\big](g_2,...,g_n)\,.
\end{aligned}
\tag{3.7}
$$

By the induction hypothesis,

$$
\big[D^{n-1}\big(x_{i_1}\cdots\widehat{x_{i_s}}\cdots x_{i_n}\big)\big](g_2,...,g_n) = \sum_{\boldsymbol{\sigma}_s} g_{2,i_{\boldsymbol{\sigma}_s(2)}}\cdots g_{n,i_{\boldsymbol{\sigma}_s(n)}}\,,
\tag{3.8}
$$

where $\boldsymbol{\sigma}_s$ runs over all one-to-one mappings $\{2,...,n\}\to\{1,...,\widehat{s},...,n\}$. It follows from (3.7) and (3.8) that

$$
\begin{aligned}
[D^nQ_n](g_1,g_2,...,g_n) &= \sum_{s=1}^{n} g_{1,i_s}\cdot\big[D^{n-1}\big(x_{i_1}\cdots\widehat{x_{i_s}}\cdots x_{i_n}\big)\big](g_2,...,g_n)\\
&= \sum_{s=1}^{n} g_{1,i_s}\cdot\sum_{\boldsymbol{\sigma}_s} g_{2,i_{\boldsymbol{\sigma}_s(2)}}\cdots g_{n,i_{\boldsymbol{\sigma}_s(n)}}\\
&= \sum_{\boldsymbol{\sigma}\in\mathbf{S}(n)} g_{1,i_{s_1}}g_{2,i_{s_2}}\cdots g_{n,i_{s_n}}\,,
\end{aligned}
$$

which concludes the induction step and proves Lemma. □

PROPOSITION 3.2. *Let a lattice $G\subset\mathbb{R}^r$ of rank r act naturally in $C(\mathbb{R}^r)$. A function $p\in C(\mathbb{R}^r)$ is a G-periodic polynomial of degree at most n if and only if $D^{n+1}p=0$. In other words,*

$$
P_n^G=\mathcal{P}_n(G,C(\mathbb{R}^r))\overset{\text{def}}{=\!=}\ker\big[D^{n+1}\colon C(\mathbb{R}^r)\to\mathcal{C}^{n+1}(G,C(\mathbb{R}^r))\big]\,.
$$

PROOF. Since all the iterated difference operators are linear, the inclusion $P_n^G\subseteq\ker D^{n+1}$ follows immediately from Lemma 3.1(a).

The proof of the opposite inclusion $\ker D^{n+1}\subseteq P_n^G$ is by induction in n. Let $p\in C(\mathbb{R}^r)$ and $D^1p=0$; that is, $p(x+g)=p(x)$ for all $x\in\mathbb{R}^r$ and $g\in G$, which means that p is a G-periodic polynomial of degree 0. This gives us the base of induction. Suppose that the inclusion $\ker D^n\subseteq P_{n-1}^G$ is already proven for some $n\geq 1$. Let $p\in C(\mathbb{R}^r)$ and $D^{n+1}p=0$. By part (a) of Main Theorem, D^np is a symmetric n-polymorphism on G with values in $C^G(\mathbb{R}^r)$. Hence, D^np may be recovered from its values $(D^np)(\mathbf{e}_{i_1},...,\mathbf{e}_{i_n})$, where $\mathbf{e}_1,...,\mathbf{e}_r$ is a free basis of G and $(i_1,...,i_n)$ runs over $\{1,...,r\}^n$. For every $\boldsymbol{\nu}=(\nu_1,...,\nu_r)\in\mathbb{Z}_+^r$ with $|\boldsymbol{\nu}|=\nu_1+...+\nu_r=n$, we denote by $\mathbf{I}_{\boldsymbol{\nu}}$ the set of all $\mathbf{i}=(i_1,...,i_n)\in\{1,...,r\}^n$ such that the number of appearences of each $j \in \{1, ..., r\}$ in the sequence $\mathbf{i} = (i_1, ..., i_n)$ is precisely ν_j. Since $D^n p$ is symmetric, the value $(D^n p)(\mathbf{e}_{i_1}, ..., \mathbf{e}_{i_n})$ depends only on $\boldsymbol{\nu} = (\nu_1, ..., \nu_r) \in \mathbb{Z}_+^r$ and does not depend on a choice of a sequence $\mathbf{i} = (i_1, ..., i_n) \in \mathbf{I}_{\boldsymbol{\nu}}$. Thus, we may define

$$a_{\boldsymbol{\nu}} \overset{\text{def}}{=\!=} \frac{1}{\boldsymbol{\nu}!} \cdot (D^n p)(\mathbf{e}_{i_1}, ..., \mathbf{e}_{i_n}),$$

where $(i_1, ..., i_n)$ is an arbitrary element of $\mathbf{I}_{\boldsymbol{\nu}}$. Clearly $a_{\boldsymbol{\nu}}$ is a continuous G-invariant function on $\mathbb{R}^r$. Let

$$p' = p - \sum_{\boldsymbol{\nu}:\ |\boldsymbol{\nu}|=n} a_{\boldsymbol{\nu}}(x)\, x_1^{\nu_1} \cdots x_r^{\nu_r}. \tag{3.9}$$

Certainly $D^{n+1}p' = 0$. Moreover, $(D^n p')(\mathbf{e}_{i_1}, ..., \mathbf{e}_{i_n}) = 0$ for all $(i_1, ..., i_n) \in \{1, ..., r\}^n$ due to Lemma 3.1(b). By part (a) of Main Theorem, this implies that $D^n p' = 0$. The induction hypothesis implies that p' is a G-periodic polynomial of degree at most $n-1$. According to (3.9), p is a G-periodic polynomial of degree at most n, which completes the proof. □

3.2. G-periodic polynomials in G-invariant subspaces. Let A be a G-invariant vector subspace of $C(\mathbb{R}^r)$. Set $A^G = A \cap C^G$ and $P_n^G(A) = A \cap P_n^G$. In other words, A^Γ consists of all G-invariant continuous functions that belong to A, and $P_n^G(A)$ consists of all G-periodic polynomials of degree at most n that belong to A. Clearly, A^G and $P_n^G(A)$ are vector spaces.

THEOREM 3.3. *Suppose that the space A^G is of finite dimension. Then every $P_n^G(A)$ is of finite dimension as well.*

PROOF. By the above definition and Proposition 3.2, we have

$$P_n^G(A) = A \cap P_n^G = A \cap \mathcal{P}_n(G, C(\mathbb{R}^r)). \tag{3.10}$$

On the other hand, it is clear that

$$\begin{aligned} A \cap \mathcal{P}_n(G, C(\mathbb{R}^r)) &= A \cap \ker\{D^{n+1} \colon C(\mathbb{R}^r) \to \mathcal{C}^{n+1}(G, C(\mathbb{R}^r))\} \\ &= \ker\{\, D^{n+1}\big|_A : A \to \mathcal{C}^{n+1}(G, C(\mathbb{R}^r))\} \\ &= \ker\{\, D^{n+1}\big|_A : A \to \mathcal{C}^{n+1}(G, A)\} = \mathcal{P}_n(G, A). \end{aligned} \tag{3.11}$$

Combining (3.10) and (3.11), we see that

$$P_n^G(A) = \mathcal{P}_n(G, A). \tag{3.12}$$

The lattice G is finitely generated and, by our assumption, $\dim A^G < \infty$. Hence, by Main Theorem, $\dim \mathcal{P}_n(G, A) < \infty$, and (3.12) implies $\dim P_n^G(A) < \infty$. □

REMARK 3.4. Any G-periodic polynomial $a \in A$ is a sum of monomials with G-invariant coefficients. In the case we know that the coefficients of all these monomials are in A (and thereby, actually, in A^G), we could prove that $\dim P_n^G(A) < \infty$ without referring to Main Theorem.

Indeed, let us denote by P_n the vector space of all polynomials in $x_1, ..., x_r$ of degree at most n with constant coefficients. The tensor product $\mathcal{T}_n = A^G \otimes P_n$ of the finite dimensional vector spaces A^G and P_n is of finite dimension. In fact, $\mathcal{T}_n$ may be represented as the space of all functions $F(y, x)$ on the direct product $\mathbb{R}_y^r \times \mathbb{R}_x^r$ of the form

$$\sum_{j_1+\ldots+j_r \le n} f_{j_1,\ldots,j_r}(y_1, ..., y_r)\, x_1^{j_1} \ldots x_r^{j_r}$$

with coefficients $f_{j_1,\dots,j_r} \in A^G$. Any G-polynomial $a \in P_n^G(A)$ with coefficients in A^G may be considered as the restriction of a certain function $F \in \mathcal{T}_n$ to the diagonal $\Delta = \{x = y\}$ of $\mathbb{R}_y^r \times \mathbb{R}_x^r$. Since $\mathcal{T}_n$ is of finite dimension, $P_n^G(A)$ is such as well.

However, the coefficients of a G-periodic polynomial $a \in P_n^G(A)$ may not be in A^G, and the above "proof" does not apply in this situation. ○

As above, let G be a full rank lattice in $\mathbb{R}^r$ and $\mathcal{D}$ be a linear partial differential operator in $\mathbb{R}^r$ with continuous G-periodic coefficients. Let $\mathcal{S} = \mathcal{S}_{\mathcal{D}}$ denote the space of all classical global solutions u of the equation $\mathcal{D}u = 0$. Clearly, $\mathcal{S}$ is a G-invariant vector subspace of $C(\mathbb{R}^r)$. Denote by $P_n^G(\mathcal{S})$ the space of all solutions $p \in \mathcal{S}$ that are G-polynomials of degree at most n:

$$P_n^G(\mathcal{S}) = \Big\{ p = \sum_{j_1+\dots+j_r \le n} f_{j_1,\dots,j_r}(x_1, \dots, x_r)\, x_1^{j_1} \dots x_r^{j_r} \mid \text{all } f_{j_1,\dots,j_r} \text{ are } G\text{-periodic}, \quad \mathcal{D}p = 0 \Big\}.$$

The following result follows immediately from Theorem 3.3:

COROLLARY 3.5. *Suppose that the space $\mathcal{S}^G$ of all G-periodic solutions of the equation $\mathcal{D}u = 0$ is of finite dimension. Then* $\dim P_n^G(\mathcal{S}) < \infty$ *for every* $n \in \mathbb{Z}_+$.

Notice that no additional restrictions to the linear partial differential operator $\mathcal{D}$ are required. One has just to assume that *the coefficients of $\mathcal{D}$* (real or complex) *are continuous and G-periodic, and the space $\mathcal{S}^G$ of all classical G-periodic solutions of the equation $\mathcal{D}u = 0$ is of finite dimension.* Furthermore, the continuity of the coefficients of $\mathcal{D}$ does not seem necessary. In this case, one can define and then obtain analogous results for certain classes of "generalized G-periodic polynomial" solutions. However, for general linear partial differential operators $\mathcal{D}$, the *apriori Liouville-type assumption* $\dim \mathcal{S}^G < \infty$ cannot be omitted, unless $\mathcal{D}$ satisfies an appropriate maximum principle.

EXAMPLE 3.6. Let $\mathcal{D}$ be an elliptic operator of second-order with real coefficients acting on functions $u \in C^2(\mathbb{R}^r)$:

$$\mathcal{D} = -\sum_{i,j=1}^{r} a_{ij}(x)\partial_i\partial_j + \sum_{i=1}^{r} b_i(x)\partial_i + c(x).$$

Here the coefficients a_{ij}, b_i, c are real, locally Hölder continuous, $\mathbb{Z}^r$-periodic functions. The matrix $A(x) := (a_{ij}(x))$ is positive definite. Also, we assume that the zeroth-order coefficient $c(x) = \mathcal{D}(\mathbf{1})$ is non-negative for each $x \in \mathbb{R}^r$, where $\mathbf{1}$ is the constant function with value 1. Then $\mathcal{D}$ satisfies the strong maximum principle (see e.g., [**11**, Lemma 3.6]).

3.3. Polynomial-like solutions of periodic differential operators on co-compact Riemannian coverings. In this subsection, we provide briefly some details as in Subsection 3.2 for the case when $\mathcal{D}$ is a periodic differential operator defined on a co-compact Riemannian covering. Let X be a connected Riemannian manifold equipped with an isometric, free, properly discontinuous and co-compact right group action of a finitely generated discrete group G (G may be non-abelian) and let $\mathcal{D}$ be a G-periodic elliptic differential operator on X, i.e., $\mathcal{D}$ commutes with the group action of G. We always assume that the principal symbol of $\mathcal{D}$ is a negative-definite quadratic form. To study G-periodic polynomials in this setting, we define the class of additive functions on the covering X as follows (see more details in [**7**, **11**]):

DEFINITION 3.7. A real continuous function u on X is said to be *additive* if there is a homomorphism $\alpha : G \to \mathbb{R}$ such that

$$u^g(x) = u(x) + \alpha(g), \quad \text{for all} \quad (g, x) \in G \times X, \tag{3.13}$$

where $u^g(x) = u(g \cdot x)$.

We also denote by $\mathcal{A}(X)$ the vector space consisting of all additive functions on X.

It is known that the vector space $\mathcal{A}(X)/C^G(X)$ is isomorphic to $\mathrm{Hom}(G, \mathbb{R})$ (see [**11**, Lemma 2.7]). Clearly, $\mathrm{Hom}(G, \mathbb{R}) = \mathrm{Hom}(\widetilde{G}, \mathbb{R})$, where $\widetilde{G} = G/[G, G]$. Hence, the dimension of $\mathcal{A}(X)/C^G(X)$ is equal to the rank r of the finitely generated abelian group $\widetilde{G}$. Let $\alpha_1, \ldots, \alpha_r$ be a vector basis of $\mathrm{Hom}(\widetilde{G}, \mathbb{R}) \cong \mathbb{R}^r$ and $h_1, \ldots, h_r$ be a corresponding basis of $\mathcal{A}(X)$ (modulo G-periodic functions) via the isomorphism between $\mathcal{A}(X)/C^G(X)$ and $\mathrm{Hom}(\widetilde{G}, \mathbb{R})$. Notice that when $G = \mathbb{Z}^r$ and $X = \mathbb{R}^r$, it is easy to see that $h_j(x) = \phi_j(x) = x_j$ for any $1 \le j \le r$ and $x \in X$; thus, we may regard these functions $h_1, \ldots, h_r$ as some analogs of Euclidean coordinate functions on the covering X (see also [**1**] for the case of co-compact abelian coverings). By misuse of language, we say that a *G-periodic monomial of degree n* is an element Q_n if it has the form $Q_n = f(x) \cdot h_1(x)^{j_1} \ldots h_r(x)^{j_r}$, where the coefficient $f \neq 0$ is G-periodic and $j_1, \ldots, j_r \in \mathbb{Z}_+$ such that $j_1 + \ldots + j_r = n$. As before, a *G-periodic polynomial* is a sum of G-periodic monomials and this representation is unique up to a G-periodic function. Let P^G (P^G_n) be the algebra of G-periodic polynomials (of order at most n). Then P^G (P^G_n) is a G-invariant subalgebra (resp. subspace) of $C(X)$.

The G-action on X induces the iterated difference operators

$$D^n : C(X) \to \mathcal{C}^n(G, C(X)).$$

Again, the subspaces $\mathcal{P}_n(G, C(X))$ of $C(X)$ of *polynomial-like elements in* $C(X)$ of order at most n is the kernel of the operator D^{n+1}. Due to (3.13), each term $[D^1 h_i](g)(x) = h_i(g \cdot x) - h_i(x)$ is independent of x $(1 \le i \le r)$. Using this fact, we can repeat the proof of Lemma 3.1(a) to see that the same statement should hold, i.e., $D^{n+1}Q{=}0$ for any G-periodic monomial Q of degree at most n. This means that

$$P^G_n \subseteq \mathcal{P}_n(G, C(X)).$$

Now suppose that A is a G-invariant vector subspace of $C(X)$. We also denote $A^G = A \cap C^G(X)$ and $P^G_n(A) = A \cap P^G_n$. Therefore, $P^G_n(A) \subseteq A \cap \mathcal{P}_n(G, C(X)) = \mathcal{P}_n(G, A)$ (see the proof of Theorem 3.3). By applying the Main Theorem again, whenever $\dim A^G < \infty$, we have $\dim P^G_n(A) \le \dim \mathcal{P}_n(G, A) < \infty$ for any $n \in \mathbb{Z}_+$.

It is worthy mentioning that when G is abelian, all of the results in Subsection 3.1 and Subsection 3.2 still hold. The proofs of these results do not require any change in this case, so we skip the details.

We finish this subsection by proving the following statement:

PROPOSITION 3.8. *Let X be a Riemannian manifold which is a Galois covering of a compact Riemannian manifold and G be its deck transformation group. Suppose* *that the abelianization $\widetilde{G}$ of G has rank r and G is of polynomial growth* [7]. *Let $h_1, \dots, h_r$ be a basis of the vector space $\mathcal{A}(X)$ (modulo G-periodic functions on X). Let $\mathcal{D}$ be a G-periodic, real elliptic operator of second-order on X such that $\mathcal{D}(\mathbf{1}) \geq 0$. Also, let $P_n^G(\mathcal{S}_{\mathcal{D}})$ be the space of all solutions u of the equation $\mathcal{D}u = 0$ on X such that*

$$u(x) = \sum_{j_1+\ldots+j_r \leq n} f_{j_1,\ldots,j_r}(x)\, h_1(x)^{j_1} \ldots h_r(x)^{j_r},$$

where each term $f_{j_1,\ldots,j_r}(x)$ in the above sum is G-periodic.

Then $\dim P_n^G(\mathcal{S}_{\mathcal{D}}) < \infty$ *for every $n \in \mathbb{Z}_+$. Furthermore, for any $n \geq 0$, we have*

(i) *If $\mathcal{D}(\mathbf{1}) \neq 0$, $P_n^G(\mathcal{S}_{\mathcal{D}}) = \{0\}$.*
(ii) *If $\mathcal{D}(\mathbf{1}) = 0$, the following estimate holds:*

$$\dim P_n^G(\mathcal{S}_{\mathcal{D}}) \leq \binom{n+r}{r}.$$

PROOF. According to our above discussion, $\dim P_n^G(\mathcal{S}_{\mathcal{D}}) < \infty$ for all $n \in \mathbb{Z}_+$ if and only if $\dim P_0^G(\mathcal{S}_{\mathcal{D}}) < \infty$. It is known (see [**11**, Theorem 6.9]) that when $\mathcal{D}(\mathbf{1}) = 0$, the dimension of the space $P_0^G(\mathcal{S}_{\mathcal{D}})$ consisting of all G-periodic (bounded) solutions on X is one. When $\mathcal{D}(\mathbf{1}) \neq 0$, [**11**, Theorem 4.5] yields that $\dim P_0^G(\mathcal{S}_{\mathcal{D}}) = 0$. In both cases, $P_0^G(\mathcal{S}_{\mathcal{D}})$ has finite dimension. This proves the first statement. The second statement then follows immediately from the fact that $\dim P_n^G(\mathcal{S}_{\mathcal{D}}) \leq \dim \mathcal{P}_n(G, \mathcal{S}_{\mathcal{D}})$ and Proposition 1.26. □

REMARK 3.9. Note that in the case $\mathcal{D}(\mathbf{1}) \neq 0$, Proposition 3.8 is still valid even if the growth of G is not polynomial.

Acknowledgments

The authors are grateful to the referees for useful comments on this manuscript. The work of the first author was partially supported by the NSF grant DMS-1517938. He expresses his gratitude to the NSF for the support.

References

[1] S. Agmon, *On positive solutions of elliptic equations with periodic coefficients in $\mathbf{R}^n$, spectral results and extensions to elliptic operators on Riemannian manifolds*, Differential equations (Birmingham, Ala., 1983), North-Holland Math. Stud., vol. 92, North-Holland, Amsterdam, 1984, pp. 7–17, DOI 10.1016/S0304-0208(08)73672-7. MR799327

[2] M. Avellaneda and F.-H. Lin, *Un théorème de Liouville pour des équations elliptiques à coefficients périodiques* (French, with English summary), C. R. Acad. Sci. Paris Sér. I Math. **309** (1989), no. 5, 245–250. MR1010728

[3] A. Brudnyi, *Holomorphic functions of polynomial growth on abelian coverings of a compact complex manifold*, Comm. Anal. Geom. **6** (1998), no. 3, 485–510, DOI 10.4310/CAG.1998.v6.n3.a3. MR1638866

[4] T. H. Colding and W. P. Minicozzi II, *Harmonic functions on manifolds*, Ann. of Math. (2) **146** (1997), no. 3, 725–747.

[5] T. H. Colding and W. P. Minicozzi II, *Weyl type bounds for harmonic functions*, Invent. Math. **131** (1998), no. 2, 257–298, DOI 10.1007/s002220050204. MR1608571

[7] Due to the celebrated work of M. Gromov, this is equivalent to the assumption that G is virtually nilpotent.

[6] T. H. Colding and W. P. Minicozzi II, *Liouville theorems for harmonic sections and applications*, Comm. Pure Appl. Math. **51** (1998), no. 2, 113–138, DOI 10.1002/(SICI)1097-0312(199802)51:2⟨113::AID-CPA1⟩3.0.CO;2-E. MR1488297

[7] M. Kha, *A short note on additive functions on co-compact Riemannian normal coverings.* arXiv:1511.00185, preprint.

[8] P. Kuchment, *An overview of periodic elliptic operators*, Bull. Amer. Math. Soc. (N.S.) **53** (2016), no. 3, 343–414, DOI 10.1090/bull/1528. MR3501794

[9] P. Kuchment and Y. Pinchover, *Liouville theorems and spectral edge behavior on abelian coverings of compact manifolds*, Trans. Amer. Math. Soc. **359** (2007), no. 12, 5777–5815, DOI 10.1090/S0002-9947-07-04196-7. MR2336306

[10] V. Ya. Lin, *Liouville coverings of complex spaces, and amenable groups* (Russian), Mat. Sb. (N.S.) **132(174)** (1987), no. 2, 202–224, DOI 10.1070/SM1988v060n01ABEH003163; English transl., Math. USSR-Sb. **60** (1988), no. 1, 197–216. MR882834

[11] V. Ya. Lin and Y. Pinchover, *Manifolds with group actions and elliptic operators*, Mem. Amer. Math. Soc. **112** (1994), no. 540, vi+78, DOI 10.1090/memo/0540. MR1230774

[12] V. Lin and M. Zaidenberg, *Liouville and and Carathéodory coverings in Riemannian and complex geometry*, Voronezh Winter Mathematical Schools, Amer. Math. Soc. Transl. Ser. 2, Adv. Math. Sci., 37 **184** (1998), 111–130.

[13] T. Lyons and D. Sullivan, *Function theory, random paths and covering spaces*, J. Differential Geom. **19** (1984), no. 2, 299–323. MR755228

[14] J. Moser and M. Struwe, *On a Liouville-type theorem for linear and nonlinear elliptic differential equations on a torus*, Bol. Soc. Brasil. Mat. (N.S.) **23** (1992), no. 1-2, 1–20.

Department of Mathematics, The University of Arizona, Tucson, Arizona, 85721, USA

Email address: minhkha@math.arizona.edu

Department of Mathematics, Technion-Israel Institute of Technology, Haifa, Israel 32000

Email address: vlalin@gmail.com

Email address: vlin@technion.ac.il

Contemporary Mathematics
Volume **734**, 2019
https://doi.org/10.1090/conm/734/14773

On two hydromechanical problems inspired by works of S. Krein

N. D. Kopachevsky, V. I. Voytitsky, and Z. Z. Sitshayeva

This paper is dedicated to the 100th anniversary of the birth of Selim G. Krein

ABSTRACT. This article deals with a problem on small oscillations of two pendula, the first of which is fixed at an immovable point and the second one is connected to the first; all connections are swiveling. Each pendulum has a cavity partially filled with an incompressible homogeneous fluid. We study an initial-boundary value problem as well as a corresponding spectral problem of normal (eigen-) oscillations of the hydromechanical system. The theorems on a correct solvability of the problem on a finite time interval are proved both for an ideal and a viscous fluids in the cavities. The properties of normal oscillations of the system are established.

The history of the problems

The work is written by a student of Selim G. Krein (first co-author), as well as student and colleague of Nikolay D. Kopachevsky. It is well known that in the study of hydromechanical problems S. Krein widely used methods of functional analysis, and he was a world-famous specialist in this topic. In particular, he dreamed that his approaches to hydrodynamic problems (see [**1**]–[**4**]) would be published as a monograph both in Russian and in English. Currently these ideas are implemented in [**5**]– [**7**].

This paper is devoted to the study of two hydromechanical problems using the approaches developed by S. Krein: problems on small motions and eigen oscillations of two joined pendula containing cavities partially filled with homogeneous ideal or viscous fluid.

We note that the first studies of the problem on dynamics of a rigid body with a cavity completely filled with a homogeneous ideal fluid were carried out by N. Zhukovsky [**8**]. In particular, he introduced auxiliary functions depending only on the shape of the cavity which are called Zhukovsky's potentials now. Using these

---

2010 *Mathematics Subject Classification.* Primary 35Q35, 35D35; Secondary 46E20, 39B42, 76B99, 76D99.

Key words and phrases. Ideal fluid, viscous fluid, initial-boundary value problem, Hilbert space, linear operator, spectral problem, strong solvability.

The first author was supported in part by the Ministry of Education and Science of the Russian Federation (grant 14.Z50.31.0037), the second author was supported by the V.I. Vernadsky Crimean Federal University development program for 2015 — 2024 within the framework of grant support for young scientists. The authors thank the reviewer for useful comments.

potentials the problem of the dynamics of a body with a cavity completely filled with an ideal fluid can be replaced by the problem of motion of an equivalent solid body with modified inertia tensor.

If fluid fills a cavity partially then the hydromechanical system has infinitely many degrees of freedom. This problem was actively studied in the middle of the 20th century due to the development of space technology (oscillations of a fluid fuel in a tank of a space rocket). The results of studies are described in numerous articles and monographs. The present work is a continuation of the first co-author research that began in [**9**]. It also uses approaches proposed in [**10**].

1. The case of ideal fluids in cavities of pendula.

1.1. The problem statement. The basic equations, boundary and initial conditions. We will study the hydromechanical system which consists of two solid bodies Ω_{01} and Ω_{02} having the densities ρ_{01} and ρ_{02} respectively. These bodies (pendula) are connected to each other by the swing (swivel) joints situated in selected points O_k of the pendula (see picture). The first body oscillates near the fixed point in space O_1, the second one oscillates near the point O_2 rigidly connected with first pendulum. Both bodies have the cavities partially filled with homogeneous ideal fluids with the densities ρ_1 and ρ_2 respectively.

We assume that a homogeneous gravitational field acts on the system, and in the state of equilibrium the suspension points O_k and mass centers C_1 and C_2 of pendula with fluids are situated on the same vertical axis (see right-hand part of the picture). In this case fluids in the cavities occupy the regions Ω_1 and Ω_2; the boundaries $\partial\Omega_1$ and $\partial\Omega_2$ of these regions consist of the solid walls S_1 and S_2 and the free horizontal surfaces Γ_1 and Γ_2, i.e. perpendicular to the direction of the gravitational field.

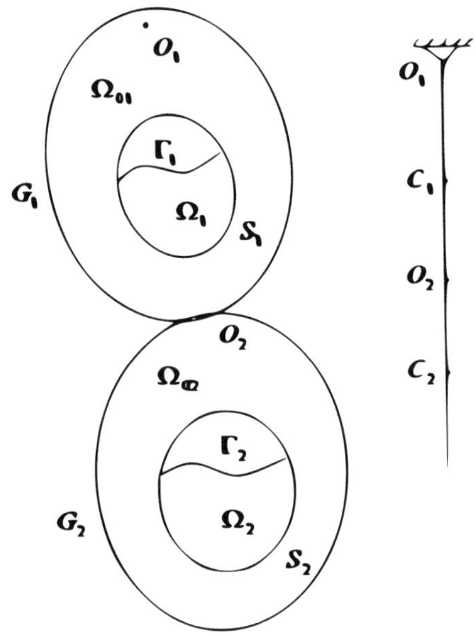

Now we give the statement of the problem on small motions of the hydromechanical system close to the equilibrium state. Let us note that in the process of oscillation regions Ω_1 and Ω_2 as well as its free boundaries Γ_1 and Γ_2 (and mass centers C_1, C_2) move in the space (see left-hand part of the picture). We suppose that this change is close enough to equilibrium state, and the linearization of the equations describing the process is reasonable. We consider both situations in the equilibrium state: hydromechanical system can be statically stable in linear approximation (see lemma 1.7 and definition 1.8) or unstable. The unstable modes of movement are connected with negative eigenvalues (and corresponding eigenfunctions) of potential energy operator (see theorem 1.17 and subsection 2.5).

To derive equations of pendula motions we introduce a fixed Cartesian coordinate system $O_1x^1x^2x^3$ with the unit vectors $\vec{e}^j$, $j = 1, 2, 3$, so that the acceleration of the gravitational field $\vec{g} = -g\vec{e}^3$, $g > 0$. We also introduce the moving coordinate systems $O_kx_k^1x_k^2x_k^3$, $k = 1, 2$, rigidly connected with bodies Ω_{0k}, $k = 1, 2$, with unit vectors $\vec{e}_k^j$, $k = 1, 2$, $j = 1, 2, 3$, respectively. We assume that in the state of equilibrium the system $O_1x_1^1x_1^2x_1^3$ coincides with $O_1x^1x^2x^3$ and the system $O_2x_2^1x_2^2x_2^3$ is obtained from $O_1x^1x^2x^3$ by parallel transfer along the vertical axis from point O_1 to O_2.

The position of the moving coordinate system $O_kx_k^1x_k^2x_k^3$, $k = 1, 2$, with respect to the fixed system $O_1x^1x^2x^3$ in the process of small motions of the hydromechanical system is given by the small angular displacement vector $\vec{\delta}_k(t) = \sum_{j=1}^{3} \delta_k^j(t)\vec{e}_k^j$, $k = 1, 2$. Then the angular velocity $\vec{\omega}_k(t)$ of the body Ω_{0k} is $\vec{\omega}_k = d\vec{\delta}_k/dt$ and the angular acceleration of this body is $d^2\vec{\delta}_k/dt^2 = d\vec{\omega}_k/dt$.

We also denote by $\vec{u}_k(t, x)$ the field of the relative velocity of the fluid in the region Ω_k, $p_k(t, x)$ is the deviation of the pressure from the equilibrium one and $\zeta_k(t, x)$, $x \in \Gamma_k$, are the perpendicular deviations of the moving surfaces $\Gamma_k(t)$ from equilibrium surfaces Γ_k, $k = 1, 2$. In addition, we introduce the notations $\vec{h}_1 = \overrightarrow{O_1O_2}$, $h_1 = |\vec{h}_1|$, $\alpha_k > 0$ are the friction coefficients in the hinges, m_k are the masses of the pendula with the fluids, $l_k = |O_kC_k|$ and $P_2\vec{\delta}_k = \sum_{j=1}^{2} \delta_k^j \vec{e}_k^j$. We will also assume that the field acting on the system deviates a little from the gravitational field, i.e., the field is $-g\vec{e}^3 + \vec{f}$, $\vec{f}_1 := \vec{f}|_{G_1}$, $\vec{f}_2 := \vec{f}|_{G_2}$, $G_k = \Omega_{0k} \cup \Omega_k$, $k = 1, 2$.

Taking into account this notation the linearized equations of pendula motion (equations of changing of kinetic moments) have the form

$$
\begin{aligned}
(1.1)\quad & \int_{G_1} \vec{r}_1 \times \left(\frac{d\vec{\omega}_1}{dt} \times \vec{r}_1 \right) dm_1 + \rho_1 \int_{\Omega_1} \vec{r}_1 \times \frac{\partial \vec{u}_1}{\partial t}\, d\Omega_1 + \\
& + \int_{G_2} \vec{h}_1 \times \left(\frac{d\vec{\omega}_1}{dt} \times \vec{h}_1 + \frac{d\vec{\omega}_2}{dt} \times \vec{r}_2 \right) dm_2 + \rho_2 \int_{\Omega_2} \vec{h}_1 \times \frac{\partial \vec{u}_2}{\partial t}\, d\Omega_1 + \\
& + \alpha_1 \vec{\omega}_1 - \alpha_2 (\vec{\omega}_2 - \vec{\omega}_1) + g (m_1 l_1 + m_2 h_1) P_2 \vec{\delta}_1 - g \rho_1 \int_{\Gamma_1} (\vec{e}_1^{\,3} \times \vec{r}_1) \zeta_1 \, d\Gamma_1 = \\
& = \int_{G_1} \vec{r}_1 \times \vec{f}_1 \, dm_1 + \int_{G_2} \vec{h}_1 \times \vec{f}_2 \, dm_2 =: \vec{M}_1(t),
\end{aligned}
$$

$$\int_{G_2} \vec{r}_2 \times \left(\frac{d\vec{\omega}_1}{dt} \times \vec{h}_1 + \frac{d\vec{\omega}_2}{dt} \times \vec{r}_2 \right) dm_2 + \rho_2 \int_{\Omega_2} \vec{r}_2 \times \frac{\partial \vec{u}_2}{\partial t} \, d\Omega_2 + \alpha_2(\vec{\omega}_2 - \vec{\omega}_1) + \\ + g m_2 l_2 P_2 \vec{\delta}_2 - g\rho_2 \int_{\Gamma_2} (\vec{e}_2^{\,3} \times \vec{r}_2)\zeta_2 \, d\Gamma_2 = \int_{G_2} (\vec{r}_2 \times \vec{f}_2) \, dm_2 =: \vec{M}_2(t). \tag{1.2}$$

We use notation

$$\int_{G_k} (\ldots) \, dm_k := \rho_{0k} \int_{\Omega_{0k}} (\ldots) \, d\Omega_k + \rho_k \int_{\Omega_k} (\ldots) \, d\Omega_k, \quad k = 1, 2.$$

The equations of the motions of fluids in cavities (Euler equations) are the following:

$$\rho_1 \left(\frac{\partial \vec{u}_1}{\partial t} + \frac{d\vec{\omega}_1}{dt} \times \vec{r}_1 \right) + \nabla p_1 = \rho_1 \vec{f}_1, \quad \operatorname{div} \vec{u}_1 = 0 \quad (\text{in } \Omega_1), \tag{1.3}$$

$$\rho_2 \left(\frac{\partial \vec{u}_2}{\partial t} + \frac{d\vec{\omega}_1}{dt} \times \vec{h}_1 + \frac{d\vec{\omega}_2}{dt} \times \vec{r}_2 \right) + \nabla p_2 = \rho_2 \vec{f}_2, \quad \operatorname{div} \vec{u}_2 = 0 \quad (\text{in } \Omega_2). \tag{1.4}$$

On the solid walls the following nonleaking conditions must be fulfilled:

$$\vec{u}_k \cdot \vec{n}_k = 0 \quad (\text{on } S_k), \quad k = 1, 2, \tag{1.5}$$

and at the equilibrium boundaries Γ_k we have the kinematic conditions

$$\frac{\partial \zeta_k}{\partial t} = u_k^3 = \vec{u}_k \cdot \vec{n}_k \quad (\text{on } \Gamma_k), \; k = 1, 2. \tag{1.6}$$

Here $\vec{n}_k$ is the outer normal to $\partial\Omega_k$. We also have the kinematic conditions

$$\frac{d}{dt}\left(P_2 \vec{\delta}_k\right) = P_2 \vec{\omega}_k, \quad \frac{d}{dt}\delta_k^3 = \omega_k^3. \tag{1.7}$$

Besides on the equilibrium free surfaces it is necessary to take into account the dynamic conditions

$$p_k = \rho_k g(\zeta_k + (P_2 \vec{\delta}_k \times \vec{r}_k) \cdot \vec{e}_k^{\,3}) \quad (\text{on } \Gamma_k), \tag{1.8}$$

as well as the conservation conditions for the volumes of fluids in the process of oscillations:

$$\int_{\Gamma_k} \zeta_k \, d\Gamma_k = 0. \tag{1.9}$$

Finally, for the complete formulation of problem (1.1)–(1.9) we imply to add the initial conditions

$$\begin{aligned} &\vec{u}_k(0, x) = \vec{u}_k^{\,0}(x), \quad x \in \Omega_k, \quad \zeta_k(0, x) = \zeta_k^0(x), \quad x \in \Gamma_k, \\ &\vec{\omega}_k(0) = \vec{\omega}_k^{\,0}, \quad \vec{\delta}_k(0) = \vec{\delta}_k^{\,0}, \quad k = 1, 2. \end{aligned} \tag{1.10}$$

1.2. The law of full energy balance. Let assume that problem (1.1)–(1.10) has a classical solution for $t \geq 0$ and we derive the law of full energy balance of the investigated hydromechanical system. It takes the differential form

$$(1.11)\quad \frac{1}{2}\frac{d}{dt}\Big\{\rho_{01}\int_{\Omega_{01}} |\vec{\omega}_1 \times \vec{r}_1|^2\,d\Omega_{01} + \rho_1\int_{\Omega_1} |\vec{\omega}_1 \times \vec{r}_1 + \vec{u}_1|^2\,d\Omega_1 +$$

$$+\rho_{02}\int_{\Omega_{02}} |\vec{\omega}_1 \times \vec{h}_1 + \vec{\omega}_2 \times \vec{r}_2|^2\,d\Omega_{02} + \rho_2\int_{\Omega_2} |\vec{\omega}_1 \times \vec{h}_1 + \vec{\omega}_2 \times \vec{r}_2 + \vec{u}_2|^2\,d\Omega_2\Big\} +$$

$$+\frac{g}{2}\frac{d}{dt}\Big\{\rho_1\Big[\int_{\Gamma_1} |\zeta_1 + \theta_1((P_2\vec{\delta}_1 \times \vec{r}_1)\cdot \vec{e}_1^{\,3})|^2\,d\Gamma_1 - \int_{\Gamma_1} |\theta_1((P_2\vec{\delta}_1 \times \vec{r}_1)\cdot \vec{e}_1^{\,3})|^2\,d\Gamma_1\Big] +$$

$$+\rho_2\Big[\int_{\Gamma_2} |\zeta_2 + \theta_2((P_2\vec{\delta}_2 \times \vec{r}_2)\cdot \vec{e}_2^{\,3})|^2\,d\Gamma_2 - \int_{\Gamma_2} |\theta_k((P_2\vec{\delta}_2 \times \vec{r}_2)\cdot \vec{e}_2^{\,3})|^2\,d\Gamma_2\Big]\Big\} +$$

$$+\frac{1}{2}\frac{d}{dt}\Big\{(m_1 l_1 + m_2 h_1)|P_2\vec{\delta}_1|^2 + m_2 l_2 |P_2\vec{\delta}_2|^2\Big\} =$$

$$= -\Big(\alpha_1|\vec{\omega}_1|^2 + \alpha_2|\vec{\omega}_2 - \vec{\omega}_1|^2\Big) + \sum_{k=1}^{2}\rho_k\int_{\Omega_k} \vec{f}_k \cdot \vec{u}_k\,d\Omega_k + \sum_{k=1}^{2}\vec{M}_k(t)\cdot\vec{\omega}_k,$$

$$\theta_k:\ L_2(\Gamma_k) \to L_{2,\Gamma_k} := L_2(\Gamma_k) \ominus \{1_{\Gamma_k}\},\quad k = 1,2,$$

θ_k are orthoprojectors.

The left-hand side of (1.11) is the sum of integral terms involving curly brackets. The first consists of a sum of kinetic energies of solids and kinetic energies of fluids in pendula cavities. Further, the second one (with the factor g) gives the potential energy of the system which corresponds to the perturbations ζ_k of free surfaces Γ_k. The last gives the change of the potential energy due to displacement of the bodies with the angles of the rotation $\vec{\delta}_1$ and $\vec{\delta}_2$.

The right-hand side of (1.11) is a sum of powers due to the frictional forces in the swivels and the external forces corresponding to action of additional field $\vec{f}$. Thus, the relation (1.11) means that the change in time of full energy of the hydromechanical system is equal to the power of internal and external forces acting on the system.

1.3. Operator approach to study of the initial-boundary problems. We introduce functional Hilbert spaces naturally related to the problem.

1°. Since the kinetic energy of fluids in cavities Ω_k at any time must be finite, the fields of relative velocities $\vec{u}_k(t,x)$ must be functions of the variable t with values in Hilbert spaces $\vec{L}_2(\Omega_k)$ with a scalar products

$$(\vec{u}_k, \vec{v}_k)_{\vec{L}_2(\Omega_k)} := \int_{\Omega_k} \vec{u}_k(x)\cdot\overline{\vec{v}_k(x)}\,d\Omega_k.$$

Taking into account the solenoidality property of $\vec{u}_k$ and the boundary conditions (see (1.3)–(1.5)) we use the orthogonal decomposition $\vec{L}_2(\Omega_k)$ onto the subspaces that naturally arise in the considered problem (see [**5**], p. 106):

$$(1.12)\qquad \vec{L}_2(\Omega_k) = \vec{G}_{0,\Gamma_k}(\Omega_k) \oplus \vec{J}_0(\Omega_k) \oplus \vec{G}_{h,S_k}(\Omega_k),$$

$$\vec{G}_{0,\Gamma_k}(\Omega_k) := \{\nabla\varphi_k \in \vec{L}_2(\Omega_k) : \ \varphi_k = 0 \quad (\text{on } \Gamma_k)\}, \tag{1.13}$$

$$\vec{J}_0(\Omega_k) := \{\vec{w}_k \in \vec{L}_2(\Omega_k) : \ \operatorname{div}\vec{w}_k = 0 \ (\text{in } \Omega_k), \ \vec{w}_k \cdot \vec{n}_k = 0 \ (\text{on } \partial\Omega)\},$$

(1.14)

$$\vec{G}_{h,\,S_k}(\Omega_k) := \Big\{\nabla\Phi_k \in \vec{L}_2(\Omega_k) : \ \Delta\Phi_k = 0 \ (\text{in } \Omega_k), \ \frac{\partial\Phi_k}{\partial n_k} = 0 \ (\text{on } S_k), \ \int_{\Gamma_k} \Phi_k d\Gamma_k = 0\Big\}.$$

Here we suppose that the boundaries $\partial\Omega_k$ are Lipschitz, S_k and Γ_k are Lipschitz pieces of these boundaries (see, for example, [**11**]).

2°. Since potential energy of fluids (as well as the whole system) must also be finite for $t \geq 0$, again in (1.11) we should assume that $\zeta_k(t,x)$, $x \in \Gamma_k$, are functions of the variable t with values in the space $L_2(\Gamma_k)$ with the scalar product

$$(\zeta_k, \eta_k)_{L_2(\Gamma_k)} := \int_{\Gamma_k} \zeta_k \overline{\eta}_k \, d\Gamma_k.$$

Then from conditions (1.9) it follows that

$$\zeta_k \in L_{2,\Gamma_k} = L_2(\Gamma_k) \ominus \{1_{\Gamma_k}\},$$

and the orthoprojectors $\theta_k : L_2(\Gamma_k) \to L_{2,\Gamma_k}$ have the form

$$\theta_k \zeta_k := \zeta_k - |\Gamma_k|^{-1} \int_{\Gamma_k} \zeta_k \, d\Gamma_k.$$

On the base of (1.13) and (1.14), the conditions (1.8) can be rewritten in the form

$$p_k = \rho_k g(\zeta_k + \theta_k((P_2\vec{\delta}_k \times \vec{r}_k) \cdot \vec{e}_k^{\,3})) \quad \text{on } \Gamma_k.$$

To the fluid motion equations (1.3), (1.4) we apply the method of orthogonal projection onto the subspaces (1.12)–(1.14) introducing the corresponding orthoprojectors P_{0,Γ_k}, $P_{0,k}$, P_{h,S_k} and representing $\vec{u}_k(t,x)$, $\nabla p_k(t,x)$ in the form

$$\begin{aligned} \vec{u}_k &= \vec{w}_k + \nabla\Phi_k, \quad \vec{w}_k \in \vec{J}_0(\Omega_k), \ \nabla\Phi_k \in \vec{G}_{h,\,S_k}(\Omega_k), \\ \nabla p_k &= \nabla\widetilde{p}_k + \nabla\varphi_k, \quad \nabla\widetilde{p}_k \in \vec{G}_{h,\,S_k}(\Omega_k), \ \nabla\varphi_k \in \vec{G}_{0,\,\Gamma_k}(\Omega_k). \end{aligned} \tag{1.15}$$

After projecting it turns out that $\nabla\varphi_k$ can be found using the other components of the solution, and the equations governing motions of the fluids and the pendula, with the exception of kinematic conditions (1.6), (1.7), can be rewritten in a vector-matrix form as a first order differential relation

$$C_1 \frac{dz_1}{dt} + A_1 z_1 + gB_{12} z_2 = f_1(t), \tag{1.16}$$

$$\begin{aligned} z_1 &:= (z_{1,1}; z_{1,2})^\tau, \ z_{1,k} = (\vec{w}_k; \nabla\Phi_k; \vec{\omega}_k)^\tau, \\ z_2 &:= (z_{2,1}; z_{2,2})^\tau, \ z_{2,k} = (\zeta_k; P_2\vec{\delta}_k)^\tau, \end{aligned} \tag{1.17}$$

where C_1, A_1, B_{12} are the operators acting respectively in

$$\mathcal{H}_1 = \bigoplus_{k=1}^{2} (\vec{J}_0(\Omega_k) \oplus \vec{G}_{h,S_k}(\Omega_k) \oplus \mathbb{C}^3),$$

(the first and the second operators) and $B_{12} : \mathcal{H}_2 \to \mathcal{H}_1$,

$$\mathcal{H}_2 = \bigoplus_{k=1}^{2} (L_{2,\Gamma_k} \oplus \mathbb{C}^2). \tag{1.18}$$

Here

$$(1.19)\quad C_1 z_1 = \Big(\rho_1\vec{w}_1 + \rho_1 P_{0,1}(\vec{\omega}_1\times\vec{r}_1);\quad \rho_1\nabla\Phi_1 + \rho_1 P_{h,S_1}(\vec{\omega}_1\times\vec{r}_1);$$

$$\rho_1\int\limits_{\Omega_1}(\vec{r}_1\times\vec{w}_1)\,d\Omega_1 + \rho_1\int\limits_{\Omega_1}(\vec{r}_1\times\nabla\Phi_1)\,d\Omega_1 + \vec{J}_1\omega_1 + \int\limits_{G_2}\vec{h}_1\times(\vec{\omega}_1\times\vec{h}_1)\,dm_2 +$$

$$+\int\limits_{G_2}(\vec{h}_1\times\vec{w}_2)\,dm_2 + \int\limits_{G_2}(\vec{h}_1\times\nabla\Phi_2)\,dm_2 + \int\limits_{G_2}\vec{h}_1\times(\vec{\omega}_2\times\vec{r}_2)\,dm_2;$$

$$\rho_2\vec{w}_2+\rho_2P_{0,2}(\vec{\omega}_1\times\vec{h}_1)+\rho_2P_{0,2}(\vec{\omega}_2\times\vec{r}_2);\ \rho_2P_{h,S_2}(\vec{\omega}_1\times\vec{h}_1)+\rho_2\nabla\Phi_2+\rho_2P_{h,S_2}(\vec{\omega}_2\times\vec{r}_2);$$

$$\int\limits_{G_2}\vec{r}_2\times(\vec{\omega}_1\times\vec{h}_1)\,dm_2 + \rho_2\int\limits_{\Omega_2}(\vec{r}_2\times\vec{w}_2)\,d\Omega_2 + \rho_2\int\limits_{\Omega_2}(\vec{r}_2\times\nabla\Phi_2)\,d\Omega_2 + \vec{J}_2\omega_2\Big)^{\tau},$$

where $\vec{J}_k$ are the inertia tensors of the pendula together with the fluids:

$$\vec{J}_k\vec{\omega}_k := \rho_{0k}\int\limits_{\Omega_{0k}}\vec{r}_k\times(\vec{\omega}_k\times\vec{r}_k)\,d\Omega_{0k} + \rho_k\int\limits_{\Omega_k}(\vec{r}_k\times(\vec{\omega}_k\times\vec{r}_k))\,d\Omega_k.$$

Further, the operator matrix B_{12} acts by the law

$$(1.20)\quad B_{12}z_2 = \Big(\vec{0};\ \rho_1V_1\zeta_1 + \rho_1V_1(\theta_1(P_2\vec{\delta}_1\times\vec{r}_1)\cdot\vec{e}_1^{\,3}));$$

$$-\rho_1\int\limits_{\Gamma_1}(\vec{e}_1^{\,3}\times\vec{r}_1)\zeta_1\,d\Gamma_1 + (m_1l_1+m_2h_1)P_2\vec{\delta}_1;$$

$$\vec{0};\ \rho_2V_2\zeta_2 + \rho_2V_2(\theta_2(P_2\vec{\delta}_2\times\vec{r}_2)\cdot\vec{e}_2^{\,3}));-\rho_2\int\limits_{\Gamma_2}(\vec{e}_2^{\,3}\times\vec{r}_2)\zeta_2\,d\Gamma_2 + m_2l_2P_2\vec{\delta}_2\Big)^{\tau}.$$

where V_k is the operator of boundary value Zaremba problem

$$(1.21)\quad \begin{aligned} &\Delta\widetilde{p}_k = 0\quad (\text{in } \Omega_k), && \frac{\partial\widetilde{p}_k}{\partial n_k} = 0\quad (\text{on } S_k),\\ &\widetilde{p}_k = \psi_k\quad (\text{on } \Gamma_k), && \int\limits_{\Gamma_k}\psi_k\,d\Gamma_k = 0; \end{aligned}$$

and by definition we have

$$(1.22)\qquad \nabla\widetilde{p}_k =: V_k\psi_k.$$

Finally, the operator matrix A_1 from (1.16) has non-zero elements only of a following type

$$(1.23)\qquad A_{1,33} = \alpha_1+\alpha_2,\quad A_{1,36} = -\alpha_2 = A_{1,63},\quad A_{1,66} = \alpha_2.$$

Using the above definitions of the operators and operator matrices we will formulate their properties and indicate their physical meaning.

LEMMA 1.1. *Zaremba problem* (1.21) *has a unique solution* $\nabla\widetilde{p}_k = V_k\psi_k$ *if and only if* $\psi_k \in H^{1/2}_{\Gamma_k} := H^{1/2}(\Gamma_k)\cap L_{2,\Gamma_k}$, *and in that case*

$$V_k \in \mathcal{L}(H^{1/2}_{\Gamma_k};\vec{G}_{h,S_k}(\Omega_k)).$$

LEMMA 1.2. *The operator matrix C_1 from (1.19) is a bounded selfadjoint and positive definite operator acting in $\mathcal{H}_1$. The quadratic form $(C_1 z_1, z_1)_{\mathcal{H}_1}$ is equal to twice kinetic energy of the hydromechanical system, i.e. C_1 is a kinetic energy operator.*

LEMMA 1.3. *The operator matrix A_1 with the elements (1.23) is a bounded selfadjoint nonnegative operator. The quadratic form of the operator A_1 is equal to*

$$(A_1 z_1, z_1)_{\mathcal{H}_1} = \alpha_1 |\vec{\omega}_1|^2 + \alpha_2 |\vec{\omega}_2 - \vec{\omega}_1|^2 \geq 0,$$

and therefore A_1 can be regarded as an energy dissipation operator corresponding to a friction in the hinges.

LEMMA 1.4. *The operator $B_{12} : \mathcal{H}_2 \to \mathcal{H}_1$, defined by the formula (1.20), is a block-diagonal unbounded operator defined on the domain*

$$\mathcal{D}(B_{12}) = (H_{\Gamma_1}^{1/2} \oplus \mathbb{C}^2) \oplus (H_{\Gamma_2}^{1/2} \oplus \mathbb{C}^2),$$

that is dense in $\mathcal{H}_2$.

The further application of the operator approach to problem (1.1)–(1.10) is based on the fact that kinematic conditions (1.6) and the first conditions in (1.7) can be rewritten in the form that allows us to introduce the potential energy operator of the system:

$$gC_2 \frac{dz_2}{dt} + gB_{21} z_1 = 0, \tag{1.24}$$

$$C_2 z_2 = \Big(\rho_1\zeta_1 + \rho_1\theta_1((P_2\vec{\delta}_1 \times \vec{r}_1)\cdot\vec{e}_1^{\,3});\ -\rho_1 \int_{\Gamma_1} (\vec{e}_1^{\,3} \times \vec{r}_1)\zeta_1\, d\Gamma_1 + (m_1 l_1 + m_2 h_1) P_2 \vec{\delta}_1;$$
$$\rho_2\zeta_2 + \rho_2\theta_2((P_2\vec{\delta}_2 \times \vec{r}_2)\cdot\vec{e}_2^{\,3});\ -\rho_2 \int_{\Gamma_2} (\vec{e}_2^{\,3} \times \vec{r}_2)\zeta_2\, d\Gamma_2 + m_2 l_2 P_2 \vec{\delta}_2\Big)^{\tau},$$

$$B_{21} z_1 = \Big(\rho_1 \gamma_{n,1} \nabla\Phi_1 - \rho_1\theta_1((P_2\vec{\omega}_1 \times \vec{r}_1)\cdot\vec{e}_1^{\,3});\tag{1.25}$$
$$\rho_1 \int_{\Gamma_1} (\vec{e}_1^{\,3} \times \vec{r}_1)\gamma_{n,1}\nabla\Phi_1\, d\Gamma_1 - (m_1 l_1 + m_2 h_1) P_2 \vec{\omega}_1;$$
$$-\rho_2\gamma_{n,2}\nabla\Phi_2 - \rho_2\theta_2((P_2\vec{\omega}_2 \times \vec{r}_2)\cdot\vec{e}_2^{\,3});\ \rho_2 \int_{\Gamma_2} (\vec{e}_2^{\,3} \times \vec{r}_2)\nabla\Phi_2\, d\Gamma_2 - m_2 l_2 P_2 \vec{\omega}_2\Big)^{\tau},$$

$$\gamma_{n,k} \nabla\Phi_k := \frac{\partial \Phi_k}{\partial n_k} \quad (\text{on } \Gamma_k). \tag{1.26}$$

LEMMA 1.5. *The operator $C_2 : \mathcal{H}_2 \to \mathcal{H}_2$ is a bounded and selfadjoint operator. Its quadratic form is equal to twice potential energy of the hydromechanical system.*

Let us introduce the notation for the inertia moments:

$$\beta_{jl}^{(k)} := \int_{\Gamma_k} x_k^j (\theta_k x_k^l)\, d\Gamma_k = \beta_{lj}^{(k)}, \quad j, l = 1, 2, \quad k = 1, 2.$$

We also introduce the determinants

$$\Delta_2^{(1)} := \det \begin{pmatrix} m_1 l_1 + m_2 h_1 - \rho_1 \beta_{22}^{(1)} & \rho_1 \beta_{21}^{(1)} \\ \rho_1 \beta_{12}^{(1)} & (m_1 l_1 + m_2 h_1) - \rho_1 \beta_{11}^{(1)} \end{pmatrix},$$

$$\Delta_2^{(2)} := \det \begin{pmatrix} m_2 l_2 - \rho_2 \beta_{22}^{(2)} & \rho_2 \beta_{21}^{(2)} \\ \rho_2 \beta_{12}^{(2)} & m_2 l_2 - \rho_2 \beta_{11}^{(2)} \end{pmatrix}.$$

LEMMA 1.6. *If conditions of general situation*

$$\Delta_2^{(1)} \neq 0, \quad \Delta_2^{(2)} \neq 0, \tag{1.27}$$

are satisfied, then relations (1.24)–(1.26) *and* (1.6), (1.7) *are equivalent.*

From now on we will assume that conditions (1.27) are fulfilled. Then the original initial-boundary value problem for small oscillations of two adjoined pendula with cavities partially filled with ideal fluids will be equivalent to the trivial problem

$$\frac{d\delta_k^3}{dt} = \omega_k^3, \quad k = 1, 2,$$

and also to a Cauchy problem for the system of equations

$$\begin{pmatrix} C_1 & 0 \\ 0 & gC_2 \end{pmatrix} \frac{d}{dt} \begin{pmatrix} z_1 \\ z_2 \end{pmatrix} + \begin{pmatrix} A_1 & 0 \\ 0 & 0 \end{pmatrix} \begin{pmatrix} z_1 \\ z_2 \end{pmatrix} + g \begin{pmatrix} 0 & B_{12} \\ B_{21} & 0 \end{pmatrix} \begin{pmatrix} z_1 \\ z_2 \end{pmatrix} = \begin{pmatrix} f_1(t) \\ 0 \end{pmatrix},$$

$$z_1(0) = z_1^0, \quad z_2(0) = z_2^0.$$

LEMMA 1.7. *The operator* C_2 *of potential energy of the system is boundedly invertible one. The conditions*

$$\begin{aligned} \Delta_1^{(1)} &:= m_1 l_1 + m_2 h_1 - \rho_1 \beta_{11}^{(1)} \geq 0, \quad \Delta_2^{(1)} \geq 0, \\ \Delta_1^{(2)} &:= m_2 l_2 - \rho_2 \beta_{11}^{(2)} \geq 0, \quad \Delta_2^{(2)} \geq 0, \end{aligned} \tag{1.28}$$

are necessary and sufficient for C_2 *to be non-negative and the conditions*

$$\Delta_1^{(k)} > 0, \quad \Delta_2^{(k)} > 0, \quad k = 1, 2, \tag{1.29}$$

are necessary and sufficient for its positive definiteness. If conditions (1.28) *or* (1.29) *are not fulfilled, then indefiniteness rank* $\kappa(C_2)$ *of the operator* C_2 *(i.e. number of the negative squares in its quadratic form) does not exceed* 4.

DEFINITION 1.8. We will say that the considered hydromechanical system is statically stable in linear approximation if operator of potential energy C_2 is positive definite.

Now we turn to study properties of the operator B_{21} from (1.25) and relationship between the operators B_{12} and B_{21}.

LEMMA 1.9. *The operator* B_{21} *is unbounded and defined on the domain*

$$\mathcal{D}(B_{21}) = (\vec{J}_0(\Omega_1) \oplus \mathcal{D}(\gamma_{n,1}) \oplus \mathbb{C}^3) \oplus (\vec{J}_0(\Omega_2) \oplus \mathcal{D}(\gamma_{n,2}) \oplus \mathbb{C}^3),$$

$$\mathcal{D}(\gamma_{n,k}) = \{\nabla \Phi_k \in \vec{G}_{h,S_k}(\Omega_k) : \ \gamma_{n,k} \nabla \Phi_k = \frac{\partial \Phi_k}{\partial n_k}\Big|_{\Gamma_k} = \varphi_k, \ \forall \varphi_k \in L_{2,\Gamma_k}\}.$$

The operators

$$V_k : \ \mathcal{D}(V_k) = H_{\Gamma_k}^{1/2} \subset L_{2,\Gamma_k} \to \vec{G}_{h,S_k}(\Omega_k),$$

$$\gamma_{n,k} : \ \mathcal{D}(\gamma_{n,k}) \subset \vec{G}_{h,S_k}(\Omega_k) \to L_{2,\Gamma_k}, \quad k = 1, 2,$$

are mutually conjugate:

$$(V_k\zeta_k, \nabla\Phi_k)_{\vec{L}_2(\Omega_k)} = (\zeta_k, \gamma_{n,k}\nabla\Phi_k)_{L_{2,\Gamma_k}},$$
$$\forall\zeta_k \in \mathcal{D}(V_k), \quad \forall\,\nabla\Phi_k \in \mathcal{D}(\gamma_{n,k}), \quad k = 1, 2.$$

LEMMA 1.10. *The operators B_{12} and B_{21} which are given by formulas* (1.20) *and* (1.25) *on their domains are skew-selfadjoint: $B_{12}^* = -B_{21}$, i.e.*

$$(B_{12}z_2, z_1)_{\mathcal{H}_1} = -(z_2, B_{21}z_1)_{\mathcal{H}_2}, \quad \forall z_1 \in \mathcal{D}(B_{21}),\ \forall\, z_2 \in \mathcal{D}(B_{12}).$$

The result of these observations is the following statement.

THEOREM 1.11. *The initial problem on small oscillations of pendula with fluids is equivalent to Cauchy problem*

$$C\frac{dz}{dt} + Az + gBz = f(t), \quad z(0) = z^0, \quad f(t) := (f_1(t); 0)^\tau, \tag{1.30}$$

in Hilbert space $\mathcal{H} = \mathcal{H}_1 \oplus \mathcal{H}_2$ (see (1.17)–(1.18)*). Here*

$$C = \mathrm{diag}(C_1; gC_2) = C^* \in \mathcal{L}(H)$$

is full energy operator of the considered system,

$$A = \mathrm{diag}(A_1; 0) \in \mathcal{L}(H)$$

is a dissipation operator and

$$B = \begin{pmatrix} 0 & B_{12} \\ B_{21} & 0 \end{pmatrix} = -B^*, \quad \mathcal{D}(B) = \mathcal{D}(B_{21}) \oplus \mathcal{D}(B_{12}), \tag{1.31}$$

is an exchange operator between kinetic and potential energies.

The properties of the operators C, A, B allow us to establish the following result.

THEOREM 1.12. *Suppose that in the problem* (1.30)–(1.31) *the following conditions*

$$z^0 \in \mathcal{D}(B), \quad f(t) \in C^1([0, T]; \mathcal{H}),$$

are satisfied. Then this problem has (both under the condition of static stability and under its absence) a strong solution in $[0, T]$, i.e. all terms of the equation in (1.30) *are continuous functions of t for $t \in [0, T]$ and the initial condition in* (1.30) *is fulfilled.*

The proof of theorem 1.12 uses the theory of contracting semigroups of the operators and similar facts of Pontryagin space theory with indefiniteness rank $\kappa = \kappa(C_2) \geq 1$, and also the theory of Volterra first order integrodifferential equations.

THEOREM 1.13. *Let, in the initial boundary-value problem* (1.1)–(1.10), *the following conditions*

$$\begin{gathered} \vec{u}_k^0 \in \vec{J}_{0,S_k}(\Omega_k), \quad P_{h,S_k}\vec{u}_k^0 =: \nabla\Phi_k^0 \in \vec{G}_{h,S_k}(\Omega_k) : \left.\frac{\partial\Phi_k^0}{\partial n_k}\right|_{\Gamma_k} \in L_{2,\Gamma_k}, \\ \zeta_k^0 \in H^{1/2}_{\Gamma_k}, \quad \vec{\omega}_k^0 \in \mathbb{C}^3, \quad \vec{\delta}_k^0 \in \mathbb{C}^3, \\ \vec{f}_k(t, x) \in C^1([0, T]; \vec{L}_2(\Omega_k)), \quad k = 1, 2. \end{gathered} \tag{1.32}$$

be satisfied. Then this problem has a unique strong solution on the interval $[0, T]$, i.e. $\vec{u}_k(t, x) \in C^1([0, T]; \vec{J}_{0,S_k}(\Omega_k))$, $\nabla p_k(t, x) \in C^1([0, T]; \vec{G}(\Omega_k))$, $\vec{G}(\Omega_k)) := \vec{G}_{0,\Gamma_k}(\Omega_k)) \oplus \vec{G}_{h,S_k}(\Omega_k)$, $\vec{\omega}_k(t) \in C^1([0, T]; \mathbb{C}^3)$, $\zeta_k(t, x) \in C^1([0, T]; L_{2,\Gamma_k})$, $x \in \Gamma_k$,

$\vec{\delta}_k(t) \in C^2([0,T];\mathbb{C}^3)$. *Further, for any* $t \in [0,T]$ *all of the following are satisfied: motion equations* (1.1), (1.2); *the Euler equations* (1.3), (1.4); *the dynamic conditions* (1.8) *(where all summands are the elements from* $C([0,T];H^{1/2}_{\Gamma_k})$*; the kinematic conditions* (1.6) *(all terms belong to* $C^1([0,T];L_{2,\Gamma_k})$*; the conditions* (1.7) *and initial conditions* (1.10).

REMARK 1.14. If instead of (1.32) the weaker conditions

$$\vec{u}_k^0 \in \vec{J}_{0,S_k}(\Omega_k), \quad \zeta_k^0 \in L_{2,\Gamma_k}, \quad \vec{\omega}_k^0 \in \mathbb{C}^3, \quad \vec{\delta}_k^0 \in \mathbb{C}^3,$$
$$\vec{f}_k(t,x) \in C([0;T];\vec{L}_2(\Omega_k)), \quad k=1,2,$$

are satisfied, then the initial problem has a generalized solution with continuous full energy, for which the law of full energy balance is fulfilled.

Now let friction in the hinges be absent and consider the problem of the eigen oscillations of the system, i.e., such solutions of homogeneous problem (1.30) for $A=0$, which depend on t according to the law

$$z(t) = e^{i\lambda t}z, \quad z \in \mathcal{H},$$

where λ is a frequency of oscillations and $z = (z_1;z_2)^\tau$ is an amplitude element. Then the spectral problem arises:

$$\begin{aligned} &gB_{12}z_2 + i\lambda C_1 z_1 = 0, \quad gB_{21}z_1 + i\lambda g C_2 z_2 = 0, \quad \omega_k^3 = i\lambda\delta_k^3, \quad k=1,2, \\ &z_1 = (\vec{w}_1;\nabla\Phi_1;\vec{\omega}_1;\vec{w}_2;\nabla\Phi_2;\vec{\omega}_2)^\tau, \quad z_2 = (\zeta_1;P_2\vec{\delta}_1;\zeta_2;P_2\vec{\delta}_2)^\tau. \end{aligned} \tag{1.33}$$

LEMMA 1.15. *Problem* (1.33) *has an infinite-multiple zero eigenvalue to which solutions of a form*

$$z_1 = (\vec{w}_1;\vec{0};\vec{0};\vec{w}_2;\vec{0};\vec{0})^\tau, \quad z_2 = (\zeta_1;P_2\vec{\delta}_1;\zeta_2;P_2\vec{\delta}_2)^\tau = 0,$$
$$\forall\, \vec{w}_k \in \vec{J}_0(\Omega_k), \quad \forall\, \delta_k^3 \in \mathbb{C}, \quad k=1,2.$$

are there corresponds. These solutions describe time-stationary motions of ideal fluids in pendulum cavities without deflection of free surfaces. In this case the pendula remain motionless and only turned along the vertical axis at arbitrary angles of rotation δ_k^3, $k=1,2$.

For nonzero oscillation frequencies λ we establish the new variational principle.

THEOREM 1.16. *If* $\lambda \neq 0$ *and the conditions for static stability of the system* ($C_2 \gg 0$) *are satisfied then the numbers* $\mu := \lambda^2/g$ *form a discrete spectrum* $\{\mu_j\}_{j=1}^\infty$ *consisting of finite positive eigenvalues* μ_j *with limit point* $\mu = +\infty$. *The corresponding system of the eigenelements* $\{\widetilde{z}_{1,j}\}_{j=1}^\infty$, $\widetilde{z}_{1,j} = (\nabla\Phi_{1,j};\vec{\omega}_{1,j};\nabla\Phi_{2,j};\vec{\omega}_{2,j})^\tau$, *forms the basis in the space* $\widetilde{\mathcal{H}}_1 = (\vec{G}_{h,S_1}(\Omega_1)\oplus\mathbb{C}^3)\oplus(\vec{G}_{h,S_2}(\Omega_2)\oplus\mathbb{C}^3)$ *which is orthogonal on the forms of potential and kinetic energies.*

The eigenvalues $\mu = \lambda^2/g$ and the eigenelements of the problem can be found by considering the successive minima of the following variational ratio

$$\mu = \Big\{\rho_1\Big[\int_{\Gamma_1}\Big|\frac{\partial\Phi_1}{\partial n_1} + \theta_1((P_2\vec{\omega}_1\times\vec{r}_1)\cdot\vec{e}_1^{\,3})\Big|^2\,d\Gamma_1 - \int_{\Gamma_1}|\theta_1((P_2\vec{\omega}_1\times\vec{r}_1)\cdot\vec{e}_1^{\,3})|^2\,d\Gamma_1\Big]+$$
$$+\rho_2\Big[\int_{\Gamma_2}\Big|\frac{\partial\Phi_2}{\partial n_2} + \theta_2((P_2\vec{\omega}_2\times\vec{r}_2)\cdot\vec{e}_2^{\,3})\Big|^2\,d\Gamma_2 - \int_{\Gamma_2}|\theta_2((P_2\vec{\omega}_2\times\vec{r}_2)\cdot\vec{e}_2^{\,3})|^2\,d\Gamma_2\Big]+$$
$$+(m_1l_l+m_2h_1)|P_2\vec{\omega}_1|^2 + m_2l_2|P_2\vec{\omega}_2|^2\Big\}\Big/\Big\{\rho_1\int_{\Omega_1}|\nabla\Phi_1+\sum_{j=1}^{3}\omega_{1,j}\nabla\psi_{1,j}|^2\,d\Omega_1+$$
$$+\rho_{01}\int_{\Omega_{01}}|\vec{\omega}_1\times\vec{r}_1|^2\,d\Omega_{01}+\rho_2\int_{\Omega_2}|\vec{\omega}_1\times\vec{h}_1+\nabla\Phi_2+\sum_{j=1}^{3}\omega_{2,j}\nabla\psi_{2,j}|^2\,d\Omega_2+$$
$$+\rho_{02}\int_{\Omega_{02}}|\vec{\omega}_1\times\vec{h}_1+\vec{\omega}_2\times\vec{r}_2|^2\,d\Omega_{02}\Big\}$$

under the conditions

$$\Delta\Phi_k = 0\ (in\ \Omega_k),\qquad \frac{\partial\Phi_k}{\partial n_k} = 0\ (on\ S_k),\qquad \int_{\Gamma_k}\Phi_k\,d\Gamma_k = 0,\quad k=1,2,$$

and also on the solutions $\psi_{k,j}$ of the equations

(1.34)
$$\Delta\psi_{k,j} = 0\ (in\ \Omega_k),\qquad \frac{\partial\psi_{k,j}}{\partial n_k} = (\vec{e}_k^{\,j}\times\vec{r}_k)\cdot\vec{n}_k\ (on\ \partial\Omega_k),\quad j=1,2,3,\quad k=1,2,$$

called Zhukovsky potentials which are depended only on a shape of the regions Ω_k, $k=1,2$.

THEOREM 1.17. *If the conditions of static stability of the system are not satisfied and $1 \le \kappa(C_2) \le 4$, then the numbers $\mu = \lambda^2/g$ form a discrete spectrum $\{\mu_j\}_{j=1}^{\infty}$ consisting of finitely multiple eigenvalues $\mu_j \in \mathbb{R}$ with limit point $\mu = +\infty$. In this case the first κ eigenvalues are negative and the rest ones are positive. Thus, the property of dynamic instability of the system follows from the property of its static instability (that is the inversion of Lagrange theorem on stability for the systems with infinite number of freedom degrees).*

2. The case of viscous fluids in cavities of pendula.

2.1. The problem statement. Now we assume that fluids in the cavities of pendula are viscous with coefficients of dynamic viscosity μ_k, $k=1,2$. In this case the motion equations (1.1), (1.2) do not change. So, below we give only the linearized equations of viscous fluids motion in cavities as well as the boundary and initial conditions.

Now instead (1.3), (1.4) we have the following equations:

(2.1)
$$\rho_1\left(\frac{\partial\vec{u}_1}{\partial t}+\frac{d\vec{\omega}_1}{dt}\times\vec{r}_1\right)+\nabla p_1-\mu_1\Delta\vec{u}_1=\rho_1\vec{f}_1,\quad \operatorname{div}\vec{u}_1=0\ (\text{in }\Omega_1),$$

(2.2)
$$\rho_2\left(\frac{\partial \vec{u}_2}{\partial t}+\frac{d\vec{\omega}_1}{dt}\times\vec{h}_1+\frac{d\vec{\omega}_2}{dt}\times\vec{r}_2\right)+\nabla p_2-\mu_2\Delta\vec{u}_2=\rho_2\vec{f}_2,\ \operatorname{div}\vec{u}_2=0\ (\text{in }\ \Omega_2),$$

where Δ is the vector Laplace operator.

In the case of viscous fluids the following adhesion conditions must be satisfied on solid walls S_k:

$$\vec{u}_k=\vec{0}\ \ (\text{on }\ S_k),\quad k=1,2. \tag{2.3}$$

Further, kinematic conditions still have the form (1.6), (1.7), and the dynamic ones are transformed as follows. The tangential stresses on Γ_k are equal to zero and the normal ones are compensated by a gravitational pressure jump:

$$\begin{aligned}&\mu_k\tau_{j3}(\vec{u}_3):=\mu_k\left(\frac{\partial u_k^3}{\partial x_k^j}+\frac{\partial u_k^j}{\partial x_k^3}\right)=0\quad (j=1,2),\\ &-p_k+2\mu_k\frac{\partial u_k^3}{\partial x_k^3}=-\rho_k g(\zeta_k+(P_2\vec{\delta}_k\times\vec{r}_k)\cdot\vec{e}_k^{\,3}),\quad k=1,2,\ \ (\text{on }\ \Gamma_k),\end{aligned} \tag{2.4}$$

where

$$\tau_{jl}(\vec{u}_k):=\frac{\partial u_k^l}{\partial x_k^j}+\frac{\partial u_k^j}{\partial x_k^l},\quad j,l=1,2,3,\quad k=1,2,$$

are the elements of the strain velocity tensor for a viscous fluid.

Thus, in the variant when the pendulum cavities are partially filled with the viscous fluids it is necessary to study initial-boundary value problem (1.1), (1.2), (2.1), (2.2), (1.6), (1.7), (1.9), (1.10).

Assuming a solution of this problem to be classical we derive (as in paragraph 1.2) the law of full energy balance of the system. In this variant we use Green's identity taking into account the fact that the solenoidal fields $\vec{v}_k$ and $\vec{u}_k$ satisfy the adhesion conditions on S_k (see (2.3)). Then we obtain the identity

$$\int_{\Omega_k}\vec{v}_k\cdot(-\mu_k\Delta\vec{u}_k+\nabla p_k)d\Omega_k=\mu_k E_k(\vec{v}_k,\vec{u}_k)-\int_{\Gamma_k}\sum_{j=1}^{3}v_k^j(\mu_k\tau_{j3}(\vec{u}_k)-p_k\delta_{j3})d\Gamma_k, \tag{2.5}$$

$$\mu_k E_k(\vec{v}_k,\vec{u}_k)=\frac{1}{2}\mu_k\int_{\Omega_k}\Big[\sum_{j,l=1}^{3}\tau_{jl}(\vec{v}_k)\tau_{jl}(\vec{u}_k)\Big]d\Omega_k. \tag{2.6}$$

Here the quadratic functional

$$\mu_k E_k(\vec{u}_k,\vec{u}_k)$$

is equal to energy dissipation velocity in viscous fluid occupying the region Ω_k.

Repeating the calculations of paragraph 1.2 and taking into account (2.5), (2.6), we arrive at the law of full energy balance in form (1.11) with the additional term in the right-hand side

$$-\sum_{k=1}^{2}\mu_k E_k(\vec{u}_k,\vec{u}_k),$$

which takes into account energy dissipation velocity in viscous fluids.

2.2. The additional choice of the functional spaces. By analogy with paragraph 1.3 we introduce the spaces $\vec{L}_2(\Omega_k)$ and L_{2,Γ_k} corresponding to finite kinetic and potential energies of the system, as well as a new space corresponding to finite energy dissipation velocity in viscous fluids. Namely, we introduce subspaces

$$\vec{J}^1_{0,S_k}(\Omega_k) := \{\vec{u}_k \in \vec{H}^1(\Omega_k) : \operatorname{div} \vec{u}_k = 0 \ \text{(in } \Omega_k), \quad \vec{u}_k = \vec{0} \ \text{(on } S_k)\}, \ k = 1, 2,$$

which are dense sets in $\vec{J}_{0,S_k}(\Omega_k)$. For these subspaces the Korn inequality holds:

$$\|\vec{u}_k\|^2_{\vec{J}^1_{0,S_k}(\Omega_k)} := E_k(\vec{u}_k, \vec{u}_k) \geq c_k \|\vec{u}_k\|_{\vec{H}^1(\Omega_k)} \geq \widetilde{c}_k \|\vec{u}_k\|^2_{\vec{J}_{0,S_k}(\Omega_k)}, \ \widetilde{c}_k > 0, \ k = 1, 2.$$

Hence it follows that any set bounded in $\vec{J}^1_{0,S_k}(\Omega_k)$ is compactly embedded in $\vec{J}_{0,S_k}(\Omega_k)$ and therefore $(\vec{J}^1_{0,S_k}(\Omega_k); \vec{J}_{0,S_k}(\Omega_k))$ is a Hilbert pair of spaces. We denote by A_{kk} an operator of this Hilbert pair and assume that it acts in an rigged Hilbert space

$$\vec{J}^1_{0,S_k}(\Omega_k) \hookrightarrow\hookrightarrow \vec{J}_{0,S_k}(\Omega_k) \hookrightarrow\hookrightarrow (\vec{J}^1_{0,S_k}(\Omega_k))^*, \tag{2.7}$$

i.e. $\mathcal{D}(A_{kk}) = \vec{J}^1_{0,S_k}(\Omega_k)$, $\mathcal{R}(A_{kk}) = (\vec{J}^1_{0,S_k}(\Omega_k))^*$.

As proved in [**11**] in this case for a domain Ω_k with Lipschitz boundary $\partial\Omega_k$ we have a generalized Green's identity for the vector Laplace operator (compare with (2.5), (2.6)):

$$\begin{gathered}
\mu_k E_k(\vec{v}_k, \vec{u}_k) = \langle \vec{v}_k, -\mu_k \Delta \vec{u}_k + \nabla p_k \rangle_{\vec{L}_2(\Omega_k)} + \\
+ \sum_{j=1}^{3} \langle \gamma_k v_k^j, \mu_k \tau_{j3}(\vec{u}_k) - p_k \delta_{j3} \rangle_{L_2(\Gamma_k)}, \quad \vec{v}_k = \sum_{j=1}^{3} v_k^j \vec{e}_k^{\,j}, \ \ \vec{v}_k, \vec{u}_k \in \vec{J}^1_{0,S_k}(\Omega_k), \\
\nabla p_k \in \vec{G}(\Omega_k) = \vec{G}_{0,\Gamma_k}(\Omega_k) \oplus \vec{G}_{h,S_k}(\Omega_k), \ \ \gamma_k \varphi := \varphi|_{\Gamma_k}, \ \ \forall \varphi \in H^1(\Omega_k).
\end{gathered} \tag{2.8}$$

Here the oblique brackets denote the functionals corresponding to rigging (2.7) and also to the corresponding riggings of the spaces L_{2,Γ_k}.

2.3. Transition to the system of differential operator equations in Hilbert space. This transition will be carried out according to the same plan that was used in paragraph 1.3, however with the corresponding changes associated with the presence of viscous fluids in the cavities of the pendula. First of all, we assume that velocity fields $\vec{u}_k(t,x)$ are functions of t with values in $\vec{J}_{0,S_k}(\Omega_k) = \vec{J}_0(\Omega_k) \oplus \vec{G}_{h,S_k}(\Omega_k)$ (see (1.12)–(1.14)). Therefore we do not need to look for these fields in form (1.15). Further, we assume $\nabla p_k = \nabla \widetilde{p}_k + \nabla \varphi_k$, $\nabla \widetilde{p}_k \in \vec{G}_{h,S_k}(\Omega_k) \subset \vec{J}_{0,S_k}(\Omega_k)$, $\nabla \varphi_k \in \vec{G}_{0,\Gamma_k}(\Omega_k)$.

We introduce orthoprojectors

$$P_{0,S_k} : \vec{L}_2(\Omega_k) \to \vec{J}_{0,S_k}(\Omega_k), \quad P_{0,\Gamma_k} : \vec{L}_2(\Omega_k) \to \vec{G}_{0,\Gamma_k}(\Omega_k),$$

and apply them to motion equations (2.1), (2.2) of the fluids.

Finally, we consider the auxiliary problem of S. Krein:

$$\begin{aligned}
&- \mu_k P_{0,S_k} \Delta \vec{u}_k + \nabla \widetilde{p}_k = \vec{f}_k, \quad \operatorname{div} \vec{u}_k = 0 \ \text{(in } \Omega_k), \\
&\vec{u}_k = \vec{0} \ \text{(on } S_k), \quad \mu_k \tau_{j3}(\vec{u}_k) = 0 \ \text{(on } \Gamma_k), \quad j = 1, 2, \\
&- \widetilde{p}_k + 2\mu_k \frac{\partial u_k^3}{\partial x_k^3} = 0 \ \text{(on } \Gamma_k).
\end{aligned} \tag{2.9}$$

Using the Green's identity (2.8) we come to a conclusion that a generalized solution of the problem (2.9) determined from the identity

$$\mu_k E_k(\vec{v}_k, \vec{u}_k) = (\vec{v}_k, \vec{f}_k)_{\vec{L}_2(\Omega_k)}, \quad \forall \vec{v}_k \in \vec{J}^1_{0,S_k}(\Omega_k),$$

has the form

$$\vec{u}_k = (\mu_k A_{kk})^{-1}\vec{f}_k \in \mathcal{D}(A_{kk}) \subset \mathcal{D}(A_{kk}^{1/2}) = \vec{J}^1_{0,S_k}(\Omega_k).$$

Now for the potential φ_k the boundary value problem arises:

$$\Delta\varphi_k = 0 \ \ (\text{in } \ \Omega_k), \qquad \frac{\partial\varphi_k}{\partial n_k} = 0 \ \ (\text{on } \ S_k),$$

$$\varphi_k = \rho_k g(\zeta_k + \theta_k(((P_2\vec{\delta}_k) \times \vec{r}_k) \cdot \vec{e}_k^{\,3})) =: \psi_k, \qquad \int_{\Gamma_k} \varphi_k d\Gamma_k = 0,$$

i.e. problem (1.21), and then $\nabla\varphi_k = V_k\psi_k$ (see (1.22)).

This allows us to treat equations (2.1), (2.2) together with boundary conditions (2.4) as an operator relation which together with (1.1), (1.2) has the form

$$C_1\frac{dz_1}{dt} + A_1 z_1 + gB_{12}z_2 = f_1(t), \quad z_1(0) = z_1^0,$$

$$z = (z_1; z_2)^\tau, \ \ z_1 = (\vec{u}_1; \vec{\omega}_1; \vec{u}_2; \vec{\omega}_2)^\tau, \ \ z_2 = (\zeta_1; P_2\vec{\delta}_1; \zeta_2; P_2\vec{\delta}_2)^\tau,$$

$$z_1 \in \mathcal{H}_1 = (\vec{J}_{0,S_1}(\Omega_1) \oplus \mathbb{C}^3) \oplus (\vec{J}_{0,S_2}(\Omega_2) \oplus \mathbb{C}^3), \tag{2.10}$$

$$z_2 \in \mathcal{H}_2 = (L_{2,\Gamma_1} \oplus \mathbb{C}^2) \oplus (L_{2,\Gamma_2} \oplus \mathbb{C}^2). \tag{2.11}$$

Here the action of the operator matrix C_1 is obtained from formulas (1.19) after substitution $\vec{w}_k + \nabla\Phi_k \mapsto \vec{u}_k$, and the operator matrix B_{12} is obtained from (1.20) respectively. Finally, the operator matrix A_1 has the form

$$A_1 = \begin{pmatrix} \mu_1 A_{11} & 0 & 0 & 0 \\ 0 & \alpha_1 + \alpha_2 & 0 & -\alpha_2 \\ 0 & 0 & \mu_2 A_{22} & 0 \\ 0 & -\alpha_2 & 0 & \alpha_2 \end{pmatrix}, \tag{2.12}$$

It takes into account not only the action of frictional forces in the hinges but also of dissipation forces in the fluids.

As in the problem of paragraph 1, the kinetic energy operator C_1 is a bounded positive definite operator acting in $\mathcal{H}_1$.

LEMMA 2.1. *The energy dissipation operator* (2.12) *given on the domain*

$$\mathcal{D}(A_1) = (\mathcal{D}(A_{11}) \oplus \mathbb{C}^3) \oplus (\mathcal{D}(A_{22}) \oplus \mathbb{C}^3),$$

dense in $\mathcal{H}_1$, is an unbounded positive definite operator with a positive compact inverse operator A_1^{-1}.

Together with relation (1.24), which takes into account the relationship between dynamic and kinematic variables, we obtain the following conclusion.

THEOREM 2.2. *The initial problem of small motions of two adjoined pendula with cavities partially filled with viscous fluids is equivalent (after separation of trivial relations) to the Cauchy problem*

$$C\frac{dz}{dt} + Az + gBz = f(t), \quad z(0) = z^0, \quad f(t) = (f_1(t); 0)^\tau, \tag{2.13}$$

in the Hilbert space $\mathcal{H} = \mathcal{H}_1 \oplus \mathcal{H}_2$ (see (2.10), (2.11)) where

$$C = diag(C_1; gC_2), \quad A = diag(A_1; 0),$$

$$B = \begin{pmatrix} 0 & B_{12} \\ B_{21} & 0 \end{pmatrix}, \quad \mathcal{D}(B) = \mathcal{D}(B_{21}) \oplus \mathcal{D}(B_{12}),$$

$$\mathcal{D}(B_{12}) = (H^{1/2}_{\Gamma_1} \oplus \mathbb{C}^2) \oplus (H^{1/2}_{\Gamma_2} \oplus \mathbb{C}^2),$$

$$\mathcal{D}(B_{12}) = \mathcal{D}(\widehat{\gamma}_n) = (\mathcal{D}(\gamma_{n,1}) \oplus \mathbb{C}^2) \oplus (\mathcal{D}(\gamma_{n,2}) \oplus \mathbb{C}^2),$$

$$\mathcal{D}(\gamma_{n,k}) = \{\vec{u}_k \in \vec{J}_{0,S_k}(\Omega_k) : \gamma_{n,k}\vec{u}_k := (\vec{u}_k \cdot \vec{n}_k)_{\Gamma_k} \in L_{2,\Gamma_k}\}.$$

Here again C is the full energy of the system operator, B is an energy exchange operator, and A is a new energy dissipation operator; as above, $B_{12} = -B^*_{21}$, and

$$\mathcal{D}(B^*_{12}B_{21}) = (\vec{J}^{\,1}_{0,S_1}(\Omega_1) \oplus \mathbb{C}^2) \oplus (\vec{J}^{\,1}_{0,S_2}(\Omega_2) \oplus \mathbb{C}^2). \tag{2.14}$$

Relying on these properties we consider the solvability questions for the Cauchy problem (2.13) .

THEOREM 2.3. *Let in the Cauchy problem (2.13) the conditions*

$$f(t) := (f_1(t); 0)^\tau \in C^\beta([0,T]; \mathcal{H}), \quad 0 < \beta \le 1,$$
$$z^0 = (z^0_1; z^0_2)^\tau, \quad z^0_1 \in \mathcal{D}(A_1), \quad z^0_2 \in \mathcal{D}(B_{12}).$$

be satisfied. Then this problem has unique strong solution on the interval $[0,T]$, i.e.

$$z(t) \in C([0,T]); \mathcal{D}(A) \cap \mathcal{D}(B)) \cap C^1([0,T]; \mathcal{H}),$$
$$\mathcal{D}(A) = \mathcal{D}(A_1) \oplus \mathcal{H}_2, \quad \mathcal{D}(B) = \mathcal{D}(B_{21}) \oplus \mathcal{D}(B_{12}).$$

The proof is carried out by transition from (2.13) to the first order Volterra integrodifferential equation for the function $z_1(t)$ and takes into account property (2.14).

On this basis we prove the theorem on correct solvability of the original problem (with viscous fluids in cavities).

THEOREM 2.4. *Let the conditions of Theorem 2.3 be satisfied, i.e.*

$$f_1(t) := (\rho_1 P_{0,S_1}\vec{f}_1; \vec{M}_1(t); \rho_2 P_{0,S_2}\vec{f}_2; \vec{M}_2(t))^\tau \in C^\beta([0,T]; \mathcal{H}), \quad 0 < \beta \le 1,$$
$$z^0_1 = (\vec{u}^{\,0}_1; \vec{\omega}^{\,0}_1; \vec{u}^{\,0}_2; \vec{\omega}^{\,0}_2)^\tau \in \mathcal{D}(A_1) = (\mathcal{D}(A_{11}) \oplus \mathbb{C}^3) \oplus (\mathcal{D}(A_{22}) \oplus \mathbb{C}^3)$$
$$\subset \mathcal{D}(A^{1/2}_1) = (\vec{J}^{\,1}_{0,S_1}(\Omega_1) \oplus \mathbb{C}^3) \oplus (\vec{J}_{0,S_2}(\Omega_2) \oplus \mathbb{C}^3),$$
$$z^0_2 = (\zeta^0_1; P_2\vec{\delta}^{\,0}_1; \zeta^0_2; P_2\vec{\delta}^{\,0}_2)^\tau \in \mathcal{D}(B_{12}) = (H^{1/2}_{\Gamma_1} \oplus \mathbb{C}^2) \oplus (H^{1/2}_{\Gamma_2} \oplus \mathbb{C}^2).$$

Then the original initial-boundary value problem (with viscous fluids) has on the interval $[0,T]$ unique strong solution which has the following properties.

1°. *Functions*

$$z_1(t) = (\vec{u}_1(t,x); \vec{\omega}_1(t); \vec{u}_2(t,x); \vec{\omega}_2(t))^\tau \in$$
$$C^1([0,T]; (\vec{J}_{0,S_1}(\Omega_1) \oplus \mathbb{C}^3) \oplus (\vec{J}_{0,S_2}(\Omega_2) \oplus \mathbb{C}^3))$$
$$\cap\, C([0,T]; (\mathcal{D}(A_{11}) \oplus \mathbb{C}^3) \oplus (\mathcal{D}(A_{22}) \oplus \mathbb{C}^3)),$$

$$z_2(t) = (\zeta_1(t,x); P_2\vec{\delta}_1(t); \zeta_2(t,x); P_2\vec{\delta}_2(t))^\tau \in$$
$$C^1([0,T]; (L_{2,\Gamma_1} \oplus \mathbb{C}^2) \oplus (L_{2,\Gamma_2} \oplus \mathbb{C}^2)) \cap C([0,T]; (H^{1/2}_{\Gamma_1} \oplus \mathbb{C}^2) \oplus (H^{1/2}_{\Gamma_2} \oplus \mathbb{C}^2)).$$

2°. *For any $t \in [0,T]$ the motion equations (2.1), (2.2) of fluids are fulfilled (with the orthoprojectors P_{0,S_k} applied on the left), and all terms of these equations are the elements from $C([0,T]; \vec{J}_{0,S_k}(\Omega_k))$, $k = 1,2$.*

3°. *For any $t \in [0,T]$ the motion equations (1.1), (1.2) of the pendula are fulfilled, and all terms of these equations are the elements from $C([0,T]; \mathbb{C}^3)$.*

4°. *For any $t \in [0,T]$ the kinematic conditions (1.6) are fulfilled, and all terms of these relations are the elements from $C([0,T]; H_{\Gamma_k}^{1/2})$; also the kinematic conditions (1.7) for the angular velocities and angular displacements are satisfied, and all terms of these relations are elements from $C([0,T]; \mathbb{C}^3)$.*

5°. *The initial conditions of the original problem are satisfied (see (1.10)).*

REMARK 2.5. If $\partial\Omega_k$ is almost smooth (see [**12**]) then a strong solution of the problem can have the property

$$\vec{u}_k(t,x) \in C([0,T]; \vec{J}^{\,1}_{0,S_k}(\Omega_k) \cap \vec{H}^2(\Omega_k)), \quad k = 1,2. \tag{2.15}$$

In this case the tangential stresses on Γ_k (see (2.4)) belong to space $C([0,T]; H^{1/2}(\Gamma_k))$, and the normal stresses are the elements of $C([0,T]; H_{\Gamma_k}^{1/2})$, $k = 1,2$.

2.4. The problem of normal oscillations. We consider solutions of the homogeneous problem (2.13) depending on t according to law

$$z_k(t) = \exp(-\lambda t) z_k, \quad \lambda \in \mathbb{C}, \quad k = 1,2, \tag{2.16}$$

where λ is a complex decrement of a damping, and $z_k \in \mathcal{H}_k$ is an amplitude element.

Then the spectral problem arises (taking into account the second relation in (1.7))

$$A_1 z_1 + g B_{12} z_2 = \lambda C_1 z_1, \quad B_{21} z_1 = \lambda C_2 z_2, \quad -\lambda \delta_k^3 = \omega_k^3. \tag{2.17}$$

LEMMA 2.6. *To an eigenvalue $\lambda = 0$ of problem (2.17) corresponds a new rest state of the hydromechanical system; it is obtained from the original rest state by a turning of the pendula to arbitrary angles $\delta_1^3 \vec{e}_1^{\,3}$ and $\delta_2^3 \vec{e}_2^{\,3}$ respectively.*

If $\lambda \neq 0$ then from (2.17) we obtain the spectral problem

$$A_1 z_1 = \lambda C_1 z_1 + g\lambda^{-1} B_{21}^* C_2^{-1} B_{21} z_1, \quad B_{21}^* = -B_{12}.$$

Carrying out substitution

$$A_1^{1/2} z_1 = \varphi_1$$

and acting on the left by the operator $A_1^{-1/2}$ we get the problem

$$L_1(\lambda)\varphi_1 := (I_1 - \lambda A_1^{-1/2} C_1 A_1^{-1/2} - g\lambda^{-1} A_1^{-1/2} B_{21}^* C_2^{-1} B_{21} A_1^{-1/2})\varphi_1 = 0. \tag{2.18}$$

LEMMA 2.7. *In problem (2.18) the operators*

$$\widetilde{A}_1 := A_1^{-1/2} C_1 A_1^{-1/2}, \quad \widetilde{B}_1 := A_1^{-1/2} B_{21}^* C_2^{-1}$$

possess the following properties.

1°. *The operator $\widetilde{A}_1 : \mathcal{H}_1 \to \mathcal{H}_1$ is compact positive one. Its eigenvalues have the asymptotic behavior*

$$\lambda_j(\widetilde{A}_1) = \Big(\frac{1}{3\pi^2} \sum_{k=1}^{2} (\rho_k/\mu_k)^{3/2} |\Omega_k|\Big)^{2/3} j^{-2/3} [1 + o(1)] \quad (j \to \infty).$$

2°. *If $C_2 \gg 0$ (the system is statically stable in linear approximation) then the operator $\widetilde{B}_1 : \mathcal{H}_1 \to \mathcal{H}_1$ is compact nonnegative and its nonzero eigenvalues have the asymptotic behavior*

$$\lambda_j(\widetilde{B}_1) = \Big(\frac{1}{16\pi}\sum_{k=1}^{2}(\rho_k/\mu_k)^2|\Gamma_k|\Big)^{1/2} j^{-1/2}[1+o(1)] \quad (j \to \infty). \tag{2.19}$$

In the general case (C_2 is possibly indefinite operator and its rank of indefiniteness $1 \le \kappa = \kappa(C_2) \le 4$) the operator $\widetilde{B}_1$ has the infinite-dimensional kernel

$$\ker \widetilde{B}_1 = \Big\{\varphi_1 = A_1^{1/2} z_1 : \ z_1 \in \mathcal{D}(A_1^{1/2}) \subset \mathcal{H}_1,$$
$$z_1 \in \ker \widehat{\gamma}_n = \{z_1 \in \mathcal{D}(\widehat{\gamma}_n) : \widehat{\gamma}_{n,k}\vec{u}_k = 0, \ \ P_2\vec{\omega}_k = \vec{0}, \ \ k = 1, 2\}\Big\},$$

and the quadratic form of the operator $\widetilde{B}_1$ is also indefinite and has exactly κ negative squares. Thus, $\widetilde{B}_1$ has κ negative eigenvalues (taking into account their multiplicities), infinite-dimensional zero eigenvalue, and a countable set of positive eigenvalues having limit point $\lambda = 0$ with asymptotic behavior (2.19).

REMARK 2.8. If $C_2 \gg 0$ then the operator pencil $L_1(\lambda)$ in (2.18) is a variant of well known S. Krein's one (see, for example, [**5**], chapter 7, paragraphs 1–3).

On the basis of this fact we give without proofs the properties of the solutions of problem (2.18). First we assume that the property of static stability by linear approximation ($C_2 \gg 0$, $\widetilde{B}_1 \ge 0$) is fulfilled.

1°. The spectrum of problem (2.18) consists of the eigenvalues of finite algebraic multiplicity located in right complex half-plane and having the limit points $\lambda = 0$ and $\lambda = \infty$. Nonreal eigenvalues (they are located symmetrically with respect to the real axis), as well as those real ones to which coprrespond not only the eigenelements but also the adjoined elements, are located in the segment

$$\Lambda := \{\lambda \in \mathbb{C} : \operatorname{Re}\lambda \ge (2\|\widetilde{A}_1\|)^{-1}, \ \ |\lambda| \le 2\|\widetilde{B}_1\|\}, \tag{2.20}$$

and there are a finite number of them.

If the rough condition of a strong damping of the system, i.e. a condition

$$4g\|\widetilde{A}_1\| \cdot \|\widetilde{B}_1\| < 1, \tag{2.21}$$

is fulfilled, then all eigenvalues are real (and positive) and problem (2.18) does not have the adjoined elements.

2°. If the condition (2.21) is satisfied then problem (2.18) has on the interval $(0, r)$, $r \in (r_-, r_+)$,

$$r_\pm = (1 \pm \sqrt{1 - 4g\|\widetilde{A}_1\| \cdot \|\widetilde{B}_1\|}/(2\|\widetilde{A}_1\|),$$

a countable set of finite eigenvalues $\{\lambda_j^-\}_{j=1}^\infty$ with limit point $\lambda = 0$. These eigenvalues correspond to the system of eigenelements $\{\varphi_{1j}^-\}_{j=1}^\infty$ which after the projecting onto the subspace $\mathcal{H}_{11} := \mathcal{H}_1 \ominus \ker \widetilde{B}_1$ forms Riesz basis in $\mathcal{H}_{11}$ and even p-basis for $p \ge p_0 = 6/7$.

3°. If condition (2.21) is satisfied then problem (2.18) has on the interval $[r_+, +\infty)$ a countable set of finite-valued eigenvalues $\{\lambda_j^+\}_{j=1}^\infty$ with limit point $\lambda = +\infty$. To these eigenvalues there are correspond the eigenelements $\{\varphi_{1j}^+\}_{j=1}^\infty$ which form Riesz basis in $\mathcal{H}_1$ and even p-basis for $p \ge p_0 = 6/7$.

4°. For the eigenvalues $\{\lambda_j^-\}_{j=1}^\infty$ and $\{\lambda_j^+\}_{j=1}^\infty$ we have the following two-sided inequalities:

$$g\lambda_j(\widetilde{B}_1) \le \lambda_j^- \le g\lambda_j(\widetilde{B}_1)/(1 - 2\lambda_j(\widetilde{B}_1)\|\widetilde{A}_1)\|), \quad j \in \mathbb{N},$$
$$1/\lambda_j(\widetilde{A}_1) - 2g\|\widetilde{B}_1)\| \le \lambda_j^+ \le 1/\lambda_j(\widetilde{A}_1), \quad j \in \mathbb{N},$$

and also the following asymptotic formulas

$$\lambda_j^- = g\lambda_j(\widetilde{B}_1)[1 + o(1)], \quad \lambda_j^+ = (\lambda_j(\widetilde{A}_1))^{-1}[1 + o(1)] \quad (j \to \infty). \tag{2.22}$$

5°. If condition (2.21) is not satisfied then problem (2.18) still has a discrete spectrum with the limit points $\lambda = 0$ and $\lambda = \infty$. The branch of eigenvalues $\{\lambda_j^-\}_{j=1}^\infty$ with limit point $\lambda = 0$ is located on the positive semiaxis and as above it has asymptotic behavior (2.22). Correspondingly, the branch of eigenvalues $\{\lambda_j^+\}_{j=1}^\infty$ with limit point $\lambda = +\infty$ also is located on the positive semiaxis and has asymptotic behavior (2.22). In addition, a basicity property of the eigen elements system for these branches is replaced by the basicity property up to a finite defect. Finally, problem (2.18) can have a finite number of the eigenvalues in sector (2.20).

2.5. The inversion of Lagrange theorem on stability. Assume that the quadratic form of the operator C_2 is indefinite, i.e. the hydromechanical system is not statically stable in linear approximation and $1 \le \kappa = \kappa(C_2) = \kappa(\widetilde{B}_1) \le 4$. We show that in this case the system is also dynamically unstable.

THEOREM 2.9. *Let condition* (2.21) *be satisfied and let* $\kappa(C_2) = \kappa$ *be positive. Then problem* (2.18) *has exactly* κ *negative eigenvalues located on the interval* $[-r_-, 0)$ *and the problem does not have other eigenvalues in the left half-plane.*

The proof is based on the use of the variational principles by Yu. Abramov (see [**13**]) applied to polynomial operator pencils in spectral problems.

THEOREM 2.10 (The inversion of Lagrange theorem on stability). *Let* $\kappa = \kappa(C_2) \ge 1$, *i.e. the considered hydromechanical system be statically unstable. Then it is also dynamically unstable and the instability is lost by exponential increase in time for normal movements, i.e. in a non-oscillatory manner.*

References

[1] S.G. Krein, *O funktsional'nykh svoystvakh operatorov vektornogo analiza i gidrodinamiki*, Reports of the Academy of Sciences of the USSR, **93** (6), 969–972.

[2] S.G. Krein, N.N. Moiseev, *O kolebaniyakh tverdogo tela, soderzhashchego zhidkost' so svobodnoy granitsey*, Applied Mathematics and Mechanics, 1957, **21** (2), 169–174.

[3] S.G. Krein, *O kolebaniyakh vyazkoy zhidkosti v sosude*, Reports of the Academy of Sciences of the USSR, 1964, **159** (2), 969–972.

[4] S.G. Krein, G.I. Laptev, *K zadache o dvizhenii vyazkoy zhidkosti v otkrytom sosude*, Functional analysis and its applications, 1968, **2** (1), 40–50.

[5] N.D. Kopachevsky, S.G. Krein, Ngo Zuy Kan, *Operatornyye metody v lineynoy gidrodinamike: Evolyutsionnyye i spektral'nyye zadachi*, Nauka, Moscow, 1989.

[6] N. D. Kopachevsky and S. G. Krein, *Operator approach to linear problems of hydrodynamics. Vol. 1*, Operator Theory: Advances and Applications, vol. 128, Birkhäuser Verlag, Basel, 2001. Self-adjoint problems for an ideal fluid. MR1860016

[7] N. D. Kopachevsky and S. G. Krein, *Operator approach to linear problems of hydrodynamics. Vol. 2*, Operator Theory: Advances and Applications, vol. 146, Birkhäuser Verlag, Basel, 2003. Nonself-adjoint problems for viscous fluids. MR2002951

[8] N.E. Zhukovsky, *O dvizhenii tverdogo tela, imeyushchego polosti, napolnennyye odnorodnoy kapel'noy zhidkost'yu*, Selected works, **1**, Gostekhizdat, Moscow, Leningrad, 1948. – pp. 31–52.
[9] N.D. Kopachevsky, *O kolebaniyakh tela s polost'yu, chastichno zapolnennoy tyazheloy ideal'noy zhidkost'yu: teoremy sushchestvovaniya, yedinstvennosti i ustoychivosti sil'nykh resheniy*, Problems of Dynamics and Stability of multidimensional systems, Institute of Mathematics of the Academy of Sciences of Ukraine (Kiev), **1** (1), 2005, 158–194.
[10] E. I. Batyr and N. D. Kopachevsky, *Small motions and normal oscillations in systems of connected gyrostats*, J. Math. Sci. (N.Y.) **211** (2015), no. 4, 441–530, DOI 10.1007/s10958-015-2615-y. Translated from Sovrem. Mat. Fundam. Napravl. 49 (2013). MR3418927
[11] N. D. Kopachevskiĭ, *On an abstract Green formula for a triple of Hilbert spaces and sesquilinear forms* (Russian, with Russian summary), Sovrem. Mat. Fundam. Napravl. **57** (2015), 71–107, DOI 10.1007/s10958-017-3470-9; English transl., J. Math. Sci. (N.Y.) **225** (2017), no. 2, 226–264. MR3540746
[12] M. S. Agranovich, *Spectral problems for second-order strongly elliptic systems in domains with smooth and nonsmooth boundaries* (Russian, with Russian summary), Uspekhi Mat. Nauk **57** (2002), no. 5(347), 3–78, DOI 10.1070/RM2002v057n05ABEH000552; English transl., Russian Math. Surveys **57** (2002), no. 5, 847–920. MR1992082
[13] Yu. Sh. Abramov, *Variatsionnye metody v teorii operatornykh puchkov* (Russian), Leningrad. Univ., Leningrad, 1983. Spektral′naya optimizatsiya. [Spectral optimization]. MR731257

Department of Mathematics and Informatics, Crimean Federal University, Vernadsky Avenue 4, Simferopol 295007

Current address: Department of Mathematics and Informatics, Crimean Federal University, Vernadsky Avenue 4, Simferopol 295007

Email address: kopachevsky@list.ru

Department of Mathematics and Informatics, Crimean Federal University, Vernadsky Avenue 4, Simferopol 295007

Email address: victor.voytitsky@gmail.com

Department of Mathematics, Crimean Engineering and Pedagogical University, Training Lane 8, Simferopol 295015

Email address: szz2008@mail.ru

Contemporary Mathematics
Volume **734**, 2019
https://doi.org/10.1090/conm/734/14774

Adjoint discrete systems with properly stated leading terms

G. A. Kurina

ABSTRACT. We study the adjointness property for explicit systems following from adjoint discrete descriptor systems with properly stated leading terms and with index less than or equal to 2 of the following forms

$$A_{i+1}B_{i+1}x_{i+1} = C_ix_i$$

and

$$B_i^*A_i^*z_i = C_i^*z_{i+1},$$

where $i = 0, 1, 2, ...$, the superscript $*$ means an adjoint operator, $x_i \in X$, $z_i \in Z$, $A_i \in L(Y, Z)$, $B_i \in L(X, Y)$, $C_i \in L(X, Z)$, X, Y and Z are real Hilbert spaces, the operators A_i, B_i are normally solvable and

$$kerA_i \oplus imB_i = Y.$$

Illustrative examples are presented.

1. Introduction

Since 1974 about 25 years Selim Grigor'evich Krein has been a head of the seminar on a small parameter in Voronezh Forest Engineering Institute (now G. F. Morozov Voronezh State University of Forestry and Technology). More than ten theses were defended by participants of this seminar. The main subject of the seminar was the equations not resolved with respect to derivative and also singular perturbations of such equations.

It is impossible to repeat the atmosphere of Krein′s seminars. Selim Grigor′evich always demonstrated sincere interest to problems discussed in the seminar and to persons who studied these problems. He had the encyclopedic breadth of interests and erudition and was always very glad to successes of seminar participants. Now the students of Krein′s students continue to work on the problems that were originally stated and studied by Selim Grigor′evich.

S. G. Krein will be the supervisor of his numerous students forever!

In the classical theory of explicit ordinary differential equations (e.g. [**H**]), the equation of the form

$$y'(t) = -C(t)^*y(t), \tag{1.1}$$

2010 *Mathematics Subject Classification.* Primary 15A06, 93B10.
Key words and phrases. Discrete-time problems, descriptor systems, adjoint systems.
The author was supported in part by the Russian Science Foundation Grant #17-11-01220.

is called the adjoint equation with respect to the equation

$$x'(t) = C(t)x(t). \tag{1.2}$$

It is well known that there exists the connection between the fundamental solution matrices of these equations and the so-called Lagrange identity

$$< x(t), y(t) >= const$$

takes place for each arbitrary pair of solutions of equations (1.1) and (1.2).

Here the prime means the derivative with respect to t, the superscript * with the notation of an operator denotes the adjoint operator, the corner brackets $< \cdot, \cdot >$ mean an inner product, $t \in J \subset \mathbb{R}$.

Adjoint equations arise in many fields, for instance, in the control theory.

A lot of problems in economics, sociology, biology is described by discrete models. Another source of appearing discrete models is a digital simulation of continuous systems where the differential equations are approximated by the corresponding difference equations. The study of sampled-data systems and computer-based systems leads in a natural way to the third source of discrete-time models.

For explicit discrete-time system

$$x_{i+1} = C_i x_i, \; i = 0, 1, 2..., \tag{1.3}$$

the adjoint system has the form

$$y_i = C_i^* y_{i+1} \tag{1.4}$$

(see, e.g., [**G**]). The analog of the Lagrange identity for each pair of solutions of systems (1.3) and (1.4) is

$$< x_i, y_i >= const.$$

The nature of differential-algebraic equations (DAEs) and their adjoint ones is much more complicated. Standard index-1 tractable DAEs and their adjoint ones have been studied, for instance, in [**BM2**].

K. Balla and R. März proposed to formulate DAEs in the so-called form of DAEs with properly stated leading form

$$A(t)(Bx)'(t) + C(t)x(t) = 0, \tag{1.5}$$

where some smoothness conditions are assumed and all space is decomposed for each t into the sum $ker A(t) \oplus im B(t)$.

Then the adjoint DAE has the following form

$$-B(t)^*(A^*y)'(t) + C(t)^*y(t) = 0 \tag{1.6}$$

and the Lagrange identity in this case is

$$< A(t)B(t)x(t), y(t) >= const.$$

Note that equations of form (1.5) arise in practice (see, for instance, [**EST**]).

The pair of adjoint DAEs (1.5) and (1.6) with tractability index less than or equal to 2 was studied in [**BLi**], [**BM1**] and [**BM3**]. In [**LM**], the research concerning these adjoint pairs of DAEs was continued for arbitrarily high tractability index.

The definitions of index-i tractable DAEs are given in the cited here papers.

The idea to use the notion of self-adjoint DAEs in the study of linear-quadratic control problems with a state equation of the form

$$A(t)(Bx)'(t) = C(t)x(t) + D(t)u(t) \tag{1.7}$$

belongs to K. Balla. For the first time, it has been made in [**BKM1**]. One section in [**BKM1**], [**BKM2**] is called "Adjoint and selfadjoint DAEs".

After some years the notion of self-adjoint DAEs was applied in [**KMS**] in the study of linear-quadratic control problems for a special case of a state equation of form (1.7) $(A(t)(Bx)'(t) = E(t)x'(t))$ in a strangeness-free formulation.

2. Problem statement

The discrete analogs of the equations not resolved with respect to derivative of type (1.7) are the discrete equations not resolved with respect to the unknown variable x_i on the next step $i+1$ of the form

$$A_{i+1}B_{i+1}x_{i+1} = C_i x_i. \tag{2.1}$$

We will consider adjoint discrete descriptor systems (2.1) and

$$B_i^* A_i^* z_i = C_i^* z_{i+1},\ i = 0, 1, 2, ..., \tag{2.2}$$

where $x_i \in X$, $z_i \in Z$, $A_i \in L(Y, Z)$, $B_i \in L(X, Y)$, $C_i \in L(X, Z)$, X, Y and Z are real Hilbert spaces, the operators A_i, B_i are normally solvable for each i, i. e. imA_i and imB_i are closed.

We will assume that the decompositions

$$kerA_i \oplus imB_i = Y,\ i = 0, 1, 2, ..., \tag{2.3}$$

hold. Then system (2.1) is called a descriptor system with a properly stated leading term.

It is easy to verify that the analog of the Lagrange identity for each pair of solutions of systems (2.1) and (2.2) is

$$< A_i B_i x_i, z_i >= const.$$

Indeed, $< A_i B_i x_i, z_i >=< C_{i-1} x_{i-1}, z_i >=< x_{i-1}, C_{i-1}^* z_i >=$
$< x_{i-1}, B_{i-1}^* A_{i-1}^* z_{i-1} >=< A_{i-1} B_{i-1} x_{i-1}, z_{i-1} >$.

It is well known that the discrete dynamic Leontief model of a multisector economy [**LA**] is a descriptor system since the capital coefficient matrix may be singular.

The solvability of discrete linear-quadratic optimal control problems with the state equation

$$A_{i+1}B_{i+1}x_{i+1} = C_i x_i + D_i u_i, \tag{2.4}$$

where $\{u_i\}$ is a control, has been studied in [**K1**], [**K2**]. In these papers it was supposed that an equation following from control optimality conditions is index-1 DAE. The regularity of the pencil of the operators from the state equation was not required in contrast to the works [**BL**], [**MK**], [**M**], devoted to linear-quadratic discrete optimal control problems for standard descriptor systems with constant coefficients in the finite-dimensional case. It was also not assumed that state equation (2.4) is causal. A linear-quadratic discrete optimal control problem for a special case of state equation (2.4) $(A_{i+1}B_{i+1} = E_{i+1})$ was considered in [**MS**] where the state equation was given in regular strangeness-free form.

The short survey of publications dealing with control problems for discrete descriptor systems is presented in [**KDN**].

Further we will study the adjointness property for explicit systems, following from (2.1) and (2.2). The paper is organized as follows. In the next section we will get from system (2.1) with index-0 an explicit system with respect to $B_i x_i$ and from system (2.2) we will have an explicit system with respect to $A_i^* z_i$. These systems are adjoint. In the fourth section we will obtain the similar result for index-1 systems. The fifth section is devoted to more complicated index-2 systems. The last section is a short conclusion.

Some results from this paper were presented in the 2015 International Conference on Scientific Computation and Differential Equations [**K3**].

3. Index-0 systems

Let $X = Y = Z$ and the operator $G_{i0} = A_i B_i : X \to X$ is invertible for each i=0,1,2,....

Then we say that system (2.1) is index-0 system.

In this case system (2.1) implies the explicit system

$$x_{i+1} = (G_{(i+1)0})^{-1} C_i x_i \tag{3.1}$$

and system (2.2) implies the explicit system

$$z_i = (G_{i0}^*)^{-1} C_i^* z_{i+1}. \tag{3.2}$$

It is easy to see that systems (3.1) and (3.2) are not adjoint.

Assume the condition

H_0. The operators A_i and B_i are invertible for each i=0,1,2,....

In the finite-dimensional case the last condition is always valid.

If we introduce the new variables

$$u_i = B_i x_i, \ v_i = A_i^* z_i \tag{3.3}$$

then the explicit systems with respect to u_i and v_i, following from (3.1) and (3.2), may be written as follows

$$u_{i+1} = A_{i+1}^{-1} C_i B_i^{-1} u_i,$$
$$v_i = (A_{i+1}^{-1} C_i B_i^{-1})^* v_{i+1}.$$

The last two systems are adjoint.

Thus the following assertion takes place.

THEOREM 3.1. *Under condition H_0 system* (2.1) *with index-0 and system* (2.2) *imply the explicit adjoint systems with respect to $B_i x_i$ and $A_i^* z_i$ respectively.*

4. Index-1 systems

4.1. Decomposition into systems in subspaces. We will use the decompositions of the spaces X, Y, Z into the orthogonal sums

$$\begin{aligned} X &= kerB_i \oplus imB_i^*, \\ Y &= kerB_i^* \oplus imB_i = kerA_i \oplus imA_i^*, \\ Z &= kerA_i^* \oplus imA_i. \end{aligned} \tag{4.1}$$

From (2.3) and from the second equality in (4.1) we obtain

$$imB_i = imA_i^*. \tag{4.2}$$

Denote by P_i, Q_i, P_{*i}, Q_{*i} the orthogonal projectors onto the subspaces $kerB_i$, $kerA_i$, $kerB_i^*$, $kerA_i^*$ respectively, corresponding to decompositions (4.1).

REMARK 4.1. It follows from (4.1) that adjoint descriptor systems (2.1) and (2.2) are descriptor systems with properly stated leading terms simultaneously.

According to the Banach Theorem the operators

$$A_i = (I - Q_{*i})A_i(I - Q_i) : imA_i^* \longrightarrow imA_i,$$

$$B_i = (I - P_{*i})B_i(I - P_i) : imB_i^* \longrightarrow imB_i$$

have the bounded inverse operators $A_i^+ = (I - Q_i)A_i^+(I - Q_{*i})$ and $B_i^+ = (I - P_i)B_i^+(I - P_{*i})$. By I, as usually, we mean identity operators.

Using decompositions (4.1), corresponding projectors, equality (4.2) and the identities

$$(I - P_i)x_i = B_i^+ B_i x_i,\ (I - Q_{*i})z_i = A_i^{*+} A_i^* z_i,,$$

we obtain from systems (2.1) and (2.2) the following systems

(4.3)

$$0 = Q_{*(i+1)}C_i(I - P_i)B_i^+ B_i x_i + Q_{*(i+1)}C_i P_i x_i,$$

$$B_{i+1}x_{i+1} = A_{i+1}^+(I - Q_{*(i+1)})C_i(I - P_i)B_i^+ B_i x_i + A_{i+1}^+(I - Q_{*(i+1)})C_i P_i x_i$$

and

(4.4)

$$0 = P_i C_i^*(I - Q_{*(i+1)})A_{i+1}^{*+} A_{i+1}^* z_{i+1} + P_i C_i^* Q_{*(i+1)} z_{i+1},$$

$$A_i^* z_i = B_i^{*+}(I - P_i)C_i^*(I - Q_{*(i+1)})A_{i+1}^{*+} A_{i+1}^* z_{i+1} + B_i^{*+}(I - P_i)C_i^* Q_{*(i+1)} z_{i+1}.$$

4.2. Index-1 systems. Let $X = Z$ and the operator

(4.5) $$G_{i1} = Q_{*(i+1)}C_i P_i : kerB_i \to \ker A_{i+1}^*$$

is invertible for each $i = 0, 1, 2, ...$

Then we say that system (2.1) is index-1 system.

In this case, we express $P_i x_i$ from the first system in (4.3) and substitute it into the second system in (4.3). After transformations we get the system

(4.6) $$B_{i+1}x_{i+1} = A_{i+1}^+(I - Q_{*(i+1)})(I - C_i P_i (G_{i1})^{-1} Q_{*(i+1)})C_i(I - P_i)B_i^+ B_i x_i.$$

Express $Q_{*(i+1)}z_{i+1}$ from the first system in (4.4) and substitute it into the second system in (4.4). We get the system

(4.7) $$A_i^* z_i = B_i^{*+}(I - P_i)C_i^*(I - Q_{*(i+1)}(G_{i1}^*)^{-1} P_i C_i^*)(I - Q_{*(i+1)})A_{i+1}^{*+} A_{i+1}^+ z_{i+1}.$$

It is not difficult to see that systems (4.6) and (4.7) with respect to variables (3.3) are adjoint.

So we have proved the following assertion.

THEOREM 4.2. *Adjoint discrete descriptor systems* (2.1) *with index-1 and* (2.2) *imply the explicit adjoint systems* (4.6) *and* (4.7) *with respect to the variables* $B_i x_i$ *and* $A_i^* z_i$ *respectively.*

REMARK 4.3. In contrast to the continuous case for DAEs with index-1 we do not need here the constancy of some subspaces in the original systems, namely, $\ker A(t)$ and $imB(t)$ (see, e. g., [**BL**], Theorem 2), to obtain the adjointness of explicit equations following from adjoint DAEs.

EXAMPLE 4.4. Consider the system of form (2.1) consisting of two equations

(4.8) $$\begin{aligned} x_{i+1}^1 &= C_i^1 x_i^1 + C_i^2 x_i^2, \\ 0 &= C_i^3 x_i^1 + C_i^4 x_i^2,\ i = 0, 1, 2, ... \end{aligned}$$

It is not difficult to verify that this system is a system with properly stated leading term. Here $X = Z = \mathbb{R}^2$, $Y = \mathbb{R}$, $x_i = \begin{pmatrix} x_i^1 \\ x_i^2 \end{pmatrix}$, $A_i = \begin{pmatrix} 1 \\ 0 \end{pmatrix}$, $B_i = \begin{pmatrix} 1 & 0 \end{pmatrix}$, $C = \begin{pmatrix} C_i^1 & C_i^2 \\ C_i^3 & C_i^4 \end{pmatrix}$, $\ker A_i = 0$, $\ker B_i = \ker A_i^* = \left\{ \begin{pmatrix} 0 \\ 1 \end{pmatrix} \right\}$, $imB_i = \mathbb{R}$, $Q_{*i} = P_i = \begin{pmatrix} 0 & 0 \\ 0 & 1 \end{pmatrix}$.

The operator $G_{i1} = Q_{*(i+1)} C_i P_i = \begin{pmatrix} 0 & 0 \\ 0 & C_i^4 \end{pmatrix} : kerB_i \to kerA_{i+1}^*$ is invertible if $C_i^4 \neq 0$.

In this case we have the index-1 system and we obtain from (4.8) the explicit equation

$$x_{i+1}^1 = (C_i^1 - C_i^2 (C_i^4)^{-1} C_i^3) x_i^1. \tag{4.9}$$

The adjoint system for (4.8) has the form

$$z_i^1 = C_i^1 z_{i+1}^1 + C_i^3 z_{i+1}^2,$$

$$0 = C_i^2 z_{i+1}^1 + C_i^4 z_{i+1}^2.$$

Assuming that $C_i^4 \neq 0$, write the corresponding explicit system for z_i^1

$$z_i^1 = (C_i^1 - C_i^3 (C_i^4)^{-1} C_i^2) z_{i+1}^1. \tag{4.10}$$

Equations (4.9) for $B_i x_i = x_i^1$ and (4.10) for $A_i^* z_i = z_i^1$ are adjoint.

5. Index-2 systems

We will consider systems (2.1) and (2.2) not assuming the invertibility of operator (4.5).

For the simplicity we will assume the condition

H_1. $G_{i1} = 0$ for all $i = 0, 1, 2, \ldots$.

Then systems (4.3) and (4.4) have respectively the forms

(5.1)
$$0 = Q_{*(i+1)} C_i (I - P_i) B_i^+ B_i x_i,$$
$$B_{i+1} x_{i+1} = A_{i+1}^+ (I - Q_{*(i+1)}) C_i (I - P_i) B_i^+ B_i x_i + A_{i+1}^+ (I - Q_{*(i+1)}) C_i P_i x_i$$

and

(5.2)
$$0 = P_i C_i^* (I - Q_{*(i+1)}) A_{i+1}^{*+} A_{i+1}^* z_{i+1},$$
$$A_i^* z_i = B_i^{*+} (I - P_i) C_i^* (I - Q_{*(i+1)}) A_{i+1}^{*+} A_{i+1}^* z_{i+1} + B_i^{*+} (I - P_i) C_i^* Q_{*(i+1)} z_{i+1}.$$

From (5.1) we obtain the relation

$$\begin{aligned} 0 = Q_{*(i+2)} C_{i+1} (I - P_{i+1}) B_{i+1}^+ (A_{i+1}^+ (I - Q_{*(i+1)}) C_i (I - P_i) B_i^+ B_i x_i \\ + A_{i+1}^+ (I - Q_{*(i+1)}) C_i P_i x_i). \end{aligned} \tag{5.3}$$

Let the operator

$$G_{i2} = Q_{*(i+2)} C_{i+1} (I - P_{i+1}) B_{i+1}^+ A_{i+1}^+ (I - Q_{*(i+1)}) C_i P_i : kerB_i \to kerA_{i+2}^*$$

is invertible for each $i = 0, 1, 2, \ldots$.

Then we say that system (2.1) is index-2 system.

In this case we can resolve equality (5.3) with respect to $P_i x_i$. Substituting the obtained expression for $P_i x_i$ into the second equality in (5.1) we get the explicit system with respect to $B_i x_i$

$$B_{i+1}x_{i+1} = (I - \mathfrak{M}_i)D_i B_i x_i, \tag{5.4}$$

where

$$\mathfrak{M}_i = A^+_{i+1}(I - Q_{*(i+1)})C_i P_i G^{-1}_{i2} Q_{*(i+2)} C_{i+1}(I - P_{i+1})B^+_{i+1},$$
$$D_i = A^+_{i+1}(I - Q_{*(i+1)})C_i(I - P_i)B^+_i.$$

It is easy to verify that the operator $\mathfrak{M}_i$ is a projector, i.e. $\mathfrak{M}_i^2 = \mathfrak{M}_i$.

In a similar way we obtain from system (5.2) the explicit system with respect to $A^*_i z_i$

$$A^*_i z_i = (I - \mathfrak{M}^*_{i-1})D^*_i A^*_{i+1} z_{i+1}. \tag{5.5}$$

Taking into account the first systems in (5.1) and (5.2) and the notation $\mathfrak{M}_i$ we get

$$\mathfrak{M}_{i-1}B_i x_i = 0 \text{ and } \mathfrak{M}^*_i A^*_{i+1} z_{i+1} = 0.$$

In view of the last equalities we obtain from (5.4) and (5.5) the explicit adjoint systems

$$B_{i+1}x_{i+1} = (I - \mathfrak{M}_i)D_i(I - \mathfrak{M}_{i-1})B_i x_i \tag{5.6}$$

and

$$A^*_i z_i = (I - \mathfrak{M}^*_{i-1})D^*_i(I - \mathfrak{M}^*_i)A^*_{i+1} z_{i+1}. \tag{5.7}$$

So we have proved the following assertion.

THEOREM 5.1. *Under the condition H_1 adjoint discrete descriptor systems with properly stated leading terms (2.1) with index-2 and (2.2) imply the explicit adjoint systems (5.6) and (5.7) with respect to the variables $B_i x_i$ and $A^*_i z_i$ respectively.*

EXAMPLE 5.2. Consider a system of the form

$$\begin{aligned} x^1_{i+1} &= C^1_i x^1_i + C^2_i x^2_i, \\ 0 &= C^3_i x^1_i. \end{aligned} \tag{5.8}$$

This system satisfies the condition H_1. Here $G_{i2} = \begin{pmatrix} 0 & 0 \\ 0 & C^3_{i+1}C^2_i \end{pmatrix}$. If the operator $C^3_{i+1}C^2_i$ is invertible for each $i = 0, 1, 2, ...$ then system (5.8) is index-2 system. We will assume that this condition is valid.

In this case the obtained from (5.8) relation

$$0 = C^3_{i+1}x^1_{i+1} = C^3_{i+1}(C^1_i x^1_i + C^2_i x^2_i)$$

can be resolved with respect to x^2_i, i. e.

$$x^2_i = -(C^3_{i+1}C^2_i)^{-1}C^3_{i+1}C^1_i x^1_i.$$

Substituting the last expression into the first equation in (5.8) we get the explicit system with respect to x^1_i of the form

$$x^1_{i+1} = (I - \Omega_i)C^1_i(I - \Omega_{i-1})x^1_i, \tag{5.9}$$

where

$$\Omega_i = C^2_i(C^3_{i+1}C^2_i)^{-1}C^3_{i+1}.$$

We took into account that $\Omega_{i-1}x_{i-1}^1 = 0$ in view of the second equation in (5.8). Here $\mathfrak{M}_i = \Omega_i$, $D_i = C_i^1$.

Similarly we obtain from the adjoint system for (5.8)

$$\begin{aligned} z_i^1 &= C_i^{1*} z_{i+1}^1 + C_i^{3*} z_{i+1}^2, \\ 0 &= C_i^{2*} z_{i+1}^1 \end{aligned} \tag{5.10}$$

the explicit system with respect to z_i^1

$$z_i^1 = (I - \Omega_{i-1}^*)C_i^{1*}(I - \Omega_i^*)z_{i+1}^1.$$

We took into account here that $\Omega_i^* z_{i+1}^1 = 0$ in view of the second equation in (5.10).

The last equation and (5.9) are adjoint.

6. Conclusion

Under some conditions adjoint explicit systems with respect to the variables $B_i x_i$ and $A_i^* z_i^*$ have been obtained from adjoint discrete descriptor systems (2.1), (2.2) with index less than or equal to 2.

Acknowledgment

The author would like to thank the reviewers for constructive remarks that have helped to improve the text of the paper.

References

[BKM1] K. Balla, G. Kurina and R. März, *Index criteria for differential algebraic equations arising from linear-quadratic optimal control problems,* Humboldt-universität zu Berlin, Mathematisch-Naturwissenschaftliche Fakultät II, Institut für Mathematik, Preprint Nr. 2003-14.

[BKM2] K. Balla, G. A. Kurina, and R. März, *Index criteria for differential algebraic equations arising from linear-quadratic optimal control problems*, J. Dyn. Control Syst. **12** (2006), no. 3, 289–311, DOI 10.1007/s10450-006-0001-2. MR2233022

[BLi] K. Balla and V. H. Linh, *Adjoint pairs of differential-algebraic equations and Hamiltonian systems*, Appl. Numer. Math. **53** (2005), no. 2-4, 131–148, DOI 10.1016/j.apnum.2004.08.015. MR2128518

[BM1] K. Balla and R. März, *A unified approach to linear differential algebraic equations and their adjoints*, Z. Anal. Anwendungen **21** (2002), no. 3, 783–802, DOI 10.4171/ZAA/1108. MR1929432

[BM2] K. Balla and R. März, *Linear differential algebraic equations of index 1 and their adjoint equations*, Results Math. **37** (2000), no. 1-2, 13–35, DOI 10.1007/BF03322509. MR1742297

[BM3] K. Balla and R. März, *Linear boundary value problems for differential algebraic equations*, Miskolc Math. Notes **5** (2004), no. 1, 3–18, DOI 10.18514/mmn.2004.20. MR2040972

[BL] D. J. Bender and A. J. Laub, *The linear-quadratic optimal regulator for descriptor systems: discrete-time case*, Automatica J. IFAC **23** (1987), no. 1, 71–85, DOI 10.1016/0005-1098(87)90119-1. MR880092

[EST] D. Estévez Schwarz and C. Tischendorf, *Structural analysis of electric circuits and consequences for MNA*, Int. J. Circ. Theor. Appl. **28** (2000), 131–162.

[G] I. V. Gajshun, *Systemy s Discretnym Vremenem (Systems with Discrete Time)*, Institut matematiki NAN Belarusi, Minsk, 2001 (in Russian).

[H] P. Hartman, *Ordinary differential equations*, John Wiley & Sons, Inc., New York-London-Sydney, 1964. MR0171038

[KMS] P. Kunkel, V. Mehrmann and L. Scholz, *Self-adjoint differential-algebraic equations*, Preprint 2011/13, Preprint-Reihe des Instituts für Mathematik Technishe Universität Berlin. `http://www.math.tu-berlin.de/preprints`.

[K1] G. A. Kurina, *Solvability of linear-quadratic discrete optimal control problems for descriptor systems in Hilbert space*, Proc. 10th IEEE Mediterranean Conf. Control and Automation. Instituto Superior Tećnico, July 9-12, 2002, Lisboa, Portugal.

[K2] G. A. Kurina, *Linear-quadratic discrete optimal control problems for descriptor systems in Hilbert space*, J. Dynam. Control Systems **10** (2004), no. 3, 365–375, DOI 10.1023/B:JODS.0000034435.80761.32. MR2070198

[K3] G. Kurina, *On adjoint discrete descriptor systems with properly stated leading terms*, SciCADE 2015, Int. Conf. Scientific Computation and Differential Equations. Book of Abstracts, University of Potsdam, Potsdam, Germany, 2015, p.68.

[KDN] G. A. Kurina, M. G. Dmitriev, and D. S. Naidu, *Discrete singularly perturbed control problems (a survey)*, Dyn. Contin. Discrete Impuls. Syst. Ser. B Appl. Algorithms **24** (2017), no. 5, 335–370. MR3708407

[LM] V. H. Linh and R. März, *Adjoint pairs of differential-algebraic equations and their Lyapunov exponents*, J. Dynam. Differential Equations **29** (2017), no. 2, 655–684, DOI 10.1007/s10884-015-9474-6. MR3651604

[LA] D. G. Luenberger and A. Arbel, *Singular dynamic Leontief systems*, Econometrica **45** (1977), 991–995.

[MK] G. P. Mantas and N. J. Krikelis, *Linear quadratic optimal control for discrete descriptor systems*, J. Optim. Theory Appl. **61** (1989), no. 2, 221–245, DOI 10.1007/BF00962798. MR996378

[M] V. L. Mehrmann, *The autonomous linear quadratic control problem*, Theory and numerical solution, Lecture Notes in Control and Information Sciences, vol. 163, Springer-Verlag, Berlin, 1991. MR1133203

[MS] V. Mehrmann and L. Scholz, *Self-conjugate differential and difference operators arising in the optimal control of descriptor systems*, Oper. Matrices **8** (2014), no. 3, 659–682, DOI 10.7153/oam-08-36. MR3257884

Voronezh State University and Institute of Law and Economics, Voronezh, Russia 394000, Federal Research Center "Computer Science and Control" of Russian Academy of Sciences, Moscow, Russia 119333

Current address: Faculty of Mathematics, Voronezh State University, Universitetskaya pl., 1, Voronezh, Russia 394018

Email address: `kurina@math.vsu.ru`

Contemporary Mathematics
Volume **734**, 2019
https://doi.org/10.1090/conm/734/14775

Solution of the initial value problem for the focusing Davey-Stewartson II system

E. Lakshtanov and B. Vainberg

Dedicated to a remarkable person and mathematician Professor S. Krein.

Abstract. We consider a focusing Davey-Stewartson system and construct the solution of the Cauchy problem in the presence of exceptional points (and/or curves).

1. Introduction

Let $q_0(z)$, $z = x + iy$, $(x, y) \in \mathbb{R}^2$, be a compactly supported, sufficiently smooth function. Consider the DSII system of equations for unknown functions $q = q(z, t)$, $\phi = \phi(z, t)$, $x, y \in \mathbb{R}^2$, $t \geq 0$:

$$
\begin{aligned}
q_t &= 2iq_{xy} - 4i\phi q, \\
\phi_{xx} + \phi_{yy} &= \pm 4|q|^2_{xy}, \\
q(z, 0) &= q_0(z).
\end{aligned} \tag{1.1}
$$

The sign plus in (1.1) corresponds to the defocusing case of DSII and sign minus corresponds to the focusing case. Even though the defocusing case is well studied, we will consider both models together to stress the universality of our approach.

The Davey-Stewartson system of equations models the shallow-water limit of the evolution of weakly nonlinear water waves that travel predominantly in one direction, but in which the wave amplitude is modulated slowly in two horizontal directions [**8**],[**10**]. The shallow-water limit means that $kh \to 0$, $a \ll kh^2$, where h denotes the depth of the bed, k is a wave number of a wavy surface and a is a characteristic amplitude of the disturbance.

Independently, Ablowitz and Haberman [**2**], Morris [**15**], and Cornille [**7**] have derived (1.1) while looking for completely integrable systems generalizing the non-linear Schrödinger equation to two spatial dimensions.

2010 *Mathematics Subject Classification.* Primary 54C40, 14E20; Secondary 46E25, 20C20.
Key words and phrases. Davey-Stewartson, $\overline{\partial}$-method, Dirac equation, exceptional point, scattering data, inverse problem.

The first author was supported by Portuguese funds through the CIDMA - Center for Research and Development in Mathematics and Applications and the Portuguese Foundation for Science and Technology ("FCT–Fundção para a Ciência e a Tecnologia"), within project UID/MAT/0416/2013.

The second author was supported by the NSF grant DMS-1714402 and the Simons Foundation grant 527180.

In [**10**], [**12**], one can find results on global uniqueness and existence of solutions of (1.1) in the defocusing case and for small initial data in the focusing case, as well as a justification of local in time well-posedness for arbitrary data in both cases.

Constructions of solutions of initial value problem (1.1) via the IST (inverse scattering transform) in the defocusing case or when initial data are small enough are given in [**4**], [**5**], [**21**]-[**23**], [**20**], [**18**]. These works used the $\overline{\partial}$-method for the Dirac inverse scattering problem. The classical $\overline{\partial}$-method fails when exceptional points are present. The latter are defined as the values of the spectral parameter k for which the homogeneous direct scattering problem has a non-trivial solution. In [**13**], [**14**], we generalized the $\overline{\partial}$-method for the Schrödinger and Dirac equations to the case when exceptional points exist. A prototype of this generalization was considered in section 8 of [**19**]. In the current paper, we apply the results of [**14**] to solve the Cauchy problem for the focusing DSII equation.

We will work with equation (1.1) rewritten in the following form (see, for example, [**9**]):

$$q_t = 2iq_{xy} \pm 4q(\overline{\varphi} - \varphi),$$
$$\partial\varphi = \overline{\partial}|q|^2,$$
$$q(z,0) = q_0(z). \tag{1.2}$$

In order to obtain (1.1) from (1.2), one needs to introduce $\phi = \pm i(\overline{\varphi} - \varphi)$, apply the Laplacian to ϕ, and use that $\partial\overline{\partial} = \overline{\partial}\partial$.

2. The main result

Denote

$$Q^0(z) = \begin{pmatrix} 0 & q_0(z) \\ \pm q_0(z) & 0 \end{pmatrix}, \quad z \in \mathbb{C}. \tag{2.1}$$

Consider the Dirac equation for the 2×2 matrix $\psi(\cdot, k)$, $k \in \mathbb{C}$:

$$\frac{\partial \psi}{\partial \overline{z}} = Q^0 \overline{\psi}, \quad \psi(z,k)e^{-i\overline{k}z/2} \to I, \quad z \to \infty. \tag{2.2}$$

The corresponding Lippmann-Schwinger equation has the following form

$$\psi(z,k) = e^{i\overline{k}z/2}I + \int_{z\in\mathbb{R}^2} G(z-z',k)Q^0(z')\overline{\psi}(z',k)dz', \quad G(z,k) = \frac{1}{\pi}\frac{e^{i\overline{k}z/2}}{z}. \tag{2.3}$$

Solutions ψ of (2.3) are called the *scattering solutions*. Let $q_0(z) \in L^p_{\text{comp}}(\mathbb{R}^2)$, $p > 2$. Here and below we use the same notation for functional spaces, irrespectively of whether those are the spaces of matrix-valued or scalar-valued functions. After the substitution

$$\mu(z,k) = \psi(z,k)e^{-i\overline{k}z/2}, \tag{2.4}$$

equation (2.3) takes the form

$$\mu(z,k) = I + \int_{z\in\mathbb{R}^2} \frac{e^{-\text{Re}(i\overline{k}z')}}{\pi(z-z')} Q^0(z')\overline{\mu}(z',k)dz', \tag{2.5}$$

where the integral operator is compact in $L^q(\mathbb{R}^2)$, $q > \frac{2p}{p-2}$, see [**6**, Th.A. iii]. The values of k such that the homogeneous equation (2.5) has a non-trivial solution are called the *exceptional points*. The set of exceptional points will be denoted by $\mathcal{E}$. Note that the operators in equations (2.3), (2.5) are not analytic in k, and $\mathcal{E}$ may contain one-dimensional components. Thus the scattering solution may not exist if $k \in \mathcal{E}$. There are no exceptional points when $|k|$ is large enough (e.g., [**21**, Lemma 2.8], [**6**, Lemma C]). Let us choose $A \gg 1$ and $k_0 \in \mathbb{C}$ such that all the exceptional points are contained in the disk

$$D = \{k \in \mathbb{C} : 0 \leq |k| < A\}, \tag{2.6}$$

and that k_0 belongs to the same disc D and is not exceptional.

The generalized scattering data are defined by the following integral

$$h_0(\varsigma, k) = \frac{1}{(2\pi)^2} \int_{R^2} e^{-i\overline{\varsigma}z/2} Q^0(z) \overline{\psi}(z, k) dx dy, \ \varsigma \in \mathbb{C}, \ k \in \mathbb{C} \backslash \mathcal{E}. \tag{2.7}$$

In fact, from the Green formula it follows that h_0 can be determined without using the potential Q^0 or the solution ψ of the Dirac equation (2.2) if the Dirichlet data at $\partial\Omega$ are known for the solution of (2.2) in a bounded region Ω containing the support of Q^0.

The inverse scattering problem of reconstructing the potential Q^0 via the given h_0 plays a crucial point in the present paper. This inverse problem has been solved in [**5, 21**] for symmetric or small antisymmetric Q^0 using the $\overline{\partial}$-method. With larger potentials, exceptional points appear, and the $\overline{\partial}$-method must be replaced [**13, 14**] by a combination of $\overline{\partial}$ and the Riemann-Hilbert methods (which we called the global Riemann-Hilbert problem). Possible appearance of exceptional points makes the focusing DSII equation much more complicated than the defocusing one: while the defocusing DSII equation deals with a symmetric inverse scattering problem, in order to solve the focusing DSII equation, one should consider a non-symmetric inverse scattering problem.

For the reconstruction procedure, we consider the space

$$\mathcal{H}^s = \left\{ u \in L^s(\mathbb{R}^2) \bigcap C(D) \right\}, \quad s > 2,$$

(recall that we use the same notation for matrices if their entries belong to $\mathcal{H}^s$) and the operator

$$T_z\phi(k) = \frac{1}{\pi} \int_{\mathbb{C}\backslash D} e^{i(\overline{\varsigma}z + \overline{z}\varsigma)/2} \overline{\phi}(\varsigma) \Pi^o h_0(\varsigma, \varsigma) \frac{d\varsigma_R d\varsigma_I}{\varsigma - k} + \tag{2.8}$$

$$\frac{1}{2\pi i} \int_{\partial D} \frac{d\varsigma}{\varsigma - k} \int_{\partial D} [e^{\frac{i}{2}(\varsigma\overline{z} + \overline{\varsigma'}z)} \overline{\phi^-(\varsigma')} \Pi^o + e^{i(\varsigma - \varsigma')\frac{\overline{z}}{2}} \phi^-(\varsigma') \Pi^d \mathbf{C}] \left[\mathrm{Ln} \frac{\overline{\varsigma'} - \overline{\varsigma}}{\overline{\varsigma'} - \overline{k_0}} h_0(\varsigma', \varsigma) d\overline{\varsigma'} \right],$$

where $z \in \mathbb{C}$, $\phi \in \mathcal{H}^s$, ϕ^- is the boundary trace of ϕ from the interior of D, $\mathbf{C}$ is the operator of complex conjugation, $\Pi^o M$ is the off-diagonal part of a matrix M, $\Pi^d M$ is the diagonal part. The logarithmic function here is well defined, see the Remark after Lemma 3.4 in [**14**].

It turns out that, after the substitution $w = v - I \in \mathcal{H}^s$, $s > 2$, the equation

$$(I + T_z)v = I$$

becomes Fredholm in $\mathcal{H}^s$, and the potential q_0 can be expressed explicitly in terms of v (see [**13, 14**]).

In order to solve the DSII problem (1.2), we apply this reconstruction procedure to specially chosen scattering data. We start with the generalized scattering data defined by q_0 and extend them in time as follows:

(2.9)

$$h(\varsigma', \varsigma, t) := e^{-t(\varsigma^2 - \overline{\varsigma'}^2)/2} \Pi^o h_0(\varsigma', \varsigma) + e^{-t(\overline{\varsigma}^2 - \overline{\varsigma'}^2)/2} \Pi^d h_0(\varsigma', \varsigma), \ \varsigma' \in \mathbb{C}, \ \varsigma \in \mathbb{C} \backslash \mathcal{E}, \ t \geq 0.$$

For $t \geq 0$, we define the operator

$$T_{z,t}\phi(k) = \frac{1}{\pi}\int_{\mathbb{C}\setminus D} e^{\frac{i}{2}(\overline{\varsigma}z+\overline{z}\varsigma)}\overline{\phi}(\varsigma)\Pi^o h(\varsigma,\varsigma,t)\frac{d\varsigma_R d\varsigma_I}{\varsigma - k}+ \tag{2.10}$$

$$\frac{1}{2\pi i}\int_{\partial D}\frac{d\varsigma}{\varsigma-k}\int_{\partial D}[e^{\frac{i}{2}(\varsigma\overline{z}+\overline{\varsigma'}z)}\overline{\phi^-(\varsigma')}\Pi^o + e^{i/2(\varsigma-\varsigma')\overline{z}}\phi^-(\varsigma')\Pi^d\mathbf{C}]\left[\mathrm{Ln}\frac{\overline{\varsigma'}-\overline{\varsigma}}{\overline{\varsigma'}-\overline{k_0}}h(\varsigma',\varsigma,t)d\overline{\varsigma'}\right].$$

Assumptions on q_0 (that will be stated later) imply that the equation

$$(I+T_{z,t})v_{z,t} = I \tag{2.11}$$

becomes Fredholm in the space $\mathcal{H}^s$, $s>2$, after the substitution $w_{z,t} = v_{z,t} - I$. Let Ω be an arbitrary region in the half space $\{(z,t) : z \in \mathbb{C}, t \geq 0\}$, where the kernel of $I + T_{z,t} : \mathcal{H}^s \to \mathcal{H}^s$, $s > 2$, is trivial. We will show that the vector $(q(z,t), \varphi(z,t))$ defined by

$$\begin{pmatrix}\varphi(z,t) & q(z,t)\\ \pm q(z,t) & \varphi(z,t)\end{pmatrix} := \frac{-i}{2}(\Pi^o + \overline{\partial}\Pi^d)\left(\frac{1}{\pi}\int_{\mathbb{C}\setminus D} e^{i(\overline{\varsigma}z+\overline{z}\varsigma)/2}\overline{v_{z,t}}(\varsigma)\Pi^o h_0(\varsigma,\varsigma,t)d\varsigma_R d\varsigma_I + \right. \tag{2.12}$$

$$\left.\frac{1}{2\pi i}\int_{\partial D}d\varsigma\int_{\partial D}[e^{\frac{i}{2}(\varsigma\overline{z}+\overline{\varsigma'}z)}\overline{v_{z,t}^-(\varsigma')}\Pi^o - e^{\frac{i}{2}(\varsigma-\varsigma')\overline{z}}v_{z,t}^-(\varsigma')\Pi^d\mathbf{C}]\left[\mathrm{Ln}\frac{\overline{\varsigma'}-\overline{\varsigma}}{\overline{\varsigma'}-\overline{k_0}}h_0(\varsigma',\varsigma,t)d\varsigma'\right]\right),$$

where $v_{z,t}$ is the solution of (2.11), solves the DSII equation in Ω.

We need a couple of definitions before we state the main result.

We'll use the word *generic* when referring to elements that belong to an open, dense subset V of a topological space S. We will assume that $q_0(z)$ belongs to the space $W^{6,\infty}_{\mathrm{comp}}(\mathbb{R}^2)$ of six times differentiable functions with bounded derivatives and with the support in a bounded region $\mathcal{O} \subset R^2$. This smoothness is used to guarantee the appropriate decay of the scattering data; we did not try to relax this assumption.

We will say that a set ω of points (z,t) in $\mathbb{R}^3_+ = \mathbb{R}^3 \bigcap \{t \geq 0\}$ is half-open if ω contains points where $t = 0$ and, for each point $(z_0, 0) \in \omega$, there is a ball B_0 centered at this point such that $B_0 \bigcap \{t \geq 0\} \subset \omega$.

THEOREM 2.1. *Let $q_0(z) \in W^{6,\infty}_{\mathrm{comp}}(\mathbb{R}^2)$. Then, for each $s > 2$, the following statements are valid.*

- *The operator $T_{z,t}$ is compact in $\mathcal{H}^s$ for all $z \in \mathbb{C}$, $t \geq 0$, and depends continuously on z and $t \geq 0$. The same property holds for its first derivative in time and all the spatial derivatives (in $\Re z$, $\Im z$) up to the third order, where the derivatives are defined in the norm convergence. When (2.10) is differentiated, the derivatives can be applied to the integrands. The function $T_{z,t}I$ belongs to $\mathcal{H}^s$ for all $t \geq 0$.*
- *Let the kernel of $I + T_{z,t}$ in the space $\mathcal{H}^s$ be trivial for (z,t) in an open or half open set $\omega \subset \mathbb{R}^3_+ = \mathbb{R}^3 \bigcap \{t \geq 0\}$. Let $v_{z,t} = w_{z,t} + I$, where $w_{z,t} \in \mathcal{H}^s$ is the solution of the equation*

$$(I + T_{z,t})w_{z,t} = -T_{z,t}I. \tag{2.13}$$

Then functions the q, φ defined in (2.12) satisfy all the relations (1.2) in the classical sense when $(z,t) \in \omega$.

- *Let us fix* $z_0, t_0 \in \mathbb{R}^2 \times \mathbb{R}^+$. *Then for generic potential* $q_0(z)$ *in* $W^{6,\infty}_{\text{comp}}(\mathbb{R}^2)$, *the kernel of* $I+T_{z,t}$ *is trivial for all* (z,t) *in some neighborhood of* (z_0, t_0). *The neighborhood may depend on* q_0.
- *Consider a set of initial data* $aq_0(z)$ *that depend on* $a \in (0,1]$. *Then equation (2.13) with* Q^0 *replaced by* aQ^0 *(*Q^0 *is fixed) is uniquely solvable for almost every* $(z,t,a) \in \mathcal{O} \times \mathbb{R}^+ \times (0,1]$. *For each* (z,t), *the unique solvability can be violated for at most finitely many values of* $a = a_j(z,t)$, $z,t \in \mathcal{O} \times \mathbb{R}^+$.

REMARK. Let us mention that results on unique solvability of the Cauchy problem for the focusing Davey-Stewartson equation with general (not small) initial data are unknown. One of the difficulties to obtain such a result relates to the fact that the corresponding functional spaces have to contain non-smooth functions in order to allow for the physically relevant solutions that blow up. The solution given by the above theorem has the following property: it exists in a neighborhood of the initial plane $t = 0$ when the initial data is $aq_0(x)$ and a is small enough. The integral equation (2.13) remains uniquely solvable for larger values of t if $|z|$ is restricted and a is decreased, if needed. The solution for larger a is understood in the sense of analytic continuation in a. More light on this topic will be given in our next publication.

The proof consists of 3 parts. In section 3, we show (Lemma 3.3) that the reconstruction procedure of the inverse scattering problem can be applied to an arbitrary regular and fast decreasing matrix-function h, which is not necessarily the scattering data h_0 of a compactly supported potential Q^0. This Lemma allows us to consider scattering data h defined by (2.9) for all $t \geq 0$, and construct operator $T_{z,t}$ and the solution $v_{z,t}$ of the equation (2.11). As a result, one gets a potential Q^t (which is not necessarily compactly supported) and a function ψ, which are defined in terms of $v_{z,t}$ for all $(z,t) \in \omega$ and are related by the Dirac equation $\frac{\partial \psi}{\partial \bar{z}} = Q^t \overline{\psi}$.

In section 4, we demonstrate that the solution $v_{z,t}$ of equation (2.11) satisfies the compatibility conditions. And, finally, in section 5, we repeat well-known arguments showing that the compatibility condition implies the validity of the first equation in the DSII problem (1.2), and complete the proof of Theorem 2.1.

3. Inverse problems for general scattering data

Consider a 2×2 matrix-function h (non necessarily defined by (2.7)) and a scalar function w that satisfy the following condition.

Condition. 1) $h^o(\varsigma, \varsigma) \in L^\infty(\mathbb{C}\backslash D) \bigcap L^2(\mathbb{C}\backslash D)$ and $h(\varsigma', \varsigma) \in L^\infty(\partial D) \times L^\infty(\partial D)$,

(3.1) 2) $|w(\varsigma', \varsigma)| \leq C\,|Ln|\varsigma - \varsigma'||\,, \varsigma', \varsigma \in \partial D,$

where $h^o = \Pi^o h$ is the off-diagonal part of h. Sometimes we will need a stronger assumption on h:

$$(3.2) \qquad |\varsigma|^3 h^o(\varsigma, \varsigma) \in L^\infty(\mathbb{C}\backslash D) \bigcap L^1(\mathbb{C}\backslash D), \quad h(\varsigma', \varsigma) \in L^\infty(\partial D) \times L^\infty(\partial D).$$

Let T_z be a slightly more general operator than those defined in (2.8). Namely, consider the following operator T_z acting in the space $\mathcal{H}^s$, $s>2$:

$$T_z\phi(k)=\frac{1}{\pi}\int_{\mathbb{C}\setminus D}e^{i(\bar\varsigma z+\bar z\varsigma)/2}\overline{\phi}(\varsigma)\Pi^o h(\varsigma,\varsigma)\frac{d\varsigma_R d\varsigma_I}{\varsigma-k}+ \tag{3.3}$$

$$\frac{1}{2\pi i}\int_{\partial D}\frac{d\varsigma}{\varsigma-k}\int_{\partial D}[e^{i(\varsigma\bar z+\bar\varsigma' z)/2}\overline{\phi^-(\varsigma')}\Pi^o+e^{i(\varsigma-\varsigma')\bar z/2}\phi^-(\varsigma')\Pi^d\mathbf{C}]\left[w(\varsigma',\varsigma)h(\varsigma',\varsigma)d\overline{\varsigma'}\right],$$

where h and w satisfy (3.1), ϕ^- is the boundary trace of ϕ from the interior of D, $\mathbf{C}$ is the operator of complex conjugation. $\Pi^o M=M^o$ is the off-diagonal part of a matrix M and $\Pi^d M=M^d$ is the diagonal part. Thus everywhere below h is an arbitrary matrix satisfying (3.1). The specific matrix $h=h_0$ defined in (2.7) via the initial data q_0 will appear only in Lemmas 3.1, 4.2 and at the very end of the last section of the paper.

LEMMA 3.1. *1) If (3.1) holds, then operator T_z is compact in $\mathcal{H}^s$, $s>2$, and*

$$\|T_z\|_{\mathcal{H}^s}\le C\|h\|_H,\quad \|h\|_H:=\|h^o(\varsigma,\varsigma)\|_{L^\infty(\mathbb{C}\setminus D)\bigcap L^2(\mathbb{C}\setminus D)}+\|h(\varsigma',\varsigma)\|_{L^\infty(\partial D)\times L^\infty(\partial D)}. \tag{3.4}$$

2) If h depends analytically on a complex parameter $\alpha\in A\subset\mathbb{C}$ or has $m\ge 0$ continuous derivatives with respect to a real parameter $\tau\in[0,T]$, where the derivatives in α,τ are understood as derivatives of elements of the Banach space H defined in (3.4), then operator T_z is analytic in α and m times differentiable in τ, and the derivatives of the right hand side of (3.3) can be moved inside the integrals.

PROOF. If $\mathcal{E}=\emptyset$ (there are no exceptional points), then D is empty, $\mathcal{H}^s=L^s$, and operator T_z can be simplified significantly (in particular, the second line in (3.3) can be dropped). In this case, the validity of the first statement of the lemma requires only $h^o\in L^2$, and the statement was proved by Nachman [**17**, Lemma 4.2] (the proof was reproduced in [**16**, Lemma 5.3]). In the general case, the proof can be found in [**14**, Lemma 4.3]. In fact, the estimate (3.4) was not stated in [**14**] explicitly, but it can be easily extracted from the proof of the compactness of T_z.

The validity of the second statement is obvious due to (3.4), and it is stated above solely for the convenience of references. □

LEMMA 3.2 (Sung [**21**],[**22**]). *If $q_0(z)\in W^{6,\infty}_{\text{comp}}(\mathbb{R}^2)$, then (3.2) holds for the scattering matrix $h=h_0$ defined in (2.7).*

REMARK. It was shown by Sung [**21**],[**22**] that condition $q_0\in W^{n,\infty}_{\text{comp}}(\mathbb{R}^2)$ implies that $h_0(\varsigma,k)$ is a bounded continuous function on $\mathbb{C}\times(\mathbb{C}\setminus D)$ and

$$\varsigma_1^{\beta_1}\varsigma_2^{\beta_2}\Pi^o h_0(\varsigma,\varsigma)\in C_0(\mathbb{C}\setminus D),\ |\beta|\le n,\ \varsigma=(\varsigma_1,\varsigma_2), \tag{3.5}$$

where $C_0(\mathbb{C}\setminus D)$ is the space of continuous matrix-functions with zero limit at infinity. In fact, this inclusion is proved (see [**21**, Lemma 2.16]) when $\mathcal{E}=\emptyset$ and $D=\emptyset$, but the proof remains the same in the presence of exceptional points if $k\in\mathbb{C}\setminus D$.

LEMMA 3.3. *Let (3.2) hold. Then*

1) $T_zI\in\mathcal{H}^s$ for each $s>2$.

2) Let z be such that the kernel of $I+T_z$ is trivial. Let $v=v(z,\cdot)$ be the unique solution of

$$(I+T_z)v=I, \tag{3.6}$$

such that $w=v-I$ satisfies $(I+T_z)w=T_zI\in\mathcal{H}^s$ and belongs to H^s.

Let

(3.7)

$$Q := \frac{1}{2i}\Pi^o \mathcal{C}v, \text{ where}$$

$$\mathcal{C}v := \frac{1}{\pi}\int_{\mathbb{C}\setminus D} e^{i(\bar\varsigma z+\bar z\varsigma)/2}\overline{v}(\varsigma)\Pi^o h(\varsigma,\varsigma)d\varsigma_R d\varsigma_I +$$

$$\frac{1}{2\pi i}\int_{\partial D} d\varsigma \int_{\partial D}\left[e^{i(\varsigma\bar z+\bar\varsigma' z)/2}\overline{v^-(\varsigma')}\Pi^o + e^{i(\varsigma-\varsigma')\bar z/2}v^-(\varsigma')\Pi^d \mathbf{C}\right]\left[w(\varsigma',\varsigma)h(\varsigma',\varsigma)d\overline{\varsigma'}\right].$$

Then

$$\bar\partial\psi = Q\overline{\psi}, \quad \psi := \Pi^d\overline{v}e^{i\bar k z/2} + e^{-i\bar z k/2}\Pi^o v, \tag{3.8}$$

and

$$\partial\Phi = \bar\partial(\overline{Q}Q), \text{ where } \Phi = \frac{1}{2i}\bar\partial\Pi^d\mathcal{C}v. \tag{3.9}$$

REMARK. 1) Note that

$$\mathcal{C}v = \lim_{k\to\infty} k(v-I), \quad Q = \lim_{k\to\infty}\frac{-ik}{2}\Pi^o v. \tag{3.10}$$

Relations (3.10) can be easily checked if one replaces $v-I$ and $\Pi^o v$ from (3.6) by $-T_z v$ and $-\Pi^o T_z v$, respectively.

2) The statements and the proof remain the same if the space L^1 in (3.2) is replaced by $L^{\frac{2s}{2+s}}$.

PROOF. The first statement is proved in [**14**, Lemmas 4.2]. Lemma 3.1 above implies that the operator T_z is compact in $\mathcal{H}^s$, and therefore $I+T_z$ is Fredholm. Both properties (the validity of the first statement of the lemma and the compactness of T_z) are valid without the factor $|\varsigma|^3$ at h in (3.2). The presence of this factor is needed to guarantee that $\partial_z^\alpha\overline{\partial}_z^\beta T_z, \alpha+\beta\le 3$, is compact in $\mathcal{H}^s$ and $\partial_z^\alpha\overline{\partial}_z^\beta T_z I\in\mathcal{H}^s$ for each $s>2$.

The operator T_z can be naturally split into two terms: $T_z = \mathcal{M}+\mathcal{D}$, where $\mathcal{M}$ involves integration over $\mathbb{C}\setminus D$ and $\mathcal{D}$ involves integration over ∂D. Thus, $T_z\phi = \mathcal{M}\overline{\phi}+\Pi^o\mathcal{D}\overline{\phi}+\Pi^d\mathcal{D}\phi$. The entries $(M^{ij})(z,k)$, $(D^{ij})(z,k)$, $i,j=1,2$, of the matrix operators $\mathcal{M}$ and $\mathcal{D}$ are

$$M^{12}\phi = \frac{1}{\pi}\int_{\mathbb{C}\setminus D}\frac{e^{i\Re(\overline{\alpha}z)}\phi(\alpha)h_{12}(\alpha,\alpha)}{\alpha-k}d\alpha_{\Re}d\alpha_{\Im},$$

$$M^{21}\phi = \frac{1}{\pi}\int_{\mathbb{C}\setminus D}\frac{e^{i\Re(\overline{\alpha}z)}\phi(\alpha)h_{21}(\alpha,\alpha)}{\alpha-k}d\alpha_{\Re}d\alpha_{\Im}, \tag{3.11}$$

$$D^{jj}\phi = \frac{1}{2\pi i}\int_{\partial D}\frac{d\zeta}{\zeta-k}\int_{\partial D}\overline{w(\varsigma,\varsigma')h_{jj}(\varsigma',\varsigma)e^{\frac{i}{2}(\varsigma-\varsigma')\bar z}}\phi(\varsigma')d\varsigma', \quad j=1,2,$$

$$D^{ij}\phi = \frac{1}{2\pi i}\int_{\partial D}\frac{d\zeta}{\zeta-k}\int_{\partial D}w(\varsigma,\varsigma')h_{ij}(\varsigma',\varsigma)e^{\frac{i}{2}(\varsigma\bar z+\bar\varsigma' z)}\phi(\varsigma')d\overline{\varsigma'}, \quad i\ne j. \tag{3.12}$$

We rewrite the matrix equation (3.6) as four equations for its components. The first row of the matrix equation is equivalent to

$$(I+D^{11})v_{11} + (M^{21}+D^{21})\overline{v_{12}} = 1, \tag{3.13}$$

$$(I+D^{22})v_{12} + (M^{12}+D^{12})\overline{v_{11}} = 0. \tag{3.14}$$

The second row of (3.6) leads to similar equations for v_{22}, v_{21}. This system of equations can be rewritten as the following four independent equations:

$$(I + D^{11})v_{11} - (M^{21} + D^{21})(I + \overline{D^{22}})^{-1}(\overline{M^{12}} + \overline{D^{12}})v_{11} = 1, \tag{3.15}$$

$$(I + \overline{D^{22}})\overline{v_{12}} - (\overline{M^{12}} + \overline{D^{12}})(I + D^{11})^{-1}[(M^{21} + D^{21})\overline{v_{12}} - 1] = 0, \tag{3.16}$$

$$(I + D^{22})v_{22} - (M^{12} + D^{12})(I + \overline{D^{11}})^{-1}(\overline{M^{21}} + \overline{D^{21}})v_{22} = 1, \tag{3.17}$$

$$(I + \overline{D^{11}})\overline{v_{21}} - (\overline{M^{21}} + \overline{D^{21}})(I + D^{22})^{-1}[(M^{12} + D^{12})\overline{v_{21}} - 1] = 0, \tag{3.18}$$

if operators $(I + \overline{D^{ii}})$ are invertible. We will prove the lemma under this additional assumption on invertibility of $(I + \overline{D^{ii}})$, and we will get rid of this assumption at the very end of the proof.

By straightforward calculation, we get the following formulas for the derivatives of the operators M^{ij}, D^{ij}:

$$\frac{2}{i}\frac{\partial}{\partial z}(M^{ij} + D^{ij}) = (M^{ij} + D^{ij})\overline{X}, \quad i \neq j, \tag{3.19}$$

$$\frac{2}{i}\frac{\partial}{\partial z}(\overline{M^{ij}} + \overline{D^{ij}}) = C_{ij} - \overline{X}(\overline{M^{ij}} + \overline{D^{ij}}), \quad i \neq j, \tag{3.20}$$

$$\frac{\partial}{\partial z}D^{jj} = 0, \tag{3.21}$$

$$\frac{2}{i}\frac{\partial}{\partial z}\overline{D^{jj}} = C_j - \overline{X}\,\overline{D^{jj}} + \overline{D^{jj}}\,\overline{X}, \tag{3.22}$$

where X is the operator of multiplication by the independent variable k, and C_{ij}, C_j are integral functionals that are closely related to the entries of the matrix $\mathcal{C}$ introduced in (3.7):

$$\begin{aligned} C_{ij}\phi &= -\frac{1}{\pi}\int_{\mathbb{C}\setminus D} e^{-i\Re(\overline{\alpha}z)}\overline{h_{ij}}(\alpha,\alpha)\phi(\alpha)d\alpha_{\Re}d\alpha_{\Im} \\ &+ \frac{1}{2\pi i}\int_{\partial D}\int_{\partial D} d\varsigma' d\varsigma \overline{w(\varsigma,\varsigma')h_{ij}(\varsigma',\varsigma)}e^{\frac{-i}{2}(\overline{\varsigma}z+\varsigma'\overline{z})}\phi(\varsigma'), \end{aligned}$$

$$C_j\phi = \frac{1}{2\pi i}\int_{\partial D}\int_{\partial D} d\overline{\varsigma'}d\overline{\varsigma} w(\varsigma,\varsigma')h_{jj}(\varsigma',\varsigma)e^{\frac{-i}{2}(\overline{\varsigma}-\overline{\varsigma'})z}\phi(\varsigma'). \tag{3.23}$$

The following relation is an immediate consequence of (3.22):

$$\frac{2}{i}\frac{\partial}{\partial z}(I + \overline{D^{22}})^{-1} = -(I + \overline{D^{22}})^{-1}C_2(I + \overline{D^{22}})^{-1} + (I + \overline{D^{22}})^{-1}\overline{X} - \overline{X}(I + \overline{D^{22}})^{-1}. \tag{3.24}$$

Lets us differentiate (3.15) using formulas (3.19)-(3.22). We get

$$\begin{aligned} &[(I + D^{11}) - (M^{21} + D^{21})(I + \overline{D^{22}})^{-1}(\overline{M^{12}} + \overline{D^{12}})]\frac{\partial v_{11}}{\partial z} = \\ &\frac{i}{2}(M^{21} + D^{21})(I + \overline{D^{22}})^{-1}[-C_2(I + \overline{D^{22}})^{-1}(\overline{M^{12}} + \overline{D^{12}})v_{11} + C_{12}v_{11}] = \\ &\frac{i}{2}(M^{21} + D^{21})(I + \overline{D^{22}})^{-1}[C_2\overline{v_{12}} + C_{12}v_{11}]. \end{aligned} \tag{3.25}$$

Note, that $c_0 := \frac{-i}{2}[C_2\overline{v_{12}} + C_{12}v_{11}]$ does not depend on k. Let us replace $\frac{\partial v_{11}}{\partial z}$ in equation (3.25) by v_{21} and omit the constant factor c_0 in its right-hand side. The resulting equation coincides with the equation obtained from (3.18) by complex conjugation. Thus, due to the assumption on the triviality of the kernel of $I+T_z$, we get that

$$\frac{\partial v_{11}}{\partial z}=\overline{\mathcal{Q}_{12}}v_{21},\quad\text{where } \mathcal{Q}_{12}=\overline{c_0}=\frac{i}{2}\overline{[C_2\overline{v_{12}}+C_{12}v_{11}]}. \tag{3.26}$$

It is easy to see that this formula for Q_{12} coincides with the one given in (3.7).

The analysis for v_{12} is similar. We will differentiate (3.16) by taking the derivatives of each term separately:

$$\begin{aligned}\frac{2}{i}\frac{\partial}{\partial z}(I+\overline{D^{22}})\overline{v_{12}}&=\frac{2}{i}(I+\overline{D^{22}})\frac{\partial\overline{v_{12}}}{\partial z}+C_2\overline{v_{12}}-\overline{X}\overline{D^{22}}\overline{v_{12}}+\overline{D^{22}}\overline{X}\overline{v_{12}}=\\ &(I+\overline{D^{22}})\left(\frac{2}{i}\frac{\partial\overline{v_{12}}}{\partial z}+\overline{X}\overline{v_{12}}\right)+C_2\overline{v_{12}}-\overline{X}(I+\overline{D^{22}})\overline{v_{12}}.\end{aligned}\tag{3.27}$$

$$\begin{aligned}&\frac{2}{i}\frac{\partial}{\partial z}\left[(\overline{M^{12}}+\overline{D^{12}})(I+D^{11})^{-1}\left[(M^{21}+D^{21})\overline{v_{12}}-1\right]\right]=\\ &\frac{2}{i}(\overline{M^{12}}+\overline{D^{12}})(I+D^{11})^{-1}(M^{21}+D^{21})\frac{\partial\overline{v_{12}}}{\partial z}-C_{12}v_{11}+\overline{X}(\overline{M^{12}}+\overline{D^{12}})v_{11}\\ &+(\overline{M^{12}}+\overline{D^{12}})(I+D^{11})^{-1}(M^{21}+D^{21})\overline{X}\overline{v_{12}}.\end{aligned}\tag{3.28}$$

Here we used the following consequence of (3.13):

$$v_{11}=-(I+D^{11})^{-1}\left[(M^{21}+D^{21})\overline{v_{12}}-1\right].$$

We multiply relations (3.27), (3.28) by $i/2$ and equate their right-hand sides (due to (3.16)). If we note that the last term in the right-hand side of (3.27) coincides with the term $\overline{X}(\overline{M^{12}}+\overline{D^{12}})v_{11}$ in (3.28) (due to (3.14)), then we arrive at the following equation

$$[(I+\overline{D^{22}})+(\overline{M^{12}}+\overline{D^{12}})(I+D^{11})^{-1}(M^{21}+D^{21})]\left(\frac{\partial\overline{v_{12}}}{\partial z}+\frac{i}{2}\overline{X}\overline{v_{12}}\right)=c_0,$$

where $c_0=\frac{-i}{2}[C_2\overline{v_{12}}+C_{12}v_{11}]$ does not depend on k (see (3.25)). Let us replace $\frac{\partial\overline{v_{12}}}{\partial z}+\frac{i}{2}\overline{X}\overline{v_{12}}$ in the latter equation by $\overline{v_{22}}$. If we also divide its right-hand side by c_0, then the resulting equation will coincide with the equation obtained from (3.17) by complex conjugation. Hence,

$$\left(\frac{\partial\overline{v_{12}}}{\partial z}+\frac{i}{2}\overline{X}\overline{v_{12}}\right)=\overline{\mathcal{Q}_{12}}\overline{v_{22}},\quad\text{where } \mathcal{Q}_{12}=\overline{c_0}=\frac{i}{2}\overline{[C_2\overline{v_{12}}+C_{12}v_{11}]}. \tag{3.29}$$

Similarly, we show that

$$\frac{\partial v_{22}}{\partial z}=\overline{\mathcal{Q}_{21}}v_{12},\quad\text{where}\quad \mathcal{Q}_{21}=\lim_{k\to\infty}\frac{-ik}{2}v_{21}, \tag{3.30}$$

and finally

$$\left(\frac{\partial\overline{v_{21}}}{\partial z}+\frac{i\overline{X}}{2}\overline{v_{21}}\right)=\overline{\mathcal{Q}_{21}}\overline{v_{11}}. \tag{3.31}$$

In order to complete the proof of (3.8), it remains only to note that equations (3.26), (3.29), (3.31), and (3.30) for v can be rewritten in the form $\overline{\partial}\psi=Q\overline{\psi}$ using the relation between v and ψ provided in the statement of the lemma.

Let us prove (3.9). From (3.26) and (3.30), it follows that

$$\partial\Pi^d(v-1)=\Pi^d(\overline{Q}v). \tag{3.32}$$

We multiply both sides of the above equation by k and pass to the limit as $k \to \infty$ using (3.10). This implies

$$\partial \Pi^d \mathcal{C} v = 2i\overline{Q}Q. \tag{3.33}$$

Applying $\overline{\partial}$ and using $\partial\overline{\partial} = \overline{\partial}\partial$, we get

$$\partial\left[\frac{1}{2i}\overline{\partial}\Pi^d\mathcal{C}v\right] = \overline{\partial}(\overline{Q}Q).$$

This completes the proof of the lemma under the condition that operators $(I + \overline{D^{ii}})$ are invertible.

In order to prove Lemma 3.3 in the general case, we consider the scattering data γh instead of h, where $\gamma \in [0,1]$. Then operators T_z and $\overline{D^{ii}}$ are analytic in γ and vanish when $\gamma = 0$. In order to prove the compactness of T_z in [**14**, Lemma 4.3], we established the compactness of its components $\mathcal{M}$ and $\mathcal{D}$, i.e., $\overline{D^{ii}}$ were shown to be compact. Due to the analytic Fredholm theorem, operators $(I + \overline{D^{ii}})$ are invertible when γ is close enough to one, $\gamma \neq 1$. Operator $I + T_z$ is invertible when $1 - \gamma \ll 1$. Hence, the relations (3.8), (3.9) hold when $0 < 1 - \gamma \ll 1$. Since all the components of equalities (3.8), (3.9) are analytic in γ, these equalities holds for $\gamma = 1$. □

4. Derivation of the compatibility condition

A symmetry of the potential Q and scattering data h will play an important role in this section. So, we start the section with two simple lemmas establishing the relation between those symmetries.

LEMMA 4.1. *Let (3.2) hold and*

$$h_{11}(\varsigma, k) = h_{22}(\varsigma, k), \quad h_{12}(\varsigma, k) = \pm h_{21}(\varsigma, k), \tag{4.1}$$

for all the pairs $(\varsigma, k) = (k, k)$, $k \in \mathbb{C}\backslash D$ or $(\varsigma, k) \in \partial D \times \partial D$. Then for all $z \in \mathbb{R}^2$ such that the kernel of $(I + T_z)$ is trivial, the following symmetry relations are valid for the matrix Q defined in (3.7) and solution v of (3.6):

$$Q_{12}(z) = \pm Q_{21}(z), \tag{4.2}$$

$$v_{11} = v_{22}, \quad v_{12} = \pm v_{21}. \tag{4.3}$$

(The converse statement is given in Lemma 4.2.)

PROOF. It is enough to prove (4.3). Then (4.2) follows from (3.10). Note that (3.11) and (3.12) imply that

$$D^{11} = D^{22}, \quad D^{12} = \pm D^{21}, \quad M^{12} = \pm M^{21}. \tag{4.4}$$

We rewrite equations (3.16), (3.18) in the form

$$[(I + \overline{D^{22}}) - (\overline{M}^{12} + \overline{D}^{12})(I + D^{11})^{-1}(M^{21} + D^{21})]\overline{v_{12}} = -(\overline{M}^{12} + \overline{D}^{12})(I + D^{11})^{-1}1,$$

$$[(I + \overline{D^{11}}) - (\overline{M}^{21} + \overline{D}^{21})(I + D^{22})^{-1}(M^{12} + D^{12})]\overline{v_{21}} = -(\overline{M}^{21} + \overline{D}^{21})(I + D^{22})^{-1}1.$$

From (4.4) it follows that the coefficients for v_{12}, v_{21} in these equations are equal to each other, and the right-hand sides differ by the factor ± 1. This justifies the second relation in (4.3). The first relation can be proved similarly using (3.15), (3.17). □

LEMMA 4.2. *Let $Q_{12} = \pm Q_{21} \in L^p_{\text{comp}}(\mathbb{R}^2)$, $p > 2$, be compactly supported functions. Let $h = h_0$ be the scattering data defined in (2.7). Then*

$$h_{11} = h_{22}, \quad h_{12} = \pm h_{21}$$

for $k \notin \mathcal{E}$, where $\mathcal{E}$ is the set of exceptional points.

PROOF. Let

$$\mathcal{L}_k\varphi(z) = \frac{1}{\pi}\int_{\mathbb{R}^2}\varphi(w)\frac{e^{-i\Re(\overline{k}w)}dw_R dw_I}{z-w}. \tag{4.5}$$

Then the Lippmann-Schwinger equation (2.3) after substitution (2.4) can be rewritten in the following form (see also [**21**, (1.6)]):

$$\mu_{11} = 1 + \mathcal{L}_k[Q_{12}(z')\overline{\mu_{21}}(z',k)], \quad \mu_{21} = \mathcal{L}_k[Q_{21}(z')\overline{\mu_{11}}(z',k)],$$
$$\mu_{22} = 1 + \mathcal{L}_k[Q_{21}(z')\overline{\mu_{12}}(z',k)], \quad \mu_{12} = \mathcal{L}_k[Q_{12}(z')\overline{\mu_{22}}(z',k)].$$

Therefore,

$$\mu_{11} = 1 + \mathcal{L}_k[Q_{12}(z')\overline{\mathcal{L}_k}[\overline{Q_{21}}(z')\mu_{11}(z',k)]],$$
$$\mu_{21} = \mathcal{L}_k\left[Q_{21}(z')\left(1 + \overline{\mathcal{L}_k}[\overline{Q_{12}}(z')\mu_{21}(z',k)]\right)\right],$$

and

$$\mu_{22} = 1 + \mathcal{L}_k[Q_{21}(z')\overline{\mathcal{L}_k}[\overline{Q_{12}}(z')\mu_{22}(z',k)]],$$
$$\mu_{12} = \mathcal{L}_k\left[Q_{12}(z')\left(1 + \overline{\mathcal{L}_k}[\overline{Q_{21}}(z')\mu_{12}(z',k)]\right)\right].$$

So, we obtain

$$\mu_{11} = \mu_{22}, \quad \mu_{12} = \pm\mu_{21}.$$

Now Lemma 4.2 follows from (2.7). □

Consider an arbitrary time-independent matrix h that satisfies (3.2) and the symmetry condition (4.1). Using this matrix h, we define the following time dependent data

$$h(\varsigma',\varsigma,t) := e^{-t(\varsigma^2-\overline{\varsigma'}^2)/2}\Pi^o h(\varsigma',\varsigma) + e^{-t(\overline{\varsigma}^2-\overline{\varsigma'}^2)/2}\Pi^d h(\varsigma',\varsigma), \ \varsigma \in \mathbb{C}\backslash\mathcal{E}, \ t \geq 0, \tag{4.6}$$

and then apply Lemma 3.3 to construct, for each $t \geq 0$, the solution $v = v(z,k,t)$ of equation (3.6) and the matrix $Q = Q^t(z)$ defined in terms of $\mathcal{C}$ in (3.7). The symmetry condition (4.1) implies that the same symmetry condition holds for function (4.6) for each $t > 0$, and therefore, from Lemma 4.1 it follows that

$$Q^t_{12}(z) = \pm Q^t_{21}(z) =: q(z,t), \tag{4.7}$$

and that (4.3) holds. We denote by φ the diagonal entries of the matrix $\Phi = \frac{1}{2i}\overline{\partial}\Pi^d\mathcal{C}v$ defined in (3.9). These entries are equal due to (4.3) and (3.10). The proof of the latter statement requires changing the order of the operator $\overline{\partial}$ and the limit in (3.10). One can give an elementary alternative proof using the definition (3.7) of matrix $\mathcal{C}$ and the symmetries of h and v. Thus (3.9) takes the form

$$\partial\varphi = \overline{\partial}|q|^2. \tag{4.8}$$

We will show that v satisfies the compatibility conditions (see [**21**, (1.26)], or Section 5):

$$\frac{\partial v}{\partial t} + 2\left[\frac{\partial^2 v}{\partial\overline{z}^2} + \frac{\partial^2 v}{\partial z^2}\right] - 2ik\frac{\partial v}{\partial\overline{z}} + A(\Pi^d v) + \overline{A}(\Pi^o v) = 0, \tag{4.9}$$

where the entries of the matrix $A = A(z,t)$ are

$$A_{12} = \pm A_{21} = -4\overline{\partial}q, \tag{4.10}$$

$$A_{11} = A_{22} = \mp 4\varphi. \tag{4.11}$$

LEMMA 4.3. *Suppose that the matrix h have properties (3.2) and (4.1), and let $T = T_{z,t}$ be the operator (3.3) with h defined in (4.6). Then (4.9) holds in the classical sense at each (z,t) for which the kernel of $I + T_{z,t}$ is trivial.*

REMARK. We will prove later that the vector (q,φ), where q is defined by (4.7) and φ is defined immediately after (4.7), satisfies the DSII equation.

PROOF. As earlier, we split operator (3.3) as follows: $T = \mathcal{M} + \mathcal{D}^o + \mathcal{D}^d$, where $\mathcal{M}$ involves integration over $\mathbb{C}\backslash D$, and $\mathcal{D} = \mathcal{D}^o + \mathcal{D}^d$ involves integration over ∂D. Similarly, we split $\mathcal{C}v$ in three natural terms

$$\mathcal{C}v = (\mathcal{C}^1 + \mathcal{C}^o + \mathcal{C}^d)v = \frac{1}{\pi}\int_{\mathbb{C}\backslash D} e^{i(\overline{\varsigma}z + \overline{z}\varsigma)/2}\overline{v}(\varsigma)\Pi^o h(\varsigma,\varsigma,t)d\varsigma_R d\varsigma_I + \tag{4.12}$$

$$\frac{1}{2\pi i}\int_{\partial D} d\varsigma \int_{\partial D} [e^{i/2(\varsigma\overline{z} + \overline{\varsigma'}z)}\overline{v^-(\varsigma')}\Pi^o + e^{i/2(\varsigma - \varsigma')\overline{z}}v^-(\varsigma')\Pi^d\mathbf{C}]\left[\mathrm{Ln}\frac{\overline{\varsigma'} - \overline{\varsigma}}{\overline{\varsigma'} - \overline{k_0}}h(\varsigma',\varsigma,t)d\varsigma'\right].$$

WARNING. We are using the standard notation $T\phi$ for the action of an operator T on a matrix ϕ. However, one must keep in mind that the products of matrices in the integrands in (3.3) are taken in an unusual order with the factor ϕ being on the left.

We will use the fact that

$$\begin{aligned}\Pi^d[(I+T)^{-1}(iI)] &= i[\Pi^d(I+T)^{-1}I] \quad \text{and}\\ \Pi^o[(I+T)^{-1}(iI)] &= -i\Pi^o[(I+T)^{-1}I].\end{aligned} \tag{4.13}$$

One can justify (4.13) by checking that the elements of the matrix $i\Pi^d v - i\Pi^o v$ satisfy (3.15)-(3.18) if the latter equations are multiplied by i.

Since $(I+T)v = I$ and matrix A does not depend on k, it follows from (4.13) and the warning above that

$$(I+T)(A(\Pi^d v) + \overline{A}(\Pi^o v)) = A.$$

Since the kernel of the operator $I+T$ is trivial, one can check that $I+T$ applied to the left-hand side of (4.9) is zero, instead of proving (4.9). Thus, in order to prove Lemma 4.3, it is enough to show that

$$(I+T)(v_t + 2Lv - 2ikv_{\overline{z}}) + A = 0, \quad Lv := v_{\overline{z}\overline{z}} + v_{zz} = \frac{1}{2}(v_{xx} - v_{yy}). \tag{4.14}$$

Applying operators $\frac{\partial}{\partial t}$ and L to the relation $(I+T)v = I$, we obtain that

$$(I+T)v_t = -T_t v, \quad (I+T)Lv = -(LT)v - (T_x v_x - T_y v_y).$$

This allows us to rewrite (4.14) in the form

$$-(T_t + 2T_{\overline{z}\overline{z}} + 2T_{zz})v - 2(T_x v_x - T_y v_y) - (I+T)(2ikv_{\overline{z}}) + A = 0. \tag{4.15}$$

We split the proof of (4.15) into several steps. Note that by a straightforward calculation (using (3.3) and formula (3.7) for $\mathcal{C}$), one can verify that

$$(T_t + 2T_{\overline{z}\overline{z}} + 2T_{zz} - 2ikT_{\overline{z}})v = 2i\mathcal{C}_{\overline{z}}v. \tag{4.16}$$

Recall that T is not linear with respect to multiplication by i, so the term $(I + T)(2ikv_z)$ in (4.15) needs an accurate treatment. We will show that

$$-2(T_x v_x - T_y v_y) - (I+T)(2ikv_{\overline{z}}) = 2ikT_{\overline{z}}v - 2i(\mathcal{C}^1 v_z + \mathcal{C}^o v_z + \mathcal{C}^d v_{\overline{z}}). \tag{4.17}$$

It will be also shown that

$$A = 2i\overline{\partial}(\mathcal{C}v). \tag{4.18}$$

Since both functions v and $\overline{v}$ are present in integrands in the right-hand side of (4.12), formula (4.18) is equivalent to $A = 2i(\mathcal{C}_{\overline{z}}v + \mathcal{C}^1 v_z + \mathcal{C}^o v_z + \mathcal{C}^d v_{\overline{z}})$. Hence, (4.16)-(4.18) justify (4.15), and therefore Lemma 4.3 will be proved as soon as (4.17), (4.18) are established.

The equality of the diagonal terms in (4.18) follows from (4.11) and the definition of φ. In order to verify the validity of (4.18) for the non-diagonal elements, we substitute the left-hand side of (4.7) for q in (4.10) and use the relation $2i\Pi^o(\mathcal{C}v) = -4\Pi^o Q$, which follows from (3.10). Thus (4.18) is proved, and it remains only to prove (4.17).

We apply operator $\overline{\partial}$ to (3.6) and then multiply the resulting equation by $-2ik$:

$$-2ikv_{\overline{z}} - 2ik((\mathcal{M} + \mathcal{D}^o)v_z + \mathcal{D}^d v_{\overline{z}}) = 2ik(\mathcal{M}_{\overline{z}}v + \mathcal{D}^o_{\overline{z}}v + \mathcal{D}^d_{\overline{z}}v). \tag{4.19}$$

Now we will rearrange each term in (4.19) separately, keeping in mind that k is an independent variable, and multiplication by k does not commute with the operators above. The rearrangement will involve an additional term $T_x v_x - T_y v_y$.

Operator $\mathcal{M}$. . We will use the fact that $i\mathcal{M}(\cdot) = \mathcal{M}(-i\cdot)$. We have

$$\mathcal{M}(2i\overline{k}v_z) = \mathcal{M}((i\Re k + \Im k)(v_x - iv_y)) = \mathcal{M}(i(\Re k v_x - \Im k v_y) + (\Im k v_x + \Re k v_y)), \tag{4.20}$$

$$\mathcal{M}(2ikv_{\overline{z}}) = \mathcal{M}(i(\Re k v_x - \Im k v_y) - (\Im k v_x + \Re k v_y)), \tag{4.21}$$

$$-\mathcal{M}_x v_x + \mathcal{M}_y v_y = \mathcal{M}(i(\Re k v_x - \Im k v_y)). \tag{4.22}$$

From (4.20)-(4.22) it follows that

$$2[-\mathcal{M}_x v_x + \mathcal{M}_y v_y] - \mathcal{M}(2ikv_{\overline{z}}) = \mathcal{M}(2i\overline{k}v_z) = -2ik\mathcal{M}v_z - 2i\mathcal{C}^1 v_z. \tag{4.23}$$

Operator $\mathcal{D}^o$. . Similar formulas hold for $\mathcal{D}^o$. We will use the fact that $i\mathcal{D}^o(\cdot) = \mathcal{D}^o(-i\cdot)$. The exponent in the term of (3.3) that corresponds to $\mathcal{D}^o$ can be rewritten as follows:

$$\frac{i}{2}(\varsigma\overline{z} + \overline{\varsigma'}z) = ix\frac{\varsigma + \overline{\varsigma'}}{2} + y\frac{\varsigma - \overline{\varsigma'}}{2}.$$

One can check that

$$2[-\mathcal{D}^o_x v_x + \mathcal{D}^o_y v_y] = -ik\mathcal{D}^o v_x + \mathcal{D}^o(ikv_x) - \mathcal{D}^o(kv_y) + k\mathcal{D}^o v_y + [-i\mathcal{C}^o v_x + \mathcal{C}^o v_y].$$

Note that

$$\mathcal{D}^o(2ikv_{\overline{z}}) = \mathcal{D}^o(ikv_x - kv_y).$$

From the last two equalities it follows that

$$\begin{aligned} 2[-\mathcal{D}^o_x v_x + \mathcal{D}^o_y v_y] - \mathcal{D}^o(2ikv_{\overline{z}}) &= [-i\mathcal{C}^o v_x + \mathcal{C}^o v_y] - ik\mathcal{D}^o v_x + k\mathcal{D}^o v_y \\ = -i[\mathcal{C}^o v_x + i\mathcal{C}^o v_y)] - ik(\mathcal{D}^o v_x + i\mathcal{D}^o v_y) &= -i\mathcal{C}^o(v_x - iv_y) - ik\mathcal{D}^o(v_x - iv_y) \\ &= -2i\mathcal{C}^o v_z - 2ik\mathcal{D}^o v_z. \end{aligned} \tag{4.24}$$

Operator $\mathcal{D}^d$. Let us rearrange the exponent in the term of (3.3) that corresponds to $\mathcal{D}^d$:

$$\frac{i}{2}(\varsigma-\varsigma')\overline{z} = ix\frac{\varsigma-\varsigma'}{2} + y\frac{\varsigma-\varsigma'}{2}.$$

Then

$$2[-\mathcal{D}^d_x v_x + \mathcal{D}^d_y v_y] = -ik\mathcal{D}^d v_x + \mathcal{D}^d(ikv_x) - \mathcal{D}^d(kv_y) + k\mathcal{D}^d v_y - [i\mathcal{C}^d v_x - \mathcal{C}^d v_y].$$

Note also that

$$\mathcal{D}^d(2ikv_{\overline{z}}) = \mathcal{D}^d(ikv_x - kv_y).$$

From the last two equalities we get

$$\begin{aligned} 2[-\mathcal{D}^d_x v_x + \mathcal{D}^d_y v_y] - \mathcal{D}^d(2ikv_{\overline{z}}) &= [-i\mathcal{C}^d v_x + \mathcal{C}^d v_y] - ik\mathcal{D}^d v_x + k\mathcal{D}^d v_y \\ = -i(\mathcal{C}^d v_x + i\mathcal{C}^d v_y) - ik(\mathcal{D}^d v_x + i\mathcal{D}^d v_y) &= -i\mathcal{C}^d(v_x + iv_y) - ik\mathcal{D}^d(v_x + iv_y)) \\ &= -2i\mathcal{C}^d v_{\overline{z}} - 2ik\mathcal{D}^d v_{\overline{z}}. \end{aligned} \tag{4.25}$$

Now (4.17) follows from (4.19) combined with (4.23), (4.24), and (4.25). □

5. Proof of Theorem 2.1

We start with a theorem showing that the compatibility condition for the solution v of (2.11) implies that the functions q, φ defined by v satisfy the first two relations of the DSII system (1.2):

$$q_t = 2(\overline{\partial}^2 - \partial^2)q \pm 4q(\overline{\varphi} - \varphi), \quad \partial\varphi = \overline{\partial}|q|^2. \tag{5.1}$$

THEOREM 5.1. *Let h be an arbitrary scattering data that depends on $t \geq 0$ and, for each t, satisfies (3.2) and the symmetry relations (4.1). Let the kernel of operator $I + T_{z,t}$ be trivial in a neighborhood ω of a point (z_0, t_0), $z_0 \in \mathbb{C}$, $t_0 \geq 0$. If the solution $v = v_{z,t}$ of equation (2.11) satisfies the compatibility condition (4.9) in ω, then the function q defined in (4.7) and the diagonal elements φ of the matrix Φ satisfy (5.1) in ω.*

PROOF. One needs to prove only the first relation in (5.1), since the second one was proved in (4.8).

We take the complex conjugate of (3.31), differentiate it in t, and replace the derivatives of v_{ij} using (4.9). We get

$$\begin{aligned} &(\overline{\partial} - \frac{ik}{2})\left(-2(\partial^2 + \overline{\partial^2})v_{21} + 2ik\overline{\partial}v_{21} \pm 4\overline{\partial}qv_{11} \pm 4\overline{\varphi}v_{21}\right) \\ &= \pm q_t v_{11} \pm q\left(-2(\partial^2 + \overline{\partial^2})v_{11} + 2ik\overline{\partial}v_{11} \pm 4\varphi v_{11} + 4\partial\overline{q}v_{21}\right), \quad k \in \mathcal{C} \setminus D. \end{aligned}$$

Hence, the theorem will be proved if we show that

$$\begin{aligned} (\overline{\partial} - \frac{ik}{2})&\left(-2(\partial^2 + \overline{\partial^2})v_{21} + 2ik\overline{\partial}v_{21} \pm 4\overline{\partial}qv_{11} \pm 4\overline{\varphi}v_{21}\right) \\ &= \pm\left(2(\overline{\partial}^2 - \partial^2)q \pm 4q(\overline{\varphi} - \varphi)\right)v_{11} \\ &\pm q\left(-2(\partial^2 + \overline{\partial^2})v_{11} + 2ik\overline{\partial}v_{11} \pm 4\varphi v_{11} + 4\partial\overline{q}v_{21}\right), \quad k \in \mathcal{C} \setminus D. \end{aligned} \tag{5.2}$$

Indeed, the difference between the last two equations is equal to the difference between the left and right hand sides in the first of equations (5.1) multiplied by $\pm v_{11}$. Thus (5.2) implies that either (5.1) holds or $v_{11} = 0$. We note that the terms in (5.1) do not depend on k, and $v_{11} \to 1$ as $|k| \to \infty$. Thus the validity of (5.2) implies (5.1).

We apply the operator $\overline{\partial} - \frac{ik}{2}$ to all the terms in the left-hand side of (5.2) and then open all the brackets except the ones in the expression $(\overline{\partial} - \frac{ik}{2})$. Then the left-hand side will have five terms, which will be denoted by A^i, and the second and third lines in (5.2) will have four terms $B^i, 1 \le i \le 4$, and five terms $C^i, 1 \le i \le 5$, respectively. Thus we need to show that

$$\sum_{i=1}^{5} A^i = \sum_{i=1}^{4} B^i + \sum_{i=1}^{5} C^i. \tag{5.3}$$

We apply the complex conjugation to equation (3.31). Using the resulting equation, we obtain

$$A^1 = -2\partial^2(\pm q v_{11}) = \mp 2(\partial^2 q)v_{11} \mp 4(\partial q)\partial v_{11} \mp 2q(\partial^2 v_{11}) =: \sum_{i=1}^{3} A^1_i.$$

$$A^2 = -2\overline{\partial}^2(\pm q v_{11}) = \mp 2(\overline{\partial}^2 q)v_{11} \mp 4(\overline{\partial} q)\overline{\partial} v_{11} \mp 2q(\overline{\partial}^2 v_{11}) =: \sum_{i=1}^{3} A^2_i.$$

$$A^3 = 2ik\overline{\partial}(\pm q v_{11}) = \pm 2ik(\overline{\partial} q)v_{11} \pm 2ikq(\overline{\partial} v_{11}) =: A^3_1 + A^3_2.$$

$$A^4 = \pm 4(\overline{\partial} - \frac{ik}{2})(\overline{\partial} q v_{11}) = \pm 4((\overline{\partial} - \frac{ik}{2})\overline{\partial} q)v_{11} \pm 4(\overline{\partial} q)(\overline{\partial} v_{11}) =: A^4_1 + A^4_2.$$

$$A^4_1 = \pm 4(\overline{\partial}^2 q)v_{11} \mp 2ik\overline{\partial} q v_{11} =: A^4_{1,1} + A^4_{1,2}.$$

$$A^5 = \pm 4(\overline{\partial} - \frac{ik}{2})(\overline{\varphi} v_{21}) = \pm 4(\overline{\partial}\overline{\varphi})v_{21} + 4\overline{\varphi} q v_{11} =: A^5_1 + A^5_2,$$

where, due to (4.8),

$$A^5_1 = \pm 4 v_{21}\partial|q|^2 =: \pm 4 v_{21}(q\partial\overline{q} + \overline{q}\partial q) = A^5_{1,1} + A^5_{1,2}.$$

From (3.26) it follows that $A^1_2 + A^5_{1,2} = 0$. Other relations below can be easily checked. Together, they prove (5.3).

$$A^1_1 = B^2, \quad A^1_3 + A^2_3 = C^1 + C^2, \quad A^2_1 + A^4_{1,1} = B^1, \quad A^2_2 + A^4_2 = 0,$$
$$A^3_1 + A^4_{1,2} = 0, \quad A^3_2 = C^3, \quad A^5_{1,1} = C^5, \quad A^5_2 = B^3, \quad B^4 + C^4 = 0.$$

□

Proof of Theorem 2.1. The compactness of $T_{z,t}$ and its derivatives stated in the first item of the theorem follows from Lemmas 3.1, 3.2. Note that condition (3.1) implies the compactness of $T_{z,t}$, and (3.2) is needed for the compactness of its derivatives. The inclusion $T_{z,t}I \in \mathcal{H}^s$, $s > 2$, is due to the Hardy-Littlewood-Sobolev inequality; more details can be found in Lemma 4.2 of [**14**].

From Lemma 4.3 and Theorem 5.1 it follows that the functions q, φ defined in (2.12) satisfy the first two relations of (1.2) when $(z,t) \in \omega$. Let us justify the validity of the initial condition (1.2). Note that $q(x,0)$ is given by (2.12) and coincides with the elements q_{12} of the matrix Q in (3.6) if the operator T_z in (3.6) is the same as in (2.8). If T_z is given by (2.8), then $q_{12} = q_0$ due to Theorem 2.1 from [**14**].

The third item of the theorem, i.e., that equation (2.12) is solvable in a neighborhood of each point (x_0, t_0) for generic potentials, is proved in Theorem 2.1 of [**14**] when $t = t_0$ is fixed. This proof remains valid when t is close to t_0. The validity of the last statement of the theorem is justified in Remark 2 after Theorem 2.1. of [**14**]. □

6. Acknowledgments

This paper would not have been written without productive consultations with R.G. Novikov. The authors are also grateful to Paula Cerejeira, Uwe Kahler, Anna Kazeykina and Jean-Claude Saut for discussions and kind support. The authors thank the reviewers for very valuable remarks.

References

[1] M. J. Ablowitz and P. A. Clarkson, *Solitons, nonlinear evolution equations and inverse scattering*, London Mathematical Society Lecture Note Series, vol. 149, Cambridge University Press, Cambridge, 1991. MR1149378

[2] M. J. Ablowitz and R. Haberman, *Nonlinear evolution equations–two and three dimensions*, Phys. Rev. Lett. **35** (1975), no. 18, 1185–1188, DOI 10.1103/PhysRevLett.35.1185. MR0427866

[3] V. A. Arkadiev, A. K. Pogrebkov, and M. C. Polivanov, *Inverse scattering transform method and soliton solutions for Davey-Stewartson II equation*, Phys. D **36** (1989), no. 1-2, 189–197, DOI 10.1016/0167-2789(89)90258-3. MR1004216

[4] R. Beals and R. R. Coifman, *Multidimensional inverse scatterings and nonlinear partial differential equations*, Pseudodifferential operators and applications (Notre Dame, Ind., 1984), Proc. Sympos. Pure Math., vol. 43, Amer. Math. Soc., Providence, RI, 1985, pp. 45–70, DOI 10.1090/pspum/043/812283. MR812283

[5] R. Beals and R.R. Coifman, *The spectral problem for the Davey- Stewartson and Ishimori hierarchies,* In Nonlinear evolution equations: Integrability and spectral methods, Manchester University Press, 1988, pp 15-23.

[6] R. M. Brown and G. A. Uhlmann, *Uniqueness in the inverse conductivity problem for nonsmooth conductivities in two dimensions*, Comm. Partial Differential Equations **22** (1997), no. 5-6, 1009–1027, DOI 10.1080/03605309708821292. MR1452176

[7] H. Cornille, *Solutions of the generalized nonlinear Schrödinger equation in two spatial dimensions*, J. Math. Phys. **20** (1979), no. 1, 199–209, DOI 10.1063/1.523942. MR517386

[8] A. Davey and K. Stewartson, *On three-dimensional packets of surface waves*, Proc. Roy. Soc. London Ser. A **338** (1974), 101–110, DOI 10.1098/rspa.1974.0076. MR0349126

[9] A. S. Fokas, D. E. Pelinovsky, and C. Sulem, *Interaction of lumps with a line soliton for the DSII equation*, Phys. D **152/153** (2001), 189–198, DOI 10.1016/S0167-2789(01)00170-1. Advances in nonlinear mathematics and science. MR1837909

[10] J.-M. Ghidaglia and J.-C. Saut, *On the initial value problem for the Davey-Stewartson systems*, Nonlinearity **3** (1990), no. 2, 475–506. MR1054584

[11] I. C. Gohberg and M. G. Kreĭn, *Introduction to the theory of linear nonselfadjoint operators*, Translated from the Russian by A. Feinstein. Translations of Mathematical Monographs, Vol. 18, American Mathematical Society, Providence, R.I., 1969. MR0246142

[12] N. Hayashi and J.-C. Saut, *Global existence of small solutions to the Davey-Stewartson and the Ishimori systems*, Differential Integral Equations **8** (1995), no. 7, 1657–1675. MR1347974

[13] E. L. Lakshtanov, R. G. Novikov, and B. R. Vainberg, *A global Riemann-Hilbert problem for two-dimensional inverse scattering at fixed energy*, Rend. Istit. Mat. Univ. Trieste **48** (2016), 21–47. MR3592436

[14] E. Lakshtanov and B. Vainberg, *On reconstruction of complex-valued once differentiable conductivities*, J. Spectr. Theory **6** (2016), no. 4, 881–902, DOI 10.4171/JST/146. MR3584188

[15] H. C. Morris, *Prolongation structures and nonlinear evolution equations in two spatial dimensions. II. A generalized nonlinear Schrödinger equation*, J. Mathematical Phys. **18** (1977), no. 2, 285–288, DOI 10.1063/1.523248. MR0432021

[16] M. Music, *The nonlinear Fourier transform for two-dimensional subcritical potentials*, Inverse Probl. Imaging **8** (2014), no. 4, 1151–1167, DOI 10.3934/ipi.2014.8.1151. MR3296519

[17] A. I. Nachman, *Global uniqueness for a two-dimensional inverse boundary value problem*, Ann. of Math. (2) **143** (1996), no. 1, 71–96, DOI 10.2307/2118653. MR1370758

[18] A. I. Nachman, I. Regev, D. I. Tataru, A Nonlinear Plancherel Theorem with Applications to Global Well-Posedness for the Defocusing Davewy-Stewartson Equation and to the Inverse Boundary Value Problem of Calderon, arXiv:1708.04759 (2017).
[19] R. G. Novikov, *The inverse scattering problem on a fixed energy level for the two-dimensional Schrödinger operator*, J. Funct. Anal. **103** (1992), no. 2, 409–463, DOI 10.1016/0022-1236(92)90127-5. MR1151554
[20] P. A. Perry, *Global well-posedness and long-time asymptotics for the defocussing Davey-Stewartson II equation in* $H^{1,1}(\mathbb{C})$, J. Spectr. Theory **6** (2016), no. 3, 429–481, DOI 10.4171/JST/129. With an appendix by Michael Christ. MR3551174
[21] L.-Y. Sung, *An inverse scattering transform for the Davey-Stewartson II equations. I*, J. Math. Anal. Appl. **183** (1994), no. 1, 121–154, DOI 10.1006/jmaa.1994.1136. MR1273437
[22] L.-Y. Sung, *An inverse scattering transform for the Davey-Stewartson II equations. II*, J. Math. Anal. Appl. **183** (1994), no. 2, 289–325, DOI 10.1006/jmaa.1994.1145. MR1274142
[23] L.-Y. Sung, *An inverse scattering transform for the Davey-Stewartson II equations. III*, J. Math. Anal. Appl. **183** (1994), no. 3, 477–494, DOI 10.1006/jmaa.1994.1155. MR1274849
[24] X. Zhou, *The Riemann-Hilbert problem and inverse scattering*, SIAM J. Math. Anal. **20** (1989), no. 4, 966–986, DOI 10.1137/0520065. MR1000732

Department of Mathematics, Aveiro University, Aveiro 3810, Portugal.
Email address: lakshtanov@ua.pt

Department of Mathematics and Statistics, University of North Carolina, Charlotte, NC 28223
Email address: brvainbe@uncc.edu

Contemporary Mathematics
Volume **734**, 2019
https://doi.org/10.1090/conm/734/14776

Vortex quantization in classical mechanics

V. P. Maslov

We were friends with Selim Grigorievich Krein. He would tell me about his mathematical preferences, which might not had shared with others. In particular, he loved totally unexpected mathematical results, which border with science fiction.

We all accustomed to the fact that quantization (parameter $\hbar$) belongs to another world, observed only under a microscope. In this text, we describe an example, where quantization can be observed "with a naked eye," namely of vorticity quantization. It is related to the Berezinskii –Kosterlitz – Thouless topological phase transition. Experimentators working with helium-4 had noticed this phenomenon quite some time ago: in low-temperature liquid helium-4 the vorticity velocity changes with a jump. This can be counted as a science fiction effect, since vortices have been observed numerous times in practice, and no one thought that their velocities can change discontinuously.

Physics derivation of this effect has been available for a long time. From the mathematical point of view, quasiclassical behaviour of this effect looks unusual, since the parameter $\hbar$ can be sent to zero and discrete things become continuous, but the discrete velocities remain. It looks like not all physicists grasped the mathematical substance of this unusual behavior.

The second quantization, introduced by Dirac, is a purely quantum effect. On the other hand, a prominent mathematician Schönberg applied it in classical mechanics in 1952 - 1953 [**1**], and did this without using the parameter $\hbar$. It was not accepted by the physics community that one can have "second quantization" without $\hbar$.

If the number of particles tends to infinity, the second quantization leads to the self-consistent Hartree-Fock equations, which are quantum equations. Now, sending $\hbar$ to zero one arrives to the classical Thomas-Fermi equation. It is rarely mentioned that this equation is in essence a classical one. Usually $\hbar$ is preserved in Thomas-Fermi equations, albeit one can eliminate it by sending only the number of particles to infinity. It is irrelevant whether $\hbar$ is kept or eliminated, since this leads to no observable effects.

This is an example of a rather frequent situation, when one can obtain the same asymptotics with respect to different parameters tending to zero or infinity. The situation with equations studied by Schönberg is different. From them, sending the number of particles to infinity, one can obtain self-consistent Vlasov equation

---

2010 *Mathematics Subject Classification.* Primary 81Qxx, 81Vxx, 82Bxx.

(see the author's work [**2**]). There is nothing fantastic here, since in both cases the quantum parameter $\hbar$ is not used and equations of the classical statistical physics are derived.

1. Asymptotic solutions of the N-particle Schrödinger equation as $N \to \infty$ and superfluidity

1.1. Asymptotic solutions of the N-particle Schrödinger equation. The one-particle Blokhintsev–Wigner density matrix is the function

$$\rho^t(p,q) = \frac{1}{(2\pi\hbar)^\nu} \int dx\, dy\, w^t(q,x) w^t(x,y) e^{-\frac{i}{\hbar}p(q-y)}. \tag{1}$$

The following equation for the function $\rho^t(p,q)$ is known as the Hartree–Wigner equation:

$$\begin{aligned} i\hbar\frac{\partial\rho^t(p,q)}{\partial t} &= \left[\frac{1}{2m}\left(p - i\hbar\frac{\partial}{\partial q}\right)^2 - \frac{1}{2m}p^2 + W^t(q) - \right. \\ &\left. W^t\left(q - i\hbar\frac{\partial}{\partial p}\right)\right]\rho^t(p,q), \end{aligned} \tag{2}$$

where

$$W^t(q) = U(q) + \int V(q,q')\rho^t(p',q')dp'dq'. \tag{3}$$

Note that equation (2) goes at the limit as $\hbar \to 0$ into the Vlasov equation.

Schrödinger equation for the system of N particles of mass m moving in the external potential $U(x)$, $x \in \mathbb{R}^n$ and interacting with one another has the form

$$i\hbar\frac{\partial}{\partial t}\psi_N(t,x_1,\dots,x_N) = (\widehat{H}\psi_N)(t,x_1,\dots,x_N), \quad x_i \in \mathbb{R}^n, \tag{4}$$

where

$$\widehat{H} = \sum_{i=1}^N \left(-\frac{\hbar^2}{2m}\Delta_i + U(x_i)\right) + \frac{1}{N}\sum_{1\le i<j\le N} V(x_i,x_j), \tag{5}$$

Δ_i is the Laplace operator with respect to the i-th variable x_i.

Let us now proceed to construction of asymptotic solutions of the multi-particle Schrödinger equation (4).

Consider the initial condition φ_0 for the Hartree equation

$$\begin{aligned} i\hbar\dot{\varphi}^t(x) &= \left(-\frac{\hbar^2}{2m}\Delta + W_t(x)\right)\varphi^t(x), \\ W_t(x) &= U(x) + \int V(x,y)|\varphi^t(y)|^2 dy, \end{aligned} \tag{6}$$

such that $\varphi_0 \in W_2^\infty(\mathbb{R}^\nu)$, $\int dx\, |\varphi_0(x)|^2 = 1$.

PROPOSITION 1. *There exists a unique solution $\varphi^t \in W_2^\infty(\mathbb{R}^\nu)$ of equation (1) satisfying the initial condition $\varphi^0 = \varphi_0$.*

Denote by $\mathcal{H}_R^\varphi$ the space of all functions $R: \mathbb{R}^{2\nu} \to \mathbb{C}$ such that

a) $\widetilde{R} \in W_2^\infty(\mathbb{R}^\nu)$, $R(x,y) = R(y,x)$;

b) $\int dy\, R(x,y)\varphi^*(y) = -\varphi(x)$;

c) the operator M in the space $L^2(\mathbb{R}^\nu)$ with the kernel

$$M(x,y) = R(x,y) + \varphi(x)\varphi(y)$$

has the norm less than unity.

Consider the following system of equations:

$$\begin{aligned} i\hbar \dot{g}_n^t(x_1,\dots,x_n) &= \sum_{i=1}^n \left(-\frac{\hbar^2}{2m}\Delta_i + W_t(x_i)\right) g_n^t(x_1,\dots,x_n) + \\ &\sum_{i=1}^n \int dy\, \varphi(x_i)V(x_i,y)g_n^t(y,x_1,\dots,x_{i-1},x_{i+1},\dots,x_n)\varphi^*(y) + \\ &\frac{(n+1)(n+2)}{2}\int dy_1 dy_2 V(y_1,y_2)\varphi^*(y_1)\varphi^*(y_2)g_{n+2}^t(y_1,y_2,x_1,\dots,x_n). \end{aligned} \tag{7}$$

Let $g_0^0,\dots,g_{n_0}^0$ be functions $g_k^0 : \mathbb{R}^{\nu k} \to \mathbb{C}$ such that $g_k^0 \in W_2^\infty(\mathbb{R}^{\nu k})$.

PROPOSITION 2. *There exists a unique solution $g_k^t \in W_2^\infty(\mathbb{R}^{\nu k})$, $k = 0,\dots,n_0$, of the Cauchy problem for system (7) with the initial conditions g_k^0.*

Denote

$$c^t = \exp\left\{-\frac{i}{2}\int_0^t dt' \int dx\,dy\,V(x,y)\varphi^{t'*}(x)\varphi^{t'*}(y)R^{t'}(x,y)\right\},$$

$$\begin{aligned} H(\varphi^*,\varphi) = \hbar^{-1}\Bigg(\int dx\,\varphi^*(x)\left(-\frac{\hbar^2}{2m}\Delta + U(x)\right)\varphi(x) + \\ \frac{1}{2}\int dx\,dy\,|\varphi(x)|^2|\varphi(y)|^2 V(x,y)\Bigg), \end{aligned}$$

$$\begin{aligned} \psi_N^{t,\mathrm{as}}(x_1,\dots,x_N) = c^t e^{iNS^t}\sum_{n=1}^{n_0}\sum_{l=0}^{[N/2]}\frac{1}{2^l N^{l+1/2} l!}\times \\ \sum_{1\le i_1\ne\dots\ne i_{2l+n}\le N} M^t(x_{i_1},x_{i_2})\dots M^t(x_{i_{2l-1}},x_{i_{2l}})g_n^t(x_{i_{2l+1}},\dots,x_{i_{2l+n}})\prod_{i\ne i_s}\varphi^t(x_i), \end{aligned} \tag{8}$$

where $M^t(x,y) = R^t(x,y) + \varphi^t(x)\varphi^t(y)$, and S^t is of the form

$$\begin{aligned} S^t &= \int_0^t d\tau \left(i\int dx \varphi^{\tau*}(x)\frac{d\varphi^\tau}{d\tau}(x) - H[\varphi^{\tau*},\varphi^\tau]\right), \\ c^\tau &= \exp\left\{-\frac{i}{2}\int_0^t d\tau \int dxdy \frac{\delta^2 H}{\delta\varphi(x)\delta\varphi(y)}R^\tau(x,y)\right\}. \end{aligned} \tag{9}$$

Similarly one can construct asymptotic as $N \to \infty$ solutions of the many-particle Schrödinger equation using the action of the germ creation and annihilation operators.

Let $F(x), G(x) \in L^2(\mathbb{R}^\nu)$. Consider the following operator $\bar{A}_{F,G} : L^2(\mathbb{R}^{\nu N}) \to L^2(\mathbb{R}^{\nu N})$,

$$(\bar{A}_{F,G}\psi_N)(x_1,\dots,x_N) = \frac{1}{\sqrt{N}} \sum_{k=1}^{N} \int dx'_k (G^*_\perp(x_k)\varphi^*(x'_k) - F^*_\perp(x'_k)\varphi(x_k))\psi_N(x_1,\dots,x_{k-1},x'_k,x_{k+1},\dots,x_N), \tag{10}$$

where $F_\perp = F - \varphi(\varphi, F)$, $G_\perp = G - \varphi(\varphi, G)$. Let us now consider the system:

$$\begin{aligned} -i\dot{G}^t_i(x) &= \left(-\frac{\hbar^2}{2m}\Delta + W_t(x)\right) G^t_i(x) + \\ &\int dy V(x,y)\varphi^{*t}(x)(\varphi^{*t}(y)F^t_i(y) + \varphi^t(y)G^t_i(y)), \\ i\dot{F}^t_i(x) &= \left(-\frac{\hbar^2}{2m}\Delta + W_t(x)\right) F^t_i(x) + \\ &\int dy V(x,y)\varphi^t(x)(\varphi^{*t}(y)F^t_i(y) + \varphi^t(y)G^t_i(y)). \end{aligned} \tag{11}$$

Let $G^0_i \in W^\infty_2(\mathbb{R}^\nu)$ for $i = 1,\dots,k$. Then consider the following initial condition for the function F^t_i: $F^0_i(x) = \int dy\, R^0(x,y)G^0_i(y)$.

Proposition 3. *Under these initial conditions, there exists a unique solution $G^t_i \in W^\infty_2(\mathbb{R}^\nu)$, $F^t_i \in W^\infty_2(\mathbb{R}^\nu)$ of the system (11).*

Denote by ψ^t_{N,M^t} the wave function (8) with $g_\alpha = 0$, $\alpha \geq 1$, $g_0 = 1$. Consider the solution $\psi^t_N \in W^\infty_2(\mathbb{R}^{\nu N})$ of equation (1) satisfying the initial condition

$$\psi^0_N = \bar{A}^{\nu_1}_{F_1,G_1}\bar{A}^{\nu_2}_{F_2,G_2}\cdots\psi^0_{N,M^0}.$$

Proposition 4. *The following relation holds:*

$$\|\psi^t_N - A_{F_1,G_1}\dots\bar{A}_{F_k,G_k}\psi^t_{N,M}\|_{L^2(\mathbb{R}^{\nu N})} \overset{N\to\infty}{\longrightarrow} 0.$$

Let $F^0_i, G^0_i \in W^\infty_2(\mathbb{R}^\nu)$, $i = 0,\dots,\infty$, be functions such that for $t = 0$ the following conditions are satisfied:

1)

$$G^t_0(x) = -\varphi^{*t}(x), \quad F^t_0(x) = \varphi^t(x); \tag{12}$$

2) $G^t_i(x)$, $i = 0,\dots,\infty$, is a complete system of functions in $L^2(\mathbb{R}^\nu)$, and

$$\sum_{i=1}^{\infty}\int |F^t_i(x)|^2 dx < \infty;$$

3)

$$\int dx(G^t_i(x)F^t_j(x) - G^t_j(x)F^t_i(x)) = 0, \quad i,j = 0,\dots,\infty; \tag{13}$$

4)

$$\int dx(G^t_\alpha(x)G^{*t}_\beta(x) - F^t_\alpha(x)F^{*t}_\beta(x)) = \delta_{\alpha\beta}, \quad \alpha,\beta = 0,\dots,\infty. \tag{14}$$

Proposition 5. *Relations 1)–4) hold at arbitrary moment of time t.*

Properties 1)–4) allow one to uniquely determine a function $R^t : \mathbb{R}^\nu \times \mathbb{R}^\nu \to \mathbb{C}$ from $L^2(\mathbb{R}^{2\nu})$ such that

$$\int dy\, R^t(x,y) G_i^t(y) = F_i^t(y) \quad \text{for all } i,$$

and to put

$$M^t(x,y) = R^t(x,y) + \varphi^t(x)\varphi^t(y).$$

PROPOSITION 6. *The following relation holds:*

$$\left\| \left(i\hbar \frac{d}{dt} - H_N \right) \bar{A}^{\nu_1}_{F_1,G_1} \bar{A}^{\nu_2}_{F_2,G_2} \cdots \psi^t_{N,M} \right\|_{L^2(\mathbb{R}^{\nu N})} \overset{N\to\infty}{\longrightarrow} 0,$$

$$\nu_1, \nu_2, \ldots \in \mathbb{N}, \quad \sum_{k=1}^\infty \nu_k < \infty.$$

COROLLARY 1. *Assume that* $\varphi^t = \varphi e^{-i\Omega t}$, *and functions* F_i^t, G_i^t *depend on* t *as follows:* $F_i^t = F_i e^{i(\beta_i - \Omega)t}$, $G_i^t = G_i e^{i(\beta_i + \Omega)t}$.

Then

$$\left\| \left(\widehat{H}_N - E_N^{(\nu)} \right) \bar{A}^{\nu_1}_{F_1,G_1} \bar{A}^{\nu_2}_{F_2,G_2} \cdots \psi^t_{N,M} \right\|_{L^2(\mathbb{R}^{\nu N})} \overset{N\to\infty}{\longrightarrow} 0,$$

$$E_N^{(\nu)} = NH(\varphi^*, \varphi) + \frac{1}{2} \operatorname{Re} \operatorname{Tr} \left(\frac{\partial^2 H}{\partial\varphi \partial\varphi} M \right) + \sum_{i=1}^\infty \beta_i \nu_i.$$

Let us now consider examples.

1.2. Bogoliubov's superfluidity theory. Let us show how the construction of the paper [**3**] is related to what was said above. Let $U = 0$, and let the set $\mathcal{X}$ be the d-dimensional torus $T^d = T \times \ldots \times T$, where T is the circle. This setting corresponds to a system of N particles in a d-dimensional box with the length of edge L, and the wave functions satisfy the periodicity condition.

Consider the function

$$\varphi(x) = L^{-d/2} e^{\frac{i}{\hbar}(px - \Omega t)}, \tag{15}$$

where $p = 2\pi n/L$, n is an integer vector. This function satisfies the Hartree equation (6) with

$$\Omega = \frac{p^2}{2m} + L^{-d} \int dx\, V(x).$$

Consider the following functions $F^{(\lambda)}(x)$ and $G^{(\lambda)}(x)$, where $\lambda = 2\pi\widetilde{n}/L$, and $\widetilde{n} \neq 0$ is an integer vector:

$$F^{(\lambda)t}(x) = L^{-d/2} \rho_\lambda e^{\frac{i}{\hbar}[(p+\lambda)x + (\beta - \Omega)t]},$$
$$G^{(\lambda)t}(x) = L^{-d/2} \sigma_\lambda e^{\frac{i}{\hbar}[(-p+\lambda)x + (\beta + \Omega)t]},$$

where

$$\begin{aligned} -\beta_\lambda \rho_\lambda &= \left(\frac{(p+\lambda)^2}{2m} - \frac{p^2}{2m} + V_\lambda \right) \rho_\lambda + V_\lambda \sigma_\lambda, \\ \beta_\lambda \sigma_\lambda &= \left(\frac{(p-\lambda)^2}{2m} - \frac{p^2}{2m} + V_\lambda \right) \sigma_\lambda + V_\lambda \rho_\lambda, \\ |\sigma_\lambda|^2 - |\rho_\lambda|^2 &= 1, \quad V_\lambda = L^{-d} \int dx\, V(x) e^{\frac{i}{\hbar}\lambda x}. \end{aligned} \tag{16}$$

From the system (16) we find

$$\beta_\lambda = -p\lambda + \sqrt{\left(\frac{\lambda^2}{2m} + V_\lambda\right)^2 - V_\lambda^2}. \tag{17}$$

In the case when the numbers V_λ rapidly decrease as $|\lambda| \to \infty$, the conditions of the Proposition hold for

$$\frac{\lambda^2}{2m} + 2V_\lambda > 0, \quad \lambda = \frac{2\pi\widetilde{n}}{L}, \tag{18}$$

and we have

$$\begin{aligned} E^{(0)} = & \frac{Np^2}{2m} + \frac{N-1}{2}\int V(x)dx - \frac{1}{2}\sum_{\substack{\kappa=2\pi k/L\\ k\in\mathbb{Z}}} \Bigg[\left(\frac{\kappa^2}{2m} + V_\kappa\right) - \\ & \sqrt{\left(\frac{\kappa^2}{2m} + V_\kappa\right)^2 - V_\kappa^2}\Bigg], \end{aligned} \tag{19}$$

$$E^{(\nu)} = E^{(0)} + \sum_{\substack{\lambda=2\pi k/L\\ \lambda\neq 0}} \beta_\lambda \nu_\lambda, \quad \nu_\lambda \in \mathbb{Z}_+, \quad \sum \nu_\lambda < \infty. \tag{20}$$

The spectrum (20) and the asymptotic eigenfunctions coincide with the ones obtained in Bogoliubov's paper [**3**].

In his paper, Bogoliubov considers another limit, $L \to \infty$, $L^d/N = v = \text{const}$. In this limit, the following *stability condition* arises in the paper:

$$\int V(x)\, dx > 0. \tag{21}$$

Let us show the relation between conditions (18) and (21). As a rule, in physics one considers the case when the coefficient before the potential V equals not $1/N$ but ε. In this case our methods are applicable for $N \sim \alpha/\varepsilon$, $\alpha = \text{const}$. The coefficient before V in all formulas in our notations is multiplied by $\varepsilon N = \alpha$. After such change of notations, formula (18) turns into the following one:

$$\frac{\kappa^2}{2} + 2\varepsilon\frac{N}{L^d}\int V(x)\, dx\, e^{i\kappa x} > 0.$$

If we consider the limit as $L = \text{const}$, $\varepsilon N \to 1$, $N \to \infty$, then this condition goes into the one given above. And if we consider the limit as $\varepsilon = \text{const}$, $N/L^d = v = \text{const}$, $N \to \infty$, then the condition goes into the following one:

$$\frac{\kappa^2}{2m} + \frac{2\varepsilon}{v}\int V(x) e^{\frac{i}{\hbar}\kappa x} dx > 0.$$

As $\varepsilon \to 0$ this condition goes to condition (21), since for $\kappa \neq 0$ it holds automatically, and for $\kappa = 0$ it coincides with condition (21).

Let us now recall the well known Landau argument [**4**] showing that in this example *superfluidity* arises.

Friction is possible if among the β_k found above there are negative numbers, $\beta_k < 0$. If

$$p^* = \inf_{\substack{\lambda=2\pi\widetilde{n}/L \\ \lambda\neq 0}} \left(\frac{1}{|\lambda|} \sqrt{\left(\frac{\lambda^2}{2m} + V_\lambda\right)^2 - V_\lambda^2} \right) \geq \frac{2\pi}{L}, \tag{22}$$

then for $|p| < p^*$ all $\beta_\lambda > 0$, hence Landau's argument implies that one has superfluidity.

If, as in the case of condition (18), one multiplies the potential V by εN and considers the limit as $L \to \infty$, $L^d/N = \text{const}$, and after that the limit as $\varepsilon \to 0$, then condition (22) will also go to the *Bogoliubov condition* (21).

1.3. Generalization of the notion of superfluidity. Let us now show that phenomena analogous to superfluidity arise also in the case when $U \neq 0$, i. e. when solutions of the system in variations (11) and of the Hartree equation (6) have the form different from const e^{ipx}. This corresponds to the physical situation when elementary excitations are not characterized by certain momentum.

To this end, consider the following example. Let $\mathcal{X} = T^1 \times \mathbb{R}^1$, assume that the potential U depends only on the coordinate x_2, and the potential V depends only on the coordinate x_1. Consider the case when the operator $\hat{A} = -\frac{1}{2}\frac{d^2}{dx_2^2} + U(x_2)$ has purely discrete spectrum. Then the Hartree equation (6) has the following solution:

$$\varphi(x) = L^{-1/2} e^{ipx_1} \chi_0(x_2), \quad p = \frac{2\pi j}{L}, \quad j \in \mathbb{Z}, \tag{23}$$

for

$$\Omega = \frac{p^2}{2m} + V_0 + E_0. \tag{24}$$

Denote by χ_n the normalized to unity eigenfunctions of the operator A corresponding to eigenvalues E_n numbered in the order of increasing.

The solutions of system (11) satisfying conditions (12)–(14) have the form $F^{(\lambda,k)}$, $G^{(\lambda,k)}$, $\lambda = 2\pi\widetilde{n}/L$, $\widetilde{n} \neq 0$, $\widetilde{n} \in \mathbb{Z}$, $k \in \mathbb{Z}_+$,

$$\begin{aligned}
F^{(\lambda,0)}(x_1, x_2) &= F^{(\lambda)0}(x_1)\chi_0(x_2) e^{\frac{i}{\hbar}(\widetilde{\beta}_{\lambda 0} - \Omega)t}, \\
G^{(\lambda,0)}(x_1, x_2) &= G^{(\lambda)0}(x_1)\chi_0^*(x_2) e^{\frac{i}{\hbar}(\widetilde{\beta}_{\lambda 0} + \Omega)t}, \\
F^{(\lambda,k)}(x_1, x_2) &= 0, \\
G^{(\lambda,k)}(x_1, x_2) &= L^{-d/2} e^{\frac{i}{\hbar}[(p+\lambda)\chi_1 + (\widetilde{\beta}_{\lambda k} + \Omega)t]} \chi_k^*(x_2), \quad k = 1, \dots, \infty,
\end{aligned}$$

here the functions $F^{(\lambda)}, G^{(\lambda)}$ coincide with the ones used in the previous example, and the numbers $\widetilde{\beta}_{\lambda k}$ equal

$$\widetilde{\beta}_{\lambda 0} = \beta_\lambda, \quad \widetilde{\beta}_{\lambda k} = E_k - E_0 + \frac{(p+\lambda)^2}{2m} - \frac{p^2}{2m}.$$

Thus, for arising of superfluidity in this case, one should require, besides condition (22), also the following condition:

$$E_1 - E_0 > \frac{p^2}{2m}.$$

This argument leads to the following definition.

DEFINITION 1. Assume that the conditions (7) hold, and one has $\beta_\alpha > 0$, $\alpha = 1, \ldots, \infty$. Then let us call the wave function $\psi_{N,M}$ by the wave function corresponding to a *superfluid state*.

The main state of the N-particle Hamiltonian also satisfies this definition. Thus, we have generalized the notion of superfluidity to the case $U \neq 0$.

2. Second quantization in the Fock's space. Schörnberg quantization. Macroscopic vorticity quantization.

As we have already mentioned, the contemporary derivation of Vlasov equation was achieved using the second quantization for the Schönberg's classical particles. Then, when $N \to \infty$, one gets the following system of equations [**5**]:

$$\dot{u}(p,q,t) = \left(\frac{\partial U}{\partial q}\frac{\partial}{\partial p} - p\frac{\partial}{\partial q}\right)u(p,q,t) +$$

$$(25) \qquad \int dp'dq'v(p',q',t)\left(\frac{\partial V(q,q')}{\partial q}\frac{\partial}{\partial p} + \frac{\partial V(q,q')}{\partial q'}\frac{\partial}{\partial p'}\right)u(p',q',t)u(p,q,t),$$

$$\dot{v}(p,q,t) = \left(\frac{\partial U}{\partial q}\frac{\partial}{\partial p} - p\frac{\partial}{\partial q}\right)v(p,q,t) +$$

$$+\int dp'dq'u(p',q',t)\left(\frac{\partial V(q,q')}{\partial q}\frac{\partial}{\partial p} + \frac{\partial V(q,q')}{\partial q'}\frac{\partial}{\partial p'}\right)v(p',q',t)v(p,q,t),$$

where $U(q_i)$ is the external field and $V(q_i, q_j)$ is the pairwise interaction.

REMARK 1. let us consider instead of u and v the creation and annihilation operators $\hat{u}, \hat{v}$ in the Fock space. After such substitution, the system (25) becomes equivalent to the N-particle problem for the Newton system. The projection from the Fock space into the $3N$-dimensional space of N particles *produces exactly the Newton system* [**6**]. This implies that "from the point of view of the operator method, their equation of characteristics is the Vlasov equation" [**7**, p. 10]. This approach allows to obtain corrections to the Vlasov equation and expansions of N-particle Newton equations into powers of $1/N$.

Let us consider a complex solution of (25). Introducing

$$(26) \qquad u(p,q,t) = \sqrt{\rho(p,q,t)}e^{i\pi(p,q,t)}, \quad v(p,q,t) = \sqrt{\rho(p,q,t)}e^{-i\pi(p,q,t)},$$

the system (25) reduces to

$$(27) \qquad \dot{\rho}(p,q,t) = \left(\frac{\partial W^t}{\partial q}\frac{\partial}{\partial p} - p\frac{\partial}{\partial q}\right)\rho(p,q,t),$$

$$\dot{\pi}(p,q,t) = \left(\frac{\partial W^t}{\partial q}\frac{\partial}{\partial p} - p\frac{\partial}{\partial q}\right)\pi(p,q,t) + \int dp'dq'\frac{\partial V(q,q')}{\partial q'}\frac{\partial \pi(p',q',t)}{\partial p'}\rho(p',q',t).$$

Here $W^t(q) = \int dq'V(q,q')\rho(p',q',t)dp'dq'$.

The first equation of (27) is the *Vlasov equation* [**8**], where ρ is the distribution function used in physics for studying many-particle systems. The second, new equation is linear with respect to the function π. Functions u and v are called semi-densities and π - phase. In the two-dimensional case, the phase in the SchrÓdinger equation plays the role of the action, which can be quantized analogously to Bohr quantization. The condition of single-valuedness of $e^{i\pi}$ brings about quantization of π, i.e. vorticity quantization, which holds also in the macroscopic classical problem.

Since the equation for π is linear, one can divide π by any constant. Dimensionality shows that, in particular, π can be divided by $\hbar$.

The quantity $\hat{p} := ih\frac{\partial}{\partial q}$ is the quantum momentum. Schönberg's quantization can be obtained as a limit of the second quantization of Schródinger equation. The latter one, in turn, leads to the Hartree equation with the parameter $\hbar$. Thus the limit when $\hbar \to 0$ in (25)–(27) leads to the limit for the average momentum $\hat{p}$.

Since (26) corresponds to a vertex, the average momentum corresponds to the vortex velocity in the Schrödinger equation.

We thus conclude that the complex solution for the classical equations has direct physical meaning, if one uses quasi-classical approach. This combination of the quantum and classical logic, when applied to the Newton system, provides a real, "visible to a naked eye" quantization of the vortex momenta. This suggests a possibility of finding a "hidden parameter" and construct an combined quantum and classical logic [**9**].

References

[1] M. Schönberg, "Application of second quantization methods to the classical statistical mechanics," Nuovo cimento, **9** (12) 1139–1182 (1952); "Application of second quantization methods to the classical statistical mechanics(II)," Nuovo cimento, **10** (4) 419–472 (1953).

[2] V. P. Maslov, "The Vlasov equation," in *Encyclopedia of Low Temperature Plasma*, Ser. B, Ed. by V. E. Fortov, Vol. VII: *Mathematical Modelling in Low-Temperature Plasma* (Yanus-K, 2008), pp. 209–242.

[3] N. Bogolubov, *On the theory of superfluidity*, Acad. Sci. USSR. J. Phys. **11** (1947), 23–32. MR0022177

[4] E. M. Lifshits and L. P. Pitaevskii, *Statistical Physics*, Part 2: *Theory of Condensed State* (Nauka, Moscow, 1978; Pergamon, Oxford, 1980).

[5] V. P. Maslov and O. Yu. Shvedov, *The Complex Germ Method in Many-Particle Problems and in Quantum Field Theory* (Editorial URSS, Moscow, 2000) [in Russian]

[6] V. P. Maslov, *Nonstandard characteristics in asymptotic problems* (Russian), Uspekhi Mat. Nauk **38** (1983), no. 6(234), 3–36. MR728722

[7] V. P. Maslov, *Kompleksnyĭ metod VKB v nelineĭnykh uravneniyakh* (Russian), Izdat. "Nauka", Moscow, 1977. Nelineĭnyĭ Analiz i ego Prilozheniya. [Monographs in Nonlinear Analysis and its Applications]. MR0503031

[8] A. A. Vlasov, "On the vibrational properties of an electronic gas," Zh. Éxper. Teoret. Fiz. **8**, 291–238 (1938).

[9] V. P. Maslov, *On the hidden parameter in quantum and classical physics*, Math. Notes **102** (2017), no. 5-6, 890–893, DOI 10.1134/S000143461711030X. MR3741186

NATIONAL RESEARCH UNIVERSITY HIGHER SCHOOL OF ECONOMICS, MOSCOW, 123458, RUSSIA; A.ISHLINSKY INSTITUTE FOR PROBLEMS IN MECHANICS OF THE RUSSIAN ACADEMY OF SCIENCES, MOSCOW, 119526, RUSSIA

Email address: `v.p.maslov@mail.ru`

Contemporary Mathematics
Volume **734**, 2019
https://doi.org/10.1090/conm/734/14777

Elliptic operators with nonstandard growth condition: Some results and open problems

Alexander Pankov

Dedicated to 100th anniversary of my teacher professor Selim Grigorievich Krein

ABSTRACT. This is a short survey of certain results and open problems on the Dirichlet problem for nonlinear monotone elliptic operators with nonstandard growth condition. In this paper we pay attention to the case when the so-called Lavrentiev phenomenon may occur. Furthermore, as a rule the growth condition accepted in the paper is more general than the well-known $p(x)$ growth condition. Also we discuss the homogenization result obtained by V. Zhikov and S. Pastukhova in the case of $p(x)$ growth and its potential extensions.

1. Introduction

In this paper we consider the Dirichlet boundary value problem

$$-\operatorname{div} a(x, \nabla u) = \operatorname{div} g(x)\,, \quad u = 0 \text{ on } \partial\Omega\,, \tag{1.1}$$

for monotone elliptic equations, with nonstandard growth, on bounded Lipschitz domains as well as their multivalued counterparts. Notice that in applications multivalued problems of this kind occur when constitutive relations are multivalued. For instance, this happens in viscoelasticity (see, *e.g.*, [**34**]). The growth condition is defined in terms of a generalized N-function $\varphi(x, t)$ which is, in general, only a measurable function of x. As a typical example we mention $\varphi(x, t) = t^{p(x)}$ ($p(x)$ growth).

Equations of such type appear in many applications such as modeling electrorheological fluids, thermistor problem and image restoration to name a few (see, *e.g.*, [**4, 5, 10, 32, 33, 38, 40**] and references therein). Notice that almost all results obtained in this direction deal with the case of $p(x)$ growth.

An interesting feature of equations with nonstandard growth is the so-called Lavrentiev phenomenon. One of its faces is that, in general, the space of smooth compactly supported functions is not dense in the naturally defined Musielak-Sobolev space $W_0^{1,\varphi}(\Omega)$. Usually, the function φ and problem (1.1) are called *regular* if smooth functions are dense in $W_0^{1,\varphi}(\Omega)$, and *non-regular* otherwise. In the case of

2010 *Mathematics Subject Classification.* Primary: 35J60; Secondary: 35J25, 35B27, 47H05.
Key words and phrases. Nonstandard growth condition, multivalued elliptic equation, monotone operator, homogenization.
The work is supported by Simons Foundation, award 410289. The author thanks the anonymous referee for valuable remarks and comments.

$p(x)$-growth a sufficient condition for regularity, the so-called log-*Hölder condition*, is obtained in [**37**] (see also [**40**]). Similar results are available for more general growth functions φ [**8**]. Notice that most of authors that study equations with non-standard growth either impose the log-Hölder condition, or look for solutions only in the closure of smooth functions in $W_0^{1,\varphi}(\Omega)$. Following V. Zhikov, we denote this closure by $H_0^{1,\varphi}(\Omega)$. To the best of our knowledge, V. Zhikov is the only author who studies problem (1.1), with $p(x)$ growth, in the situation when the Lavrentiev phenomenon may occur (see [**36, 37, 40**] and references therein).

The notion of weak solution of problem (1.1) can be introduced in the standard way. However, it is not so useful in the presence of Lavrentiev phenomenon. Therefore, we consider a subclass of weak solutions, the so-called variational solutions. More precisely, if V is a closed subspace of $W_0^{1,\varphi}(\Omega)$ such that $H_0^{1,\varphi}(\Omega) \subset V$, then we define V-solutions (see Section 3) that belong to V and satisfy the usual integral identity with test functions in the space V. In the case of $p(x)$ growth such solutions are introduced by V. Zhikov (see, *e.g.*, [**40**]). Among these solutions the most important are two extreme cases when $V = H_0^{1,\varphi}(\Omega)$ (H-solutions) and $W_0^{1,\varphi}(\Omega)$ (W-solutions).

The contents of the paper is the following. In Section 2 we introduce basic functional spaces and quickly discuss their main properties. Section 3 is devoted to the existence of V-solutions to the multivalued version of problem (1.1). Here the existence problem is treated from the point of view of monotone operators. In Section 4 we consider approximation of solutions to problem (1.1) by solutions of certain approximate problems of similar form. Finally, in Section 5 we discuss the homogenization result by V. Zhikov and S. Pastukhova [**41**] for single-valued strictly monotone problems with $p(x)$ growth.

2. Generalized N-functions and functional spaces

In this section we collect some basic properties of Musielak-Orlicz and Musielak-Orlicz-Sobolev spaces. For more details we refer to [**15, 16, 26**].

Let Ω be an open, bounded subset of $\mathbb{R}^n$ with Lipschitz boundary. A function $\varphi : \Omega \times [0, +\infty) \to [0, +\infty)$ is called a *generalized N-function* if it satisfies the following conditions:

(a) *for almost all $x \in \Omega$, $\varphi(x, \cdot)$ is a N-function, i.e., convex, nondecreasing function such that, $\varphi(x, 0) = 0$, $\varphi(x, t) > 0$ for all $t > 0$, and*

$$lim_{t\to 0} \frac{\varphi(x,t)}{t} = 0, \qquad \lim_{t\to\infty} \frac{\varphi(x,t)}{t} = +\infty.$$

(b) *$\varphi(\cdot, t)$ is a measurable function on Ω for all $t \geq 0$.*

The set of all generalized N-functions is denoted by $N(\Omega)$.

The *conjugate function* $\varphi^* : \Omega \times [0, +\infty) \to [0, +\infty)$ to φ is defined by

$$\varphi^*(x, s) = \sup_{t\geq 0}(st - \varphi(x, t)) \text{ for every } s \geq 0 \text{ and } x \in \Omega. \tag{2.1}$$

It is well known that $\varphi^* \in N(\Omega)$ and φ is the conjugate function to φ^*, and the following Young inequality holds

$$st \leq \varphi(x, t) + \varphi^*(x, s) \text{ for } x \in \Omega \text{ and } t, s \in \mathbb{R}. \tag{2.2}$$

We say that $\varphi \in N(\Omega)$ satisfies the Δ_2-condition if there exists a positive constant $K > 1$ such that

$$\varphi(x, 2t) \leq K\varphi(x, t) \text{ for every } t > 0 \text{ and } x \in \Omega. \tag{2.3}$$

Throughout the paper we assume that all generalized N-functions satisfy the following assumptions.

(N_1) *For all $t \geq 0$ the functions $\varphi(\cdot, t)$ and $\varphi^*(\cdot, t)$ belong to $L^\infty(\Omega)$.*
(N_2) *Both functions φ and φ^* satisfy Δ_2-condition.*
(N_3) *For every $t_0 > 0$ there exists a constant $\gamma(t_0)$ such that*

$$t^{-1}\varphi(\cdot, t) \geq \gamma(t_0) \quad \text{and} \quad t^{-1}\varphi^*(\cdot, t) \geq \gamma(t_0)$$

for all $t \geq t_0$.

Under Assumptions (N_1)–(N_3) there exist p and q, with $1 < p \leq q < \infty$, such that

$$\underline{c}_0|t|^p - \underline{c}_1 \leq \varphi(x, t) \leq \bar{c}_0|t|^q + \bar{c}_1 \tag{2.4}$$

and

$$\underline{c}_0|t|^{q'} - \underline{c}_1 \leq \varphi^*(x, t) \leq \bar{c}_0|t|^{p'} + \bar{c}_1 \tag{2.5}$$

for almost all $x \in \Omega$, all $t \geq 0$, and some positive constants $\underline{c}_0$, $\underline{c}_1$, $\bar{c}_0$ and $\bar{c}_1$, where p' and q' are conjugate exponents defined by $(p')^{-1} + p^{-1} = 1$. Indeed, Δ_2-condition for φ implies that $\varphi(x, ts) \leq ct^q\varphi(x, s)$ for all $t \geq 1$ and $s \geq 0$ (see, *e.g.*, [**23**]). Together with (Φ) this yields the upper bound in (2.4). Similarly, Δ_2-condition for φ^* yields the upper bound in (2.5). The lower bounds follow by duality.

EXAMPLE 2.1. Let $p(x)$ be a measurable real valued function such that $1 < p \leq p(x) \leq q < \infty$. Then the function $\varphi(x, t) = t^{p(x)}$ satisfies Assumptions (N_1)–(N_3).

Let $\varphi \in N(\Omega)$. We define the convex functional

$$\varrho_\varphi(u) := \int_\Omega \varphi(x, |u(x)|)dx\,.$$

This functional is called the *modular*. Then, the *Musielak-Orlicz space*, also called *generalized Orlicz space*, $L^\varphi(\Omega)$ is defined by

$$L^\varphi(\Omega) = \{u : \Omega \to \mathbb{R} \text{ is measurable and } \varrho_\varphi(u) < +\infty\}\,.$$

We equip this space with the Luxemburg norm

$$\|u\|_{L^\varphi(\Omega)} = \|u\|_\varphi = \inf\{\lambda > 0 : \varrho_\varphi(u/\lambda) \leq 1\}.$$

If φ is the function from Example 2.1, we use the notation $L^{p(\cdot)}(\Omega)$. Notice that the classical space $L^p(\Omega)$ is a particular case.

In our context the following properties of L^φ spaces are particular cases of general results on Musielak-Orlicz spaces (see, *e.g.*, [**15, 26, 32**]).

PROPOSITION 2.1. *Let φ and ψ be generalized N-functions satisfying* (N_1)–(N_3). *Then the following statements hold true.*

(*i*) *The set $L^\varphi(\Omega)$ is a separable, reflexive Banach space.*
(*ii*) *The modular convergence is equivalent to the norm convergence,* i.e. *$\varrho_\varphi(u_n - u) \to 0$ if and only if $\|u_n - u\|_\varphi \to 0$.*

(*iii*) *The space* $L^\varphi(\Omega)$ *is embedded into* $L^\psi(\Omega)$ *densely and continuously if and only if there exists* $\gamma > 0$ *and* $h \in L^1(\Omega)$ *with* $\|h\|_{L^1(\Omega)} \le 1$ *such that*

$$\psi(x, \gamma t) \le \varphi(x,t) + h(x) \tag{2.6}$$

for almost all x *and all* $t \ge 0$. *In this case the embedding constant does not exceed* $2\gamma^{-1}$.

(*iv*) *In particular,* $L^q(\Omega) \subset L^\varphi(\Omega) \subset L^p(\Omega)$, *where* p *and* q *are the constants from inequality* (2.4).

In the spirit of [**40**], we introduce two types of *Musielak-Sobolev spaces* with zero boundary traces: $W_0^{1,\varphi}(\Omega)$ and $H_0^{1,\varphi}(\Omega)$. The first one is defined by

$$W_0^{1,\varphi}(\Omega) = \{u \in W_0^{1,1}(\Omega) : \nabla u \in L^\varphi(\Omega)^n\}$$

and is equipped with the norm

$$\|u\|_{W_0^{1,\varphi}(\Omega)} = \|u\|_W = \|\nabla u\|_\varphi\,. \tag{2.7}$$

Here $W_0^{1,1}(\Omega)$ stands for the standard Sobolev space of L^1 functions with L^1 gradients and zero boundary values. The second space is defined by

$$H_0^{1,\varphi}(\Omega) = \text{closure of } C_0^\infty(\Omega) \text{ in } W_0^{1,\varphi}(\Omega)\,.$$

We use the notation $W_0^{1,p(\cdot)}$ and $H_0^{1,p(\cdot)}$ when the function φ is of the form given in Example 2.1.

The operator ∇ maps isometrically the spaces $W_0^{1,\varphi}(\Omega)$ and $H_0^{1,\varphi}(\Omega)$ onto certain closed subspaces of $L^\varphi(\Omega)^n$. Then Proposition 2.1 implies the following.

PROPOSITION 2.2. *The spaces* $W_0^{1,\varphi}(\Omega)$ *and* $H_0^{1,\varphi}(\Omega)$ *are separable, reflexive Banach spaces.*

In general, $H_0^{1,\varphi}(\Omega) \ne W_0^{1,\varphi}(\Omega)$. We say that $\varphi \in N(\Omega)$ is a *regular* N-function if these two spaces coincide.

EXAMPLE 2.2. Let $\Omega \subset \mathbb{R}^2$ be the unit disk and $1 < p_1 < 2 < p_2$. Define the function $p(x)$ by $p(x) = p_1$ if $x_1x_2 > 0$ and $p(x) = p_2$ if $x_1x_2 < 0$. Then $H_0^{1,p(\cdot)}(\Omega)$ has codimension 1 in $W_0^{1,p(\cdot)}(\Omega)$ (see [**40**]).

The following theorem is a consequence of main result of [**8**].

THEOREM 2.1. *Suppose that there is a constant* $c > 0$ *such that*

$$|\ln\varphi(x,t) - \ln\varphi(y,t)| \le \frac{ct}{\ln(1/|x-y|)} \tag{2.8}$$

for all $t \ge 1$ *and* $x, y \in \Omega$ *with* $|x-y| \le 1/2$. *Then* $H_0^{1,\varphi}(\Omega) = W_0^{1,\varphi}(\Omega)$.

Assumption 2.8 is a direct extension of the so-called log-Hölder condition widely used in the case of variable exponent spaces (see [**15**–**18**, **40**] and references therein).

PROBLEM 1. Notice that in the variable exponent case there is a more general sufficient condition for regularity, the so-called double-log-condition ([**40**, **41**]). It would be interesting to extend that condition to Musielak-Sobolev spaces with φ satisfying (N_1)–(N_3).

A closed subspace $V \subset W_0^{1,\varphi}(\Omega)$ is called an *admissible subspace* if $H_0^{1,\varphi}(\Omega) \subset V$. We denote by V' the dual space to V. We use the notation $H^{-1,\varphi^*}(\Omega)$ and $W^{-1,\varphi^*}(\Omega)$ for the dual spaces to $H_0^{1,\varphi}(\Omega)$ and $W_0^{1,\varphi}(\Omega)$, respectively. The space $H^{-1,\varphi^*}(\Omega)$ is a space of distributions. However, in the case when φ is not regular and $V \neq H_0^{1,\varphi}(\Omega)$, the space V' is not a space of distributions anymore. Nevertheless, for any $g \in L^{\varphi^*}(\Omega)$ the functional f_g defined by

$$\langle f_g, v\rangle = -\int_\Omega g(x)\cdot\nabla v(x)\,dx\,, \quad v \in V \tag{2.9}$$

is an element of V' for every admissible subspace V.

3. Elliptic operators and Dirichlet problems

Let X be a reflexive Banach space and let X^* be its dual. The duality pairing on $X \times X^*$ is denoted by $\langle\cdot,\cdot\rangle$. A *multivalued operator* $A : X \to X^*$ is a mapping which takes every point $x \in X$ to a set $Ax \subset X^*$. The *graph* and the *domain* of the operator A are the sets

$$gr(A) = \{(x,y) \in X \times X^* : y \in Ax\} \text{ and } D(A) = \{x \in X : Ax \neq \emptyset\},$$

respectively. The operator A is single-valued if for every $x \in X$, Ax contains exactly one element. For operators A and B, we write $A \subseteq B$ if $gr(A) \subseteq gr(B)$.

An operator $A : X \to X^*$ is a *monotone operator* if

$$\langle y_1 - y_2, x_1 - x_2\rangle \geq 0.$$

whenever $y_i \in Ax_i$, $i = 1,2$. A monotone operator A is a *maximal monotone operator* if for every monotone operator $B : X \to X^*$ such that $A \subseteq B$ we have $A = B$. Notice that the values Ax, $x \in X$, of a maximal monotone operator are closed convex subsets of X^*.

A multivalued operator A from X into X^* is *coercive* if there exists a real valued function $c(s)$ such that $c(s) \to \infty$ as $s \to \infty$, and for every $x \in X$ and every $y \in Ax$,

$$\langle y, x\rangle \geq c(\|x\|_X)\|x\|_X. \tag{3.1}$$

THEOREM 3.1. *Let X be a reflexive Banach space, and $A : X \to X^*$ be a maximal monotone map. If A is coercive, then it is surjective in the sense that*

$$\bigcup_{x\in X} Ax = X^*.$$

For the proof of Theorem 3.1 we refer to [**9, 29**] (see also [**22, 28, 35**]).

Under suitable assumptions, we consider the Dirichlet problem for multivalued elliptic equations of the form

$$-\operatorname{div} a(x,\nabla u) \ni \operatorname{div} g\,, \quad u|_{\partial\Omega} = 0\,, \tag{3.2}$$

where g is a given vector function in certain Musielak-Orlicz space.

Let $\mathfrak{B}(\mathbb{R}^n)$ be the σ-algebra of all Borel subset of $\mathbb{R}^n$, $\mathfrak{L}(\Omega)$ be the σ-algebra of all Lebesgue measurable subsets of Ω and $\otimes$ stand for the tensor product of σ-algebras. The symbols $|\cdot|$ and $\cdot$ are reserved for the Euclidean norm and the inner product in $\mathbb{R}^n$, respectively.

DEFINITION 3.1. Denote by M_Ω the set of all multivalued functions

$$a : \Omega \times \mathbb{R}^n \to \mathbb{R}^n$$

which satisfy the following conditions:

(i) for almost all $x \in \Omega$, the function $a(x, \cdot) : \mathbb{R}^n \to \mathbb{R}^n$ is maximal monotone;

(ii) a is measurable with respect to $\mathfrak{L}(\Omega) \otimes \mathfrak{B}(\mathbb{R}^n)$, *i.e.*

$$a^{-1}(C) = \{(x, \xi) \in \Omega \times \mathbb{R}^n : a(x, \xi) \cap C \neq \emptyset\} \in \mathfrak{L}(\Omega) \otimes \mathfrak{B}(\mathbb{R}^n)$$

for any closed subset $C \subset \mathbb{R}^n$;

(iii) there exist positive constants c_1, c_2 and a nonnegative function $h \in L^1(\Omega)$ such that for almost all $x \in \Omega$, the inequalities

$$\varphi^*(x, |\eta|) \leq c_1 \varphi(x, |\xi|) + h(x) \tag{3.3}$$

$$\eta \cdot \xi \geq c_2 \varphi(x, |\xi|) - h(x) \tag{3.4}$$

hold true for any $\xi \in \mathbb{R}^n$ and $\eta \in a(x, \xi)$.

Let $V \subset W_0^{1,\varphi}(\Omega)$ be an admissible subspace with the dual V'. To a function $a \in M_\Omega$ we associate the (multivalued) operator A_V, acting from V into V', by

$$f \in A_V u \Leftrightarrow \exists g \in L^{\varphi^*}(\Omega)^n : g(x) \in a(x, \nabla u) \text{ a.e. in } \Omega \text{ and} \tag{3.5}$$
$$\langle f, v \rangle = \int_\Omega g(x) \cdot \nabla v(x) dx, \ \forall v \in V .$$

The operator A_V has the following useful representation. Restricting the gradient operator ∇, understood in the sense of distributions, to the space V, we obtain the operator

$$\nabla_V : V \to L^\varphi(\Omega)^n .$$

This is a linear bounded operator with the adjoint operator

$$\nabla_V^* : L^{\varphi^*}(\Omega)^n \to V' .$$

Given $a \in M_\Omega$, we introduce the (multivalued) operator $\mathbf{a}$ acting from $L^\varphi(\Omega)^n$ into $L^{\varphi^*}(\Omega)^n$ as follows. For any $h \in L^\varphi(\Omega)^n$,

$$\mathbf{a}(h) = \{g \in L^{\varphi^*}(\Omega)^n : g(x) \in a(x, h(x)) \text{ for almost all } x \in \Omega\} .$$

Now it is easily seen that

$$A_V u = \nabla_V^* \mathbf{a}(\nabla_V u) , \quad u \in V .$$

It turns out to be that A_V is a maximal monotone operator from V into V', while $\mathbf{a}$ is a maximal monotone operator from $L^\varphi(\Omega)^n$ into $L^{\varphi^*}(\Omega)^n$. In what follows we use indices H and W instead of V if $V = H_0^{1,\varphi}(\Omega)$ and $V = W_0^{1,\varphi}(\Omega)$, respectively.

Notice that the operator ∇_H^* acts as $-\text{div}$ in the sense of distributions, while this is not so for ∇_V^* if $V \neq H_0^{1,\varphi}(\Omega)$.

In the case when $\varphi(x, t) = t^{p(x)}$ and the function a is single-valued, the following result is essentially well known (see, *e.g.*, [**4, 5, 40**]).

THEOREM 3.2. *Let $a \in M_\Omega$ and V be an admissible subspace of $W_0^{1,\varphi}(\Omega)$. Then, for every $f \in V'$ the equation $A_V u \ni f$ has a solution $u \in V$. The solution u is unique if the function $a(x, \cdot)$ is strictly monotone for almost all $x \in \Omega$.*

This result is a minor extension of Theorem 3.2 in [**7**]. The proof based on Theorem 3.1 is essentially the same as in that paper. Notice that inequality (3.4) implies an obvious lower bound in terms of the modular. It is not trivial to derive the coerciveness condition (3.1) from that lower bound, and the key analytic ingredient is the following lemma (see [**7**]).

LEMMA 3.1. *The function*

$$c_\varphi(s) = \operatorname*{ess\,inf}_{x\in\Omega} \inf_{t>0} \frac{\varphi(x, st)}{s\varphi(x,t)}\,.$$

satisfies the following inequality

$$c_\varphi(s) \geq \alpha(s^{p-1} - 1)\,,$$

where the constant $\alpha > 0$ depends only on the constant in the (Δ_2) condition for φ^*.

The following problem makes sense. In particular, it is related to the homogenization problem (see Section 5).

PROBLEM 2. Study the existence of solutions in the case when the generalized N-function $\varphi(x,t)$ in M_Ω is replaced by an anisotropic N-function $\varphi(x,\xi)$, $\xi \in \mathbb{R}^n$, satisfying appropriate assumptions that reduce to (N_1)–(N_3) in the isotropic case.

The key point in such a hypothetic extension of Theorem 3.2 is the coerciveness property of the operator under consideration (*cf.* Lemma 3.1).

Now we give an interpretation of Theorem 3.2 in terms of equation (3.2) with $g \in L^{\varphi^*}(\Omega)^n$. Assuming that $a \in M_\Omega$, a function $u \in W_0^{1,\varphi}(\Omega)$ is a *weak solution* of problem (3.2) if there exists a function $w \in L^{\varphi^*}(\Omega)^n$ such that $w(x) \in a(x, \nabla u(x))$ for almost all $x \in \Omega$ and

$$\int_\Omega w(x)\cdot\nabla v(x)dx = -\int_\Omega g(x)\cdot\nabla v(x)dx \tag{3.6}$$

for all $v \in C_0^\infty(\Omega)$. The function w in equation (3.6) is considered as the *flux* corresponding to the solution u. If V is an admissible subspace, a function $u \in V$ is a V*-solution* if equation (3.6) holds for all $v \in V$. Finally, a function $u \in W_0^{1,\varphi}(\Omega)$ is a *variational solution* if u is a V-solution for some admissible subspace V. If $V = W_0^{1,\varphi}(\Omega)$ and $V = H_0^{1,\varphi}(\Omega)$, we use the terms W-solution and H-solution, respectively. The concepts of weak solution and variational solutions were introduced by V. Zhikov in the case of $p(x)$ growth condition (see, *e.g.*, [**40**]).

Every variational solution obviously satisfies the following *energy identity*

$$\int_\Omega w(x)\cdot\nabla u(x)dx = -\int_\Omega g(x)\cdot\nabla u(x)dx\,. \tag{3.7}$$

On the other hand, if a weak solution u satisfies (3.7), then it is a variational solution. Indeed, it is not difficult to verify that u is a V-solution, where $V = \operatorname{span}\{u, H_0^{1,\varphi}(\Omega)\}$. The role of the energy identity is pointed out by V. Zhikov still for $p(x)$ growth condition.

Now, making use of Theorem 3.1 with $f = f_g \in V'$, we obtain the existence of V-solution for every admissible subspace V. If $a(x,\xi)$ is strictly monotone for almost all $x \in \Omega$, then there is a unique V-solution.

EXAMPLE 3.1. Consider the problem

$$-\operatorname{div}(|\nabla u|^{p(x)-2}\nabla u) = \operatorname{div} g\,, \quad u|_{\partial\Omega} = 0\,,$$

where Ω and $p(x)$ are defined in Example 2.2. Obviously, Theorem 3.2 applies. Furthermore, the equation is strictly monotone. In [**40**, Theorem 1.4] it is shown that there exists $g \in C_0^\infty(\Omega)^n$ with the following properties:

- the unique H-solution is Hölder continuous on $\bar{\Omega}$;
- the unique W-solution is discontinuous at $x = 0$;
- there is the continuum of distinct week solutions.

Let us turn to the case of regular problems when there is no difference between weak, H-, and W-solutions. The following result on the high integrability of the gradient of solutions is obtained in [**40**, **41**].

THEOREM 3.3. *Let $\varphi(x,t) = t^{p(x)}$, where $p(x)$ is as in Example 2.1 and satisfies the* log-*Hölder condition (2.8). Suppose that $a \in M_\Omega$ is a single valued function and $g \in L^{p'}$ Then any weak solution u of problem (3.2) satisfies the estimate*

$$\int_\Omega |\nabla u|^{(1+\delta)p(x)}dx \le c\int_\Omega |g|^{(1+\delta)p'(x)}dx + c\,,$$

where $c > 0$ and $\delta > 0$ depend on $\|g\|_{L^{p'}}$.

*If the function $p(x)$ satisfies the double-*log-*condition*

$$|p(x) - p(y)| \le \frac{k\ln\ln(1/|x-y|)}{\ln(1/|x-y|)}\,, \quad |x-y| \le 1/2\,,$$

where $0 < k \le p/n$, then

$$\int_\Omega |\nabla u|^{p(x)}\ln^\delta(|\nabla u| + 2)dx \le c\int_\Omega |g|^{p'(x)}\ln(|g| + 2)dx + c\,.$$

In the regular case there are also certain results on the Hölder continuity of solutions [**1**, **17**, **18**].

PROBLEM 3. Extend the result of Theorem 3.3 to the case when $a \in M_\Omega$ is multivalued, and φ satisfies (N_1)–(N_3) and (2.8) (or a hypothetic extension of the double-log-condition mentioned in Problem 1).

4. Convergence of solutions

Along with equation (3.2), we consider a sequence of approximate problems

$$(4.1) \qquad -\operatorname{div} a_k(x,\nabla u) \ni \operatorname{div} g\,, \quad u|_{\partial\Omega} = 0\,,$$

assuming that the sequence a_k converges to a in certain sense. The question is whether solutions of problem (4.1) converge to solutions of problem (3.2).

Recall that a sequence A_k of maximal monotone operators from X into X^* *converges* to a maximal monotone operator A from X into X^* if for every $x \in X$ and $y \in X^*$ such that $y \in Ax$ there exist sequences $x_k \in X$ and $y_k \in X^*$ such that $y_k \in A_k x_k$, $x_k \to x$ strongly in X and $y_k \to y$ strongly in X^*. For a detailed discussion of this concept we refer to [**6**] (see also [**14**]).

First we consider the case when the functions a and a_k are of the same class M_Ω and satisfy the following assumption.

(A) *The functions $a_k(x,\cdot) \in M_\Omega$ converge to $a(x,\cdot) \in M_\Omega$ in the sense of maximal monotone operators a.e. in Ω.*

THEOREM 4.1 ([**7**]). *Let V be an admissible subspace of $W_0^{1,\varphi}(\Omega)$ and $g \in L^{\varphi^*}(\Omega)$. Under assumption (A), if $u_k \in V$ is a sequence of V-solutions of problems (4.1), then, along a subsequence, it converges to a V-solution of problem (3.2) weakly in V.*

This result is obtained in [**7**] for H- and W-solutions. The general case is similar. The main technical point of the proof is the following.

LEMMA 4.1. *Under assumption (A), the operators $\mathbf{a}_k : L^{\varphi}(\Omega)^n \to L^{\varphi^*}(\Omega)^n$ converge to $\mathbf{a} : L^{\varphi}(\Omega)^n \to L^{\varphi^*}(\Omega)^n$ in the sense of maximal monotone operators.*

In view of Theorem 4.1, the following problem naturally arises.

PROBLEM 4. Under which condition(s) for every $a \in M_\Omega$ there exists a sequence of single valued functions $a_k \in M_\Omega$, maybe with different constants c_1 and c_2 and function h, such that $a_k(x,\cdot)$ converges to $a(x,\cdot)$ in the sense of maximal monotone operators for almost all $x \in \Omega$.

In the simplest case of $p(x)$-growth (see Example 2.1) the answer is positive without any additional assumptions. Let

$$j(x,\xi) = |\xi|^{p(x)-2}\xi\,.$$

For a.e. $x \in \Omega$, this is the duality mapping [**22, 28, 29**] for the $l^{p(x)}$-norm on $\mathbb{R}^n$. For any $\lambda > 0$ we denote by $a_\lambda(x,\cdot)$ the Yosida approximation of $a(x,\cdot)$. More precisely, if $\xi \in \mathbb{R}^n$, then there exists a unique $\xi_\lambda \in \mathbb{R}^n$ such that

$$j(x,\xi_\lambda - \xi) + \lambda a(x,\xi_\lambda) \ni 0\,.$$

We set

$$a_\lambda(x,\xi) = \lambda^{-1} j(x,\xi - \xi_\lambda)\,.$$

Then, for a.e. $x \in \Omega$, $a_\lambda(x,\cdot)$ converges to $a(x,\cdot)$ as $\lambda \to 0$ in the sense of maximal monotone operators (see, *e.g.*, [**22**, Theorem 7.1]). Notice that the Yosida approximation is a standard tool in the theory of monotone multivalued operators.

In general, we expect that the answer is positive if there is a generalized N-function $\psi(x,t)$ which is C^1 with respect to t and such that

$$\lambda(\varphi(x,t) - 1) \le \psi(x,t) \le \Lambda(\varphi(x,t) + 1)$$

for some positive constants λ and Λ. In addition, ψ might need to have some kind of uniform convexity (see, *e.g.*, [**15, 45**]).

Now we turn to another type of approximations for a single valued Dirichlet problem

$$-\operatorname{div} a(x,\nabla u) = \operatorname{div} g\,, \quad u|_{\partial\Omega} = 0\,. \tag{4.2}$$

Approximate problems are of the same form

$$-\operatorname{div} a_k(x,\nabla u) = \operatorname{div} g\,, \quad u|_{\partial\Omega} = 0\,, \tag{4.3}$$

but with growth conditions different from that in problem (4.2).

Together with the generalized N-function φ we consider a sequence of generalized N-functions φ_k. We assume that

(A_1) *the functions φ and φ_k satisfy assumptions (N_1)–(N_3) uniformly with respect to k, and $\varphi_k(x,t) \to \varphi(x,t)$ and $\varphi_k^*(x,t) \to \varphi^*(x,t)$ for all $t \ge 0$ almost everywhere in Ω.*

This assumption implies that inequalities (2.4) and (2.5) hold for the functions φ and φ_k with all the constants independent of k.

We accept the following assumptions for a and a_k.

(A_2) *The functions $a(x,\xi)$ and $a_k(x,\xi)$ are monotone with respect to $\xi \in \mathbb{R}^n$ and satisfy the Carathéodory condition,* i.e. *they are measurable in $x \in \Omega$ for all $\xi \in \mathbb{R}^n$ and continuous in ξ for almost all $x \in \Omega$.*

(A_3) *There exist positive constants c_1 and c_2 and a nonnegative function $h \in L^1(\Omega)$ such that*

$$\varphi^*(x, a(x,\xi)) \le c_1\varphi(x,|\xi|) + h(x)\,,$$

$$a(x,\xi)\cdot\xi \ge c_c\varphi(x,|\xi|) - h(x)\,,$$

$$\varphi_k^*(x, a_k(x,\xi)) \le c_2\varphi_k(x,|\xi|) + h(x)$$

and

$$a_k(x,\xi)\cdot\xi \ge c_2\varphi_k(x,|\xi|) - h(x)$$

for all $\xi \in \mathbb{R}^n$ almost everywhere on Ω.

THEOREM 4.2 ([**7**]). *Suppose that assumptions (A_1)–(A_3) hold.*

(a) *Assume that*

$$\varphi_k(x,\gamma t) \le \varphi(x,t) + h_k(x) \tag{4.4}$$

for all k and $t \ge 0$ almost everywhere in Ω, where $\gamma > 0$ is independent of k, $h_k \in L^1(\Omega)$, with $\|h_k\|_{L^1} \le 1$. Let $g \in L^{p'}(\Omega)^n$. Then, along a subsequence, W-solutions $u_{W,k} \in W_0^{\varphi_k}(\Omega)$ of problem (4.3) converge weakly in $W_0^{1,p}(\Omega)$ to a W-solution of problem (4.2).

(b) *Assume that*

$$\varphi(x,\gamma t) \le \varphi_k(x,t) + h_k(x) \tag{4.5}$$

for all k and $t \ge 0$ almost everywhere in Ω, where $\gamma > 0$ is independent of k, $h_k \in L^1(\Omega)$, with $\|h_k\|_{L^1} \le 1$. Let $g \in L^{\varphi^*}(\Omega)^n$. Then, along a subsequence, H-solutions $u_{H,k} \in H_0^{\varphi_k}(\Omega)$ of problem (4.3) converge weakly in $H_0^{1,\varphi}(\Omega)$ to an H-solution of problem (4.2).

Under the assumptions of Theorem 4.2, the existence of solutions follows from Theorem 3.2.

Now we consider a special choice of functions a_k. Let $a(x,\xi)$ be as in Theorem 4.2 and

$$a_k(x,\xi) = a(x,\xi) + \varepsilon_k|\xi|^{q-2}\xi\,,$$

where $\varepsilon_k > 0$ and q is introduced in inequality (2.4). A natural choice of the growth function for a_k is

$$\varphi_k(x,t) = \varphi(x,t) + q^{-1}\varepsilon_k|t|^q\,.$$

If $g \in L^{\varphi^*}(\Omega)^n \subset L^{q'}(\Omega)^n$, then problem (4.3) has a unique solution

$$u_k \in H_0^{1,q}(\Omega) = W_0^{1,q}(\Omega) \subset H_0^{1,\varphi}(\Omega)\,.$$

PROPOSITION 4.1 ([**7**]). *Along a subsequence, solutions u_k converge weakly in $H_0^{1,\varphi}(\Omega)$ to an H-solution of problem (4.2) as $\varepsilon_k \to 0$.*

Theorem 4.2 and Proposition 4.1 apply to the case $\varphi(x,t) = t^{p(x)}$ and $\varphi_k(x,t) = t^{p_k(x)}$, where p and p_k are as in Example 2.1. In this case the weak convergence of W-solutions (respectively, H-solutions) takes place provided that $p_k(x) \leq p(x)$ (respectively, $p_k(x) \geq p(x)$) a. e. in Ω. For $a(x,\xi) = |\xi|^{p(x)-2}\xi$ and $a_k(x,\xi) = |\xi|^{p_k(x)-2}\xi$ this result is obtained in [**40**].

PROBLEM 5. Extend the results of Theorem 4.2 and Proposition 4.1 to the case of multivalued problems of the form (4.2) and (4.3).

The main point is to give an appropriate meaning to the convergence of maximal monotone operators defined on a variable sequence of Banach spaces so that an extension of Lemma 4.1 holds true. Notice that the problem is open even in the case of $p(x)$ and $p_k(x)$ growth assumptions.

5. Homogenization

Now we turn to the homogenization problem according to [**44**]. Let $p(y) = p(y_1, y_2, \ldots, y_n)$ be an L^∞ function on $\mathbb{R}^n$ such that $p(y)$ is 1-periodic in each y-coordinate (1-periodic in y for shortness) and $1 < p_0 \leq p(y) \leq p_1 < \infty$ for almost all $y \in \mathbb{R}^n$. Let $a(y,\xi)$ be a Carathéodory function from $\mathbb{R}^n \times \mathbb{R}^n$ into $\mathbb{R}^n$. We accept the following assumptions.

(H_1) *The function $a(y,\xi)$ is 1-periodic in y, and $a(y,0) = 0$.*

(H_2) *The function $a(y,\xi)$ is strictly monotone with respect to ξ,* i.e. *for almost all $y \in \mathbb{R}^n$*

$$(a(y,\xi) - a(y,\eta)) \cdot (\xi - \eta) > 0$$

for all $\xi \in \mathbb{R}^n$ and $\eta \in \mathbb{R}^n$, $\xi \neq \eta$.

(H_3) *There exist positive constants c_0 and c_1 such that for almost all $y \in \mathbb{R}^n$*

$$a(y,\xi) \cdot \xi \geq c_0 |\xi|^{p(y)} - c_1$$

and

$$|a(y,\xi)|^{p'(y)} \leq c(|\xi|^{p(y)} + 1)$$

for all $\xi \in \mathbb{R}^n$.

As usual, $p'(\cdot)$ stands for the conjugate exponent to $p(\cdot)$. For simplicity of the notation we set $p_\varepsilon(x) = p(x/\varepsilon)$ and $a_\varepsilon(x,\xi) = a(x/\varepsilon, \xi)$, where $\varepsilon > 0$.

Let $\Omega \subset \mathbb{R}^n$ be a bounded, Lipschitz domain. We consider the Dirichlet problem

$$-\operatorname{div} a_\varepsilon(x, \nabla u_\varepsilon) = \operatorname{div} g\,, \quad u_\varepsilon|_{\partial\Omega} = 0\,, \tag{5.1}$$

where $g \in L^\infty(\Omega)^n$. Theorem 3.2 implies that problem (5.1) possesses a unique H-solution $u_{\varepsilon,H} \in H_0^{p_\varepsilon(\cdot)}(\Omega)$ as well as a unique W-solution $u_{\varepsilon,W} \in W_0^{p_\varepsilon(\cdot)}(\Omega)$. Furthermore, the norms $\|u_{\varepsilon,H}\|_{H_0^{p_\varepsilon(\cdot)}}$ and $\|u_{\varepsilon,W}\|_{W_0^{p_\varepsilon(\cdot)}}$ are bounded by a constant that depends on the constants c, c_0 and c_1, and the norm $\|g\|_{L^{p_0'}}$.

We are interesting in the limit behavior of solutions $u_{\varepsilon,H}$ and $u_{\varepsilon,W}$ as $\varepsilon \to 0$ from the point of view of homogenization. Notice that each of the two types of solutions requires its own homogenized equation. According to the general idea of periodic homogenization, these equations are expressed in terms of auxiliary problems often called *cell* problems.

Let $\square$ denote the unit cube in $\mathbb{R}^n_y$ centered at the origin. The space $W^{1,p(\cdot)}_{per}$ consists of all 1-periodic functions $v \in W^{1,1}_{loc}(\mathbb{R}^n_y)$ such that

$$\int_\square v(y)dy = 0$$

and

$$\int_\square |v(y)|^{p(y)}dy < \infty \,.$$

The closure in $W^{1,p(\cdot)}_{per}$ of the subspace of smooth functions is denoted by $H^{1,p(\cdot)}_{per}$.

For each $\xi \in \mathbb{R}^n$, let $v_{\xi,H} \in H^{1,p(\cdot)}_{per}$ and $v_{\xi,W} \in W^{1,p(\cdot)}_{per}$ be solutions of the following problems

$$\int_\square a(y, \xi + v_{\xi,H}(y)) \cdot \nabla w(y)dy = 0 \quad \forall w \in H^{1,p(\cdot)}_{per} \tag{5.2}$$

and

$$\int_\square a(y, \xi + v_{\xi,W}(y)) \cdot \nabla w(y)dy = 0 \quad \forall w \in W^{1,p(\cdot)}_{per} \,. \tag{5.3}$$

As in the single-valued $t^{p(x)}$-case of Theorem 3.2, the unique solvability of both problems (5.2) and (5.3) is sufficiently simple, and can be obtained as in [**4, 5**] with the help of elementary Lemma 3.2.5 in [**15**].

In terms of solutions to problems (5.2) and (5.3) we introduce two homogenized functions $a_\alpha(\xi)$ ($\alpha = H, W$) of $\xi \in \mathbb{R}^n$ with values in $\mathbb{R}^n$ as follows

$$a_\alpha(\xi) = \int_\square a(y, \xi + \nabla v_\xi(y))dy \,.$$

In addition, the growth condition has to be homogenized as follows. We introduce the functions

$$\varphi_W(\xi) = \inf_{v \in W^{1,p(\cdot)}_{per}} \int_\square \frac{1}{p(y)} |\xi + \nabla_y v(y)|^{p(y)}dy$$

and

$$\varphi_H(\xi) = \inf_{v \in H^{1,p(\cdot)}_{per}} \int_\square \frac{1}{p(y)} |\xi + \nabla_y v(y)|^{p(y)}dy \,.$$

Notice that the infima above are attained [**21**].

The functions φ_H and φ_W are convex and satisfy the two-sided power estimate

$$\lambda_0|\xi|^{p_0} - 1 \le \varphi_W(\xi) \le \varphi_H(\xi) \le \lambda_1|\xi|^{p_1}$$

with some positive constants λ_0 and λ_1. Furthermore, these functions and their conjugate (in the sense of convex analysis) functions φ^*_H and φ^*_W satisfy (Δ_2) condition, *i.e.* there exists a constant $K > 0$ such that

$$\psi(2\xi) \le K\psi(\xi)\,, \quad \xi \in \mathbb{R}^n\,,$$

where ψ stands for either of those functions. Notice that there exist examples when $\varphi_W < \varphi_H$ [**21**]. These facts on homogenization of integral functionals can be found in [**21**] (see also [**44**]). Also we mention paper [**31**] which is worth the reader's attention.

If φ is one of the functions φ_H or φ_W, we denote by $L^\varphi(\Omega)$ the vector Orlicz space

$$\{v \in L^1(\Omega)^n : \varphi(v) \in L^1(\Omega)\}\,.$$

Equipped with the Luxemburg norm, this is a separable, reflexive Banach space, and

$$L^{p_1}(\Omega)^n \subset L^{\varphi}(\Omega) \subset L^{p_0}(\Omega)^n$$

(see, *e.g.*, [**26**] and references therein). The associated Orlicz-Sobolev space is defined by

$$W_0^{1,\varphi}(\Omega) = \{u \in W_0^{1,1} : \nabla u \in L^{\varphi}(\Omega)\}$$

and is a separable, reflexive Banach space.

The function $a_\alpha(\xi)$, $\alpha = H, W$, is continuous, strictly monotone, and satisfies the following inequalities

$$a_\alpha(\xi) \cdot \xi \geq c_0' \varphi_\alpha(\xi) - c_1' ,$$
$$\varphi_\alpha^*(a_\alpha(\xi)) \leq c'(\varphi_\alpha(\xi) + 1) ,$$

with some positive constants c_0', c_1' and c'.

Thus, there are two homogenized problems

$$- \operatorname{div} a_H(\nabla u) = \operatorname{div} g , \quad u \in W_0^{\varphi_H}(\Omega) , \tag{5.4}$$

and

$$- \operatorname{div} a_W(\nabla u) = \operatorname{div} g , \quad u \in W_0^{\varphi_W}(\Omega) . \tag{5.5}$$

We have the following homogenization theorem.

THEOREM 5.1 ([**44**]). *Assume that conditions* (H_1)–(H_3) *hold. Then the solutions* $u_{\varepsilon,H}$ *(respectively,* $u_{\varepsilon,W}$*) converge weakly in* $W_0^{1,p_0}(\Omega)$ *to a unique solution* u_h *(respectively,* u_w*) of problem* (5.4) *(respectively,* (5.5)*). Furthermore, the fluxes* $a_\varepsilon(u_{\varepsilon,H})$ *and* $a_\varepsilon(u_{\varepsilon,W})$ *converge weakly in* $L^{p_1'}(\Omega)^n$ *to the fluxes* $a_H(u_h)$ *and* $a_W(u_w)$*, respectively.*

The proof given in [**44**] is based on a version of compensated compactness. In the case of constant exponent this approach is introduced by Murat and Tartar (see [**25**]). The existence of solutions u_h and u_w is obtained in the proof of Theorem 5.1. The uniqueness follows immediately from strong monotonicity of the functions a_H and a_W.

To the best of our knowledge, Theorem 5.1 is the only result on homogenization of *non-variational* elliptic problems with nonstandard growth that takes into account the Lavrentiev phenomenon. Therefore, this area is open to further contributions. As the first step we suggest the following problem.

PROBLEM 6. Extend Theorem 5.1 to the case when the function a is multivalued, monotone (not necessarily strictly monotone) and satisfies growth and coerciveness conditions expressed in terms of anisotropic generalized N-function.

Notice that there are examples of non-strictly monotone single-valued problems with standard growth such that their homogenized problems are multi-valued [**12**, **27**].

We expect that a suitable extension of compensated compactness may lead to a positive result in Problem 6. The results from [**30, 39, 42, 43**] may be useful in this context. Let us mention other potential approaches such as two-scale convergence [**2, 46, 47**] and unfolding [**13, 14, 24**]. The approach developed in [**19, 20**] is of interest as well.

Especially interesting would be to put homogenization problems with nonstandard growth into the framework of an appropriate G-convergence theory as it is done in the case of regular power growth (see [**11, 12, 27**] and references therein). Such a hypothetic G-convergence theory should deal with sequences a_k of multivalued monotone functions from $\Omega \times \mathbb{R}^n$ into $\mathbb{R}^n$ that have properties similar to those given in Definition 3.1, with anisotropic generalized N-function $\varphi = \varphi_k$ depending on k. As in Section 3, a_k is expected to define two operators A_H^k and A_W^k acting between properly defined Sobolev spaces of H and W types, respectively. As consequence, there should be two type of G-convergence: for operators A_H^k and A_W^k, respectively. In contrast to the standard case, the main challenge on this way is caused by the fact that these operators act between variable spaces.

References

[1] Yu. A. Alkhutov, *The Harnack inequality and the Hölder property of solutions of nonlinear elliptic equations with a nonstandard growth condition* (Russian, with Russian summary), Differ. Uravn. **33** (1997), no. 12, 1651–1660, 1726; English transl., Differential Equations **33** (1997), no. 12, 1653–1663 (1998). MR1669915

[2] G. Allaire, *Homogenization and two-scale convergence*, SIAM J. Math. Anal. **23** (1992), no. 6, 1482–1518, DOI 10.1137/0523084. MR1185639

[3] B. Amaziane and L. Pankratov, *Homogenization in Sobolev spaces with nonstandard growth: brief review of methods and applications*, Int. J. Differ. Equ., posted on 2013, Art. ID 693529, 16, DOI 10.1155/2013/693529. MR3046764

[4] S. Antontsev and S. Shmarev, *Elliptic equations and systems with nonstandard growth conditions: existence, uniqueness and localization properties of solutions*, Nonlinear Anal. **65** (2006), no. 4, 728–761, DOI 10.1016/j.na.2005.09.035. MR2232679

[5] S. N. Antontsev and S. I. Shmarev, Elliptic equations with anisotropic nonlinearity and nonstandard growth, *Handbook of Differential Equations, Stationary Partial Differential Equations*, Vol. **3**, Elsevier, Amsterdam, 2006, 1–100.

[6] H. Attouch, *Variational convergence for functions and operators*, Applicable Mathematics Series, Pitman (Advanced Publishing Program), Boston, MA, 1984. MR773850

[7] M. Avci and A. Pankov, *Multivalued elliptic operators with nonstandard growth*, Adv. Nonlinear Anal. **7** (2018), no. 1, 35–48, DOI 10.1515/anona-2016-0043. MR3757454

[8] A. Benkirane and M. Ould Mohamedhen Val, *Some approximation properties in Musielak-Orlicz-Sobolev spaces*, Thai J. Math. **10** (2012), no. 2, 371–381. MR3001860

[9] F. E. Browder, *Nonlinear operators and nonlinear equations of evolution in Banach spaces*, Nonlinear functional analysis (Proc. Sympos. Pure Math., Vol. XVIII, Part 2, Chicago, Ill., 1968), Amer. Math. Soc., Providence, R. I., 1976, pp. 1–308. MR0405188

[10] Y. Chen, S. Levine, and M. Rao, *Variable exponent, linear growth functionals in image restoration*, SIAM J. Appl. Math. **66** (2006), no. 4, 1383–1406, DOI 10.1137/050624522. MR2246061

[11] V. Chiadò Piat, G. Dal Maso, and A. Defranceschi, *G-convergence of monotone operators* (English, with French summary), Ann. Inst. H. Poincaré Anal. Non Linéaire **7** (1990), no. 3, 123–160, DOI 10.1016/S0294-1449(16)30298-0. MR1065871

[12] V. Chiadò Piat and A. Defranceschi, *Homogenization of monotone operators*, Nonlinear Anal. **14** (1990), no. 9, 717–732, DOI 10.1016/0362-546X(90)90102-M. MR1049117

[13] D. Cioranescu, A. Damlamian, and G. Griso, *Periodic unfolding and homogenization* (English, with English and French summaries), C. R. Math. Acad. Sci. Paris **335** (2002), no. 1, 99–104, DOI 10.1016/S1631-073X(02)02429-9. MR1921004

[14] A. Damlamian, N. Meunier, and J. Van Schaftingen, *Periodic homogenization of monotone multivalued operators*, Nonlinear Anal. **67** (2007), no. 12, 3217–3239, DOI 10.1016/j.na.2006.10.007. MR2350881

[15] L. Diening, P. Harjulehto, P. Hästö, and M. Růžička, *Lebesgue and Sobolev spaces with variable exponents*, Lecture Notes in Mathematics, vol. 2017, Springer, Heidelberg, 2011. MR2790542

[16] X. Fan, *Differential equations of divergence form in Musielak-Sobolev spaces and a sub-supersolution method*, J. Math. Anal. Appl. **386** (2012), no. 2, 593–604, DOI 10.1016/j.jmaa.2011.08.022. MR2834769

[17] X. Fan and D. Zhao, *A class of De Giorgi type and Hölder continuity*, Nonlinear Anal. **36** (1999), no. 3, Ser. A: Theory Methods, 295–318, DOI 10.1016/S0362-546X(97)00628-7. MR1688232

[18] X. Fan and D. Zhao, *The quasi-minimizer of integral functionals with $m(x)$ growth conditions*, Nonlinear Anal. **39** (2000), no. 7, Ser. A: Theory Methods, 807–816, DOI 10.1016/S0362-546X(98)00239-9. MR1736389

[19] G. Francfort, F. Murat, and L. Tatar, *Monotone operators in divergence form with x-dependent multivalued graphs* (English, with English and Italian summaries), Boll. Unione Mat. Ital. Sez. B Artic. Ric. Mat. (8) **7** (2004), no. 1, 23–59. MR2044260

[20] G. A. Francfort, F. Murat, and L. Tartar, *Homogenization of monotone operators in divergence form with x-dependent multivalued graphs*, Ann. Mat. Pura Appl. (4) **188** (2009), no. 4, 631–652, DOI 10.1007/s10231-009-0094-9. MR2533960

[21] V. V. Jikov, S. M. Kozlov, and O. A. Oleĭnik, *Homogenization of differential operators and integral functionals*, Springer-Verlag, Berlin, 1994. Translated from the Russian by G. A. Yosifian [G. A. Iosif′yan]. MR1329546

[22] N. Kenmochi, *Monotonicity and compactness methods for nonlinear variational inequalities*, Handbook of differential equations: stationary partial differential equations. Vol. IV, Handb. Differ. Equ., Elsevier/North-Holland, Amsterdam, 2007, pp. 203–298, DOI 10.1016/S1874-5733(07)80007-6. MR2569333

[23] M. A. Krasnosel′skiĭ and Ja. B. Rutickiĭ, *Convex functions and Orlicz spaces*, Translated from the first Russian edition by Leo F. Boron, P. Noordhoff Ltd., Groningen, 1961. MR0126722

[24] N. Meunier and J. Van Schaftingen, *Periodic reiterated homogenization for elliptic functions* (English, with English and French summaries), J. Math. Pures Appl. (9) **84** (2005), no. 12, 1716–1743, DOI 10.1016/j.matpur.2005.08.003. MR2180388

[25] F. Murat, *Compacité par compensation* (French), Ann. Scuola Norm. Sup. Pisa Cl. Sci. (4) **5** (1978), no. 3, 489–507. MR506997

[26] J. Musielak, *Orlicz spaces and modular spaces*, Lecture Notes in Mathematics, vol. 1034, Springer-Verlag, Berlin, 1983. MR724434

[27] A. Pankov, *G-convergence and homogenization of nonlinear partial differential operators*, Mathematics and its Applications, vol. 422, Kluwer Academic Publishers, Dordrecht, 1997. MR1482803

[28] N. S. Papageorgiou and S. Th. Kyritsi-Yiallourou, *Handbook of applied analysis*, Advances in Mechanics and Mathematics, vol. 19, Springer, New York, 2009. MR2527754

[29] D. Pascali and S. Sburlan, *Nonlinear mappings of monotone type*, Martinus Nijhoff Publishers, The Hague; Sijthoff & Noordhoff International Publishers, Alphen aan den Rijn, 1978. MR531036

[30] S. E. Pastukhova and A. S. Khripunova, *Some versions of the compensated compactness principle* (Russian, with Russian summary), Mat. Sb. **202** (2011), no. 9, 135–160, DOI 10.1070/SM2011v202n09ABEH004192; English transl., Sb. Math. **202** (2011), no. 9-10, 1387–1412. MR2884368

[31] S. E. Pastukhova and A. S. Khripunova, *Gamma-closure of some classes of nonstandard convex integrands*, J. Math. Sci. (N.Y.) **177** (2011), no. 1, 83–108, DOI 10.1007/s10958-011-0449-9. Problems in mathematical analysis. No. 59. MR2838988

[32] V. D. Rădulescu and D. D. Repovš, *Partial differential equations with variable exponents*, Monographs and Research Notes in Mathematics, CRC Press, Boca Raton, FL, 2015. Variational methods and qualitative analysis. MR3379920

[33] M. Růžička, *Electrorheological fluids: modeling and mathematical theory*, Lecture Notes in Mathematics, vol. 1748, Springer-Verlag, Berlin, 2000. MR1810360

[34] E. Zeidler, *Nonlinear functional analysis and its applications. IV*, Springer-Verlag, New York, 1988. Applications to mathematical physics; Translated from the German and with a preface by Juergen Quandt. MR932255

[35] E. Zeidler, *Nonlinear functional analysis and its applications. II/B*, Springer-Verlag, New York, 1990. Nonlinear monotone operators; Translated from the German by the author and Leo F. Boron. MR1033498

[36] V. V. Zhikov, *The Lavrent′ev effect and averaging of nonlinear variational problems* (Russian), Differentsial′nye Uravneniya **27** (1991), no. 1, 42–50, 180; English transl., Differential Equations **27** (1991), no. 1, 32–39. MR1133509

[37] V. V. Zhikov, *On Lavrentiev's phenomenon*, Russian J. Math. Phys. **3** (1995), no. 2, 249–269. MR1350506

[38] V. V. Zhikov, *Solvability of the three-dimensional thermistor problem* (Russian, with Russian summary), Tr. Mat. Inst. Steklova **261** (2008), no. Differ. Uravn. i Din. Sist., 101–114, DOI 10.1134/S0081543808020090; English transl., Proc. Steklov Inst. Math. **261** (2008), no. 1, 98–111. MR2489700

[39] V. V. Zhikov, *On the technique of the passage to the limit in nonlinear elliptic equations* (Russian, with Russian summary), Funktsional. Anal. i Prilozhen. **43** (2009), no. 2, 19–38, DOI 10.1007/s10688-009-0014-1; English transl., Funct. Anal. Appl. **43** (2009), no. 2, 96–112. MR2542272

[40] V. V. Zhikov, *On variational problems and nonlinear elliptic equations with nonstandard growth conditions*, J. Math. Sci. (N.Y.) **173** (2011), no. 5, 463–570, DOI 10.1007/s10958-011-0260-7. Problems in mathematical analysis. No. 54. MR2839881

[41] V. V. Zhikov and S. E. Pastukhova, *On the improved integrability of the gradient of solutions of elliptic equations with a variable nonlinearity exponent* (Russian, with Russian summary), Mat. Sb. **199** (2008), no. 12, 19–52, DOI 10.1070/SM2008v199n12ABEH003980; English transl., Sb. Math. **199** (2008), no. 11-12, 1751–1782. MR2489687

[42] V. V. Zhikov and S. E. Pastukhova, *The compensated compactness principle* (Russian), Dokl. Akad. Nauk **433** (2010), no. 5, 590–595, DOI 10.1134/S1064562410040241; English transl., Dokl. Math. **82** (2010), no. 1, 590–595. MR2766572

[43] V. V. Zhikov and S. E. Pastukhova, *Lemmas on compensated compactness in elliptic and parabolic equations* (Russian, with Russian summary), Tr. Mat. Inst. Steklova **270** (2010), no. Differentsial′nye Uravneniya i Dinamicheskie Sistemy, 110–137, DOI 10.1134/S0081543810030089; English transl., Proc. Steklov Inst. Math. **270** (2010), no. 1, 104–131. MR2768940

[44] V. V. Zhikov and S. E. Pastukhova, *Homogenization of monotone operators under conditions of coercitivity and growth of variable order* (Russian, with Russian summary), Mat. Zametki **90** (2011), no. 1, 53–69, DOI 10.1134/S0001434611070078; English transl., Math. Notes **90** (2011), no. 1-2, 48–63. MR2908169

[45] V. V. Zhikov and S. E. Pastukhova, *Uniform convexity and variational convergence*, Trans. Moscow Math. Soc., posted on 2014, 205–231, DOI 10.1090/s0077-1554-2014-00232-6. MR3308610

[46] V. V. Zhikov and S. E. Pastukhova, *Homogenization and two-scale convergence in a Sobolev space with an oscillating exponent* (Russian, with Russian summary), Algebra i Analiz **30** (2018), no. 2, 114–144. MR3790734

[47] V. V. Zhikov and G. A. Yosifian, *Introduction to the theory of two-scale convergence*, J. Math. Sci. (N.Y.) **197** (2014), no. 3, 325–357, DOI 10.1007/s10958-014-1717-2. Translation of Tr. Semin. im. I. G. Petrovskogo No. 29 (2013), Part II, 281–332. MR3392863

Mathematics Department, Morgan State University
Email address: `alexander.pankov@morgan.edu`

Contemporary Mathematics
Volume **734**, 2019
https://doi.org/10.1090/conm/734/14778

Essential spectrum of Schrödinger operators with δ and δ'-interactions on systems of unbounded smooth hypersurfaces in $\mathbb{R}^n$

Vladimir Rabinovich

Dedicated to the centenary of the birth of Professor Selim Krein is dedicated

Abstract. The purpose of this paper is to study of the essential spectra of Schrödinger operators on $\mathbb{R}^n$ with surface δ and δ'-type interactions on systems $\cup_{k=1}^{N}\Gamma_k$ of simply connected unbounded C^2-hypersurfaces in $\mathbb{R}^n$ ($\dim\Gamma_k = n-1$) such that $dist(\Gamma_k, \Gamma_l) > 0$ for $k \neq l$. We consider the formal Schrödinger operators

$$(1) \qquad H_{\delta_\Gamma,\alpha} = -\Delta + W + \sum_{j=1}^{N}\alpha_j\delta_{\Gamma_j}, H_{\delta'_\Gamma,\beta} = -\Delta + W + \sum_{j=1}^{N}\beta_j\delta'_{\Gamma_j},$$

where δ_{Γ_j} are the delta-functions supported on the hypersurfaces Γ_j, δ'_{Γ_j} are the normal derivatives of δ_{Γ_j}. We show in the paper that the addition to the potential $W \in L^\infty(\mathbb{R}^n)$ of delta-type potentials supported on unbounded hypersurfaces significantly changes the essential spectrum of the Schrödinger operators $-\Delta + W$, and we give an effective description of the essential spectra of Schrödinger operators (1) taking into account a behavior at infinity the hypersurfaces Γ_j, the potentials W, and the strength interaction coefficients α_j, β_j , $j = 1, ..., N$. We apply for the study of the essential spectra of operators $H_{\delta_\Gamma,\alpha}, H_{\delta'_\Gamma,\beta}$ the limit operators method adapted for investigations of unbounded operators.

1. Introduction

Let $\{\Gamma_k\}_{k=1}^{N}$ be a family of simply connected unbounded C^2-hypersurfaces ($\dim\Gamma_j = n-1$) such that $\Gamma_k \cap \Gamma_l = \varnothing$ if $k \neq l$. We suppose that the hypersurfaces Γ_k have the following description at infinity: there exists $R > 0$ such that

$$(2) \qquad \Gamma_k = \{x = (x', x_n) \in \mathbb{R}^n, x' = (x_1, x_2, ..., x_{n-1}) : x_n = F_k(x'), |x'| > R\}$$

with $\partial_{x_j}F_k \in C_b^1(\mathbb{R}^{n-1}), j = 1, ..., n-1$ where $C_b^1(\mathbb{R}^{n-1})$ is the space of differentiable functions on $\mathbb{R}^{n-1}$ bounded with all first partial derivatives. Moreover we suppose

---

2010 *Mathematics Subject Classification.* Primary 35J10, 47A10.

Key words and phrases. Delta-interactions on hypersurfaces, Fredholm theory, essential spectrum.

This work is partially supported by the National System of Investigators of Mexico (SNI), and the Conacyt Project SB-179872.

that

$$\liminf_{x' \to \infty} (F_{k+1}(x') - F_k(x')) > 0 \tag{3}$$

for every $k = 1, 2, ..., N-1$.

The main purpose of the paper is to study the essential spectra of Schrödinger operators on $\mathbb{R}^n$ with δ and δ'-type interactions on the system of hypersurfaces $\{\Gamma_k\}_{k=1}^N$. That is we consider the formal Schrödinger operators of the form

$$H_{\delta_\Gamma, \alpha} = -\Delta + W + \sum_{j=1}^N \alpha_j \delta_{\Gamma_j}, H_{\delta'_\Gamma, \beta} = -\Delta + W + \sum_{j=1}^N \beta_j \delta'_{\Gamma_j}, \tag{4}$$

where $W \in L^\infty(\mathbb{R}^n)$, δ_{Γ_j} are delta-functions supported on the hypersurfaces Γ_j, δ'_{Γ_j} are the normal derivatives of δ_{Γ_j}, α_j, β_j have a physical sense the strength interaction coefficients on the hypersurfaces $\Gamma_j, j = 1, ...N$.

The Schrödinger operators with δ-type potentials supported on hypersurfaces are important in Mathematical Physics and have attracted a lot of attention: for instance they are used for a description of quantum particles interacting with charged hypersurfaces, in approximations of Hamiltonians of propagation of the electrons through thin barriers. There is an extensive literature devoted to Schrödinger operators with singular potentials (see, [**1**], [**2**], [**5**], [**23**],[**6**]-[**20**]),[**35**], and the references cite there.)

We note also the recent papers [**7**], [**8**], [**9**] devoted to Schrödinger operators with δ and δ'-interactions on a union of *compact* Lipschitz hypersurfaces. It was proved in these papers that the essential spectra of the operators $H_{\delta_\Gamma,\alpha}, H_{\delta'_\Gamma,\beta}$ coincide with the essential spectra of the operator $-\Delta + W$, that is the addition to the potential W of potentials of delta-interactions on compact hypersurfaces (possibly with intersections) does not change the essential spectrum of the Schrödinger operators.

However, as we will show in the paper that the addition to the potential W of delta-type potentials supported on unbounded hypersurfaces significantly changes the essential spectrum of the Schrödinger operators $-\Delta + W$. We will give an effective description of the essential spectra of Schrödinger operators (4) taking into account the behavior at infinity of the hypersurfaces Γ_j, the potentials W, and the strength interaction coefficients $\alpha_j, \beta_j, j = 1, ..., N$.

Let Ω_0 be a semi-bounded domain in $\mathbb{R}^n$ with the boundary Γ_1, Ω_k be layers between Γ_k and $\Gamma_{k+1}, k = 1, ..., N-1$, and Ω_N be a semi-bounded domain in $\mathbb{R}^n$ with the boundary Γ_N. We set $\Gamma = \bigcup_{k=1}^N \Gamma_k$, $\Omega = \bigcup_{j=0}^N \Omega_j = \mathbb{R}^n \backslash \Gamma$, and $H^s(\mathbb{R}^n \backslash \Gamma) = H^s(\Omega) = \oplus_{j=0}^N H^s(\Omega_j)$ where $H^s(\Omega_j)$ are the Sobolev spaces of order s on the domains Ω_j. The norm in $H^s(\Omega) = H^s(\mathbb{R}^n \backslash \Gamma)$ is introduced as

$$\|u\|_{H^s(\mathbb{R}^n \backslash \Gamma)} = \left(\sum_{j=0}^N \|u_j\|^2_{H^s(\Omega_j)} \right)^{1/2}$$

where u_j are the restrictions of $u \in H^s(\mathbb{R}^n \backslash \Gamma)$ on the domain Ω_j.

It is well-known (see for instance [**7**], [**8**], [**9**]) that one can associate with operators (4) unbounded operators $\mathcal{H}_{\delta_\Gamma,\alpha}, \mathcal{H}_{\delta'_\Gamma,\beta}$ in $L^2(\mathbb{R}^n)$ of transmission problems

defined as
(5)

$$\mathcal{H}_{\delta_\Gamma,\alpha}u(x)=\begin{cases}(-\Delta+W(x))\,u(x),x\in\mathbb{R}^n\setminus\Gamma,\\ dom\mathcal{H}_{\delta_\Gamma,\alpha}=D_{\mathcal{H}_{\delta_\Gamma,\alpha}}=\\ =\{u\in H^2(\mathbb{R}^n\backslash\Gamma):[u]_{\Gamma_j}=0,[\partial_\nu u]_{\Gamma_j}=\alpha_j u_j\mid_{\Gamma_j},j=1,...,N\}\end{cases},$$

(6)

$$\mathcal{H}_{\delta'_\Gamma,\beta}u(x)=\begin{cases}(-\Delta+W(x))\,u(x),x\in\mathbb{R}^n\setminus\Gamma,\\ dom\mathcal{H}_{\delta'_\Gamma,\beta}=D_{\mathcal{H}_{\delta'_\Gamma,\beta}}=\\ \{u\in H^2(\mathbb{R}^n\backslash\Gamma):[\partial_\nu u]_{\Gamma_j}=0,[u]_{\Gamma_j}=\beta_j\partial_\nu u_j\mid_{\Gamma_j},j=1,...,N\}\end{cases},$$

where $[u]_{\Gamma_j}$ is a jump of the function $u\in H^2(\mathbb{R}^n\backslash\Gamma)$ on the surface $\Gamma_j,[\partial_\nu u]_{\Gamma_j}$ is a jump on Γ_j of the normal derivative. We suppose that $W\in L^\infty(\mathbb{R}^n)$, and $\alpha_j,\beta_j\in C_b^1(\Gamma_j),j=1,...N$ where $C_b^1(\Gamma_j)$ is the space of differentiable functions on Γ_j bounded with their first partial derivatives.

We will prove that if $W\in L^\infty(\mathbb{R}^n)$ and $\alpha_j,\beta_j\in C_b^1(\Gamma_j),j=1,...N$ are real-valued functions then the operators $\mathcal{H}_{\delta_\Gamma,\alpha},\mathcal{H}_{\delta'_\Gamma,\beta}$ are self-adjoint in $L^2(\mathbb{R}^n)$.

We recall that the essential spectrum of an unbounded closed operator $\mathcal{A}$ in a Hilbert space $\mathcal{H}$ with a dense in $\mathcal{H}$ domain $dom\mathcal{A}$ is defined as

$$sp_{ess}\mathcal{A}=\{\lambda\in\mathbb{C}:\mathcal{A}-\lambda I\text{ is not Fredholm operator as unbounded operator}\}.$$

The unbounded operator $\mathcal{A}$ can be considered as a bounded operator $\mathcal{A}$ acting from $dom\mathcal{A}$ with the graph norm into the space $\mathcal{H}$. Note that the operator $\mathcal{A}-\lambda I$ is a Fredholm operator as an unbounded operator if and only if $\mathcal{A}-\lambda I$ is a Fredholm operator as a bounded operator acting from $dom\mathcal{A}$ into $\mathcal{H}$ (see for instance [**4**]). Hence for the study of the essential spectra of the operators $\mathcal{H}_{\delta_\Gamma,\alpha},\mathcal{H}_{\delta'_\Gamma,\beta}$ one can consider the Fredholm property of bounded operators $\mathcal{H}_{\delta_\Gamma,\alpha}:D_{\mathcal{H}_{\delta_\Gamma,\alpha}}\to L^2(\mathbb{R}^n),\mathcal{H}_{\delta'_\Gamma,\beta}:D_{\mathcal{H}_{\delta'_\Gamma,\beta}}\to L^2(\mathbb{R}^n)$ of transmission problems (5),(6).

The transmission problems for two domains D_+ and $D_-=\mathbb{R}^n\backslash\bar{D}_+$ where D_+ is a bounded domain with a smooth boundary Γ have been considered in [**36**], [**34**], see also [**4**], page 70-72, [**22**], page 141-144. Transmission problems for two unbounded domains with a common smooth unbounded boundary has been considered in [**33**] by means of the limit operators method. Applying ideas and methods of this paper we obtain necessary and sufficient conditions for transmission problems (5),(6) to be Fredholm what allows us to obtain an effective description of essential spectra of the operators $\mathcal{H}_{\delta_\Gamma,\alpha},\mathcal{H}_{\delta'_\Gamma,\beta}$.

Note that the limit operators method is an effective tool for a study of essential spectra of different problems of Mathematical Physics in unbounded domains. The general theory of limit operators method for bounded operators has been presented in the book [**24**]. In the papers [**26**], [**27**] this method was applied to an investigation of the essential spectra of electromagnetic Schrödinger and Dirac operators on $\mathbb{R}^n$ for wide classes of potentials, and in the papers [**28**],[**30**],[**29**],[**31**],[**32**] for an investigation of essential spectra of discrete operators on $\mathbb{Z}^n$, on periodic combinatorial and quantum graphs, and quantum waveguides.

The paper is organized as follows. In Section 2 we give some auxiliary material and prove self-adjointness of operators $\mathcal{H}_{\delta_\Gamma,\alpha},\mathcal{H}_{\delta'_\Gamma,\beta}$. In Section 3 we give a description of the essential spectra of the operators $\mathcal{H}_{\delta_\Gamma,\alpha},\mathcal{H}_{\delta'_\Gamma,\beta}$ for a system $\{\Gamma_k\}_{k=1}^N$ of parallel hyperplanes in $\mathbb{R}^n$ applying the limit operators method. For a case of slowly oscillating at infinity potentials W and strength interaction coefficients $\alpha_j, \beta_j, j = 1, ..., N$ the limit operators are enough simple and their spectra can be found in explicit forms. In Section 3 we consider the essential spectra of the operators $\mathcal{H}_{\delta_\Gamma,\alpha}, \mathcal{H}_{\delta'_\Gamma,\beta}$ for C^2-hypersurfaces $\Gamma_k, k = 1, ..., N$ slowly oscillating at infinity.

2. Self-adjointness of the operators $\mathcal{H}_{\delta_\Gamma,\alpha}, \mathcal{H}_{\delta'_\Gamma,\beta}$

- We use the following notations: $C_b(\mathbb{R}^n)$ is a C^*-algebra of bounded continuous functions on $\mathbb{R}^n$, $C_b^m(\mathbb{R}^n)$ is a sub-algebra of $C_b(\mathbb{R}^n)$ consisting of functions f such that $\partial^\alpha f \in C_b(\mathbb{R}^n)$ for all multi-indices $\alpha : |\alpha| \leq m, C_b^\infty(\mathbb{R}^n) = \bigcap_{m=0}^\infty C_b^m(\mathbb{R}^n)$. In the natural way the space $C_b^m(\Gamma_j)$ of functions bounded with all derivatives of the order smaller or equal m on Γ_j is defined. If a is a function as usual we denote by the same letter a an operator of multiplication by this function acting in an admissible functional space.
- If X, Y are Banach spaces then we denote by $\mathcal{B}(X, Y)$ the space of bounded linear operators acting from X into Y with the uniform operator topology, and by $\mathcal{K}(X, Y)$ the subspace of $\mathcal{B}(X, Y)$ of all compact operators. In the case $X = Y$ we write shortly $\mathcal{B}(X)$ and $\mathcal{K}(X)$.
- We recall that $A \in \mathcal{B}(X, Y)$ is a Fredholm operator if

$$ker A = \{x \in X : Ax = 0\}, coker A = Y/\operatorname{Im} A$$

are finite dimensional spaces. Let $\mathcal{A}$ be a closed unbounded operator in a Hilbert space $\mathcal{H}$ with a dense in $\mathcal{H}$ domain $D_\mathcal{A}$. Then $\mathcal{A}$ is called a Fredholm operator if $\{x \in D_\mathcal{A} : \mathcal{A}x = 0\}$ and $\mathcal{H}/\operatorname{Im}\mathcal{A}$ where $\operatorname{Im}\mathcal{A} = \{y \in \mathcal{H} : y = \mathcal{A}x, x \in D_\mathcal{A}\}$ are finite dimensional spaces.

Note that $\mathcal{A}$ is a Fredholm operator as unbounded operator in $\mathcal{H}$ if and only if $\mathcal{A} : D_\mathcal{A} \to \mathcal{H}$ is a Fredholm operator as a bounded operator where $D_\mathcal{A}$ is equipped by the graph norm

$$\|u\|_{D_\mathcal{A}} = \left(\|u\|_\mathcal{H}^2 + \|\mathcal{A}u\|_\mathcal{H}^2\right)^{1/2}, u \in D_\mathcal{A}$$

(see for instance [**4**]).
- The essential spectrum $sp_{ess}\mathcal{A}$ of an unbounded operator $\mathcal{A}$ is a set of $\lambda \in \mathbb{C}$ such that $\mathcal{A} - \lambda I$ is not Fredholm operator as unbounded operator.

We consider the unbounded operators $\mathcal{H}_{\delta_\Gamma,\alpha}, \mathcal{H}_{\delta'_\Gamma,\beta}$ in $L^2(\mathbb{R}^n)$ defined as:

(7)
$$\mathcal{H}_{\delta_\Gamma,\alpha}u(x) = \left\{ \begin{array}{c} (-\Delta + W(x))\, u(x), x \in \mathbb{R}^n \setminus \Gamma, \\ D_{\mathcal{H}_{\delta_\Gamma,\alpha}} = \left\{u \in H^2(\mathbb{R}^n \backslash \Gamma) : [u]_{\Gamma_j} = 0, [\partial_\nu u]_{\Gamma_j} = \alpha_j u\,|_{\Gamma_j}, j = 1, ..., N\right\} \end{array} \right.,$$

$$(8)\quad \mathcal{H}_{\delta'_\Gamma,\beta}u(x) = \left\{ \begin{array}{c} (-\Delta + W(x))\, u(x), x \in \mathbb{R}^n \setminus \Gamma, \\ D_{\mathcal{H}_{\delta'_\Gamma,\beta}} = \left\{u \in H^2(\mathbb{R}^n \backslash \Gamma) : [\partial_\nu u]_{\Gamma_j} = 0, [u]_{\Gamma_j} = \beta_j \partial_\nu u\,|_{\Gamma_j}\right\} \end{array} \right.,$$

where $([u]_{\Gamma_j} = \gamma_{\Gamma_j}^+ u_j - \gamma_{\Gamma_j}^- u_{j-1}$, $[\partial_\nu u]_{\Gamma_j} = \gamma_{\Gamma_j}^+ \partial_\nu u_j - \gamma_{\Gamma_j}^- \partial_\nu u_{j-1}$, and ∂_ν is the normal derivative to Γ_j defined by the unit normal vector ν to Γ_j directed to the domain $\Omega_j, j = 0, 1, ..., N-1$, $\gamma_{\Gamma_j}^- : H^s(\Omega_{j-1}) \to H^{s-1/2}(\Gamma_j)$, $\gamma_{\Gamma_j}^+ : H^s(\Omega_j) \to H^{s-1/2}(\Gamma_j), s > 1/2$ are the trace operators. We suppose that $W \in L^\infty(\mathbb{R}^n)$, and $\alpha_j, \beta_j \in C_b^1(\Gamma_j), j = 1, ..., N$.

We denote by $\mathcal{A}_{\delta_\Gamma,\alpha}(\mu), \mathcal{B}_{\delta'_\Gamma,\beta}(\mu)$ families of operators of transmission problems depending on the parameter $\mu^2, \mu>0$

$$\mathcal{A}_{\delta_\Gamma,\alpha}(\mu)u(x)=\left\{\begin{array}{c}(-\Delta+\mu^2)\,u(x), x\in\mathbb{R}^n\setminus\Gamma, \mu>0,\\ [u]_{\Gamma_j}=0, [\partial_\nu u]_{\Gamma_j}=\alpha_j u\mid_{\Gamma_j}, j=1,...,N\end{array}\right.$$

$$\mathcal{B}_{\delta'_\Gamma,\beta}(\mu)=\left\{\begin{array}{c}(-\Delta+\mu^2)\,u(x), x\in\mathbb{R}^n\setminus\Gamma, \mu>0,\\ [\partial_\nu u]_{\Gamma_j}=0, [u]_{\Gamma_j}=\beta_j\partial_\nu u\mid_{\Gamma_j}\end{array}\right. .$$

Let $H^2_{\delta_\Gamma,\alpha}(\mathbb{R}^n\backslash\Gamma)=D_{\mathcal{H}_{\delta_\Gamma,\alpha}}, H^2_{\delta'_\Gamma,\beta}(\mathbb{R}^n\backslash\Gamma)=D_{\mathcal{H}_{\delta'_\Gamma,\beta}}$. Then we consider $\mathcal{A}_{\delta_\Gamma,\alpha}(\mu):H^2_{\delta_\Gamma,\alpha}(\mathbb{R}^n\backslash\Gamma)\to L^2(\mathbb{R}^n)$, $\mathcal{B}_{\delta'_\Gamma,\beta}(\mu):H^2_{\delta'_\Gamma,\beta}(\mathbb{R}^n\backslash\Gamma)\to L^2(\mathbb{R}^n)$ as bounded operators.

PROPOSITION 1. *Let $\alpha_j,\beta_j\in C^1_b(\Gamma_j)$ and $\inf_{s\in\Gamma_j}|\beta_j(s)|>0, j=1,...,N$. Then there exists $\mu_0\geq 1$ such that the operators $\mathcal{A}_{\delta_\Gamma,\alpha}(\mu_0):H^2_{\delta_\Gamma,\alpha}(\mathbb{R}^n\backslash\Gamma)\to L^2(\mathbb{R}^n)$ and $\mathcal{B}_{\delta'_\Gamma,\beta}(\mu_0):H^2_{\delta'_\Gamma,\beta}(\mathbb{R}^n\backslash\Gamma)\to L^2(\mathbb{R}^n)$ are isomorphisms.*

PROOF. For the proof of Proposition 1 we use an approach of the well-known paper [**3**] where the authors studied general elliptic boundary value problems depending on parameters in bounded domains in $\mathbb{R}^n$. Just like in this paper for the proof of Proposition 1 we use a partition of unity and construct local parametrizes depending on the parameters at every element of this partition of unity, and then we gluing a global inverse operator from these parametrizes for enough large values of the parameter. But unlike of the paper [**3**] we use an infinite partition of the unity (see for instance [**12**], [**24**].) □

THEOREM 2. *Let $W\in L^\infty(\mathbb{R}^n), \alpha_j,\beta_j\in C^1_b(\Gamma_j), j=1,...,N$ are real-valued functions, and $\inf_{s\in\Gamma_j}|\beta_j(s)|>0$. Then $\mathcal{H}_{\delta_\Gamma,\alpha},\mathcal{H}_{\delta'_\Gamma,\beta}$ are self-adjoint operators in $L^2(\mathbb{R}^n)$.*

PROOF. We prove the statement of Theorem 2 for $\mathcal{H}_{\delta_\Gamma,\alpha}$ only. For $\mathcal{H}_{\delta'_\Gamma,\beta}$ the proof is same. Let $\mu_0\geq 1$ be such that $\mathcal{A}_{\delta_\Gamma,\alpha}(\mu_0):H^2_{\delta_\Gamma,\alpha}(\mathbb{R}^n\backslash\Gamma)\to L^2(\mathbb{R}^n)$ is an isomorphism. Applying the Green formula we obtain that

$$((-\Delta+\mu_0)u,v)=((u,(-\Delta+\mu_0)v)$$
$$+\sum_{j=1}^N\int_{\Gamma_j}\left([(\partial_\nu u)\,\bar v]_{\Gamma_j}-[u\,(\partial_\nu\bar v)]_{\Gamma_j}\right)ds_j,\ u,v\in H^2_{\delta_\Gamma,\alpha}(\mathbb{R}^n\backslash\Gamma)$$

where (u,v) is the standard scalar product in $L^2(\mathbb{R}^n), ds_j$ is the surface Lebesgue measure on Γ_j. Taking into account conditions $[u]_{\Gamma_j}=0, [\partial_\nu u]_{\Gamma_j}=\alpha_j u\mid_{\Gamma_j}$, and the fact that α_j are real-valued functions, we obtain that $[(\partial_\nu u)\,\bar v]_{\Gamma_j}-[u\,(\partial_\nu\bar v)]_{\Gamma_j}=0$. Hence $-\Delta+\mu_0$ is a symmetric operator. To prove that $-\Delta+\mu_0$ with domain $H^2_{\delta_\Gamma,\alpha}(\mathbb{R}^n\backslash\Gamma)$ is a self-adjoint operator we have to show that $D_{(-\Delta+\mu_0)^*}=D_{-\Delta+\mu_0}$. Since $-\Delta+\mu_0$ is a symmetric operator $D_{-\Delta+\mu_0}\subset D_{(-\Delta+\mu_0)^*}$. Suppose that $u\in D_{(-\Delta+\mu_0)^*}$, then $(-\Delta+\mu_0)^*\,u=f\in L^2(\mathbb{R}^n)$ and Proposition 1 implies that there exists $v\in D_{-\Delta+\mu_0}$ such that $(-\Delta+\mu_0)\,v=f$. Since $D_{-\Delta+\mu_0}\subset D_{(-\Delta+\mu_0)^*}$ we obtain that $(-\Delta+\mu_0)^*\,v=f$. Hence $u-v\in\ker(-\Delta+\mu_0)^*=(\mathrm{Im}(-\Delta+\mu_0))^\perp=\{0\}$. Therefore $u=v\in D_{-\Delta+\mu_0}$ and $D_{-\Delta+\mu_0}=D_{(-\Delta+\mu_0)^*}$. Thus $-\Delta+\mu_0$ is a self-adjoint operator in $L^2(\mathbb{R}^n)$ with domain $H^2_{\delta_\Gamma,\alpha}(\mathbb{R}^n\backslash\Gamma)$. Note that bounded operators in $L^2(\mathbb{R}^n)$ are strongly dominated by the operator $-\Delta$ with domain $H^2_{\delta_\Gamma,\alpha}(\mathbb{R}^n\backslash\Gamma)$ (see [**10**], page 73). Hence $-\Delta$ with domain $H^2_{\delta_\Gamma,\alpha}(\mathbb{R}^n\backslash\Gamma)$ is a self-adjoint operator and since $W\in L^\infty(\mathbb{R}^n)$ is a real-valued function the operators $-\Delta$

and $-\Delta + W$ with domains $H^2_{\delta_\Gamma,\alpha}(\mathbb{R}^n\backslash\Gamma)$ are self-adjoint operators simultaneously (see, [**10**], page 100, Theorem 9). □

3. Delta-interactions on systems of parallel hyperplanes

3.1. Fredholm property of operators in the algebra $\mathfrak{A}(\mathbb{R}^n)$. We give here in a short form some definitions and results from the papers [**25**], [**21**] and book [**24**], Sec. 3.1.4. concerning the local invertibility at infinity and limit operators.

- We denote by $C_{b,u}(\mathbb{R}^n)$ the C^*-algebra of bounded uniformly continuous functions on $\mathbb{R}^n$. If $a \in C(\mathbb{R}^n)$, then we set

$$a_t(x) = a(t_1x_1, t_2x_2, ..., t_nx_n); t = (t_1, t_2, ..., t_n) \in \mathbb{R}^n.$$

  We say that an operator $A \in \mathcal{B}(L^2(\mathbb{R}^n))$ belongs to the class $\mathfrak{A}(\mathbb{R}^n)$ if for every function $a \in C_{b,u}(\mathbb{R}^n)$

$$\lim_{t\to 0} \|[A, a_t]\|_{\mathcal{B}(L^2(\mathbb{R}^n))} = \lim_{t\to 0} \|Aa_t - a_tA\|_{\mathcal{B}(L^2(\mathbb{R}^n))} = 0.$$

  Note that $\mathfrak{A}(\mathbb{R}^n)$ is a C^*-algebra. It implies in particular that if $A \in \mathfrak{A}(\mathbb{R}^n)$ is an invertible operator in $L^2(\mathbb{R}^n)$ then $A^{-1} \in \mathfrak{A}(\mathbb{R}^n)$ also.
- We introduce a compactification $\tilde{\mathbb{R}}^n$ of $\mathbb{R}^n$ obtained by joining to an every ray emanating from the origin and intersecting the unit sphere S^{n-1} at a point ω an infinitely distant point η_ω. A topology in $\tilde{\mathbb{R}}^n$ is introduced such that $\tilde{\mathbb{R}}^n$ is isomorphic to the closed unit ball $B_1 = \{x \in \mathbb{R}^n : |x| \leq 1\}$. Fundamental systems of neighborhoods of points $x \in \mathbb{R}^n$ are formed by standard systems of neighborhoods of this point, and fundamental systems of neighborhoods of points η_ω are formed by conical sets

$$U_{\eta_\omega} = \{x \in \mathbb{R}^n : x = \lambda\theta, \theta \in \Omega_\omega, \lambda > R > 0\}$$

  where Ω_ω are neighborhoods of the point ω on the unit sphere S^{n-1}.
- Let $\varphi \in C_0^\infty(\mathbb{R}^n), 0 \leq \varphi(x) \leq 1,\ \varphi(x) = \begin{cases} 1, |x| \leq 1 \\ 0, |x| \geq 2 \end{cases}, \varphi_R(x) = \varphi(x/R)$, $R > 0$, and $\psi_R(x) = 1 - \varphi_R(x)$. We say that an operator $A \in \mathcal{B}(L^2(\mathbb{R}^n))$ is locally invertible at infinity if there exists $R > 0$ and operators $\mathcal{L}_R, \mathcal{R}_R \in \mathcal{B}(L^2(\mathbb{R}^n))$ such that

$$\mathcal{L}_RA\psi_R = \psi_R, \psi_RA\mathcal{R}_R = \psi_R.$$

- Let a sequence $h = \{h_m\}_{m=1}^\infty$ be such that $h_m \in \mathbb{Z}^n$ and $\lim_{m\to\infty} h_m = \infty$. We denote the set of such sequences by $\mathfrak{M}(\mathbb{R}^n)$. Let an operator $A \in \mathcal{B}(L^2(\mathbb{R}^n))$. We say that $A^h \in \mathcal{B}(L^2(\mathbb{R}^n))$ is a limit operator of A defined by a sequence $h \in \mathfrak{M}(\mathbb{R}^n)$ if for every function $\varphi \in C_0^\infty(\mathbb{R}^n)$

$$\begin{aligned}&\lim_{m\to\infty} \left\|(V_{-h_m}AV_{h_m} - A^h)\varphi\right\|_{\mathcal{B}(L^2(\mathbb{R}^n))} \\ &= \lim_{m\to\infty} \left\|\varphi(V_{-h_m}AV_{h_m} - A^h)\right\|_{\mathcal{B}(L^2(\mathbb{R}^n))} = 0\end{aligned}$$

  where $V_gu(x) = u(x-g), g \in \mathbb{R}^n$ is the unitary in $L^2(\mathbb{R}^n)$ shift operator.
- We say that an operator $A \in \mathcal{B}(L^2(\mathbb{R}^n))$ is *rich* if every sequence $g \in \mathfrak{M}(\mathbb{R}^n)$ has a subsequence $h \in \mathfrak{M}(\mathbb{R}^n)$ defining a limit operator A^g. We denote by $Lim_\infty A$ the set of all limit operators of A defined by the sequences $h = \{h_m\}_{m=1}^\infty \in \mathfrak{M}(\mathbb{R}^n)$.

THEOREM 3. *(see for instance* [**25**], [**21**]*) Let an operator* $A \in \mathfrak{A}(\mathbb{R}^n)$ *and be rich. Then* A *is a locally invertible at infinity operator if and only if all limit operators* $A^h \in Lim_\infty A$ *are invertible.*

REMARK 4. Note that in the paper [**25**] the following result has been obtained. Let an operator $A \in \mathfrak{A}(\mathbb{R}^n)$ and be rich. Then A is a locally invertible at infinity operator if and only if all limit operators $A^h \in Lim_\infty A$ are invertible and the norms of their inverses are *uniformly bounded.* But in the paper [**21**] it was proved that the invertibility of all limit operators implies the uniform boundedness of their inverses.

3.2. Fredholm properties and essential spectrum of operators of interactions on a system of parallel hyperplanes. Let

$$\begin{aligned}
&\Gamma_k = \{x = (x', x_n) \in \mathbb{R}^n : x_n = d_k\}, k = 1, ..., N,\\
&d_k > d_{k-1}, k = 2, ..., N;\\
&\Omega_0 = \{x = (x', x_n) \in \mathbb{R}^n : x_n < d_1\}, ...,\\
&\Omega_k = \{x = (x', x_n) \in \mathbb{R}^n : d_k < x_n < d_{k+1}\}, k = 1, ..., N-1,\\
&\quad \Omega_N = \{x = (x', x_n) \in \mathbb{R}^n : x_n > d_N\},\\
&\Gamma = \cup_{k=1}^N \Gamma_k, \mathbb{R}^n \backslash \Gamma = \cup_{k=1}^N \Omega_k.
\end{aligned}$$

We denote by $\mathcal{H}_{\delta_\Gamma,\alpha}, \mathcal{H}_{\delta'_\Gamma,\beta}$ unbounded operators in $L^2(\mathbb{R}^n)$ defined by the Schrödinger operators $S = -\Delta + W, W \in L^\infty(\mathbb{R}^n)$ with domains

$$H^2_{\delta_\Gamma,\alpha}(\mathbb{R}^n\backslash\Gamma) = \left\{ u \in H^2(\mathbb{R}^n\backslash\Gamma) : [u]_{\Gamma_k} = 0, \left[\frac{\partial u}{\partial x_n}\right]_{\Gamma_k} = \alpha_k u|_{\Gamma_k}, k = 1, ..., N \right\},$$

and

$$H^2_{\delta'_\Gamma,\beta}(\mathbb{R}^n\backslash\Gamma) = \left\{ u \in H^2(\mathbb{R}^n\backslash\Gamma) : \left[\frac{\partial u}{\partial x_n}\right]_{\Gamma_k} = 0, [u]_{\Gamma_k} = \beta_k \left(\frac{\partial u_k}{\partial x_n}\right)|_{\Gamma_k}, k = 1, ..., N \right\},$$

respectively.

We will suppose that $\alpha_k, \beta_k \in C^1_b(\Gamma_k)$. It means that $\tilde{\alpha}_k(x') = \alpha(x', d_k), \tilde{\beta}_k(x') = \beta(x', d_k)$ belong to $C^1_b(\mathbb{R}^{n-1})$. Hence the equality $\left[\frac{\partial u}{\partial x_n}\right]_{\Gamma_k} = \alpha_k u_k|_{\Gamma_k}$, and $[u]_{\Gamma_k} = \beta_k \frac{\partial u}{\partial x_n}|_{\Gamma_k}$ have a sense in $H^{1/2}(\Gamma_k)$.

We consider also the unbounded operators $\mathcal{H}_{\delta_\Gamma,\alpha}, \mathcal{H}_{\delta'_\Gamma,\beta}$ as bounded operators of transmission problems acting from $H^2_{\delta_\Gamma,\alpha}(\mathbb{R}^n\backslash\Gamma)$ into $L^2(\mathbb{R}^n)$, and $H^2_{\delta'_\Gamma,\beta}(\mathbb{R}^n\backslash\Gamma)$ into $L^2(\mathbb{R}^n)$, respectively.

Let $h = (h', h_n) \in \mathbb{R}^n$, and $V_h u(x) = u(x-h)$ be the shift operator. Then

$$\begin{aligned}
(9)\qquad & V_{-h}(-\Delta + W)V_h = -\Delta + W(x+h),\\
& V_{-h}\left[V_h u\right]_{\Gamma_k} = [u]_{\Gamma_k}, V_{-h}\left[\frac{\partial (V_h u)}{\partial x_n}\right]_{\Gamma_k} = \left[\frac{\partial u}{\partial x_n}\right]_{\Gamma_k},\\
& V_{-h}\alpha_k (V_h u)|_{\Gamma_k} = \tilde{\alpha}_k(\cdot + h')u_k|_{\Gamma_k},\\
& V_{-h}\beta_k \frac{\partial (V_h u)}{\partial x_n}|_{\Gamma_k} = \tilde{\beta}(\cdot + h'_m)\frac{\partial u_{k'}}{\partial x_n}|_{\Gamma_k}.
\end{aligned}$$

DEFINITION 5. We say that a function $f \in L^{\infty}(\mathbb{R}^n)$ is *rich* if every sequence $\mathbb{Z}^n \in g_m \to \infty$ has a subsequence $h_m \to \infty$ defining a limit function f^h such that for every compact set $K \subset \mathbb{R}^n$.

$$\lim_{m\to\infty} \sup_{x\in K} \left| f^h(x) - f(x+h_m) \right| = 0. \tag{10}$$

Note that $\|f^h\|_{L^\infty(\mathbb{R}^n)} \leq \|f\|_{L^\infty(\mathbb{R}^n)}$.

In what follows we suppose that the potential $W \in L^{\infty}(\mathbb{R}^n)$ is rich.

Let a sequence $\mathbb{Z}^n \ni g_m \to \infty$. Since $\tilde{\alpha}_k, \tilde{\beta}_k \in C_b^1(\mathbb{R})$ the sequences $\tilde{\alpha}_k(x' + g'_m), \tilde{\beta}_k(x' + g'_m)$, $g'_m \in \mathbb{Z}^{n-1}$ are uniformly bounded and equicontinuous. Hence by the Arzela-Ascoli Theorem every sequence $g'_m \to \infty$ has a subsequence h'_m defining limits functions a_k^h, β_k^h such that for every compact set $K' \subset \mathbb{R}^{n-1}$

$$\lim_{m\to\infty} \sup_{x'\in K'} \left| \alpha_k^h(x') - \tilde{\alpha}_k(x'+h'_m) \right| = 0, \tag{11}$$
$$\lim_{m\to\infty} \sup_{x'\in K'} \left| \beta_k^h(x') - \tilde{\beta}_k(x'+h'_m) \right| = 0.$$

That is $\tilde{a}_k, \tilde{b}_k$ are rich functions on $\mathbb{R}^{n-1}$.

DEFINITION 6. We say that $\mathcal{H}^h_{\delta_\Gamma,\alpha}, \mathcal{H}^h_{\delta'_\Gamma,\beta}$ are *limit operators for the operators* $\mathcal{H}_{\delta_\Gamma,\alpha}, \mathcal{H}_{\delta'_\Gamma,\beta}$ if for every $R > 0$

$$\begin{aligned} &\lim_{m\to\infty} \left\| \left(V_{-h_m}\mathcal{H}_{\delta_\Gamma,\alpha}V_{h_m} - \mathcal{H}^h_{\delta_\Gamma,\alpha} \right)\psi_R \right\|_{\mathcal{B}(H^2_{\delta_\Gamma,\alpha}(\mathbb{R}^n\setminus\Gamma),L^2(\mathbb{R}^n))} \\ &= \lim_{m\to\infty} \left\| \psi_R \left(V_{-h_m}\mathcal{H}_{\delta_\Gamma,\alpha}V_{h_m} - \mathcal{H}^h_{\delta_\Gamma,\alpha} \right) \right\|_{\mathcal{B}(H^2_{\delta_\Gamma,\alpha}(\mathbb{R}^n\setminus\Gamma),L^2(\mathbb{R}^n))} = 0, \end{aligned} \tag{12}$$

$$\begin{aligned} &\lim_{m\to\infty} \left\| \left(V_{-h_m}\mathcal{H}_{\delta'_\Gamma,\beta}V_{h_m} - \mathcal{H}^h_{\delta'_\Gamma,\beta} \right)\psi_R \right\|_{\mathcal{B}(H^2_{\delta'_\Gamma,\beta}(\mathbb{R}^n\setminus\Gamma),L^2(\mathbb{R}^n))} \\ &= \lim_{m\to\infty} \left\| \psi_R \left(V_{-h_m}\mathcal{H}_{\delta'_\Gamma,\beta}V_{h_m} - \mathcal{H}^h_{\delta'_\Gamma,\beta} \right) \right\|_{\mathcal{B}(H^2_{\delta'_\Gamma,\beta}(\mathbb{R}^n\setminus\Gamma),L^2(\mathbb{R}^n))} = 0. \end{aligned} \tag{13}$$

Applying formulas (9) we obtain that: (i) if $h_m \to \eta_\omega, \omega = (\omega', \omega_n), \omega_n \neq 0$, then

$$\mathcal{H}^h_{\delta_\Gamma,\alpha} = \mathcal{H}^h_{\delta'_\Gamma,\beta} = -\Delta + W^h : H^2(\mathbb{R}^n) \to L^2(\mathbb{R}^n), \tag{14}$$

where W^h is defined by formula (10), and (ii) if $\omega = (\omega', 0)$ and a sequence $\mathbb{R}^n \ni h_m \to \eta_\omega$, then

$$\mathcal{H}^h_{\delta_\Gamma,\alpha}u(x) = \left\{ \begin{array}{c} (-\Delta + W^h)u(x), x \in \mathbb{R}^n\setminus\Gamma, \\ [u]_{\mathbb{R}^{n-1}} = 0, \left[\frac{\partial u}{\partial x_n}\right]_{\Gamma_k} = \alpha_k^h u\mid_{\Gamma_k}, k = 1, ..., N \end{array} \right., \tag{15}$$

$$\mathcal{H}^h_{\delta'_\Gamma,\beta}u(x) = \left\{ \begin{array}{c} (-\Delta + W^h)u(x) = \varphi, x \in \mathbb{R}^n\setminus\Gamma, \\ \left[\frac{\partial u}{\partial x_n}\right]_{\Gamma_k} = 0, [u]_{\Gamma_k} = \beta_k^h \frac{\partial u}{\partial x_n}\mid_{\Gamma_k}, k = 1, ..., N \end{array} \right., \tag{16}$$

where W^h is defined by formula (10) and α_k^h, β_k^h are defined by formulas (11).

Note that if the potential $W \in L^{\infty}(\mathbb{R}^n)$ is rich the operators $\mathcal{H}_{\delta_\Gamma,\alpha}, \mathcal{H}_{\delta'_\Gamma,\beta}$ are rich: that is every sequence $\mathbb{Z}^n \ni g_m \to \infty$ has a subsequence $h_m \to \infty$ defining by formulas (12),(13) the limit operators $\mathcal{H}^h_{\delta_\Gamma,\alpha}, \mathcal{H}^h_{\delta'_\Gamma,\beta}$.

We will consider also the limit operators $\mathcal{H}^h_{\delta_\Gamma,\alpha}, \mathcal{H}^h_{\delta'_\Gamma,\beta}$ as unbounded operators in $L^2(\mathbb{R}^n)$ generates by the Schrödinger operator $-\Delta + W^h$ with domains $H^2(\mathbb{R}^n)$

if $h_m \to \eta_\omega, \omega = (\omega', \omega_n), \omega_n \neq 0$, and if $h_m \to \eta_\omega, \omega = (\omega', 0)$ the domains of the operators $\mathcal{H}^h_{\delta_\Gamma,\alpha}, \mathcal{H}^h_{\delta'_\Gamma,\beta}$ are defined, as

$$H^2_{\delta_\Gamma,\alpha^h}(\mathbb{R}^n\backslash\Gamma) = \left\{ u \in H^2(\mathbb{R}^n\backslash\Gamma) : [u]_{\Gamma_k} = 0, \left[\frac{\partial u}{\partial x_n}\right]_{\Gamma_k} = \alpha^h_k u|_{\Gamma_k}, k = 1, ..., N, \right\},$$

$$H^2_{\delta'_\Gamma,\beta^h}(\mathbb{R}^n\backslash\Gamma) = \left\{ u \in H^2(\mathbb{R}^n\backslash\Gamma) : \left[\frac{\partial u}{\partial x_n}\right]_{\Gamma_k} = 0, [u]_{\Gamma_k} = \beta^h_k \frac{\partial u}{\partial x_n}|_{\Gamma_k}, k = 1, ..., N, \right\},$$

respectively.

THEOREM 7. *Let the potential $W \in L^\infty(\mathbb{R}^n)$ be rich, the strength interaction coefficients $\alpha_k, \beta_k \in C^1_b(\Gamma_k)$ and $\inf_{s\in\Gamma_k} |\beta_k| > 0, k = 1, ..., N$. Then $\mathcal{H}_{\delta_\Gamma,\alpha} : H^2_{\delta_\Gamma,\alpha}(\mathbb{R}^n\backslash\Gamma) \to L^2(\mathbb{R}^n)$, $\mathcal{H}_{\delta'_\Gamma,\beta} : H^2_{\delta'_\Gamma,\beta}(\mathbb{R}^n\backslash\Gamma) \to L^2(\mathbb{R}^n))$ are Fredholm operators if and only if all limit operators $\mathcal{H}^h_{\delta_\Gamma,\alpha} : H^2_{\delta_\Gamma,\alpha^h}(\mathbb{R}^n\backslash\Gamma) \to L^2(\mathbb{R}^n)$, $\mathcal{H}^h_{\delta'_\Gamma,\beta} : H^2_{\delta_\Gamma,\beta^h}(\mathbb{R}^n\backslash\Gamma) \to L^2(\mathbb{R}^n))$ are invertible.*

PROOF. We prove the theorem for $\mathcal{H}_{\delta_\Gamma,\alpha}$ only. For $\mathcal{H}_{\delta'_\Gamma,\beta}$ the proof is similar. *Sufficiency.* Since $-\Delta + W$ is an elliptic operator on $\mathbb{R}^n$ and for every point $x \in \Gamma_k$ the Shapiro-Lopatisky conditions for the transmission problems are satisfied (see for instance [**34**],[**36**],[**33**]) the operators $\mathcal{H}_{\delta_\Gamma,\alpha}, \mathcal{H}_{\delta'_\Gamma,\beta}$ are locally Fredholm. For $\mathcal{H}_{\delta_\Gamma,\alpha}$ it means that for every $R > 0$ there exist operators $L^R_{\delta_\Gamma,\alpha}$ and $M^R_{\delta_\Gamma,\alpha}$ such that

$$L^R_{\delta_\Gamma,\alpha}\mathcal{H}_{\delta_\Gamma,\alpha}\varphi_R I = \varphi_R + T^R_{\delta_\Gamma,\alpha}, \varphi_R \mathcal{H}_{\delta_\Gamma,\alpha} M^R_{\delta_\Gamma,\alpha} = \varphi_R + \acute{T}^R_{\delta_\Gamma,\alpha},$$

where $T^R_{\delta_\Gamma,\alpha} \in \mathcal{K}(H^2_{\delta_\Gamma,\alpha}(\mathbb{R}^n\backslash\Gamma)), \acute{T}^R_{\delta_\Gamma,\alpha} \in \mathcal{K}(L^2(\mathbb{R}^n))$. Then $\mathcal{H}_{\delta_\Gamma,\alpha} : H^2_{\delta_\Gamma,\alpha^h}(\mathbb{R}^n\backslash\Gamma) \to L^2(\mathbb{R}^n)$ is a Fredholm operator if and only if $\mathcal{H}_{\delta_\Gamma,\alpha}$ is a locally invertible at infinity operator: that is there exists $R > 0$ and operators $\mathcal{L}^R_{\delta_\Gamma,\alpha}, \mathcal{M}^R_{\delta_\Gamma,\alpha} \in \mathcal{B}(L^2(\mathbb{R}^n), H^2_{\delta_\Gamma,\alpha}(\mathbb{R}^n\backslash\Gamma))$ such that

$$\mathcal{L}^R_{\delta_\Gamma,\alpha}\mathcal{H}_{\delta_\Gamma,\alpha}\psi_R = \psi_R, \psi_R\mathcal{H}_{\delta_\Gamma,\alpha}\mathcal{M}^R_{\delta_\Gamma,\alpha} = \psi_R. \tag{17}$$

We investigate the local invertibility at infinity of $\mathcal{H}_{\delta_\Gamma,\alpha}$ applying Theorem 3. Let $\mathcal{A}_{\delta_\Gamma,\alpha}(\mu_0) : H^2_{\delta_\Gamma,\alpha}(\mathbb{R}^n\backslash\Gamma) \to L^2(\mathbb{R}^n)$ be the isomorphism introduced in Proposition 1. We set $\tilde{\mathcal{H}}_{\delta_\Gamma,\alpha} = \mathcal{H}_{\delta_\Gamma,\alpha}\mathcal{A}^{-1}_{\delta_\Gamma,\alpha}(\mu_0)$. Then $\tilde{\mathcal{H}}_{\delta_\Gamma,\alpha}$ is a bounded operator in $L^2(\mathbb{R}^n)$. We will prove that the operator $\tilde{\mathcal{H}}_{\delta_\Gamma,\alpha}$ belongs to the algebra $\mathfrak{A}(\mathbb{R}^n)$. Let $a \in C_{b,u}(\mathbb{R}^n)$ and $a_t(x) = a(t_1x_1, ..., t_nx_n), t = (t_1, ..., t_n)$. We have to prove that

$$\lim_{t\to 0} \left\| \left[\tilde{\mathcal{H}}_{\delta_\Gamma,\alpha}, a_t\right] \right\|_{\mathcal{B}(L^2(\mathbb{R}^n))} = 0. \tag{18}$$

Since the space $C^\infty_b(\mathbb{R}^n)$ is dense in $C_{b,u}(\mathbb{R}^n)$ (see [**25**]) it is enough to prove (18) for $a \in C^\infty_b(\mathbb{R}^n)$. Note that

$$\lim_{t\to 0} \|[\mathcal{A}_{\delta_\Gamma,\alpha}(\mu_0), a_t]\|_{\mathcal{B}(H^2_{\delta_\Gamma,\alpha}(\mathbb{R}^n\backslash\Gamma), L^2(\mathbb{R}^n))} = 0, a \in C^\infty_b(\mathbb{R}^n). \tag{19}$$

Since

$$\left[a_t, \mathcal{A}^{-1}_{\delta_\Gamma,\alpha}(\mu_0)\right] = \mathcal{A}^{-1}_{\delta_\Gamma,\alpha}(\mu_0)\left[\mathcal{A}_{\delta_\Gamma,\alpha}(\mu_0), a_t\right]\mathcal{A}^{-1}_{\delta_\Gamma,\alpha}(\mu_0)$$

we obtain from (19) that

$$\lim_{t\to 0} \left\| \left[\mathcal{A}^{-1}_{\delta_\Gamma,\alpha}(\mu_0), a_t\right] \right\|_{\mathcal{B}(L^2(\mathbb{R}^n), H^2_{\delta_\Gamma,\alpha}(\mathbb{R}^n\backslash\Gamma))} = 0. \tag{20}$$

Since

$$\left[\tilde{\mathcal{H}}_{\delta_\Gamma,\alpha}, a_t\right] = \mathcal{H}_{\delta_\Gamma,\alpha}\left[\mathcal{A}^{-1}_{\delta_\Gamma,\alpha}(\mu_0), a_t\right] + \left[\mathcal{H}_{\delta_\Gamma,\alpha}, a_t\right]\mathcal{A}^{-1}_{\delta_\Gamma,\alpha}(\mu_0), \tag{21}$$

formulas (20) and (21) yield (18). Hence $\tilde{\mathcal{H}}_{\delta_\Gamma,\alpha} \in \mathfrak{A}(\mathbb{R}^n)$. Now we can apply Theorem 3 to the investigation of local invertibility of the operator $\tilde{\mathcal{H}}_{\delta_\Gamma,\alpha}$ at infinity. First we note that the operator $\tilde{\mathcal{H}}_{\delta_\Gamma,\alpha}$ is rich since the operator $\mathcal{H}_{\delta_\Gamma,\alpha}$ is rich according the assumptions of Theorem 7. Next step is to consider the limit operators for $\tilde{\mathcal{H}}_{\delta_\Gamma,\alpha}$. Let a sequence $h_m \to \eta_\omega, \omega = (\omega', \omega_n), \omega_n \neq 0$ define the limit operators for $\mathcal{H}_{\delta_\Gamma,\alpha}$ and $\mathcal{A}_{\delta_\Gamma,\alpha}(\mu_0)$. The conditions of Theorem 7 imply that $\mathcal{H}^h_{\delta_\Gamma,\alpha} = -\Delta + W^h : H^2(\mathbb{R}^n) \to L^2(\mathbb{R}^n)$ are invertible operators. Then the formula

$$V_{-h_m}\tilde{\mathcal{H}}_{\delta_\Gamma,\alpha}V_{h_m} = \left(V_{-h_m}\mathcal{H}_{\delta_\Gamma,\alpha}V_{h_m}\right)\left(V_{-h_m}\mathcal{A}^{-1}_{\delta_\Gamma,\alpha}(\mu_0)V_{h_m}\right) \tag{22}$$

yields that

$$\tilde{\mathcal{H}}^h_{\delta_\Gamma,\alpha} = (-\Delta + W^h)(-\Delta + \mu_0^2)^{-1}.$$

Hence $\tilde{\mathcal{H}}^h_{\delta_\Gamma,\alpha}$ is an invertible operator in $L^2(\mathbb{R}^n)$. Let a sequence $h_m \to \eta_\omega, \omega = (\omega', 0)$. Then (22) yields that

$$\tilde{\mathcal{H}}^h_{\delta_\Gamma,\alpha} = \mathcal{H}^h_{\delta_\Gamma,\alpha}\left[\mathcal{A}^{-1}_{\delta_\Gamma,\alpha}(\mu_0)\right]^h. \tag{23}$$

By assumptions of Theorem 7 all operators $\mathcal{H}^h_{\delta_\Gamma,\alpha}$ are invertible from $H^2_{\delta_\Gamma,\alpha^h}(\mathbb{R}^n\backslash\Gamma)$ into $L^2(\mathbb{R}^n)$, and because the operator $\mathcal{A}_{\delta_\Gamma,\alpha}(\mu_0)$ is invertible all limit operators $\mathcal{A}^h_{\delta_\Gamma,\alpha}(\mu_0)$ are invertible from $H^2_{\delta,\alpha^h}(\mathbb{R}^n\backslash\Gamma)$ into $L^2(\mathbb{R}^n)$. Moreover $\left[\mathcal{A}^{-1}_{\delta_\Gamma,\alpha}(\mu_0)\right]^h = \left[\mathcal{A}^h_{\delta_\Gamma,\alpha}(\mu_0)\right]^{-1}$ hence $\tilde{\mathcal{H}}^h_{\delta_\Gamma,\alpha^h} : L^2(\mathbb{R}^n) \to L^2(\mathbb{R}^n)$ are invertible operators. Applying Theorem 3 we obtain that the operator $\tilde{\mathcal{H}}_{\delta_\Gamma,\alpha}$ is locally invertible at infinity that is there exists $R_0 > 0$ and operators $\mathcal{L}^{R_0}_{\delta_\Gamma,\alpha}, \mathcal{M}^{R_0}_{\delta_\Gamma,\alpha} \in \mathcal{B}(L^2(\mathbb{R}^n))$ such that

$$\mathcal{L}^{R_0}_{\delta_\Gamma,\alpha}\tilde{\mathcal{H}}_{\delta_\Gamma,\alpha}\psi_{R_0} = \psi_{R_0}, \quad \psi_{R_0}\tilde{\mathcal{H}}_{\delta_\Gamma,\alpha}\mathcal{M}^{R_0}_{\delta_\Gamma,\alpha} = \psi_{R_0}.$$

Hence

$$\mathcal{L}^{R_0}_{\delta_\Gamma,\alpha}\mathcal{H}_{\delta_\Gamma,\alpha}\mathcal{A}^{-1}_{\delta_\Gamma,\alpha}(\mu_0)\psi_{R_0} = \psi_{R_0}, \tag{24}$$

$$\psi_{R_0}\mathcal{H}_{\delta_\Gamma,\alpha}\mathcal{A}^{-1}_{\delta_\Gamma,\alpha}(\mu_0)\mathcal{M}^{R_0}_{\delta_\Gamma,\alpha} = \psi_{R_0}. \tag{25}$$

Equality (25) yields that $\mathcal{A}^{-1}_{\delta_\Gamma,\alpha}(\mu_0)\mathcal{M}^{R_0}_{\delta_\Gamma,\alpha}$ is a right locally inverse operator at infinity for $\mathcal{H}_{\delta_\Gamma,\alpha}$. Let $R_1 > 2R_0$. Then $\psi_{R_0}\psi_{R_1} = \psi_{R_1}$, and (25) yields that

$$\mathcal{L}^{R_0}_{\delta_\Gamma,\alpha}\mathcal{H}_{\delta_\Gamma,\alpha}\mathcal{A}^{-1}_{\delta_\Gamma,\alpha}(\mu_0)\psi_{R_1} = \psi_{R_1} \tag{26}$$

for every $R_1 > 2R_0$. Note that the equality

$$\lim_{R\to\infty}\left\|\left[\mathcal{A}_{\delta_\Gamma,\alpha}(\mu_0), \psi_R\right]\right\|_{\mathcal{B}(H^2(\mathbb{R}^n\backslash\Gamma), L^2(\mathbb{R}^n))} = 0,$$

implies that

$$\lim_{R\to\infty}\left\|\left[\mathcal{A}^{-1}_{\delta_\Gamma,\alpha}(\mu_0), \psi_R\right]\right\|_{\mathcal{B}(L^2(\mathbb{R}^n), H^2(\mathbb{R}^n\backslash\Gamma))} = 0. \tag{27}$$

Hence

$$\mathcal{A}^{-1}_{\delta_\Gamma,\alpha}(\mu_0)\mathcal{L}^{R_0}_{\delta_\Gamma,\alpha}\mathcal{H}_{\delta_\Gamma,\alpha}\psi_{R_1} = \psi_{R_1} + T_{R_1}, \tag{28}$$

where $\lim_{R_1\to\infty}\|T_{R_1}\|_{\mathcal{B}(L^2(\Gamma))} = 0$. Let $R_2 > 2R_1 > 0$, and R_1 be such that $\|T_{R_1}\|_{\mathcal{B}(L^2(\Gamma))} < 1$. Then

$$\mathcal{A}^{-1}_{\delta_\Gamma,\alpha}(\mu_0)\mathcal{L}^{R_0}_{\delta_\Gamma,\alpha}\mathcal{H}_{\delta_\Gamma,\alpha}\psi_{R_2} = (I + T_{R_1}\psi_{R_1})\,\psi_{R_2},$$

and

$$(I + T_{R_1}\psi_{R_1})^{-1}\mathcal{A}^{-1}_{\delta_\Gamma,\alpha}(\mu_0)\mathcal{L}^{R_0}_{\delta_\Gamma,\alpha}\mathcal{H}_{\delta_\Gamma,\alpha}\psi_{R_2} = \psi_{R_2}.$$

Hence $(I + T_{R_1}\psi_{R_1})^{-1}\mathcal{A}^{-1}_{\delta_\Gamma,\alpha}(\mu_0)\mathcal{L}^{R_0}_{\delta_\Gamma,\alpha}$ is a left inverse operator of $\mathcal{H}_{\delta_\Gamma,\alpha}$ at infinity. Thus $\mathcal{H}_{\delta_\Gamma,\alpha} : H^2_{\delta_\Gamma,\alpha}(\mathbb{R}^n\backslash\Gamma) \to L^2(\mathbb{R}^n)$ is a locally invertible at infinity operator. Together with its local Fredholmness this statement proves that $\mathcal{H}_{\delta_\Gamma,\alpha}$ is a Fredholm operator.

Necessity. Note that Fredholmness of $\mathcal{H}_{\delta_\Gamma,\alpha} : H^2_{\delta_\Gamma,\alpha}(\mathbb{R}^n\backslash\Gamma) \to L^2(\mathbb{R}^n)$ yields Fredholmness of $\tilde{\mathcal{H}}_{\delta_\Gamma,\alpha} : L^2(\mathbb{R}^n) \to L^2(\mathbb{R}^n)$. Thus Theorem 3 implies that all limit operators $\tilde{\mathcal{H}}^h_{\delta_\Gamma,\alpha} : L^2(\mathbb{R}^n) \to L^2(\mathbb{R}^n)$ are invertible. Taking into account formula (23) we obtain that all limit operators $\mathcal{H}^h_{\delta_\Gamma,\alpha} : H^2_{\delta_\Gamma,\alpha^h}(\mathbb{R}^n\backslash\Gamma) \to L^2(\mathbb{R}^n)$ are invertible. □

THEOREM 8. *Let the potential $W \in L^\infty(\mathbb{R}^n)$ be rich, the strength interaction coefficients $\alpha_k, \beta_k \in C^1_b(\Gamma_k)$ and $\inf_{s\in\mathbb{R}}|\beta_k| > 0, k = 1, ..., N$. Then*

$$sp_{ess}\mathcal{H}_{\delta_\Gamma,\alpha} = \bigcup_{\mathcal{H}^h_{\delta_\Gamma,\alpha}\in Lim\mathcal{H}_{\delta_\Gamma,\alpha}} sp\mathcal{H}^h_{\delta_\Gamma,\alpha},$$

$$sp_{ess}\mathcal{H}_{\delta'_\Gamma,\beta} = \bigcup_{\mathcal{H}^h_{\delta_\Gamma,\beta}\in Lim\mathcal{H}_{\delta'_\Gamma,\beta}} sp\mathcal{H}^h_{\delta'_\Gamma,\beta},$$

PROOF. Theorem 8 follows from Theorem 7. Indeed, the operators $\mathcal{H}_{\delta_\Gamma,\alpha} - \lambda, \mathcal{H}_{\delta'_\Gamma,\beta} - \lambda, \lambda \in \mathbb{C}$ are not Fredholm operators if and only if there exist limit operators $\mathcal{H}^h_{\delta_\Gamma,\alpha}$, $\mathcal{H}^h_{\delta'_\Gamma,\beta}$ such that $\lambda \in sp\mathcal{H}^h_{\delta_\Gamma,\alpha}, \lambda \in sp\mathcal{H}^h_{\delta'_\Gamma,\beta}$, respectively. □

3.3. Slowly oscillating at infinity electric potentials and strength interaction coefficients.

- We say that a function $a \in C^1_b(\mathbb{R}^n)$ is slowly oscillating at infinity and belongs to the class $SO(\mathbb{R}^n)$ if

(29)
$$\lim_{x\to\infty}\frac{\partial a(x)}{\partial x_j} = 0, j = 1, ..., n.$$

  Let $a \in SO(\mathbb{R}^n)$ and a sequence $g = (g_m) \to \infty$. Then there exists a subsequence $h = (h_m)$ of g and a constant $a^g \in \mathbb{C}$ such that

$$\lim_{m\to\infty}\sup_{x\in K}|a(x + h^m) - a^g| = 0$$

  for every compact set $K \subset \mathbb{R}^n$ (see for instance [**24**], Chap.3). Moreover

$$\lim_{m\to\infty} a(h_m) = a^g.$$

- Let the potential $W \in SO(\mathbb{R}^n)$, and the strength interaction coefficient $\alpha_k, \beta_k \in C^1_b(\mathbb{R}^n)$. It implies that every sequence $\mathbb{R}^n \ni g_m \to \infty$ has a subsequence h_m such that

(30)
$$\lim_{m\to\infty} W(x + h_m) = W^h \in \mathbb{C}$$

uniformly on every compact set $K\subset\mathbb{R}^n$, and

(31) $$\lim_{m\to\infty}\tilde{\alpha}_k(x'+h'_m)=\alpha_k^h\in\mathbb{C} \text{ and } \lim_{m\to\infty}\tilde{\beta}_k(x'+h'_m)=\beta_k^h\in\mathbb{C}$$

uniformly on every compact set $K\subset\mathbb{R}^{n-1}_{x'}$. Hence the operators $\mathcal{H}_{\delta_\Gamma,\alpha}$, $\mathcal{H}_{\delta'_\Gamma,\beta}$ are rich and the limit operators of $\mathcal{H}^h_{\delta_\Gamma,\alpha}$, $\mathcal{H}^h_{\delta'_\Gamma,\beta}$ are of the form:

- If $\mathbb{R}^n\ni h_m\to\eta_\omega, S^{n-1}\in\omega=(\omega',\omega_n),\omega_n\neq 0$ then
$$\mathcal{H}^h_{\delta_\Gamma,\alpha}=\mathcal{H}^h_{\delta'_\Gamma,\beta}=-\Delta+W^h,W^h\in\mathbb{C}$$
with domain $H^2(\mathbb{R}^n)$. Note if the potential W is real-valued, then

(32) $$sp(-\Delta+W^hI)=\left[W^h,+\infty\right).$$

- If $\mathbb{R}^n\ni h_m\to\eta_\omega$ where $S^{n-1}\in\omega=(\omega',0)$, then $\mathcal{H}^h_{\delta_\Gamma,\alpha},\mathcal{H}^h_{\delta'_\Gamma,\beta}$ are the Schrödinger operators $-\Delta+W^h,W^h\in\mathbb{C}$ with domains
$$H^2_{\delta_\Gamma,\alpha^h}(\mathbb{R}^n\backslash\Gamma)=\Big\{u\in H^2(\mathbb{R}^n\backslash\Gamma):[u]_{\Gamma_k}=0,\left[\frac{\partial u}{\partial x_n}\right]_{\Gamma_k}=\alpha_k^h u\mid_{\Gamma_k},$$
$$\alpha_k^h\in\mathbb{C},k=1,...,N\Big\}$$
$$H^2_{\delta'_\Gamma,\beta^h}(\mathbb{R}^n\backslash\Gamma)=\Big\{u\in H^2(\mathbb{R}^n\backslash\Gamma):\left[\frac{\partial u}{\partial x_n}\right]_{\Gamma_k}=0,[u]_{\Gamma_k}=\beta_k^h\frac{\partial u}{\partial x_n}\mid_{\Gamma_k},$$
$$\beta_k^h\in\mathbb{C},k=1,...,N\Big\}$$

The operators $\mathcal{H}^h_{\delta_\Gamma,\alpha},\mathcal{H}^h_{\delta'_\Gamma,\beta}$ can be realized as
$$\mathcal{H}^h_{\delta_\Gamma,\alpha}=-\Delta_{x'}+W^h+\mathcal{L}_{\delta,\alpha^h},\mathcal{H}^h_{\delta'_\Gamma,\beta}=-\Delta_{x'}+W^h+\mathcal{L}_{\delta',\beta^h}$$
where $\mathcal{L}_{\delta_\Gamma,\alpha^h},\mathcal{L}_{\delta'_\Gamma,\beta^h}$ are unbounded operators generated by the operator $-\frac{d^2}{dx_n^2}$ acting in $L^2(\mathbb{R})$ with domains
$$D_{\mathcal{L}_{\delta,\alpha^h}}=\left\{v\in H^2\left(\mathbb{R}\backslash\{d_1,...,d_N\}\right):[v]_{x_n=d_k}=0,\left[\frac{dv}{dx_n}\right]_{x_n=d_k}=\alpha_k^h v(d_k),k=1,...,N,\right\},$$
$$D_{\mathcal{L}_{\delta',\beta^h}}=\left\{v\in H^2\left(\mathbb{R}\backslash\{d_1,...,d_N\}\right):\left[\frac{dv}{dx_n}\right]_{x_n=d_k}=0,[v]_{x_n=d_k}=\beta_k^h\frac{dv(d_k)}{dx_n},k=1,...,N\right\}$$
where $H^2\left(\mathbb{R}\backslash\{d_1,...,d_N\}\right)=H^2\left((-\infty,d_1)\right)\cup H^2\left((d_1,d_2)\right)\cup...\cup H^2\left((d_N,+\infty)\right)$. Let the strength interaction coefficients $\alpha_k,\beta_k,k=1,,,,.N$ are real-valued. Then $\mathcal{L}_{\delta_\Gamma,\alpha^h},\mathcal{L}_{\delta'_\Gamma,\beta^h}$ are self-adjoint operators with essential spectra $[0,+\infty)$ and a possible finite negative discrete spectrum. Then
$$sp\mathcal{H}^h_{\delta,\alpha}=\left[W^h+\inf sp\mathcal{L}_{\delta_\Gamma,\alpha^h},+\infty\right),sp\mathcal{H}^h_{\delta'_\Gamma,\beta}=\left[W^h+\inf sp\mathcal{L}_{\delta'_\Gamma,\beta^h},+\infty\right).$$

Thus Theorem 8 yields the following description of the essential spectra of operators $\mathcal{H}_{\delta_\Gamma,\alpha}$, $\mathcal{H}_{\delta'_\Gamma,\beta}$ with slowly oscillating data.

THEOREM 9. *Let the potential $W\in SO(\mathbb{R}^n)$, the strength interaction coefficients $\alpha_k,\beta_k\in SO(\mathbb{R}^{n-1})$, $k=1,,,,.N$ and be real-valued, and $\inf_{s\in\mathbb{R}}|\beta_k|>0$. Then the operators $\mathcal{H}_{\delta_\Gamma,\alpha},\mathcal{H}_{\delta'_\Gamma,\beta}$ are self-adjoint, and*

(33) $$sp_{ess}\mathcal{H}_{\delta_\Gamma,\alpha}=[m_1,+\infty),sp_{ess}\mathcal{H}_{\delta'_\Gamma,\beta}=[m_2,+\infty),$$

where

$$(34)\quad m_1 = \min\left\{ \inf_{\omega=(\omega',\omega_n)\in S^{n-1},\omega_n\neq 0}\ \inf_{h_m\to\eta_\omega} W^h,\ \inf_{\omega=(\omega',0)\in S^{n-1}}\ \inf_{h_m\to\eta_\omega} sp\mathcal{H}^h_{\delta,\alpha} \right\},$$

$$m_2 = \min\left\{ \inf_{\omega=(\omega',\omega_n)\in S^{n-1},\omega_n\neq 0}\ \inf_{h_m\to\eta_\omega} W^h,\ \inf_{\omega=(\omega',0)\in S^{n-1}}\ \inf_{h_m\to\eta_\omega} sp\mathcal{H}^h_{\delta',\beta} \right\},$$

and $\inf_{h_m\to\eta_\omega} W^h$ *is taken with respect to all sequences* $h_m \to \eta_\omega$ *such that* $\lim_{m\to\infty} W(h_m)$ *exists,* $\inf_{h_m\to\eta_\omega} sp\mathcal{H}^h_{\delta,\alpha}$*, and* $\inf_{h_m\to\eta_\omega} sp\mathcal{H}^h_{\delta',\beta}$ *are taken with respect to all sequences* $h_m \to \eta_\omega$ *defining the limit operators* $\mathcal{H}^h_{\delta,\alpha}, \mathcal{H}^h_{\delta',\beta}$.

EXAMPLE 10. Let the space $\mathbb{R}^n$ be divided by the subspace $\Gamma = \mathbb{R}^{n-1}_{x'}$ on two half-spaces $\mathbb{R}^n_\pm = \{x \in \mathbb{R}^n : x_n \gtrless 0\}$. We set $H^2(\mathbb{R}^n\backslash\Gamma) = H^2(\mathbb{R}^n_+) \oplus H^2(\mathbb{R}^n_-)$ and we introduce the unbounded operators $\mathcal{H}_{\delta_\Gamma,\alpha}, \mathcal{H}_{\delta'_\Gamma,\beta}$ generated by the Schrödinger operator $-\Delta + W$ in $L^2(\mathbb{R}^n)$ with real-valued potential $W \in SO(\mathbb{R}^n)$ and domains

$$D_{\mathcal{H}_{\delta_\Gamma,\alpha}} = \left\{ u \in H^2(\mathbb{R}^n\backslash\Gamma) : [u]_\Gamma = 0, \left[\frac{\partial u}{\partial x_n}\right]_\Gamma (x',0) = \alpha(x')u(x',0), x' \in \Gamma \right\},$$

$$D_{\mathcal{H}_{\delta'_\Gamma,\beta}} = \left\{ u \in H^2(\mathbb{R}^n\backslash\Gamma) : \left[\frac{\partial u}{\partial x_n}\right]_\Gamma = 0, [u]_\Gamma (x',0) = \beta(x')\frac{\partial u(x',0)}{\partial x_n}, x' \in \Gamma \right\},$$

where $\alpha, \beta \in SO(\mathbb{R}^{n-1})$ are real-valued functions. Moreover we suppose that $\inf_{x'} |\beta(x')| > 0$.

The simple calculations show that $\mathcal{L}_{\delta,\alpha^h}$ has an unique negative eigenvalue $-\frac{(\alpha^h)^2}{4}$ if $\alpha^h < 0$, and $\mathcal{L}_{\delta',\beta^h}$ has an unique negative eigenvalue $-\frac{4}{(\beta^h)^2}$ if $\beta^h < 0$. Suppose that $\liminf_{x'\to\infty} \alpha(x') < 0, \limsup_{x'\to\infty} \beta(x') < 0$. Then

$$sp_{ess}\mathcal{H}_{\delta_\Gamma,\alpha} = [m_1, +\infty), sp_{ess}\mathcal{H}_{\delta'_\Gamma,\beta} = [m_2, +\infty),$$

where

$$m_1 = \min\left\{ \inf_{\omega=(\omega',\omega_n)\in S^{n-1},\omega_n\neq 0}\ \inf_{h_m\to\eta_\omega} W^h, \liminf_{x'\to\infty}\left(W(x',0) - \frac{\alpha(x')^2}{4} \right) \right\},$$

$$m_2 = \min\left\{ \inf_{\omega=(\omega',\omega_n)\in S^{n-1},\omega_n\neq 0}\ \inf_{h_m\to\eta_\omega} W^h, \liminf_{x'\to\infty}\left(W(x',0) - \frac{4}{\beta(x')^2} \right) \right\}.$$

4. Interaction on a system of slowly oscillating at infinity hypersurfaces

Let $\{\Gamma_k\}_{k=1}^N$ be a family of the above introduced simply connected unbounded C^2-hypersurface ($\dim \Gamma_k = n-1$) such that $\Gamma_k \cap \Gamma_l = \varnothing$ if $k \neq l$. We suppose that the system $\{\Gamma_k\}_{k=1}^N$ is slowly oscillating at infinity. It means that

$$(35)\quad \Gamma_k \cap \Pi_R = \left\{ \begin{array}{c} x = (x',x_n) \in \mathbb{R}^n : x_n = F_k(x') = F(x') + d_k, k = 1,...,N, x' \in B'_R \\ 0 \leq d_1 < d_2 < .... < d_N, \end{array} \right\}$$

for some $R > 0$ where

$$B'_R = \left\{ x' = (x_1,...x_{n-1}) \in \mathbb{R}^{n-1} : |x'| > R \right\}, \Pi_R = \{x = (x',x_n) : x' \in B'_R\}$$

, and F is such that $\frac{\partial F}{\partial x_j} \in SO(\mathbb{R}^{n-1})$, that is

$$(36)\quad \lim_{x\to\infty} \frac{\partial^2 F}{\partial x_i \partial x_j} = 0 \text{ for all } i,j = 1,...,n-1.$$

Condition (36) implies that the Gaussian curvature of Γ_k equels zero at infinity.

Important examples of functions F defining slowly oscillating at infinity hypersurfaces are:

1) $F\in C_b^2(B_R')$ and it is such that $\lim_{x\to\infty}\frac{\partial F(x')}{\partial x_j}=0, j=1,...,n-1$;

2) $F(x')=|x'|^{\delta} f\left(\frac{x'}{|x'|}\right), \delta\in[0,1], f\in C^2(S^{n-2}), x'\in B_R'$. If $\delta=1$ we obtain a conic at infinity hypersurface.

3) $F(x')=|x'|^{\delta} f\left(\frac{x'}{|x'|}\right)\sin\log|x'|, x'\in B_R', \delta\in[0,1]$.

We suppose that the potential $W\in SO(\mathbb{R}^n)$ and the strength interaction coefficients α_k,β_k are slowly oscillating at infinity It means that $\alpha_k,\beta_k\in C^1(\Gamma_k)$ and the functions $\tilde{\alpha}_k(x')=\alpha_k(x',F_k(x')), \tilde{\beta}_k(x')=\beta_k(x',F_k(x'))$ belong $C_b^1(B_R')$, and

$$\lim_{x'\to\infty}\frac{\partial\tilde{\alpha}_k(x')}{\partial x_j}=\lim_{x'\to\infty}\frac{\partial\tilde{\beta}_k(x')}{\partial x_j}=0. \tag{37}$$

Let $\mathcal{H}_{\delta_\Gamma,\alpha},\mathcal{H}_{\delta_\Gamma',\beta}$ be the unbounded operators defined by formulas (5), (6) where the system $\{\Gamma_k\}_{k=1}^N$ satisfies (35), (36). Let $\Phi:\Pi_R\to\Pi_R$ be the C^2-diffeomorphism given as

$$y=(y',y_n)=\Phi(x',x_n)=(x',x_n-F(x')), x=(x',x_n)\in\Pi_R, \tag{38}$$

and $\Phi^*: L^2(\Pi_R)\to L^2(\Pi_R), \Phi_*: H^2_{\delta_\Gamma,\alpha}(\Pi_R\backslash\Gamma)\to H^2_{\delta_{\tilde{\Gamma}},\tilde{\alpha}}(\Pi_R\backslash\tilde{\Gamma})$ $(H^2_{\delta_{\tilde{\Gamma}}',\beta}(\Pi_R\backslash\Gamma)\to H^2_{\delta_{\tilde{\Gamma}}',\tilde{\beta}}(\Pi_R\backslash\Gamma))$ are defined as

$$\Phi_* f(x)=u(\Phi(x)), x\in\Pi_R,$$
$$\Phi^* f(y)=f(\Phi^{-1})y), y\in\Pi_R.$$

We denote by $\mathcal{H}^R_{\delta_\Gamma,\alpha},\mathcal{H}^R_{\delta_\Gamma',\beta}$ the restrictions of the operators $\mathcal{H}_{\delta_\Gamma,\alpha},\mathcal{H}_{\delta_\Gamma',\beta}$ on Π_R. Then we set

$$\left(\mathcal{H}^R_{\delta_\Gamma,\alpha}\right)_\Phi=\Phi^*\mathcal{H}^R_{\delta,\alpha}\Phi_*, \left(\mathcal{H}^R_{\delta',\beta}\right)_\Phi=\Phi^*\mathcal{H}^R_{\delta',\beta}\Phi_*$$

Note the change of variables (38) transforms the hypersurfaces $\Gamma_k^R=\Gamma_k\cap\Pi_R$ into the hyperplanes

$$\tilde{\Gamma}_k^R=\Phi(\Gamma_k^R)=\{y=(y',y_n)\in\Pi_R: y_n=d_k\},$$

and $\Omega_k^R=\Omega_k\cap\Pi_R, k=0,...,N$ into the layers:

$$\tilde{\Omega}_0^R=\{y\in\mathbb{R}^n\cap\Pi_R: y_n<d_1\},...\tilde{\Omega}_k=\{y\in\mathbb{R}^n\cap\Pi_R: d_k<y_n<d_{k+1}\},$$
$$k=1,...,N-1, \tilde{\Omega}_N=\{y\in\mathbb{R}^n\cap\Pi_R: y_n>d_N\}.$$

The Schrödinger operator $-\Delta+W$ is transformed to the operator

$$L_\Phi=-\sum_{j=1}^{n-1}\left(\partial_{y_j}-\left(\partial_{y_j}F\right)\partial_{y_n}\right)^2-\partial^2_{y_n}+\tilde{W}$$

where $\tilde{W}(y)=W(y',y_n+F(y')), y\in\Pi_R$, and the operator $\frac{\partial}{\partial\nu}$ of the normal derivative to Γ_k^R is transformed to the operator

$$M_\Phi=\sum_{j=1}^{n-1}-\left(\partial_{y_j}F\right)\left(\partial_{y_j}-\left(\partial_{y_j}F\right)\partial_{y_n}\right)+\partial_{y_n}.$$

It implies that

$$\left(\mathcal{H}^R_{\delta_\Gamma,\alpha}\right)_\Phi = \left\{ \begin{array}{c} L_\Phi u(y), y \in \Pi_R \backslash \tilde{\Gamma}^R, \tilde{\Gamma}^R = \cup_{k=1}^N \tilde{\Gamma}^R_k, \\ D_{\left(\mathcal{H}^R_{\delta_\Gamma,\alpha}\right)_\Phi} = \left\{ u \in H^2(\Pi_R \backslash \tilde{\Gamma}^R) : [u]_{\tilde{\Gamma}^R_k} = 0, \left[M^h_\Phi u\right]_{\tilde{\Gamma}^R_k} = \tilde{\alpha}^h_k u \mid_{\tilde{\Gamma}^R_k}, k = 1, ..., N \right\}, \end{array} \right.$$

$$\left(\mathcal{H}^R_{\delta'_\Gamma,\beta}\right)_\Phi = \left\{ \begin{array}{c} L_\Phi u(y), y \in \Pi_R \backslash \tilde{\Gamma}^R, \\ D_{\left(\mathcal{H}^R_{\delta'_\Gamma,\beta}\right)_\Phi} = \left\{ u \in H^2(\Pi_R \backslash \tilde{\Gamma}^R) : \left[M^h_\Phi\right]_{\tilde{\Gamma}^R_k} = 0, [u]_{\tilde{\Gamma}^R_k} = \tilde{\beta}_k \frac{\partial u}{\partial y_n} \mid_{\tilde{\Gamma}^R_k}, k = 1, ..., N \right\}. \end{array} \right.$$

Following to Chap.3.3 we define the limit operators for $\left(\mathcal{H}^R_{\delta_\Gamma,\alpha}\right)_\Phi$, $\left(\mathcal{H}^R_{\delta'_\Gamma,\beta}\right)_\Phi$ as follows:

- Let $\mathbb{R}^n \ni h_m \to \eta_\omega, S^{n-1} \in \omega = (\omega', \omega_n), \omega_n \neq 0$. Then

$$\left(\mathcal{H}^R_{\delta_\Gamma,\alpha}\right)^h_\Phi = \left(\mathcal{H}^R_{\delta'_\Gamma,\beta}\right)^h_\Phi = L^h_\Phi = \sum_{j=1}^{n-1} \left(\partial_{y_j} - \left(\partial_{y_j} F\right)^h \partial_{y_n}\right)^2 - \partial^2_{y_n} + \tilde{W}^h,$$

where $\left(\partial_{y_j} F\right)^h = \lim_{m\to\infty} \partial_{y_j} F(h'_m), \tilde{W}^h = \lim_{m\to\infty} \tilde{W}(h_m)$. We consider L^h_Φ as unbounded operator in $L^2(\mathbb{R}^n)$ with domain $H^2(\mathbb{R}^n)$. Note that the nondegenerate linear change of the variables

$$(39) \qquad z = \Psi_h(y) = \left\{ \begin{array}{c} z_j = y_j, j = 1, ..., n-1, \\ z_n = y_n + \sum_{j=1}^{n-1} \left(\partial_{y_j} F\right)^h y_j \end{array} \right.$$

transforms the operator L^h_Φ to the Schrödinger operator $-\Delta_z + \tilde{W}^h I$. Therefore

$$(40) \qquad sp\left(\mathcal{H}_{\delta_\Gamma,\alpha}\right)^h_\Phi = sp\left(\mathcal{H}_{\delta'_\Gamma,\beta}\right)^h_\Phi = sp\left(-\Delta_z + \tilde{W}^h I\right) = [\tilde{W}^h, +\infty).$$

- Let $\mathbb{R}^n \ni h_m \to \eta_\omega, S^{n-1} \in \omega = (\omega', 0)$. Then $\left(\mathcal{H}^R_{\delta_\Gamma,\alpha}\right)^h_\Phi$ and $\left(\mathcal{H}^R_{\delta'_\Gamma,\beta}\right)^h_\Phi$ are defined by the operator L^h_Φ with domains

$$D_{\left(\mathcal{H}^R_{\delta_\Gamma,\alpha}\right)^h_\Phi} = \left\{ u \in H^2(\mathbb{R}^n \backslash \mathbb{L}\right\} : [u]_{\mathbb{L}_k} = 0, \left[M^h_\Phi u\right]_{\mathbb{L}_k} = \tilde{\alpha}^h_k u \mid_{\mathbb{L}_k}, \tilde{\alpha}^h_k = \lim_{m\to\infty} \tilde{\alpha}_k(h_m) \in \mathbb{C}\},$$

$$\mathbb{L}_k = \left\{ y = (y', y_n) \in \mathbb{R}^n : y_n = d_k \right\}, k = 1, .., n, \mathbb{L} = \cup_{k=1}^N \mathbb{L}_k$$

and

$$D_{\left(\mathcal{H}^R_{\delta'_\Gamma,\beta}\right)^h_\Phi} = \left\{ u \in H^2(\mathbb{R}^n \backslash \mathbb{L}\right\} : \left[M^h_\Phi u\right]_{\mathbb{L}_k} = 0,$$

$$[u]_{\mathbb{L}_k} = \tilde{\beta}^h_k M^h_\Phi u \mid_{\mathbb{L}_k}, \tilde{\beta}^h_k = \lim_{m\to\infty} \tilde{\beta}(h_m) \in \mathbb{C}, k = 1, ..., N\}$$

where

$$M^h_\Phi = \sum_{j=1}^{n-1} -\left(\partial_{y_j} F\right)^h \left(\partial_{y_j} - \left(\partial_{y_j} F\right)^h \partial_{y_n}\right) + \partial_{y_n}.$$

Change of variables (39) transforms the hyperplanes $\mathbb{L}_k$ into the hyperplanes

$$\tilde{\mathbb{L}}_k = \left\{ z = (z', z_n) \in \mathbb{R}^n : z_n = \sum_{j=1}^{n-1} \left(\partial_{y_j} F\right)^h z_j + d_k, z' = (z_1, ..., z_{n-1}) \in \mathbb{R}^{n-1} \right\}$$

Note that the operator M_{Φ}^{h} is transformed into the operator of normal derivative $\frac{\partial}{\partial \nu^h}$ to the hyperplanes $\tilde{\mathbb{L}}_k$ by change of variables (39). Hence $sp\left(\mathcal{H}^{R}_{\delta_\Gamma,\alpha}\right)^{h}_{\Phi} = spA^{h}_{\delta_\Gamma,\alpha}, sp(\mathcal{H}^{R}_{\delta'_\Gamma,\beta\Phi})^{h} = spA^{h}_{\delta'_\Gamma,\beta}$ where

$$A^{h}_{\delta_\Gamma,\alpha} = \begin{cases} \left(-\Delta + \tilde{W}^{h}\right) u(z), z \in \mathbb{R}^n \setminus \cup_{k=1}^{N} \tilde{\mathbb{L}}_k \\ u \in H^2(\mathbb{R}^n \setminus \cup_{k=1}^{N} \tilde{\mathbb{L}}_k : [u]_{\tilde{\mathbb{L}}_k} = 0, \left[\frac{\partial u}{\partial \nu^h}\right]_{\tilde{\mathbb{L}}_k} = \tilde{\alpha}_k^h u \mid_{\tilde{\mathbb{L}}_k}, k = 1, ..., N \end{cases},$$

$$A^{h}_{\delta'_\Gamma,\beta} = \begin{cases} \left(-\Delta + \tilde{W}^{h}\right) u(z), z \in \mathbb{R}^n \setminus \cup_{k=1}^{N} \tilde{\mathbb{L}}_k \\ u \in H^2(\mathbb{R}^n \setminus \cup_{k=1}^{N} \tilde{\mathbb{L}}_k) : \left[\frac{\partial u}{\partial \nu^h}\right]_{\tilde{\mathbb{L}}_k} = 0, [u]_{\tilde{\mathbb{L}}_k} = \beta_k^h \frac{\partial u}{\partial \nu^h} \mid_{\tilde{\mathbb{L}}_k}, k = 1, ..., N \end{cases}.$$

- We introduce an orthonormal base $\mathfrak{e} = (\mathfrak{e}_1, \mathfrak{e}_2, ..., \mathfrak{e}_n)$ in $\mathbb{R}^n$ such that $\mathfrak{e}_n = \frac{\nu^h}{|\nu^h|}$. Let $\mathfrak{x} = (\mathfrak{x}', \mathfrak{x}_n)$, $\mathfrak{x}' = (\mathfrak{x}_1, ..., \mathfrak{x}_{n-1})$ be coordinates of a point $\mathfrak{x} \in \mathbb{R}^n$ in the base $\mathfrak{e}$. Then the Laplacian $-\Delta$ can be written as $-\Delta = -\Delta_{\mathfrak{x}'} - \frac{\partial^2}{\partial \mathfrak{x}_n}$, and as in Chapter 3

$$A^{h}_{\delta_\Gamma,\alpha} = -\Delta_{\mathfrak{x}'} + \tilde{W}^{h} + L_{\delta,\alpha^h}, A^{h}_{\delta'_\Gamma,\beta} = -\Delta_{\mathfrak{x}'} + \tilde{W}^{h} + L_{\delta',\beta^h},$$

where the operators $L_{\delta,\alpha^h}, L_{\delta',\beta^h}$ are unbounded operators defined by the operators $-\frac{d^2}{d\mathfrak{x}_n^2}$ with domains

$$D_{L_{\delta,\alpha^h}} = \left\{\varphi \in H^2\left(\mathbb{R} \setminus \{d'_1, ..., d'_N\}\right) : [\varphi]_{\mathfrak{x}_n = d'_k} = 0, \left[\frac{d\varphi}{d\mathfrak{x}_n}\right]_{\mathfrak{x}_n = d'_k} = \alpha_k^h \varphi(d'_k), k = 1, ..., N, \right\},$$

$$D_{L_{\delta',\beta^h}} = \left\{\varphi \in H^2\left(\mathbb{R} \setminus \{d'_1, ..., d'_N\}\right) : \left[\frac{d\varphi}{d\mathfrak{x}_n}\right]_{\mathfrak{x}_n = d'_k} = 0, [\varphi]_{\mathfrak{x}_n = d'_k} = \beta_k^h \frac{d\varphi(d'_k)}{d\mathfrak{x}_n}, k = 1, ..., N\right\},$$

where $d'_k = |(e_n, \mathfrak{e}_n)|\, d_k$. Note that the operators $L_{\delta_\Gamma,\alpha^h}, L_{\delta'_\Gamma,\beta^h}$ are self-adjoint with the essential spectra $[0, +\infty)$ and possible finite negative discrete spectra. Let $\gamma\left(L_{\delta_\Gamma,\alpha^h}\right), \gamma\left(L_{\delta'_\Gamma,\beta^h}\right)$ be the minimal negative eigenvalues of the operators $L_{\delta_\Gamma,\alpha^h}, \mathcal{L}_{\delta'_\Gamma,\beta^h}$ and $\gamma\left(L_{\delta_\Gamma,\alpha^h}\right) = \gamma\left(L_{\delta'_\Gamma,\beta^h}\right) = 0$ if the operators $L_{\delta_\Gamma,\alpha^h}, L_{\delta'_\Gamma,\beta^h}$ do not have the discrete spectra. Then

$$spA^{h}_{\delta,\alpha} = \left[\tilde{W}^{h} + \gamma\left(L_{\delta,\alpha^h}\right), +\infty\right), spA^{h}_{\delta'_\Gamma,\beta} = \left[\tilde{W}^{h} + \gamma\left(L_{\delta',\beta^h}\right), +\infty\right). \tag{41}$$

THEOREM 11. *Let the hypersurfaces $\Gamma_k, k = 1, ..., N$ satisfy conditions (35), (36), the potential W and the strength interaction coefficients $\alpha_k, \beta_k, k = 1, ..., N$ be real-valued and slowly oscillating at infinity, and $\inf_{s\in\Gamma_k} |\beta_k| > 0$. Then the operators $\mathcal{H}_{\delta_\Gamma,\alpha}, \mathcal{H}_{\delta'_\Gamma,\beta}$ are self-adjoint, and*

$$sp_{ess}\mathcal{H}_{\delta_\Gamma,\alpha} = [m_1, +\infty), sp_{ess}\mathcal{H}_{\delta'_\Gamma,\beta} = [m_2, +\infty),$$

where

$$m_1 = \min\left\{\inf_{\omega=(\omega',\omega_n)\in S^{n-1},\omega_n\neq 0}\ \inf_{h_m\to\eta_\omega} \tilde{W}^{h}, \inf_{\omega=(\omega',0)\in S^{n-1}}\ \inf_{h_m\to\eta_\omega} spA^{h}_{\delta,\alpha}\right\},$$

$$m_2 = \min\left\{\inf_{\omega=(\omega',\omega_n)\in S^{n-1},\omega_n\neq 0}\ \inf_{h_m\to\eta_\omega} W^{h}, \inf_{\omega=(\omega',0)\in S^{n-1}}\ \inf_{h_m\to\eta_\omega} spA^{h}_{\delta',\beta}\right\}.$$

PROOF. Self-adjointness of $\mathcal{H}_{\delta_\Gamma,\alpha}, \mathcal{H}_{\delta'_\Gamma,\beta}$ follows from Theorem 2. Conditions of Theorem 11 yield that the operators $\mathcal{H}_{\delta_\Gamma,\alpha} : D_{\mathcal{H}_{\delta_\Gamma,\alpha}} \to L^2(\mathbb{R}^n), \mathcal{H}_{\delta'_\Gamma,\beta} : D_{\mathcal{H}_{\delta'_\Gamma,\beta}} \to L^2(\mathbb{R}^n)$ are locally Fredholm operators. Hence essential spectra of the operators

$\mathcal{H}_{\delta_\Gamma,\alpha}, \mathcal{H}_{\delta'_\Gamma,\beta}$ are sets of $\lambda \in \mathbb{R}$ such that the operators $\mathcal{H}_{\delta_\Gamma,\alpha} - \lambda : D_{\mathcal{H}_{\delta_\Gamma,\alpha}} \to L^2(\mathbb{R}^n), \mathcal{H}_{\delta'_\Gamma,\beta} - \lambda : D_{\mathcal{H}_{\delta'_\Gamma,\beta}} \to L^2(\mathbb{R}^n)$ are not locally invertible at infinity. Above it has been proven that the operators $\mathcal{H}_{\delta_\Gamma,\alpha}, \mathcal{H}_{\delta'_\Gamma,\beta}$ are locally invertible at infinity if and only if the operators $\left(\mathcal{H}^R_{\delta_\Gamma,\alpha}\right)_\Phi, \left(\mathcal{H}^R_{\delta'_\Gamma,\beta}\right)_\Phi$ are such. Hence $sp_{ess}\mathcal{H}_{\delta_\Gamma,\alpha}$, $sp_{ess}\mathcal{H}_{\delta'_\Gamma,\beta}$ are the unions of spectra of limit operators of $\left(\mathcal{H}^R_{\delta_\Gamma,\alpha}\right)_\Phi, \left(\mathcal{H}^R_{\delta'_\Gamma,\beta}\right)_\Phi$, respectively. Then we obtain the statement of the theorem applying formulas (40), (41). □

References

[1] S. Albeverio, F. Gesztesy, R. Høegh-Krohn, and H. Holden, *Solvable models in quantum mechanics*, 2nd ed., AMS Chelsea Publishing, Providence, RI, 2005. With an appendix by Pavel Exner. MR2105735

[2] S. Albeverio and P. Kurasov, *Singular perturbations of differential operators*, London Mathematical Society Lecture Note Series, vol. 271, Cambridge University Press, Cambridge, 2000. Solvable Schrödinger type operators. MR1752110

[3] M. S. Agranovič and M. I. Višik, *Elliptic problems with a parameter and parabolic problems of general type* (Russian), Uspehi Mat. Nauk **19** (1964), no. 3 (117), 53–161. MR0192188

[4] M. S. Agranovich, *Elliptic boundary problems*, Partial differential equations, IX, Encyclopaedia Math. Sci., vol. 79, Springer, Berlin, 1997, pp. 1–144, 275–281, DOI 10.1007/978-3-662-06721-5_1. Translated from the Russian by the author. MR1481215

[5] F. A. Berezin and L. D. Faddeev, *Remark on the Schrödinger equation with singular potential* (Russian), Dokl. Akad. Nauk SSSR **137** (1961), 1011–1014. MR0129309

[6] J. F. Brasche, P. Exner, Yu. A. Kuperin, and P. Šeba, *Schrödinger operators with singular interactions*, J. Math. Anal. Appl. **184** (1994), no. 1, 112–139, DOI 10.1006/jmaa.1994.1188. MR1275948

[7] J. Behrndt, P. Exner,V. Lotoreichik, *Essential spectrum of Schrödinger operators with δ-interactions on the union of compact Lipschitz hypersurfaces*, PAMM Proc. Appl. Math. Mech. **13**(2013), 523-524 .

[8] J. Behrndt, M. Langer, and V. Lotoreichik, *Schrödinger operators with δ and δ′-potentials supported on hypersurfaces*, Ann. Henri Poincaré **14** (2013), no. 2, 385–423, DOI 10.1007/s00023-012-0189-5. MR3028043

[9] J. Behrndt, P. Exner, and V. Lotoreichik, *Schrödinger operators with δ- and δ′-interactions on Lipschitz surfaces and chromatic numbers of associated partitions*, Rev. Math. Phys. **26** (2014), no. 8, 1450015, 43, DOI 10.1142/S0129055X14500159. MR3256859

[10] M. Sh. Birman and M. Z. Solomjak, *Spectral theory of selfadjoint operators in Hilbert space*, Mathematics and its Applications (Soviet Series), D. Reidel Publishing Co., Dordrecht, 1987. Translated from the 1980 Russian original by S. Khrushchëv and V. Peller. MR1192782

[11] M. Sh. Birman, T. A. Suslina, and R. G. Shterenberg, *Absolute continuity of the two-dimensional Schrödinger operator with delta potential concentrated on a periodic system of curves* (Russian, with Russian summary), Algebra i Analiz **12** (2000), no. 6, 140–177; English transl., St. Petersburg Math. J. **12** (2001), no. 6, 983–1012. MR1816514

[12] F. E. Browder, *On the spectral theory of elliptic differential operators. I*, Math. Ann. **142** (1960/1961), 22–130, DOI 10.1007/BF01343363. MR0209909

[13] P. Exner, *Bound states of infinite curved polymer chains*, Lett. Math. Phys. **57** (2001), no. 2, 87–96, DOI 10.1023/A:1017923426674. MR1856902

[14] P. Exner and K. Yoshitomi, *Band gap of the Schrödinger operator with a strong δ-interaction on a periodic curve*, Ann. Henri Poincaré **2** (2001), no. 6, 1139–1158, DOI 10.1007/s00023-001-8605-2. MR1877236

[15] P. Exner and S. Kondej, *Curvature-induced bound states for a δ interaction supported by a curve in* $\mathbb{R}^3$, Ann. Henri Poincaré **3** (2002), no. 5, 967–981, DOI 10.1007/s00023-002-8644-3. MR1937610

[16] P. Exner and S. Kondej, *Bound states due to a strong δ interaction supported by a curved surface*, J. Phys. A **36** (2003), no. 2, 443–457, DOI 10.1088/0305-4470/36/2/311. MR1960070

[17] P. Exner, *Spectral properties of Schrödinger operators with a strongly attractive δ interaction supported by a surface*, Waves in periodic and random media (South Hadley, MA, 2002), Contemp. Math., vol. 339, Amer. Math. Soc., Providence, RI, 2003, pp. 25–36, DOI 10.1090/conm/339/06097. MR2042529

[18] P. Exner, *An isoperimetric problem for leaky loops and related mean-chord inequalities*, J. Math. Phys. **46** (2005), no. 6, 062105, 10, DOI 10.1063/1.1914728. MR2149809

[19] P. Exner and K. Yoshitomi, *Asymptotics of eigenvalues of the Schrödinger operator with a strong δ-interaction on a loop*, J. Geom. Phys. **41** (2002), no. 4, 344–358, DOI 10.1016/S0393-0440(01)00071-7. MR1888470

[20] P. Exner and K. Yoshitomi, *Eigenvalue asymptotics for the Schrödinger operator with a δ-interaction on a punctured surface*, Lett. Math. Phys. **65** (2003), no. 1, 19–26, DOI 10.1023/A:1027367605285. MR2019388

[21] M. Lindner and M. Seidel, *An affirmative answer to a core issue on limit operators*, J. Funct. Anal. **267** (2014), no. 3, 901–917, DOI 10.1016/j.jfa.2014.03.002. MR3212726

[22] W. McLean, *Strongly elliptic systems and boundary integral equations*, Cambridge University Press, Cambridge, 2000. MR1742312

[23] B. S. Pavlov, *The theory of extensions, and explicitly solvable models* (Russian), Uspekhi Mat. Nauk **42** (1987), no. 6(258), 99–131, 247. MR933997

[24] V. Rabinovich, S. Roch, and B. Silbermann, *Limit operators and their applications in operator theory*, Operator Theory: Advances and Applications, vol. 150, Birkhäuser Verlag, Basel, 2004. MR2075882

[25] V. S. Rabinovich, S. Roch, and B. Silbermann, *Band-dominated operators with operator-valued coefficients, their Fredholm properties and finite sections*, Integral Equations Operator Theory **40** (2001), no. 3, 342–381, DOI 10.1007/BF01299850. MR1840187

[26] V. S. Rabinovich, *Essential spectrum of perturbed pseudodifferential operators. Applications to the Schrödinger, Klein-Gordon, and Dirac operators*, Russ. J. Math. Phys. **12** (2005), no. 1, 62–80. MR2179335

[27] V. S. Rabinovich and S. Roch, *Essential spectrum and exponential decay estimates of solutions of elliptic systems of partial differential equations. Applications to Schrödinger and Dirac operators*, Georgian Math. J. **15** (2008), no. 2, 333–351. MR2428476

[28] V. S. Rabinovich and S. Roch, *The essential spectrum of Schrödinger operators on lattices*, J. Phys. A **39** (2006), no. 26, 8377–8394, DOI 10.1088/0305-4470/39/26/007. MR2238507

[29] V. S. Rabinovich and S. Roch, *Essential spectra of difference operators on $\mathbb{Z}^n$-periodic graphs*, J. Phys. A **40** (2007), no. 33, 10109–10128, DOI 10.1088/1751-8113/40/33/012. MR2371282

[30] V. S. Rabinovich and S. Roch, *Essential spectra and exponential estimates of eigenfunctions of lattice operators of quantum mechanics*, J. Phys. A **42** (2009), no. 38, 385207, 21, DOI 10.1088/1751-8113/42/38/385207. MR2540394

[31] V. Rabinovich, *On the essential spectrum of quantum graphs*, Integral Equations Operator Theory **88** (2017), no. 3, 339–362, DOI 10.1007/s00020-017-2386-6. MR3682194

[32] V. S. Rabinovich, R. Castillo-Pérez, and F. Urbano-Altamirano, *On the essential spectrum of quantum waveguides*, Math. Methods Appl. Sci. **36** (2013), no. 7, 761–772, DOI 10.1002/mma.2623. MR3056934

[33] V. Rabinovich, *Transmission problems for conical and quasi-conical at infinity domains*, Appl. Anal. **94** (2015), no. 10, 2077–2094, DOI 10.1080/00036811.2014.964911. MR3384034

[34] Ja. A. Roĭtberg and Z. G. Šeftel′, *On equations of elliptic type with discontinuous coefficients* (Russian), Dokl. Akad. Nauk SSSR **146** (1962), 1275–1278. MR0142877

[35] T. A. Suslina and R. G. Shterenberg, *Absolute continuity of the spectrum of the Schrödinger operator with the potential concentrated on a periodic system of hypersurfaces* (Russian, with Russian summary), Algebra i Analiz **13** (2001), no. 5, 197–240; English transl., St. Petersburg Math. J. **13** (2002), no. 5, 859–891. MR1882869

[36] M. Schechter, *A generalization of the problem of transmission*, Ann. Scuola Norm. Sup. Pisa (3) **14** (1960), 207–236. MR0131063

Insituto Politécnico Nacional of Mexico, ESIME Zacatenco
Email address: vladimir.rabinovich@gmail.com